Advances in Complex Electromagnetic Materials

NATO ASI Series

Advanced Science Institutes Series

A Series presenting the results of activities sponsored by the NATO Science Committee, which aims at the dissemination of advanced scientific and technological knowledge, with a view to strengthening links between scientific communities.

The Series is published by an international board of publishers in conjunction with the NATO Scientific Affairs Division

A Life Sciences **B Physics**	Plenum Publishing Corporation London and New York
C Mathematical and Physical Sciences **D Behavioural and Social Sciences** **E Applied Sciences**	Kluwer Academic Publishers Dordrecht, Boston and London
F Computer and Systems Sciences **G Ecological Sciences** **H Cell Biology** **I Global Environmental Change**	Springer-Verlag Berlin, Heidelberg, New York, London, Paris and Tokyo

PARTNERSHIP SUB-SERIES

1. Disarmament Technologies	Kluwer Academic Publishers
2. Environment	Springer-Verlag / Kluwer Academic Publishers
3. High Technology	Kluwer Academic Publishers
4. Science and Technology Policy	Kluwer Academic Publishers
5. Computer Networking	Kluwer Academic Publishers

The Partnership Sub-Series incorporates activities undertaken in collaboration with NATO's Cooperation Partners, the countries of the CIS and Central and Eastern Europe, in Priority Areas of concern to those countries.

NATO-PCO-DATA BASE

The electronic index to the NATO ASI Series provides full bibliographical references (with keywords and/or abstracts) to more than 50000 contributions from international scientists published in all sections of the NATO ASI Series.
Access to the NATO-PCO-DATA BASE is possible in two ways:

– via online FILE 128 (NATO-PCO-DATA BASE) hosted by ESRIN,
Via Galileo Galilei, I-00044 Frascati, Italy.

– via CD-ROM "NATO-PCO-DATA BASE" with user-friendly retrieval software in English, French and German (© WTV GmbH and DATAWARE Technologies Inc. 1989).

The CD-ROM can be ordered through any member of the Board of Publishers or through NATO-PCO, Overijse, Belgium.

3. High Technology – Vol. 28

Advances in Complex Electromagnetic Materials

edited by

A. Priou

University of Paris X,
Paris, France

A. Sihvola

Helsinki University of Technology,
Espoo, Finland

S. Tretyakov

St Petersburg State Technical University,
St Petersburg, Russia

and

A. Vinogradov

SCAPE, Russian Academy of Sciences,
Moscow, Russia

Kluwer Academic Publishers

Dordrecht / Boston / London

Published in cooperation with NATO Scientific Affairs Division

Proceedings of the NATO Advanced Research Workshop on
Electromagnetics of Chiral, Bi-isotropic, and Bi-anisotropic Media (Chiral '96)
St Petersburg – Moscow, Russia
23–30 July, 1996

A C.I.P. Catalogue record for this book is available from the Library of Congress.

ISBN 0-7923-4503-7

Published by Kluwer Academic Publishers,
P.O. Box 17, 3300 AA Dordrecht, The Netherlands.

Kluwer Academic Publishers incorporates the publishing programmes of
D. Reidel, Martinus Nijhoff, Dr W. Junk and MTP Press.

Sold and distributed in the U.S.A. and Canada
by Kluwer Academic Publishers,
101 Philip Drive, Norwell, MA 02061, U.S.A.

In all other countries, sold and distributed
by Kluwer Academic Publishers Group,
P.O. Box 322, 3300 AH Dordrecht, The Netherlands.

Printed on acid-free paper

TABLE OF CONTENTS

IV. Scattering and Diffraction from Bi-Anisotropic Structures

PREFACE

This volume presents the Proceedings of the NATO Advanced Research Workshop on the electromagnetics of chiral, bi-isotropic and bi-anisotropic media, *Chiral'96*, which was held in Russia from July 23 to July 30, 1996. The main organizer of the event was the Scientific Center for Applied Problems in Electromagnetics (SCAPE) of the Russian Academy of Sciences, Moscow. The Director of SCAPE, Professor A.N. Lagarkov, was co-chairman of the workshop, and the SCAPE staff provided excellent working environment for the participants. The other co-chairman was Professor A. Priou from the University of Paris X.

This NATO ARW, *Chiral'96,* followed the line of the specialist meetings held in Espoo, Finland (*Bi-isotropics'93,* February 1993), Gomel, Belarus (*Bianisotropics'93*, October 1993), Périgueux, France (*Chiral'94,* May 1994), and State College, Pennsylvania, USA (*Chiral'95,* October 1995). The meeting with its cross-disciplinary nature was devoted to discussions of the latest developments in electromagnetics, materials science, and applications of novel composite materials for microwave and optical technology. The emphasis of the meeting was on bi-anisotropic materials whose most interesting feature is the magnetoelectric interaction of the fields. Bi-anisotropic materials are expected to provide useful applications in radar technology, aerospace, microwave engineering, manufacturing technology, etc.

The workshop was held on board the river-boat *A. Suvorov*, on a trip from St. Petersburg to Moscow through the Ladoga and Onega lakes, the Neva, Volga, and Moskva rivers, and the Volgo-Balt channel system. The event was held parallel to the fourth *ETOPIM* meeting (International conference on electric transport and optical properties of inhomogeneous media). The proceedings of that conference will be published as a special issue of the journal *Physica A.*

The following persons served in the Organizing Committee for *Chiral'96*:

- Prof. A.N. Lagarkov (Russia), Co-chairman
- Prof. A. Priou (France), Co-chairman
- Dr. I.V. Semchenko (Belarus)
- Dr. A. Sihvola (Finland)

– Prof. S.A. Tretyakov (Russia)
– Dr. U. Unrau (Germany)
– Prof. V.V. Varadan (USA)
– Dr. A.P. Vinogradov (Russia), Workshop Co-ordinator

56 contributions were submitted to the Workshop. These all were reviewed by the organising committee. After selection, 33 presentations were given and discussed during the 8 days of the workshop. Before the meeting, a Book of Abstracts was published as Report 219 in the Electromagnetics Laboratory Series of the Helsinki University of Technology, Finland.

The present volume contains the proceedings of *Chiral'96*. It consists of detailed expositions of the contributions presented in the workshop. The only exception is the invited review paper *Fedorov's covariant methods in the electromagnetic field theory*. All the papers have been reviewed twice: first by external referees and then by the editors, who have also made the final editing of the manuscripts (with a few exceptions of papers whose final versions were not available in electronic form). The editors apologise for the inevitable misprints which will be found in this book.

Acknowledgements

The editors have the pleasure to acknowledge Professor A.N. Lagarkov of the Scientific Center for Applied Problems in Electrodynamics for providing the platform for the workshop, and SCAPE personnel for technical assistance during the event. We very much appreciate the financial support from the NATO Scientific Affairs Division, the French Chapter of the Electromagnetics Academy and the IEEE ED/MTT/AP Chapter of St. Petersburg. Our gratitude also extends to the Electromagnetics Laboratory (Helsinki University of Technology, Finland) which published the Book of Abstracts, and to the Department of Electromagnetic Theory (Lund Institute of Technology, Sweden) where the final editing of the proceedings was made. Many persons provided valuable assistance. We should especially mention Dr. L. Arnaut for his careful editing of the invited review paper and Dr. A. Sochava and Mr. B. Stjernberg for their technical help. It has been a pleasure to work with the contributors to this volume, who were always very co-operative and patient during the several rounds of review process.

Moscow - St. Petersburg - Paris - Lund
November 1996 *Editors*

INTRODUCTION

Structure of the Proceedings

The contents of this Proceedings of the NATO Advanced Research Workshop on the electromagnetics of chiral, bi-isotropic, and bi-anisotropic media is divided into six chapters.

The first chapter contains papers of review nature devoted to general and broad issues within the field of the workshop. It starts with a historical review paper by L.M. Barkovski and G.N. Borzdov. These authors represent the Belarussian school of scientists founded by Academician F.I. Fedorov. That famous person gave birth to many new ideas in the physics of complex media, and in particular, chiral (or *gyrotropic*) crystals. The next review paper, written by the Moscow scientists A.F. Konstantinova, A.Yu. Tronin, and B.V. Nabatov, is also related to activities of F.I. Fedorov, and it presents new developments in crystallo-optics which have their roots in Fedorov's ideas. Thirdly, following the traditions of previous specialist meetings, the book includes a literature survey with an extensive list of recent publications in the field. This review on complex media electromagnetics has been authored by S. Bolioli.

Chapter 2 contains theoretical papers of general and fundamental nature. Here, such issues of electromagnetic theory are treated as the consistency of constitutive relations for bi-anisotropic media, direct and inverse problems of reflection and transmission in bi-anisotropic slabs, and the classification of possible magnetoelectric effects in general reciprocal media.

The third chapter is devoted to various aspects of modelling complex composite materials as effective continuum media. The chapter starts with a general discussion of the physical concept of effective permeability in composite materials with non-magnetic inclusions (complex-shaped conductive particles, for example). Further, homogenisation principles for composites with bi-anisotropic inclusions are analysed as well as multilayered structures and superlattices.

Chapter 4 treats classical and canonical electromagnetic problems in the presence of bi-anisotropic media. These include scattering and diffraction from homogeneous, layered, and inhomogeneous bi-anisotropic structures. Also wave-guiding problems and surface waves that propagate along boundaries of non-reciprocal bi-anisotropic samples are analysed.

Bi-anisotropic materials for microwave applications are composite media with complex-shaped inclusions. In Chapter 5 the reader finds papers which study various aspects of the electromagnetic response of such particles. Inclusions used in the design of artificial magnetics and chiral composites are considered. The topics also include analyses of planar arrays of helices and other complex inclusions which kind of structures bear relevance to engineering applications.

Finally, in Chapter 6, the latest experimental results of the performance of chiral materials are presented. The papers deal also with measurement principles and application aspects.

Comments on Mathematical Notations

Because of obvious historical reasons, scientists working in the Western countries and in the former Soviet Union countries often use different systems of notations in their writings. For people working in the field of electromagnetics of complex materials there is still another reason for understanding and mastering different scientific languages: research in this field demands understanding of many related disciplines, such as electromagnetics, optics, physics of crystals, materials science, etc. It is natural that in the present volume the reader will find different notations, especially for vector and dyadic operators. Also, in some papers of physical nature, the authors find it preferable to use the Gaussian physical units instead of the SI system. Here we offer some comments on these issues.

First of all, it is inevitable that the character types and fonts that are used to denote vectors, tensors, matrices and dyadics vary from one paper to another. It would have required persuasion beyond reasonable amount from the editors' side to convince all the authors about the benefits of such uniformity in the book that every contribution would follow the notation the editors prefer.

Secondly, someone studying the literature of theoretical electromagnetics finds soon that there are at least three more or less widely accepted ways to denote operations between vectors and dyadics. To illustrate these systems we refer to Table 1 where products between vectors (denoted by bold-face characters $\mathbf{a}, \mathbf{b}$) are illustrated.

The system A is nearly universally adopted in the modern Western lit-

TABLE 1. Vector and dyadic algebra notations

Operation	System A	System B	System C
Scalar multiplication	$\mathbf{a}\cdot\mathbf{b}$	$\mathbf{ab}$	$(\mathbf{a},\mathbf{b})$
Vector multiplication	$\mathbf{a}\times\mathbf{b}$	$\mathbf{a}\times\mathbf{b}$	$[\mathbf{a},\mathbf{b}]$ or $[\mathbf{a}\ \mathbf{b}]$
Dyadic multiplication	$\mathbf{ab}$	$\mathbf{a}\cdot\mathbf{b}$	$\mathbf{a}\otimes\mathbf{b}$

erature on electromagnetics. In this volume, one finds these notations in papers by A. Sihvola, S. Tretyakov, A. Sochava, and in some others.

In papers written by F.I. Fedorov and his students notations according to the system B are often used. In these papers, dyadics are usually not distinguished in notation from scalars, and similarly, there is no distinction between scalar products of vectors and dyadics on one hand, and products of two numbers on the other. The reader is supposed to understand from the context what type of quantities the analysis deals with. Furthermore in the Fedorov school writings, the special notation $\mathbf{a}^\times$ is used for "tensor (or dyadic), dually conjugate to vector **a**." This means an antisymmetric dyadic and translated to the notations adopted in the Western literature, it equals $\mathbf{a}\times\overline{\overline{I}}$, where $\overline{\overline{I}}$ is the unit dyadic. For example, $\nabla^\times\mathbf{E}$ (scalar product of the dyadic $\nabla^\times$ and the vector $\mathbf{E}$) means $\nabla\times\overline{\overline{I}}\cdot\mathbf{E}=\nabla\times\mathbf{E}$.

The system C is usual in old Soviet literature and in some Russian mathematical books.

In the papers of more mathematical nature that appear in the present volume, authored by Minsk scientists L.M. Barkovski, G.N. Borzdov, and A.N. Borzdov, more complex dyadic operations are utilized. These include, in particular, exterior and interior products. In general, these authors follow system C for vector algebra notations.

The authors and editors have tried their best to add definitions into each paper for operations and quantities that are not obvious.

I.

General

FEDOROV'S COVARIANT METHODS IN THE ELECTROMAGNETIC FIELD THEORY

Creation, progress and trends of development

L.M. BARKOVSKII AND G.N. BORZDOV
Department of Theoretical Physics
Byelorussian State University
Fr. Skaryny avenue, 4
220050 Minsk, Belarus

Fedor Ivanovich Fedorov, the well-known theoretical and mathematical physicist, Professor of the Byelorussian State University (BSU), Academician and Member of the Presidium of Academy of Sciences of the Republic of Belarus and President of the Byelorussian Physical Society has died on October 13th, 1994 in Minsk, Belarus. He devoted more than 60 years of his life to scientific work, pedagogical and scientific organizational activities.

This article describes covariant methods developed by Academician Fedorov, which are now widely used in different branches of electrodynamics, acoustics and quantum physics. We present a brief overview of the history of the development of these methods and results, obtained by their application to various problems, and discuss some trends of their further development.

1. The Beginning of Fedorov's Scientific Activity

F.I. Fedorov was born on June 19th, 1911 in the village of Turetz, Grodno region in the family of a teacher, the then well-known Byelorussian writer Yanka Mavr. In 1931, after graduating from the Physics and Mathematics Department of the Pedagogical Faculty of BSU, he worked as a tutor in a pedagogical technical school in Krichev. In 1933 he became a postgraduate of V.A. Fock, a brilliant theoretical physicist of Leningrad University. In this very favourable scientific atmosphere, Fedorov became very much interested in the theory of relativity. This theory appeared first at the end of the war and shook the scientific world. There had been seven years since the publication of the well-known articles by Fock, Schrödinger, Klein and Gordon at the time when Fock met the young postgraduate Fedorov. These scientists discovered independently the first relativistic scalar wave equa-

A. Priou et al. (eds.), Advances in Complex Electromagnetic Materials, 3–18.

tion. Five years had passed since Dirac discovered the relativistic spinor wave equation. It took Fedorov a comparatively short time (1933–1936) to get acquainted with the new ideas. Using the method of Fock's functionals, he examined a set of principal questions on the theory of electromagnetic interactions between elementary particles. He described these findings in his candidate dissertation, which was defended in Leningrad in 1936. Later in BSU he continued his research devoted to the application of Fock's functionals in the quantum theory of radiation. He founded the department of theoretical physics in 1938.

Fedorov's studies in Leningrad with Fock greatly influenced the direction of his further activity. This gave him the opportunity to reveal his great talent, extraordinary wide scientific interests, owing such traits as self-discipline, very high working ability, very strong will in research work and favourable circumstances. He was head of the Deparment of Theoretical Physics of BSU for 24 years. In the toughest period for the university (1943–1950) he worked as a Dean of the Physics-Mathematics Faculty. In the post-war period he carried out research work on general relativistic theory of first-order equations for particles with arbitrary spins. In these years his interests expanded and he started to work on the problem of electromagnetic theory of light in anisotropic and gyrotropic media as well as crystal acoustics. We should mention one interesting feature in his scientific work: he never gave up an idea until it was realized, even if it took him several years. Many years later (1982) he discovered a universal system of relativistic non-linear first-order equations, describing all kinds of fundamental interactions in general. All well-known linear and non-linear equations of mathematical physics are derived from it.

In 1954 Fedorov brilliantly defended his doctoral dissertation "Invariant methods in optics of anisotropic media" and became the first Doctor of Physics and Mathematics in the Republic of Belarus. One of Fedorov's characteristics in his research work is the covariant approach; development and application of direct methods in analysis, covering a wide range of similar dynamics problems. The specific point is the use of specially developed and adapted algebraic and geometric techniques for solving these problems. One can do without using the explicit matrix forms — bases of spaces for presentation of dynamic equations – their solutions and symmetries by application of properties of minimal polynomials. This concept has obvious advantages. It raises the well-known works of Hamilton, Cayley, Grassman on non-commutative algebras, algebraic-geometric studies of Lie and famous Erlangen program of Klein. Investigations of these mathematicians inspired Maxwell, Schrödinger, Dirac, Einstein and Fock in doing research on dynamics of physical fields and were to a large extent the key to their success. Fedorov continued his studies in this direction during the second half of 20th century. At the time of wide application of methods of group theory and corresponding dynamics analogies, geometro-dynamics, topological phases, band-gap photonic structures, quaternion quantum mechanics and the discovery of the role of the Sturm-Liouville operator in soliton solutions, the fact is that Fedorov's approach forestalled that time. The significance of this fact was greatly recognized in a number of scientific communities. Periodical and monographical literature on classical and quantum electrodynamics and optics, acoustics, radiophysics and gravitation have proven this fact. This conclusion has been strengthened during international conferences of recent years: "Bi-isotropics'93" (Espoo, 1993), "Bi-anisotropics'93" (Gomel, 1993), "Quantum Systems'94" (Minsk, 1994), "Applied Optics" (St.-Petersburg, 1994), "Chiral'94" (Périgueux, 1994), "Complex Systems" (Novopolotsk, 1994), "Quantum Systems'96" (Minsk, 1996) and "Chiral'96" (St.-Petersburg–Moscow, 1996).

During the hard post-war years, Fedorov continued his scientific work in spite of being occupied with teaching and administrative work. While being a young scientist Fedorov realized the great importance of Maxwell theory, not only for the technical development of world community but also for the creation of a scientific picture of the world. He spent many years on the study of anisotropic media.

2. Intrinsic Tensor Techniques in Electromagnetic Theory

2.1. THEORY OF ANISOTROPY

The theory of anisotropic properties of transparent and, in particular, absorbing and gyrotropic crystals was originated in the works of Drude, Voigt and others at the turn of this century. Pockels' monograph [1] and the extensive work by Szivessy [2] reviewed the development of crystal optics. Only the theory of transparent crystals had been completed by that time; for absorbing and gyrotropic crystals the corresponding theory was far from finalized. Because of difficulties in connection with the necessity of taking into consideration either the dielectric tensor or the conductivity tensor (in absorbing crystals) and the tensor of gyrotropy (in gyrotropic crystals), the theory was developed only for elementary cases of isotropic media and the most symmetric uniaxial crystals. In other cases the theory was very approximative and, as it was realized later, based on erroneous assumptions in a number of cases. One can state undoubtedly that at that time, a rigorous theory of optical properties of absorbing and gyrotropic crystals was not complete. Considering magnetic crystals, the theory of optical properties which requires the tensor of permeability to be considered, was not worked out at all.

In the world literature a specific coordinate system was commonly used when considering optical properties of crystals — the system of principle axes of the tensor of dielectric permittivity. In such a system this tensor takes its simplest (diagonal) form, which avoids complicated expressions in the optics of transparent crystals. However, this approach fails for absorbing, gyrotropic and magnetic crystals as it is necessary to take into account other tensors which in general exhibit a different system of principal axes. Fedorov used a fundamentally new approach to describe crystal properties. Instead of using a specific co-ordinate system, Fedorov performed all calculations in covariant form, on the basis of general algebraic properties of vectors and tensors that characterize either a wave or the medium in which the wave propagates. This resulted into a new approach to crystal optics and to the solution of many problems that were impractical because of complexity and cumbrousity. In order to apply the new method it is necessary to describe all characteristics of both media and electromagnetic waves in a covariant manner. The principal property of a light wave is its polarization. Previously, the polarization of light was described only in noncovariant way using the phase plane of wave as a co-ordinate plane. The well-known Stokes parameters associated with this plane are applied for describing polarization states. Poincare's sphere representing some geometrical interpretation of Stokes parameters is also connected with this plane.

As an alternative to these classical methods of describing polarization which appeared in the course of the last century, the new covariant method [3] was proposed, which made use of the full complex vector of electric (or magnetic) wave field without decomposition into components. The description of polarization in such a formalism acquires an unusual form. For example, the condition of circular polarization means that the squared complex vector equals zero. However it is rather convenient and general for it permits to determine the wave polarization state directly from Maxwell's equations in which field vectors appear. The mentioned approach was further elaborated upon in the works appearing in 1954–1955. Later (1965–1966), the covariant method was generalized to the case of partially polarized light with the aid of the so-called beam tensor [4, 5]. Thus it was shown that, instead of using Stokes parameters for the complete characterization of partially polarized light, invariants of the beam tensor are considerably more convenient [4]. This approach allows one to solve a number of complex problems of polarization characteristics of light in scattering processes, luminescence etc. in an simple way [5].

A very important topic in electromagnetic theory is the concept of so-called inhomogeneous waves, i.e. waves which have non-coinciding directions of phase velocity and attenuation. Inhomogeneous waves are widespread in a nature and arise in any absorbing medium, as well as in reflections and refractions of transparent media. The knowledge of their properties is necessary in the construction of many optical devices. However, their properties are considerably more complex than those of homogeneous waves ones and their theory had not been developed earlier. Complex angles were used to describe these waves [6] but these have no direct physical meaning. Hence the not so simple theory became more complex. In 1954 Fedorov introduced the refraction vector defined as the product of the refractive index and the unit phase normal. This refraction vector is generalized for inhomogeneous waves in a natural manner if one assigns complex values to it. Its real part represents the normal (phase) refraction vector, and its imaginary part signifies an extinction vector.

Within two years (1955–1956) the general theory of inhomogeneous waves was developed [7, 8] (see also [9, 10, 11]) and several principal facts were discovered. For example, it was shown that in an isotropic medium electric and magnetic vectors of inhomogeneous wave are always differently polarized in different planes; the special case where the polarizations and planes of electric and magnetic fields coincide is the case of circular polarization. In a fixed point, the vector of energy flux of inhomogeneous wave traces out an elliptical cone, directed outward to the phase normal. It is shown that the direction associated with the average flux of energy depends on the polarization of the wave with fixed normal. A number of other unusual properties of inhomogeneous waves were established.

Meanwhile the theory of inhomogeneous waves which appear in the total reflection from the interface of a transparent isotropic medium was developed [7, 8]. It was shown that, for general polarization of the incident wave, the average flux of energy in the second medium is in the direction of a certain angle to the plane of incidence. The consequence of this fact was the prediction of a lateral displacement of the reflected beam, caused by its exit from the plane of incidence. This interesting optical phenomenon was confirmed experimentally about 10 years later and was called "Fedorov shift" (see [12]). Nowadays many works are devoted to the theoretical and experimental study of this phenomenon.

Another case of considerable interest, treated by Fedorov, is when a circularly polarized inhomogeneous wave arises at total reflection. In this case, both the averaged and instanteneous energy fluxes are absent, although the field in the second medium does exist. This means that the classical explanation which dated from the time of Eichenwald, i.e. that energy leaks into the medium but originates from another point, fails. The experimental study of total reflection under these conditions is of great interest.

In a series of works by F.I. Fedorov and L.M. Tomilchik (1956–1957) on the basis of covariant methods, a general and rigorous theory for transparent media with arbitrary anisotropy of magnetic permeability was derived. Earlier attempts of constructing such a theory failed due to difficulties connected with the use of common co-ordinate systems. Later, in works by G.S. Krinchik and M.V. Chetkin [13], transparent crystals were obtained that had noticeable magnetic anisotropy to which this theory is applicable. In works by Fedorov and Tomilchik the complete analysis and general classification of transparent magnetic crystals through their properties is given. It is shown that the optical properties of a magnetic crystal are mainly defined by the tensor $\gamma = \mu^{-1}\varepsilon$, where ε and μ are tensors of permittivity and permeability, respectively. It appeared that magnetic crystals of lowest syngonies (rhombic, monoclinic and triclinic) may be optically uniaxial, and their properties (unlike those of common uniaxial crystals) have no rotational symmetry. The previously unknown type of optical anisotropy in magnetic crystal, which is present in monorefrigent crystals, was predicted. In the latter case the velocity of waves depends on direction. However in any direction only one wave propagates with arbitrary polarization. In this way birefringence is absent, in spite of the anisotropy. The crystals of every syngony[1] may be monorefringent, except for the cubic one. A fundamental property follows from this: the separation of light waves in crystals into two waves with different polarizations and velocities (i.e. birefringence) is not a necessary consequence of anisotropy of ε and μ but takes place because ε and μ differ in anisotropy, i.e. they exhibit different sets of eigenvectors and eigenvalues in the general case.

The optical properties of absorbing crystals were considered separately. Such crystals are found in abundance in nature and have several practical applications. These media constantly attract the interest of researchers. However it is impossible to study their optical properties by the classical methods that are applicable to transparent media, because these substances are opaque even in thin layers, due to the strong absorption. At optical frequencies, one can deduce its dielectric permittivity and absorbing properties (the electric conductivity) of such media only by observation of reflected light. For this purpose it is necessary to get a complete solution of the boundary value problem, which is possible only when using an exact theory of optical properties of absorbing crystals. Fedorov developed such a theory on the basis of covariant methods [9, 10, 11]. Previously, in the optics of absorbing crystals, one used to think that all theoretical conclusions might be derived from appropriate results of optics for transparent crystals through the simple replacement of real quantities by complex ones. The associated statement received a special name: "transfer principle" (see [1]). From this principle it follows in particular that all absorbing crystals, like transparent ones, are subdivided according to their optical properties into three types — optically isotropic, uniaxial and quadroaxial. Concerning

[1]The term "syngony" is used in the Russian literature for a crystallographic system.

the latter, to which crystals of lowest syngonies belong, all of these must exhibit four "screw" optical axes, along which only circular polarized light can propagate.

In Fedorov's work [14] it was proven that the "transfer principle," which is included in classical monographs, is incorrect. It was found that in crystals of lowest syngonies with screw (or "circular") axes, axes may exist along which light with arbitrary polarization can propagate. Such axes were named "isotropic optical axes". It appeared that rhombic crystals can exhibit two isotropic or four circular optical axes. Monoclinic and triclinic crystals may have between one and four optical axes, one of which may be isotropic. In particular, absorbing monoclinic crystals, having only one circular axis, are possible. Also triclinic crystals with one isotropic optical axis are possible. Finally it was established that instead of three types of absorbing crystals, as stated in previous theories, 16 types of such crystals may exist, distinguished by the number and character of optical axes and by other optical properties as well. Monoclinic and triclinic syngonies each have 6 various types of crystals.

In all vast practical applications of crystal optics the major attention is given to the behaviour of a light beam on the crystal interface, i.e. the problem of reflection and refraction of light. The special significance that this problem has acquired is connected with the development of solid state laser systems and optical networks. A series of works (1956–1971) is devoted to a detailed treatment of this question for transparent and absorbing crystals and isotropic media. Until this moment simple and foreseeable relations were absent, even for the simplest case of transparent uniaxial crystals, that expressed the amplitudes of reflected and refracted light waves on its interface. Only particular cases were considered. Fedorov derived general expressions in compact covariant vector form. He also gave a principal solution to the problem of determining all optical parameters of a transparent or absorbing medium on the basis of the observation of properties of light reflected from an arbitrarily oriented interface. In spite of the well-known role of the surface layer, this method is now the only one with which the optical parameters of strongly absorbing media can be determined.

In 1958 the monograph *Optics of anisotropic media* [9] was published, in which results were systematized and generalized (see also [10, 11]). Original results that had not been published before can also be found in this monograph. In particular, the theory surrounding the phenomenon of complete polarization of light reflected from transparent crystals was developed here, and the notion of principal vectors of reflected and refracted waves was introduced, which considerably simplifies the solution of boundary value problems. A range of other results was also obtained.

2.2. THEORY OF GYROTROPY

Academician Fedorov made a considerable contribution to the theory of gyrotropic (optically active) crystals [10, 15, 16, 17, 18]. In gyrotropic media, the theory of wave propagation is even more complicated then in anisotropic media, and Fedorov's covariant methods prove to be especially useful. The electrodynamic law of energy conservation was firstly used by Fedorov as a condition that limits the type of tensors and characterizes the phenomenon of optical activity. It enables one to get a rather general phenomenological derivation of constitutive equations (material equations) for the electrodynamics of gyrotropic media with either dielectric or magnetic anisotropy. Covariant expressions for tensors describing the gyrotropy in all crystal classes, where applicable, were found. It was established that crystals of planar classes of middle syngonies, that were not included previously in the class of optically active media, do possess this property, though they do not rotate the plane of polarization. In one of the important works (1959) it was shown that such crystals differ from non-active crystals in the sense that light reflected from them is elliptically polarized in spite of the absence of absorption. Thus, it was proved that the number of crystallographic classes in which optical activity is possible is 18, and not 15, as it was considered earlier [1, 2].

The number of variants describing properties of optically active crystals shows the complexity of the theory. In the joint work with B.V. Bokut and A.N. Serdukov in 1972 they analyzed different variants and it was noted that all of these were approximations of a general unified method, characterized by a redefinition of electromagnetic field vectors. A discussion of the form of material equations for electrodynamics of gyrotropic media (see [16] and also [19, 20]) took place. This discussion came to an end when opponents admitted the correctness of the concept defended in [16] (see [21] and also [22]).

It was shown in [17] that the theory of homogeneous gyrotropic crystals, possessing arbitrary magnetic anisotropy, can be reduced to the theory of nonmagnetic crystals by a simple redefinition of field quantities. It is important to notice that, in the optical region, the tensors of gyrotropy and magnetic susceptibility are small but of the same order of magnitude. The existence of three material tensors, i.e. ε, μ and α, complicates the theory drastically. However, from the results in [17] it follows that magnetic anisotropy can always be excluded from equations, thereby considerably simplifying the theory. All relations derived for nonmagnetic crystals can be easily generalized to the case of crystals with arbitrary homogeneous magnetic anisotropy. In this way general equations were obtained for normals in covariant form, i.e. the basic equations of crystal optics for magnetic gyrotropic crystals of any symmetric class.

There did not exist an exact and general theory of optical properties of gyrotropic crystals, even when neglecting magnetic anisotropy. All of the authors (Pockels, Szivessy, Born etc) restricted themselves to the simplest case, in which not only terms with the squared gyrotropic parameters were neglected, but also the products of parameters of gyrotropy and anisotropy of crystals. In [17, 18] this question was analyzed and it was shown that in some cases the generally accepted theory lead to conclusions that were fundamentally wrong. So, in [17] it was shown that if the equations of electric and magnetic fields of plane waves in gyrotropic crystals were limited to terms of first order of gyrotropy, the electric field equation appears to be incompatible with the magnetic field equation. Moreover, different physical conclusions follow from them. For example, in the case of planar crystals of middle syngonies, it follows from the approximate (first-order) equation for electric field that there is an ordinary wave in the crystal whose velocity does not depend on either the direction of propagation nor the presence of gyrotropy. Meanwhile, from the approximate equation of the same order for the magnetic field it follows that there are no ordinary waves in crystals. Instead, the wave propagates with a velocity that depends weakly on either the direction or the gyrotropy. Thus an internal contradiction was found in the approximate theory if limited to first-order terms. In [17] it was shown that, if one proceeds from equations for electric and magnetic fields taking into account second-order terms of gyrotropy, this contradiction disappears and the theory becomes completely self-consistent.

In [18] the general but approximate theory of optical properties of gyrotropic media in covariant form is developed on the basis of such self-consistent and non-contradictional equations. In this theory the gyrotropy terms of the third and higher order are neglected. As a result of this consideration one more statement of ordinary theory was disproved (see [2, 3]), namely that two isonormal waves in any gyrotropic crystal have similar but perpendicularly arranged ellipses of polarization. In fact, the axial ratios of these two ellipses are not equal, but differ to first order in gyrotropy, a difference which cannot be neglected.

General relations obtained in papers [17, 18] enable one to calculate correctly all properties of light waves propagating in any direction in crystals of arbitrary symmetry, possessing simultaneously dielectric and magnetic anisotropy and gyrotropy. Special attention should be paid to the fact that at present the demand for such calculations continuously increases in connection with developments in laser technology.

Thus a number of basic problems of theory of optical properties of crystals has been solved. Whereas previously the more or less complete characterization applied to the simplest part of crystal optics — the theory of optical properties of transparent nonmagnetic nongyrotropic crystals — now the same degree of completeness was achieved in a considerably more complicated theory for crystals with other possible types of anisotropy: magnetic, absorbing anisotropic and gyrotropic. A number of conventional statements in the world literature were found to be incorrect. A novel mathematical approach allowed for clarification and completion in this area of physics.

Simultaneously, F.I. Fedorov became interested in questions of theoretical acoustics of crystals. He advanced an original method for the description of elastic waves in crystals, i.e. the method of comparison with isotropic medium, and applied it to the calculation of Debye temperatures of solid states. After an invitation to Lomonosov Moscow State University he taught a course of lectures on acoustics of crystals to students and post-graduates specializing in crystallographic physics. This lecture material was used by Fedorov when writing a new a book on acoustics. This book was published by "Nauka" in 1965,[2] and afterwards in English [23]. In the foreword to the English edition Prof. H.B. Huntington noted: "In many respects his approach will appear quite novel as his treatment introduces several concepts and parameters previously unfamiliar to the literature of the West."

In 1961 F.I. Fedorov suggested a new parameterization of the Lorentz group, i.e. complex vector parameterization. In this approach, the matrix of the Lorentz transformation takes a very convenient form, and the complex vector parameter of the group has a very simple composition law [24]. This parameterization has found many applications in group theory (Lorentz group, the three-dimensional orthogonal group and several others), in particle and field theory, as well as in optics of anisotropic media.

3. Evolution Operators

The main contents of Fedorov's monographs and articles on crystal optics and crystal acoustics can be summarized as the theory of eigenwaves. The fundamental character of eigenwaves, i.e. plane harmonic waves, is caused by the fact that they are the most simple and convenient way to analyze solutions of wave equations. These permit one not only to determine the basic laws of light propagation in an anisotropic medium, but also to investigate the more complex fields with the help of spectral expansions. The relative simplicity in the analysis of properties of eigenwaves is explained mainly by the fact that wave equations (which involve partial derivatives) then reduce to algebraic equations. At present the properties of eigenwaves in various linear media (transparent and absorbing, nonmagnetic and magnetic, biisotropic, chiral, bianisotropic) are known in detail due, to a great

[2]The name of the publishing house, "Nauka" is the Russian word for "science."

extent, to the application of Fedorov's covariant methods. However, from the time of Voigt it is well-known, that in anisotropic media eigenwaves do not form a complete system of plane wave solutions. For example, along singular directions in absorbing crystals, waves with linear dependence of amplitude on coordinate (Voigt waves) [10, 11, 25]

$$\boldsymbol{E} = [\boldsymbol{E}_0 + i(\boldsymbol{k} \cdot \boldsymbol{r})\boldsymbol{E}_1] \exp [i(\boldsymbol{k} \cdot \boldsymbol{r} - \omega t)]. \tag{1}$$

can propagate.

In 1963, Fedorov and Petrov found a new type of waves with linear amplitude dependence on coordinates [26, 27]. They showed that, in the case of oblique incidence onto a transparent or absorbing uniaxial nongyrotropic crystal, an eigenwave can excite the inhomogeneous wave [26, 27]

$$\boldsymbol{E} = [\boldsymbol{E}_0 + ik_0(\boldsymbol{q} \cdot \boldsymbol{r})\boldsymbol{E}_1] \exp [i(\boldsymbol{k} \cdot \boldsymbol{r} - \omega t)], \tag{2}$$

where k_0 is the wave number in vacuum, $\boldsymbol{q}$ is the unit normal to the interface. Later, in 1985, Fedorov and coauthors showed [28] that such waves also can propagate in transparent biaxial crystals. The similar elastic waves in crystals have also been found.

The evolution operator is a key concept in the construction and analysis of the complete system of basis functions for electromagnetic and elastic fields in anisotropic media. The exponential evolution operator which connects field vectors in different points of medium (at different instants of time) is the operational generalization of the scalar phase factor $\exp [i(\boldsymbol{k} \cdot \boldsymbol{r} - \omega t)]$. The relation between eigenwaves and evolution operators is, in fact, a relation between partial and fundamental solutions of a system of ordinary differential equations.

A plane monochromatic wave, propagating in a homogeneous anisotropic or bianisotropic medium in the direction of the unit vector $\boldsymbol{q}$ consists, in general, of two isonormal waves with different refractive indices n_+ and n_- and different polarizations. The total wave is described by the relation

$$\boldsymbol{E}(\boldsymbol{r}, t) = \exp (-i\omega t) \exp (ik_0 z N)\boldsymbol{E}(0), \tag{3}$$

where $z = \boldsymbol{q} \cdot \boldsymbol{r}$ and N is the tensor of refractive indices [29, 30, 31, 32, 33, 34, 35]. Its nonzero eigenvalues and corresponding eigenvectors define refractive indices and polarizations of isonormal waves ($N\boldsymbol{E}_\pm = n_\pm \boldsymbol{E}_\pm$). The total refracted wave, excited in an anisotropic or bianisotropic medium for oblique incidence of time-harmonic waves, can be described in a similar manner [36, 37]

$$\boldsymbol{E}(\boldsymbol{r}, t) = \exp [i(k_0 \boldsymbol{b} \cdot \boldsymbol{r} - \omega t)] \exp(ik_0 z N)\boldsymbol{E}(0), \tag{4}$$

where $\boldsymbol{b}$ is the tangential component of the refraction vector $\boldsymbol{m}_{in} = \boldsymbol{k}_{in}/k_0$ of the incident wave, N is the tensor of normal refraction and the eigenvalues $\eta_\pm$ define the normal components of the refraction vectors $\boldsymbol{m}_\pm = \boldsymbol{b}+\eta_\pm \boldsymbol{q}$ of the refracted eigenwaves ($N\boldsymbol{E}_\pm = \eta_\pm \boldsymbol{E}_\pm$). In the particular case when $\boldsymbol{b} = 0$, the tensor of normal refraction coincides with the tensor of refractive indices. Voigt waves $\boldsymbol{E}$ (1) and Fedorov–Petrov waves $\boldsymbol{E}$ (3) are described by degenerate evolution operators, for which N has a two-fold eigenvalue with one-dimensional eigensubspace.

The tensors N in (3) and N in (4) are defined by the corresponding tensor dispersion equations. It is essential that even in an isotropic medium with scalar permittivity ε and permeability μ this equation retains its tensor nature [38] and has the form $N^2 = \varepsilon\mu I$, where $I = 1 - \boldsymbol{n} \otimes \boldsymbol{n}$ and $\boldsymbol{n}$ is the unit wave normal, 1 is the unit tensor.[3] It has two sets of solutions:

$$N = \pm\sqrt{\varepsilon\mu} I, \tag{5}$$
$$N = \pm\sqrt{\varepsilon\mu}(I - 2\boldsymbol{s} \otimes \boldsymbol{c}), \tag{6}$$

where $\boldsymbol{s}$ and $\boldsymbol{c}$ are arbitrary complex vectors satisfying the conditions $\boldsymbol{s} \cdot \boldsymbol{c} = 1$, $\boldsymbol{s} \cdot \boldsymbol{n} = \boldsymbol{c} \cdot \boldsymbol{n} = 0$. The solutions N (5) describe eigenwaves propagating in $\pm\boldsymbol{n}$ directions. The solutions for N (6) describe the superpositions of two eigenwaves which travel in opposite directions and which have the polarization projection operators $\rho_1 = \boldsymbol{s} \otimes \boldsymbol{c}$ and $\rho_2 = I - \boldsymbol{s} \otimes \boldsymbol{c}$.

The evolution operators characterize the solutions of the Cauchy problem for wave equations and are the mathematical expression of Huygens' principle [39]. When studying eigenwaves the solution consists in finding a set of harmonic functions, for which the wave equations reduce to algebraic equations. However, when studying the evolution operators the solution consists in finding a set of functions, for which the wave equations reduce to a system of ordinary differential equations. This leads to a system with constant coefficients for homogeneous media and a system with variable coefficients for planar stratified media. It is clear that spectral analysis of the principal solution of this system, i.e. the exponential evolution operator in the case of homogeneous medium, permits all types of basic functions to be retrieved. Hence, the possible types of waves are specified by the medium itself, rather than being chosen a priori as being harmonic waves. In particular, here the necessity originates for using evolution operators in the analysis and synthesis of nonharmonic fields in crystals.

The operator of refractive indices is a generalization of the refraction vector. Fedorov clearly realized that the transition from the research of eigenwaves to the research of evolution operators is not only natural and

[3]The symbol $\otimes$ denotes a dyadic or tensor product. –*(Editors)*

useful, but necessary as well. Let us illustrate this by the example of a dispersive medium.

Dielectric properties of a linear homogeneous stationary dispersive medium are described by the integral constitutive equation

$$\boldsymbol{D}(\boldsymbol{r}, t) = \int \varepsilon(\boldsymbol{R}, \tau)\boldsymbol{E}(\boldsymbol{r} - \boldsymbol{R}, t - \tau)d\boldsymbol{R}\,d\tau, \tag{7}$$

i.e. the induction vector $\boldsymbol{D}$ at the point $\boldsymbol{r}$ and at the time t is defined by the values of the strength vector $\boldsymbol{E}$ in some neighbourhood of this point (spatial dispersion) at earlier instants of time (temporal dispersion). For plane harmonic waves, $\boldsymbol{E} = \boldsymbol{E}_0 \exp[i(\boldsymbol{k} \cdot \boldsymbol{r} - \omega t)]$, this equation reduces to a local (algebraic) constitutive equation

$$\boldsymbol{D}(\boldsymbol{r}, t) = \varepsilon \boldsymbol{E}(\boldsymbol{r}, t), \tag{8}$$

where the tensor of dielectric permittivity is a Fourier transformation of the tensor function $\varepsilon(\boldsymbol{R}, \tau)$ and depends on $\boldsymbol{k}$ and ω.

The description of a dispersive medium with the help of the tensor $\varepsilon(\boldsymbol{k}, \omega)$ permits one to consider only harmonic waves and their superpositions. However the tensor $\varepsilon(\boldsymbol{k}, \omega)$ is not a tensor of dielectric permittivity for Voigt waves $\boldsymbol{E}$ (1), i.e. $\boldsymbol{D}(\boldsymbol{r}, t) \neq \varepsilon(\boldsymbol{k}, \omega)\boldsymbol{E}(\boldsymbol{r}, t)$. When introducing the function $\varepsilon(\boldsymbol{k}, \omega)$, the wave vector $\boldsymbol{k}$ and frequency ω are considered as independent. However, this function cannot be measured, since for any harmonic wave excited in the medium, parameters $\boldsymbol{k}$ and ω are related by some dispersive equation. Moreover, in anisotropic media, a wave with given $\boldsymbol{k}$ and ω has a certain polarization state. As a result of these natural restrictions on the experiment only some of the components of tensor $\varepsilon(\boldsymbol{k}, \omega)$ for some values of $\boldsymbol{k}$ and ω can be measured. Another essential disadvantage that appears when solving boundary value problems is the necessity of using a separate dielectric tensor for each partial wave.

The above disadvantages can be eliminated by introducing the dielectric permittivity tensor on the set of evolution operators F [35, 37, 40], defined by the formula

$$\boldsymbol{E}(\boldsymbol{r} + \boldsymbol{R}, t + \tau) = F(\boldsymbol{R}, \tau)\boldsymbol{E}(\boldsymbol{r}, t). \tag{9}$$

For the fields satisfying this relation, the integral constitutive equation (7) reduces [40] to the equation (8), where

$$\varepsilon = \varepsilon(F) = \int \varepsilon(\boldsymbol{R}, t)F(-\boldsymbol{R}, -\tau)d\boldsymbol{R}\,d\tau. \tag{10}$$

This approach enables the use of material equations (8) for the complete system of plane wave solutions of Maxwell's equations in a dispersive medium, including Voigt waves $\boldsymbol{E}$ (1). It is important to note that superpositions of isonormal waves with different wave numbers, frequencies and polarization states are described by a single generalized tensor $\varepsilon(F)$ (10), since the dielectric permittivities of the medium for these eigenwaves are given by different dyad components of the tensor $\varepsilon(F)$. It is also very important to note, that this tensor can be uniquely expressed through refractive indices and polarization parameters of partial waves, i.e. it is a measurable quantity unlike the tensor $\varepsilon(\boldsymbol{k}, \omega)$.

Advantages of Fedorov's covariant approach are most conspicuous in solving boundary value problems [29, 30, 31, 36, 41, 42, 43, 44]. The method makes it possible to expand the methods of surface impedances and characteristic matrices, which are widely and effectively used in the theory of isotropic layered media, to various kinds of anisotropic and bianisotropic media. These operator methods and the obtained solutions considerably widen the possibilities for theoretical investigation of wave properties in various anisotropic, chiral and bianisotropic media as well as in anisotropic elements of various radiotechnical, optical and acoustical systems [30, 31, 36, 38, 41, 42, 44]. They can also be used for the development of new methods for material parameters measurements [9, 41, 42].

In [45, 46] the covariant methods were generalized to the case of geometro-acoustics and geometro-optics of anisotropic media. In continuously inhomogeneous anisotropic media the evolution operators, being the geometrical solutions of the tensor Helmholtz equation for monochromatic fields, are described by multiplicative integrals of normal refraction tensors. These integrals generalize the known scalar optical path lengths. In the framework of this approach, the solutions are obtained without separation of fields into normal waves, by making use of tensor eikonals. It is shown that, even in isotropic media, the eikonal and the refractive index are tensors that are directly associated with the polarization states of the waves.

In this review, we have not aimed to elaborate on the achievements of Academician Fedorov in all areas of the theoretical physics, as it is impossible to accomplish this in one paper. Therefore, we further note only some of the achievements of Fedorov, his pupils and collaborators in the field of physics of elementary particles and fundamental interactions. The general method of projective operators in the theory of relativistic wave equations was developed, the original vector parameterization of Lorentz group and the method of universal nonlinear equations was proposed, the general method of finite transformations of geometric, dynamic and quantum symmetries was advanced, etc.

A great body of supplementary information on works by Fedorov and various applications of his covariant methods to the research of chiral and bianisotropic media can be found in the recent review paper [47].

About Fedorov's contribution in physics one may quote the ancient Roman consuls: "Feci, quod potui, faciant meliora potentes" (Made what I could; let those who can do better), which were pronounced after the report of their activities.

Acknowledgements

The authors are grateful to Dr. L.R. Arnaut for careful reading of the manuscript and various corrections, Prof. S.A. Tretyakov for his assistance and valuable suggestions, Drs. A.V. Lavrinenko and T.M. Lavrinenko for translating the text into English, and A.N. Furs and T.A. Alekseyeva for the assistance in computer processing of the text.

References

1. Pockels, F. (1906) *Lehrbuch der Kristalloptik*, Teubner, Leipzig and Berlin.
2. Szivessy, G. (1928) Kristalloptik, in *Handbuch der Physik* **20**, Springer, Berlin.
3. Fedorov, F.I. (1955) On the polarization of electromagnetic waves, *Dokl. Akad. Nauk SSSR* **102**, 69–71.
4. Fedorov, F.I. (1965) Covariant description of properties of light beams, *Zh. Prikl. Spektrosk.* **2**, 523–533.
5. Fedorov, F.I. (1966) On the beam transformation at the interaction of light with matter, *Zh. Prikl. Spektrosk.* **4** 58–63.
6. Born, M. (1933) *Optik*, Springer, Berlin.
7. Fedorov, F.I. (1955) On the theory of total reflection, *Dokl. Akad. Nauk SSSR* **105**, 465–468.
8. Fedorov, F.I. (1956) The inhomogeneous waves and total reflection, *Trudy Inst. Fiz. Mat. Akad. Nauk BSSR* No. 1, 11–31.
9. Fedorov, F.I. (1958) *Optics of Anisotropic Media*, Izd. Akad. Nauk BSSR, Minsk.
10. Fedorov, F.I. (1976) *Theory of Gyrotropy*, Nauka i Tekhnika, Minsk.
11. Fedorov, F.I. and Filippov, V.V. (1976) *Reflection and Refraction of Light by Transparent Crystals*, Nauka i Tekhnika, Minsk.
12. Tamir, T. (1972) Inhomogeneous wave types at planar structures: I. The lateral wave, *Optik*, **36**, 209–269.
13. Krinchik, G.S. and Chetkin, M.V. (1961) Exchange interaction and magnetooptical effects in ferrite garnets, *Zh. Eksperim. i Teor. Fiz.* **41**, 673–680.
14. Fedorov, F.I. (1956) "Transfer principle" and the general theory of optical properties of absorbing crystals, *Opt. Spektrosk.* **1**, 807–808.
15. Fedorov, F.I. (1959) On the theory of optical activity of crystals. I, II, *Opt. Spektrosk.* **6**, 85–93, 377–383.
16. Fedorov, F.I. (1972) The theory of optical activity of crystals, *Usp. Fiz. Nauk* **108**, 762–764.
17. Fedorov, F.I. (1973) On the optics of magnetic gyrotropic crystals, *Vestsi Akad. Nauk BSSR, Ser. Fiz. Mat. Nauk* No. 5, 70–79.
18. Fedorov, F.I. (1974) Plane waves in gyrotropic crystals, *Vestsi Akad. Nauk BSSR, Ser. Fiz. Mat. Nauk* No. 1, 48–54.
19. Ginzburg, V.L. (1972) Development of crystal optics with taking into account of spatial dispersion, *Usp. Fiz. Nauk* **108**, 749–752.
20. Agranovich, V.M. and Ginzburg, V.L. (1972) On the phenomenological electrodynamics of gyrotropic media, *Zh. Eksperim. i Teor. Fiz.* **63**, 838–843.
21. Agranovich, V.M. and Yudson, V.I. (1973) On the phenomenological electrodynamics of gyrotropic media, *Opt. Commun.* **9**, 58–60.
22. Bokut, B.V., Serdyukov, A.N., and Fedorov, F.I. (1973) On the occasion of the article of V.M. Agranovich "On the phenomenological electrodynamics of gyrotropic media", *Zh. Prikl. Spektrosk.* **19**, 377–380.

23. Fedorov, F.I. (1968) *Theory of Elastic Waves in Crystals*, Plenum Press, New York.
24. Fedorov, F.I. (1979) *Lorentz Group*, Nauka, Moscow.
25. Voigt, W. (1902) Weiteres zur Aufklärung der Eigenschaften pleochroitischer Krystalle, *Gött. Nachr.* **5**, 269–277.
26. Fedorov, F.I. and Petrov, N.S. (1963) Special case of inhomogeneous electromagnetic waves in transparent crystals, *Opt. Spektrosk.* **14**, 256–261.
27. Petrov, N.S. and Fedorov, F.I. (1963) New type of plane electromagnetic waves in absorbing crystals, *Opt. Spektrosk.* **15** , 792–796.
28. Fedorov, F.I., Petrov, N.S. and Filippov, V.V. (1985) Special waves in transparent biaxial crystals, *Zh. Prikl. Spektrosk.* **42**, 844–849.
29. Barkovskii, L.M., Borzdov, G.N. and Fedorov, F.I. (1975) Impedance of a plane electromagnetic waves in optically active media, *Dokl. Akad. Nauk BSSR* **19**, 305–308.
30. Barkovskii, L.M. and Borzdov, G.N. (1975) Tensor impedance and transformation of light beams by systems of anisotropic layers. I. Normal incidence, *Opt. Spektrosk.* **39**, 150–154.
31. Barkovskii, L.M. (1975) One-dimensional wave equation of electrodynamics of inhomogeneous anisotropic media and an operator method for its solution, *Zh. Prikl. Spektrosk.* **23**, 304–309.
32. Barkovskii, L.M. (1976) On the tensor of refractive indices in crystal optics, *Kristallografiya* **21**, 445–449.
33. Barkovskii, L.M. (1979) Spectral expansions of operators of refractive indices in crystals, *Zh. Prikl. Spektrosk.* **30**, 115–123.
34. Fedorov, F.I., Barkovskii, L.M. and Borzdov, G.N. (1982) Tensor dispersive equations, *Dokl. Akad. Nauk BSSR* **26**, 684–687.
35. Barkovskii, L.M., Borzdov, G.N. and Fedorov, F.I. (1983) Operator phases and non harmonic optical fields in crystals, *Zh. Prikl. Spektrosk.* **39**, 996–1000.
36. Barkovskii, L.M., Borzdov, G.N. and Lavrinenko, A.V. (1987) Fresnel's reflection and transmission operators for stratified gyroanisotropic media, *J. Phys.* **A20**, 1095–1106.
37. Borzdov, G.N. (1990) Evolution operators of electromagnetic waves in crystals. I–III, *Kristallografiya* **35**, 535–558.
38. Barkovskii, L.M. and Sharapaeva, V. (1996) Local and global involutive algebras in the problem of light propagation, *Opt. Spektrosk.* **80**, 789–798.
39. Courant, R. (1962) *Partial Differential Equations*, New York-London.
40. Barkovskii, L.M., Borzdov, G.N. and Fedorov, F.I. (1990) Evolution operators in the electrodynamics of spatially dispersive media, *J. Mod. Opt.* **37**, 85–97.
41. Borzdov, G.N. (1992) Lorentz-covariant surface impedance and characteristic matrix methods with applications to measurements of material parameters of linear media, *Opt. Commun.* **94**, 159–174.
42. Borzdov, G.N. (1993) An intrinsic tensor technique in Minkowski space with applications to boundary value problems, *J. Math. Phys.* **34**, 3162–3196.
43. Borzdov, G.N. (1996) Waves with linear, quadratic and cubic coordinate dependence of amplitude in crystals, *Pramana-journal of physics* **46**, 245–257.
44. Barkovskii, L.M., Borzdov, G.N., Zhilko, V.V., Lavrinenko, A.V., Borzdov, A.N., Fedorov, F.I. and Kamach, Yu.E. (1996) The impedance operator method for calculation of electro-optic systems for control laser radiation. I–III, *J. Phys.* **D29**, 289–306.
45. Barkovskii, L.M. and Hang, F.T.N. (1990) Tensor of eikonal in geometrooptical approximations for anisotropic media, *Opt. Spektrosk.* **68**, 670–674.
46. Barkovskii, L.M. and Hang, F.T.N. (1991) Cauchy operators and ray operators in geometrooptics of anisotropic media, *Opt. Spektrosk.* **70**, 61–64.
47. Semchenko, I.V., Tretyakov, S.A. and Serdyukov, A.N. (1996) Research on chiral and bianisotropic media in Byelorussia and Russia in the last ten years, *Progress in Electromagnetic Research* **PIER12**, 335–370.

DEVELOPMENTS OF FEDOROV COVARIANT METHODS AND APPLICATIONS TO OPTICALLY ACTIVE CRYSTALS

A.F. KONSTANTINOVA, A.YU. TRONIN, B.V. NABATOV
Institute of Crystallography
Russian Academy of Sciences
Leninsky pr. 59, 117333 Moscow, Russia

Abstract. The report presents a review of crystal optical activity studies performed at the Institute of Crystallography either in direct collaboration with F.I. Fedorov and his students or influenced by his ideas. These studies give a detailed consideration of the boundary crystal-optics problems accounting for anisotropy, gyrotropy and absorption, the most complicated case of the oblique incidence on the uniaxial gyrotropic layer with arbitrary optical axis orientation and gyration tensor among them. The effects of manifestation of gyrotropy and anisotropy are discussed. Various experimental ellipsometric, spectrophotometric and spectropolarimetric methods for the determination of optical parameters are reviewed as well. The theoretical and experimental approaches for the studies of some gyrotropic crystals and crystalline thin films are described.

1. Introduction

Early in 1958 F.I. Fedorov published his book *Optics of anisotropic media* [1] in which he put forward the covariant (coordinate free) methods for solving various crystal-optics problems. In 1978 F.I. Fedorov published the book *Theory of gyrotropy* [2] in which he presented a detailed consideration of the optical activity (gyrotropy) of crystals. A detailed comparison of covariant and coordinate approaches is given in [3].

According to Fedorov [2], the constitutive equations for gyrotropic media are given as:

$$\begin{aligned} \mathbf{D} &= \epsilon\mathbf{E} + ia\mathbf{H} \\ \mathbf{B} &= \mu\mathbf{H} - i\tilde{a}\mathbf{E} \end{aligned} \qquad (1)$$

A. Priou et al. (eds.), Advances in Complex Electromagnetic Materials, 19–32.

where a is the gyration pseudotensor (further we shall omit the prefix "pseudo"); the tilde stands for transposed operation. In the case of a non-magnetic absorbing crystal, the tensor of dielectric constant ϵ and the gyration tensor a are complex:

$$\epsilon = \epsilon' + i\epsilon'', \qquad a = a' + ia'' \tag{2}$$

a' and a'' describe circular birefringence and circular dichroism, respectively. For uniaxial crystals the expression for a in the covariant form can be written as [2]:

$$a = \alpha_1 + \alpha_2 \mathbf{c} \cdot \mathbf{c} + \alpha_0 \mathbf{c}^\times \tag{3}$$

where $\mathbf{c}$ is the unit vector in the optical axis direction, $\mathbf{c} \cdot \mathbf{c}$ is a dyadic, $\mathbf{c}^\times$ stands for the antisymmetric tensor, dual to the vector $\mathbf{c}$. The coefficients α_i are related to the often used "gyration tensor" as:

$$\begin{aligned} g_{11} &= (\alpha_1 + \alpha_2)\overline{n} \\ g_{33} &= 2\alpha_1 \overline{n} & (4) \\ g_{12} &= -g_{21} = \alpha_0 \overline{n} & (5) \end{aligned}$$

where $\overline{n} = 0.5(n_o + n_e)$ is the mean refraction index of crystal, n_o, n_e are the ordinary and extraordinary refraction indices. Obviously, n_e depends on the direction of wave propagation.

Equations (1)–(2) combined with the Maxwell equations successfully describe the properties of optically active crystals. In particular they state that, in the optically active crystals eigenwaves are two elliptically polarized waves. Their ellipticities k_1, k_2 are different both in value and sign. Moreover, the main axes of the ellipses are non-orthogonal. The deviation of the angle between the ellipse axes from $\pi/2$ is denoted by θ. For transparent, optically active crystals it is usually assumed that $k_1 = -k_2 = k$ and $\theta = 0$.

Let us consider some principal results of the studies of optics of gyrotropic media.

2. Ellipticity of Eigenwaves in Gyrotropic Crystals

2.1. DIFFERENCE BETWEEN THE ELLIPTICITIES OF THE EIGENWAVES IN TRANSPARENT CRYSTALS

As it was shown in [2], the ellipticities of the different eigenwaves even in non-absorbing crystals are not equal, although in many cases this difference is not essential. This difference was experimentally observed for the first time in a quartz crystal by precise measurements with a dichrograph [4].

The signal, registered by a dichrograph from an optically active plate which is cut parallel to the optical axis is represented as

$$D = \frac{n(\overline{n}^2 - n^2)}{\overline{n}^4 + 6\overline{n}^2 + n^4 + n^2}[(2\alpha_1 + \alpha_2)(\cos\Delta - 1)] \tag{6}$$

where $\Delta = 2\pi d\Delta n/\lambda$ (phase retardation), Δn is birefringence of crystal plate, n is the refraction index of the external medium, d is the plate thickness, λ is the wavelength of the incident light. For such an orientation of optical axis the birefringence of the crystal is maximal, giving the greatest difference between the ellipticities of the eigenwaves and the maximum influence of the reflection from the boundaries of the crystal. For inactive crystals the signal registered by the dichrograph should be equal to zero.

The measurements were carried out for two specimens. The first one was made of optically active crystal (quartz) whereas the second was made of inactive crystal (leucosapphire) and played the role of a test sample. The specimens were placed in a dichrograph holder which was rotated at 60 rev/sec. For the gyrotropic crystal the signal registered by the dichrograph has risen by almost an order of magnitude and can not be explained by the residual influence of the polarization effect of the photomultiplier. For verification D was calculated by means of expression (5), starting from the values g_{11}, measured in [5], which showed good agreement of the theoretical calculations and the results of the experiment.

2.2. THE ELLIPTICITIES OF THE EIGENWAVES IN ABSORBING LOW SYMMETRIC CRYSTALS

The principal axes of the ellipses of eigenwave polarizations in absorbing low symmetric crystals are not orthogonal, and neither are the corresponding ellipticities equal (see for details [2]). Analysis of the eigenwave behavior near the optical axis was given in [6]. In the case of absorbing optically active crystals the expressions for eigenwave ellipticities are very cumbersome and cannot be present here. If the wave propagates in the direction at an appreciable angle to the optical axes than the eigenwaves may by considered as orthogonal, the principle of superposition for eigenwave ellipticities is valid, and the cumbersome formulae reduce to (see [6]):

$$\begin{aligned} k_1 &= k + s \\ k_2 &= -k + s \end{aligned} \tag{7}$$

where k is the ellipticity originating from optical activity and s is the ellipticity which is due to absorption in crystal. The calculation results for a

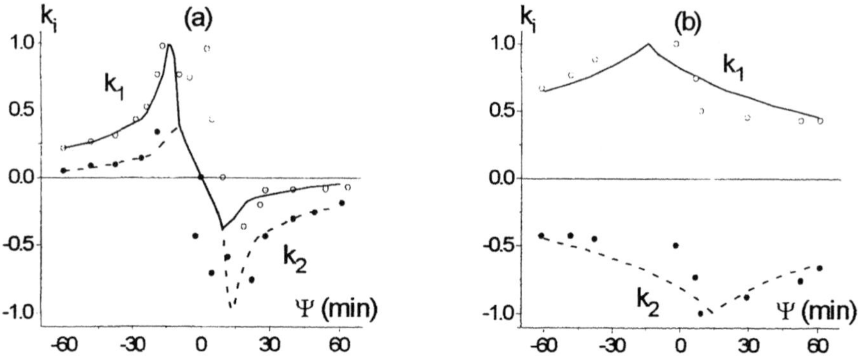

Figure 1. Evaluations of ellipticity of the eigenwave propagating at the angle ψ to the optical axis of non-absorbing crystal of class 222. Solid lines are obtained by exact formulae [6], dots correspond to approximation (6). Fig. 1(a) presents the crystal with the optical parameters: $N_1 = 1.64 + i0.004$, $N_2 = 1.65 + i0.00394$, $N_3 = 1.67 + i0.00404$; $g_{11} = 4.95 \times 10^{-5} + i8.25 \times 10^{-7}$, $g_{22} = 2.47 \times 10^{-5} + i4.95 \times 10^{-7}$, $g_{33} = 6.60 \times 10^{-5} + i3.30 \times 10^{-7}$. Fig. 1(b) presents the crystal with the same refractivity but with the components of gyration tensor enlarged in 10 times

crystal of orthorhombic system of class 222 near the optical axis are given in Figure 1 [6].

Analyzing Fig. 1, we can see that in a gyrotropic crystal the ellipticity of different eigenwaves may be of the same or opposite signs depending on the relation between the values characterizing the optical activity and absorption. The width of the region of substantial non-orthogonality, where the evaluations carried out by the exact formulae in [6] and by (6) disagree, depends on the same relation. One can see that the discrepancy between these calculations is noticeable in the direction close to optical axes.

3. Determination of Optical Parameters in Gyrotropic Crystals at Normal Incidence

3.1. POLARIZATION MEASUREMENTS IN THE PRINCIPAL PLANE

The covariant method appears to be very effective. The problem of light propagation through an optically active plate at normal or oblique incidence accounting for multiple reflections can be easily solved with the help of this method [2]. Using constitutive equations (2) and the Maxwell equations one can write down expressions for two eigenwaves, propagating in the crystal plate. Making use of the boundary conditions at the top and bottom of the plate one can obtain a simultaneous system of eight complex equations for the vector amplitudes of reflected, propagating, and transmitted waves. The boundary problem for the case of an optically active uniaxial transparent crystal was solved in [7–9], for biaxial transparent crystals in [10], for absorbing crystals in [11,12], for biaxial absorbing crystals without multiple reflections in [13]. Experimental determination of the optical parameters of absorbing optically active crystals in directions far away from the optical axis is a problem which is not completely solved yet. The analysis of the boundary problem solution has lead to proposed approaches for the optically active crystals investigation and the simultaneous determination of the anisotropic optical parameters: birefringence, dichroism and optical activity (gyrotropy) [5,12,14,15]. Let us consider some of them.

In the general case, the expressions for the azimuth of the transmitted light (χ) from the azimuth incident light (α) has the following form:

$$\tan 2(\alpha - \chi) = \frac{a_4 \tan^4 \alpha + a_3 \tan^3 \alpha + a_1 \tan \alpha + a_0}{b_4 \tan^4 \alpha + b_3 \tan^3 \alpha + b_1 \tan \alpha + b_0} \qquad (8)$$

where a_i and b_i depend on the parameters $\Delta = 2\pi d \Delta n/\lambda$ (phase retardation), $\delta = 2\pi d \Delta\kappa/\lambda$, the eigenwave ellipticities k_1 and k_2, $\Delta n = n_2 - n_1$ (the elliptical birefringence), and $\Delta\kappa = \kappa_2 - \kappa_1$ (the elliptical dichroism). $N_{1,2} = n_{1,2} + i\kappa_{1,2}$, $n_{1,2}$ are the refractive indices, and $\kappa_{1,2}$ are the absorption coefficients.

For the polarization of the incident light polarized in the main plane ($\alpha = 0$) or perpendicular to that ($\alpha = 90°$) in transparent crystals $k = s = 0$ and $\tan 2\chi = \sin 2\chi = 0$, i.e., the light goes out linearly polarized, without changing the azimuth of the incident light polarization. In the case of active absorbing crystals, the situation is different. For the same incident light polarization one has:

$$\begin{aligned} \tan 2\chi_{\|} &= -2k_1 e^{-\delta} \sin \Delta \\ \tan 2\chi_{\perp} &= -2k_2 e^{\delta} \sin \Delta \end{aligned} \qquad (9)$$

The peculiarities of the optical activity and absorption in directions different from the optical axis manifest themselves most strikingly. In these particular cases of the incident light polarization, the phase difference oscillates like $\sin \Delta$, while k_i changes smoothly. Hence, by recording the dependence of the transmitted light azimuth χ on wavelength or temperature, one can measure the eigenwave ellipticity and the phase retardation. The corresponding dependencies for a crystal of $La_3Ga_5SiO_{14}$ cut parallel to the optical axis are given in Figure 2 [16]. The enveloping curve for the oscillation function is defined by the eigenwave ellipticity k_i (Figure 2(a)). The gyration tensor component g_{11} for a given direction is calculated as

$g_{11} = k2\bar{n}\Delta n$. The other component g_{33} is determined from a change of polarization plane rotation. Figure 2(b) shows the dispersion of components g_{11} and g_{33} as well as the birefringence dispersion for a crystal $La_3Ga_5SiO_{14}$.

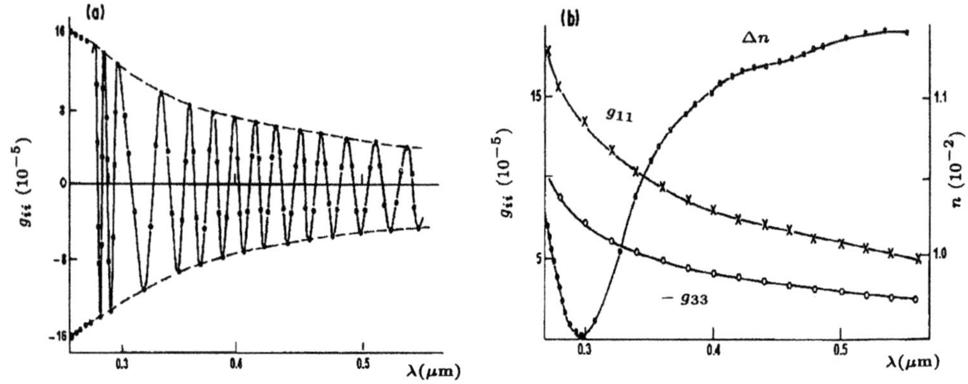

Figure 2. (a) Dependence of the transmitted light azimuth polarization on the wave length λ for the crystal $La_3Ga_5SiO_{14}$ ($d = 1.1$ mm); (b) Dispersion of the gyration tensor components g_{11} and g_{33}, and dispersion of birefringence for the same crystal

The same technique was applied to the study of the benzyl crystal [17]. In the orthorhombic crystal HIO_3 the dispersion of the ellipticities for the three normal waves and the dispersions of all three components g_{ii} were measured [18].

3.2. POLARIZATION MEASUREMENTS FOR DIFFERENT AZIMUTH OF INCIDENT LIGHT

In another variant of the polarization method the transmitted light azimuth χ is measured for various values of the incident light α [19]. Using the relationship (7), one can calculate the values of optical parameters in crystals of any symmetry possessing the optical activity and absorption.

We applied this method for investigation of yttrium, erbium and holmium formiate dihydrate crystals [20–22]. The most characteristic dependence $(\alpha - \chi) = f(\alpha)$ obtained for the plate made of $Er(HCOO)_3 2H_2O$ is shown in Figure 3. The legend for Figure 3(a) contains the computed values Δ, δ, k_1, k_2. As a result of computation, the non-orthogonality angle θ proved to be smaller than 1.5°, even in the strongest absorption bands, despite the fact the measurements were made near the optical axis. Calculated dependence of the values $k(\lambda)$ ellipticities of eigenwaves are given in Figure 3(b) [23].

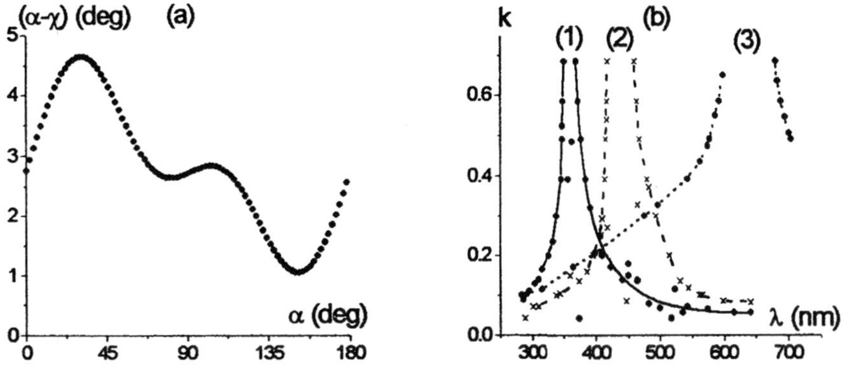

Figure 3. (a) Dependence $(\alpha - \chi) = f(\alpha)$ for the plate made of the $Er(HCOO)_3 2H_2O$ crystal (d = 0.135 mm) at wavelength λ = 0.448 μm and the values of $\Delta = 19.137°$, $\delta = 0.0431$, $k_1 = 0.1532$, $k_2 = -0.1388$ obtained by treating the experimental data; (b) Experimental values of ellipticity $k(\lambda)$ for formiate dihydrate erbium (1) formiate dihydrate holmium (2), and formiate dihydrate yttrium (3) crystals. The curves correspond to theoretical evaluations of $k(\lambda)$, evaluated without taking into account the influence of absorption

From Figure 3 we can see that the best coincidence of the direction of the wave normal of the incident light with the optical axis of the crystal occurs for specimens of yttrium, holmium, and erbium formiate dihydrates at $\lambda = 650$, 440, and 350 nm, respectively. The change in wavelength leads, owing to dispersion of the optic axis, to reduction in the values of k. It can also be seen that the smaller the value of the ratio of the linear to the circular birefringence, the greater is the half-width of the curve of $k(\lambda)$. This ratio is smaller for yttrium formiate dihydrate and greater for erbium formiate dihydrate.

The polarization method has proved to be convenient when the measurements are done under changing external conditions, such as external electric and magnetic fields, temperature and some other. A number of crystals were studied by means of this technique, magnetically ordered [24] and crystals at phase transitions [25] among them.

3.3. THE SPECTROPHOTOMETRIC METHOD

Another approach for determining the optical parameters of crystals with various combinations of optical properties was proposed in [26]. This method is based on the measurement of the light intensity transmitted through the sample placed between a polarizer and an analyzer. The dependence of the intensity $(I/I_0)(\alpha)$ on the rotation of the sample with crossed and parallel polarizers is measured and analyzed. In this case the expression has the form:

$$I/I_0 = a + b_1 \cos 2\alpha + b_2 \sin 2\alpha + c_1 \cos 4\alpha + c_2 \sin 4\alpha \tag{10}$$

where a, b_j, and c_j depend on the mutual orientations of the axes of the polarizer and analyzer and on the optical parameters of the plate [26].

The theoretical basis of this approach was developed in [26], where the general case of non-orthogonal eigenwaves was considered. The consideration was based on Müller matrix formalism. From the measurements of the light intensity for crossed polarizer and analyzer one can find out the ellipticity and the non-orthogonal angle between the eigenwaves. Then, from measurements of light intensity for parallel polarizer and analyzer one can calculate the birefringence and dichroism. This technique was applied to various crystals [27].

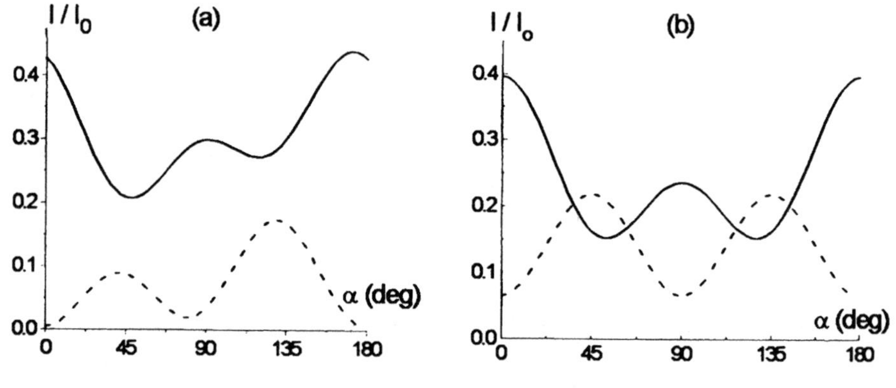

Figure 4. (a) The calculated dependence of the intensity $(I/I_0)(\alpha)$ on the rotation angle of the plate in the crossed (dashed curve), parallel (solid curve) polarizer and analyzer for low symmetrical absorbing gyrotropic crystal: $\Delta = 70°$, $\delta = 0.15$, $k_1 = 0.1$, $k_2 = -0.2$, $\theta = 10$; (b) The measured dependence of the intensity $(I/I_0)(\alpha)$ for the plate from brown quartz: $d = 2.48$ mm at $\lambda = 0.55$ μm and calculated optical parameters: $\Delta = 98.68° + 360°$, $\delta = 0.256$, $k_1 = -k_2 = 0.296$, $\theta = 0$

It is seen from Figure 4(a) that in the low-symmetrical absorbing gyrotropic crystal the maximum and minimum values $(I/I_0)(\alpha)$ are different for crossed polarizer and analyzer in comparison with a uniaxial absorbing gyrotropic crystal.

4. Oblique Incidence of Light on Crystal

Up to now we have described the studies of crystals at the normal incidence. However, in many cases one encounters the problem of oblique incidence. The covariant methods make it possible to solve the problem of light reflection and propagation in this case as well. Let us consider the light incident at an arbitrary angle on a plate of uniaxial gyrotropic crystal with arbitrary axis orientation. The electromagnetic field of incident, reflected and transmitted beams can be decomposed into two components: one parallel and the other perpendicular to the plane of incidence:

$$\begin{aligned}\mathbf{E}_l &= A_{pl}\mathbf{a} + A_{sl}[\mathbf{n}_l\ \mathbf{a}] \\ \mathbf{H}_l &= A_{pl}[\mathbf{m}_l\ \mathbf{a}] - A_{sl}n_l\mathbf{a}\end{aligned} \qquad (11)$$

where p, s stand for the parallel and perpendicular polarizations, $l = i, r, t$ denotes the incident, reflected and transmitted waves.[1]

The expressions for the refracted waves were obtained in [28]. The same relations can be obtained in a somewhat different form, which is more convenient because of the obvious reducibility to the non-gyrotropic case [29]:

$$\begin{aligned}\mathbf{E}_o &= A_o[\mathbf{m}_o\ \mathbf{c}] - ik_o\mathbf{c} + iw_o\mathbf{m}_o \\ \mathbf{H}_o &= A_o[\mathbf{m}_o[\mathbf{m}_o\ \mathbf{c}]] - ik_{o1}[\mathbf{m}_o\ \mathbf{c}] - i\alpha_o[\mathbf{c}[\mathbf{m}_o\ \mathbf{c}]] \qquad (12) \\ \mathbf{E}_e &= A_e z_e\mathbf{c} - t_e\mathbf{m}_e + ik_e[\mathbf{m}_e\ \mathbf{c}] + iw_e\mathbf{m}_e \\ \mathbf{H}_e &= A_e z_e[\mathbf{m}_e\ \mathbf{c}] - if_e\mathbf{c} + it_e k_{e1}\mathbf{m}_e - it_e\alpha_o[\mathbf{m}_e\ \mathbf{c}] \qquad (13)\end{aligned}$$

where A_o and A_e are the amplitudes of the refracted waves; indices 'o' and 'e' correspond to the ordinary and extraordinary beams. $\mathbf{m}_l$, $\mathbf{m}_{o,e}$ are the refraction vectors, k_o, w_o, k_{o1}, z_e, t_e, k_e, w_e, k_{e1}, f_e are some coefficients which depend on optical parameters of the system.

Substituting the expressions (10–12), together with similar expressions for the waves reflected from the lower interface, into the boundary conditions [1], one obtains the system of eight complex equations for the amplitudes of all the waves. This system was solved by the Gauss technique and through the singlular-value decomposition [30,31]. On the amplitudes of reflected waves have been calculated, one can readily determine such parameters of reflection as: azimuth χ' and ellipticity k' of reflected wave, diagonal (R_{pp} and R_{ss}) and off-diagonal (R_{ps} and R_{sp}) reflection coefficients, etc.

[1] [a b] denotes vector (cross) product of vectors a and b. –*(Editors)*.

Typical dependencies of azimuth and ellipticity of reflected light on rotation along plate normal are shown in Figure 5. Calculations were done for $n_{\text{medium}} = 1$ and $n_{\text{medium}} = 1.495$. One can see that due to the optical activity the symmetry of the curves changes. The curve corresponding to the inactive film is antisymmetric as a function of $(\rho - 180°)$, while that for the optically active film does not posses any property of symmetry. This effect is more pronounced when the refractive index of the external medium approaches that of the film (curve 4). Breaking of the symmetry due to the optical activity is also observed for the azimuth of the reflected light (Fig. 5a).

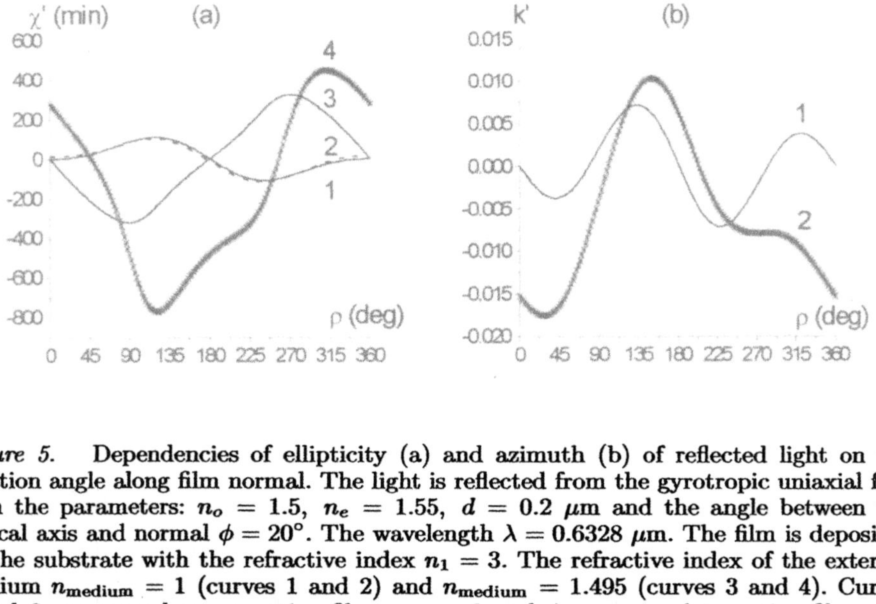

Figure 5. Dependencies of ellipticity (a) and azimuth (b) of reflected light on the rotation angle along film normal. The light is reflected from the gyrotropic uniaxial film with the parameters: $n_o = 1.5$, $n_e = 1.55$, $d = 0.2\ \mu$m and the angle between the optical axis and normal $\phi = 20°$. The wavelength $\lambda = 0.6328\ \mu$m. The film is deposited on the substrate with the refractive index $n_1 = 3$. The refractive index of the external medium $n_{\text{medium}} = 1$ (curves 1 and 2) and $n_{\text{medium}} = 1.495$ (curves 3 and 4). Curves 1 and 3 correspond to nonactive film, curves 2 and 4 correspond to active film with $\alpha_1 = 0.003$, $\lambda_2 = -0.01$, $\alpha_0 = 0$. Incident light is p-polarized

Before Fedorov developed the detailed analysis of the phenomenon of optical activity, the optical activity had been regarded as intrinsic only to the crystals with a symmetric gyration tensor, and, thus, 15 symmetry classes had been considered to be gyrotropic. Fedorov showed that crystals with an antisymmetric tensor reveal optical activity as well, and, thus, 3m, 4mm and 6mm classes should be considered as gyrotropic [2]. The light reflection from such crystals were studied in [32] and the peculiarities of the manifestation of optical activity in this case were shown. The appearance of an additional phase change in the reflected waves due to the optical activity was shown in [33]. This effect is enhanced near the Brewster angle.

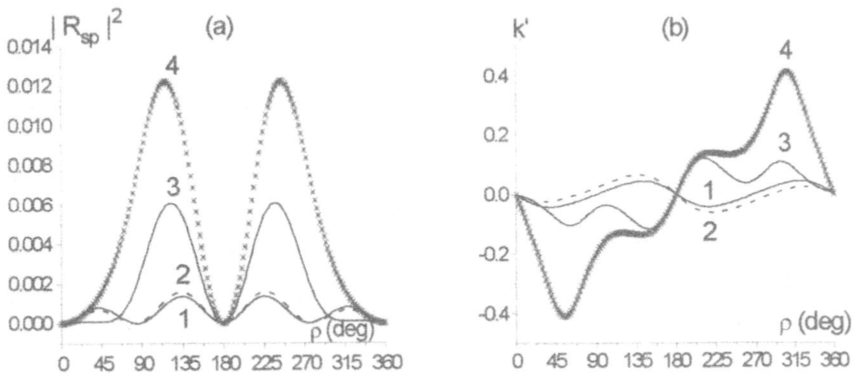

Figure 6. Dependencies of off-diagonal intensity reflection coefficient (a) and ellipticity of reflected light (b) on the rotation angle along film normal. The light is reflected from the gyrotropic uniaxial film with the parameters: $n_o = 1.50$, $n_e = 1.55$, $d = 0.2\ \mu$m and angle between optical axis and normal $\phi = 75°$. The wavelength $\lambda = 0.6328\ \mu$m. The film is deposited on the substrate with refractive index $n_1 = 3$. The refractive index of the external medium $n_{\mathrm{medium}} = 1$ (curves 1 and 2) and $n_{\mathrm{medium}} = 1.495$ (curves 3 and 4). Curves 1 and 3 correspond to nonactive film, Curves 2 and 4 correspond to active film with $\alpha_1 = 0$, $\alpha_2 = 0$, $\alpha_0 = 0.03$. Incident light is p-polarized

Reflection from crystal film was studied recently [34]. The influence of optical activity of the crystal film is demonstrated in Figure 6. There, Figure 6(a) presents dependencies of off-diagonal intensity reflection coefficient on the rotation along the normal to the film. The calculations were done for $n_{\mathrm{medium}} = 1$ and $n_{\mathrm{medium}} = 1.495$. The notation of the curves is the same as in Figure 5. One can see that the effect of optical activity is enhanced when the refractive index of the external medium is close to that of the film. A similar behavior is noticed for the reflected light ellipticity (Figure 6(b)). The influence of the external medium is such pronounced, that for some particular values of n_o, n_e, n_{medium}, and ρ one can observe almost circular polarization of the reflected light.

5. Conclusion

The ideas of Academician F.I. Fedorov have had a great impact on the development of theoretical as well as experimental methods for the studies in optics of anisotropic media in general and the optics of active crystals in particular. The main achievements in this field can be summarized as follows:

- New material equations for gyrotropic anisotropic media were proposed and proved.

- The phenomenon of optical activity was generalized to embrace not only rotation of polarization, but many other effects of changing the polarization state originating from non-zero gyration tensor.
- Crystals with antisymmetric tensor were shown to display optical activity and thus, the total number of gyrotropic classes is equal to 18.
- Covariant methods make it possible to solve very many boundary problems for various anisotropic systems.
- An extended analysis of the boundary tasks solutions give rise to a broad variety of experimental techniques for studying the effects of gyrotropy in absorbing anisotropic media.

The reference list given below is not complete. Many papers of the Fedorov school and papers of other scientific schools could not be included. We have tried to review only the studies done at the Institute of Crystallography of the Russian Academy of Sciences in Moscow. An interested reader can find more complete lists of references in the cited literature.

Acknowledgment

Authors are sincerely grateful to the co-authors of the original papers for assistance and helpful discussions.

References

1. Fedorov, F.I. (1958) Optics of Anisotropic media, AN BSSR (Academy of Sciences of the Belarus Soviet Socialist Republic), Minsk.
2. Fedorov, F.I. (1976) Theory of gyrotropy, Nauka i Tekhnika, Minsk.
3. Konstantinova, A.F., Grechushnikov, B.N., Bokut', B.V., and Valjashko, E.G. (1995) Optical Properties of Crystals, Nauka i Tekhnika, Minsk.
4. Baturin, N.A., and Konstantinova, A.F. (1987) Influence of difference in ellipticities of proper waves and multiple reflections on signal registered by a dichrograph, *Sov. Phys. Crystallogr.* **32**(5), 697-699.
5. Konstantinova, A.F., Ivanov, N.R., and Grechushnikov, B.N. (1969) Optical activity of crystals in directions different from that of the optic axis. I. Uniaxial crystals, *Sov. Phys. Crystallogr.* **14**(2), 222-229.
6. Okorochkov, A.I., and Konstantinova, A.F. (1984) Characteristic waves in absorbing gyrotropic crystals of the orthorhombic system, *Sov. Phys. Crystallogr.* **29**(5), 497-502.
7. Fedorov, F.I., and Konstantinova, A.F. (1962) Propagation of light through a plate of uniaxial optically active crystals, *Optics Spectrosc.* **12**(3), 407-411.
8. Fedorov, F.I., and Konstantinova, A.F. (1962) Propagation of light through a plate of uniaxial optically active crystals. II. Plates are parallel to the optic axis, *Optics Spectrosc.* **12**(4), 505-509.
9. Bokut', B.V., Konstantinova, A.F., and Serdyukov, A.N. (1972) Propagation of light in optically active uniaxial crystals, *Sov. Phys. Crystallogr.* **17**(4), 711-713.
10. Konstantinova, A.F., Okorochkov, A.I., and Filippov, V.V. (1983) Transmission of light by absorbent optically active crystals of the orthorhombic system, *Sov. Phys. Crystallogr.* **28**(5), 502-504.

11. Konstantinova, A.F., Shepelevich, V.V., Bokut', B.V., Grechushnikov, B.N., Kaldybaev, K.A., Perekalina, Z.B., and Serdyukov, A.N. (1976) Characteristics of the appearance of optical activity in absorbing crystals, *Sov. Phys. Crystallogr.* **21**(6), 641-643.
12. Konstantinova, A.F., Ulukhanov, I.T., and Grechushnikov, B.N. (1991) Spectrophotometric method of determining the optical parameters of gyrotropic crystals, *Sov. Phys. Crystallogr.* **36**(3), 381-384.
13. Okorochkov, A.I., and Konstantinova, A.F. (1985) Influence of nonorthogonal characteristic waves in a crystal on the polarization of transmitted light, *Sov. Phys. Crystallogr.* **30**(1), 57-62.
14. Ivanov, N.R., and Konstantinova, A.F. (1970) Optical activity of crystals in directions different from an optic axis direction. II. Biaxial crystals, *Sov. Phys. Crystallogr.* **15**(3), 416-422.
15. Grechushnikov, B.N., and Konstantinova, A.F. (1988) Crystal optics of absorbing and gyrotropic media, *Comput. Math. Applic.* **16**(5-8), 637-655.
16. Kaminskii, A.A., Mill, B.V., Khodzhabagyan, G.G., Konstantinova, A.F., and et.al. (1983) Investigation of trigonal $(La_{1-x}Nd_x)_3Ga_3SiO_{14}$ crystals. I. Growth and optical properties, *Phys. Stat. Sol.* **80**(a), 387-398.
17. Kaldybaev, K.A., Konstantinova, A.F., Perekalina, Z.B., Grechushnikov, B.N., and Kalinkina, I.N. (1978) Optical activity and circular dichroism in benzil, *Sov. Phys. Crystallogr.* **23**(4), 438-442.
18. Ivanov, N.R., Chikhladze, O.A. (1976) Measurement of the gyration tensor for orthorhombic α-HIO_3, *Sov. Phys. Crystallogr.* **21**(1), 65-68.
19. Grechushnikov, B.N., Konstantinova, A.F., Lomako, I.D., and Kalinkina, I.N. (1980) Optical activity and absorption in birefringent crystal for various azimuths of the incident light polarization, *Sov. Phys. Crystallogr.* **25**(3), 346-348.
20. Okorochkov, A.I., Konstantinova, A.F., Perekalina, Z.B., Grechushnikov, B.N., Soboleva, L.V., and Khapaeva, L.I. (1983) Measurement of optical activity near an optic axis in transparent biaxial crystals, *Sov. Phys. Crystallogr.* **28**(4), 436-438.
21. Okorochkov, A.I., Konstantinova, A.F., Soboleva, L.V., and Khapaeva, L.I. (1984) Optical activity of the biaxial absorbent crystal of $Er(HCOO)_3 2H_2O$, *Sov. Phys. Crystallogr.* **29**(6), 645-649.
22. Okorochkov, A.I., Konstantinova, A.F., Soboleva, L.V., and Khapaeva, L.I. (1985) Optical activity of crystals of holmium, erbium, and yttrium formiate, *Sov. Phys. Crystallogr.* **30**(3), 347-348.
23. Baturin, N.A., Konstantinova, A.F., Okorochkov, A.I., Grechushnikov, B.N., and Perekalina, Z.B. (1986) Measuring the circular dichroism of biaxial crystals, *Sov. Phys. Crystallogr.* **31**(2), 237-239.
24. Lomako, I.D., Konstantinova, A.F., and Ges', A.P. (1990) Determination of the optical parameters of samarium-terbium orthoferrites by the polarization method, *Sov. Phys. Crystallogr.* **35**(2), 240-242.
25. Baturina, O.A., Brezina, B., Bogachek, P., Perekalina, Z.B., and Konstantinova, A.F. (1987) Investigation of optical properties of crystals at phase transitions by means of the polarization method, *Sov. Phys. Crystallogr.* **32**(6), 847-849.
26. Konstantinova, A.F., Evdischenko, E.A., and Ulukhanov, I.T. (1994) Determination of Mueller matrix elements and optical parameters of absorbing gyrotropic low-symmetric crystals, *Crystallogr. Rep.* **39**(5), 712-719.
27. Knab, G.G., Konstantinova, A.F., Ulukhanov, I.T., and et.al. (1990) Influence of activators on the optical and mechanical properties of $ZrO_2 - Y_2O_3$ crystals, *Sov. Phys. Crystallogr.* **35**(6), 923-929.
28. Filippov, V.V., Tronin, A.Yu., and Konstantinova, A.F. (1994) Ellipsometry of anisotropic media, *Sov. Phys. Crystallogr.* **39**(2), 313-335.
29. Konstantinova, A.F., Nabatov, B.V., and Tronin, A.Yu. (1995) Reflection and propagation of light through an uniaxial optically active slab, in A. Sihvola, S. Tretyakov, U. Unrau, V.V. Varadan, V.K. Varadan, and K. Whites (eds.), Proceedings of *Chi-*

ral'95, 4th Int. Conf. on Chiral, Bi-isotropic and Bi-anisotropic Media, Penn State University, USA, 101-104.

30. Tronin, A.Yu., Konstantinova, A. F. (1989) Ellipsometric study of optical anisotropy of lead arachidate langmuir films, *Thin Solid Films* **177**, 305-314
31. Nabatov, B.V., Konstantinova, A.F., and Tronin, A.Yu. (1996) Application of ellipsometry studies of anisotropic and gyrotropic films, *Proceeding SPIE* (in press).
32. Fedorov, F.I., Bokut', B.V., and Konstantinova, A.F. (1963) Optical activity of crystals of the planar classes of the middle groups, *Sov. Phys. Crystallogr.* **7**(6), 738-741.
33. Asimova, V.D., Konstantinova, A.F., Filippov, V.V., and Sender, N.N. (1985) Reflection of light from boundaries with crystals of the planar classes of the middle systems, *Sov. Phys. Crystallogr.* **30**(5), 546-547.
34. Konstantinova, A.F., and Nabatov, B.V. (1995) Manifestation of the optical activity in uniaxial crystals of the mirror-reflection classes, *Crystallogr. Rep.* **40**(2), 197-200.

BI-ISOTROPIC AND BI-ANISOTROPIC MEDIA

General Review

S. BOLIOLI
CERT-ONERA - Microwave Department
2, Avenue Edouard Belin
31055 Toulouse Cedex
FRANCE

1 Introduction

Bi-anisotropic media have been of great interest for many years in the scientific community dealing with the electromagnetic properties of materials. Because of their diversity, they have found many potential applications, in a very large wavelength spectrum, from microwaves to optical frequencies.

The particular properties of these media are due to a coupling between electric and magnetic fields which can be described by some constitutive relations. In the most general case, the permittivity, permeability, and coupling factor are full 3x3 terms tensors; hence, 36 complex parameters are required to describe completely these properties. Moreover, dispersion characteristics may emphasise the complexity of the problem. As an example, a set of homogeneous constitutive relations can be expressed as follows :

$$\begin{cases} \vec{D} = \tilde{\varepsilon}.\vec{E} + \tilde{\xi}.\vec{H} \\ \vec{B} = \tilde{\zeta}.\vec{E} + \tilde{\mu}.\vec{H} \end{cases} \tag{1}$$

Where $\vec{D}$, $\vec{B}$ are the electric and magnetic inductions, $\vec{E}$, $\vec{H}$ are the electric and magnetic fields, $\tilde{\varepsilon}$, $\tilde{\mu}$, $\tilde{\xi}$, $\tilde{\zeta}$ are the permittivity, permeability and coupling tensors. This formalism is due to Lindell and Sihvola. Of course, one can find some other different ways to express the magneto-electric coupling, like the Post-Jaggard's or the Drudde-Born-Fedorov's formalisms. In the particular case of anisotropic media, there are no coupling tensors. Another case of great interest is the bi-isotropic medium, where all tensors reduce to scalars. The chiral media belong to this last class of materials.

These chiral media, well known for a very long time in optics, present optical activity and circular dichroïsm in the case of lossy medium. In the microwave range, optical activity is achieved by embedding chiral 3D objects in a host medium. A recent and very extensive overview on these media, their properties and applications has been proposed by Corry [1].

Up to now, no industrial application of chiral, or more generally bi-anisotropic media has been carried out, particularly in the microwave frequency range. As a matter of fact, there is no chiral medium that could be manufactured according to industrial constraints (cost, weight, mechanical properties,...). So should this topic be considered as purely

A. Priou et al. (eds.), Advances in Complex Electromagnetic Materials, 33–51.

academic science? One should better think that it is not! Nowadays, the available technology is not yet suitable for mass production. But the advances in composite and polymer chemistry science, and in the reduction of the size of chiral particles make it reasonable to expect to dispose of such kind of materials for engineering science in the next future. This justifies the large number of publications or communications that have been presented in different workshops these last years, and especially those in Espoo, Gomel, Perigueux, and Pennsylvania. In this paper, the author would like to give a review of different works published since the beginning of 1995, on bi-isotropic and bi-anisotropic media. A rough overview pointed out four different classes of problems. At first, when dealing with complex media, it is worthwhile knowing how to describe their intrinsic properties by developing new models or bringing out some new aspects in constitutive relations. Some kind of new particle can also be investigated. The second class of problems consists in the realisation of the corresponding media, and their characterisation. To solve this last challenge, one has often to refer to the third class, which concerns the propagation through waveguides or transmission lines containing these media. The last class is more connected to applications like stealth coatings, antennas, or polarisers. It requires to solve the problems in terms of reflection, diffraction, or radiation.

2 Constitutive parameters

2.1 TELLEGEN MEDIA : AN OPEN QUESTION

Non-reciprocal bi-isotropic media, so called Tellegen media, can be described by using the following set of constitutive relations :

$$\begin{cases} \vec{D} = \varepsilon.\vec{E} + \chi.\sqrt{\varepsilon_0\mu_0}.\vec{H} \\ \vec{B} = \chi.\sqrt{\varepsilon_0\mu_0}.\vec{E} + \mu.\vec{H} \end{cases} \qquad (2)$$

The possible existence of Tellegen media has been discussed for a long time on the « CHIRAL-L » network.

The main arguments against this existence are developed by Weiglhofer and Lakhtakia [6], [69], proving the uniformity constraint using the mathematical structure of Maxwell's curl equations. It can be expressed through $\mathrm{Trace}(\tilde{\alpha} - \tilde{\beta}) = 0$, where $\tilde{\alpha}$ and $\tilde{\beta}$ are the magneto-electric coupling dyadics in the Post-Jaggard formalism.

The opposite point of view is defended by Sihvola, Raab *et al.* [19], [38], [98], [106], by pointing out some natural examples of cubic magnetic crystal classes, or by giving a way to manufacture Tellegen media by imbedding some pairs of « glued » dielectric and magnetic dipoles in a dielectric host.

Until now, no definite answer has been given to this question which still remains open.

2.2 NEW INCLUSIONS AND COMPOSITE MEDIA

The chirality, or more generally the bi-anisotropic property in composites is achieved by using special inclusions such as the well known metallic helices. Sauviac, Mariotte *et al.* [92], and Guerin [122] developed new composites based on dielectric multi-turn helices. The shape of the scatterer is also of great importance. Kostin, Shevchenko *et al.* [93], [97], introduced a new concept of helically conductive sphere, imbedded in a space cubical lattice. Some other kind of new inclusion is the chiral « hook » described by Theron [100] to get a non-magnetic uniaxial chiral crystal, or the knotted media by Tretyakov *et al.* [64]. Arnaut [72] presented some results on planar chiral structures, introducing new considerations on the notion of chirality and wave propagation in a multi-dimensional context.

Typical racemic composites have been investigated by Vinogradov *et al.* [114] to get an artificially magnetic medium. Semchenko *et al.* considered the case of heterogeneous structures like a plasma in a magnetic field [109], when Slepyan *et al.* interested themselves in media described by periodically inhomogeneous constitutive parameters [21], as well as in non-linear media [14].

One can see the diversity of interests, plasmas and composites, various inclusions, random distributions in a dielectric, magnetic, or even chiral medium, or regular arrays. The same diversity is found with the models used to describe the electromagnetic properties.

2.3 MODELS BASED ON THE ANALYSIS OF THE DIFFRACTION OF THE SINGLE INCLUSION

One way to deal with composite media including metallic objects is analysing first the electromagnetic properties of a single scatterer before considering the medium as effective, or applying a multi-diffusion theory or some kind of purely numerical technique.

Tretyakov *et al.* [34] have developed an analytical approach to describe the scattering of Omega-shaped conductive particles, based on an antenna model. A full numerical method has been developed by F. Mariotte *et al.* [33], [91], [92] using the classical Method of Moments. An improvement to this technique has been reported by Yung and Wong [5], who introduced some curved segments, more suitable to describe structures like helices.

The case of regular arrays has been treated by Peters and Newman [41] applying the Periodic Method of Moments to wire scatterers in a 3D homogeneous medium and by Chuprin *et al.* to linear arrays of multifilar helices [25].

2.4 DEVELOPMENT ON BASIC EQUATIONS

An important contribution is proposed by Arnaut [101], [104], [120], on the computation of dipolarisability dyadics for chiral or bianisotropic composite media based on helices. The case of high density is developed, taking into account the mutual coupling effects. Lindell and Olyslager have generalised the TE/TM waves decomposition using duality transformations to axially chiral anisotropic media [49],[50]. Dyadic Green's functions are also of great interest; some asymptotic expressions are given by Jakoby and Olyslager [46], when Weiglhofer presents some one-dimensional forms [45], [71].

To describe the macroscopic properties of these media, Oberschmidt and Jacob examined two ways to calculate the effective behaviour of randomly distributed chiral particles [40], the first one is based on the Vector Radiative Transfer Equation, and the other one on the Lorenz-Lorentz mixing formulas. Whites considered this problem by

using a full wave Monte-Carlo scattering solution [9]. A generalised Maxwell Garnett modelling is presented by Tretyakov and Mariotte [31].
In order to improve the compacity in writing Maxwell's equations when dealing with bianisotropic media, Lindell and Sihvola introduced the six-vector formalism, derived from Heaviside [30].
Some time domain formulations have been also developed, like the description of the propagation of a harmonic pulse in a chirotropic medium proposed by Hillion [108].
All these publications show that the problem of the interaction between an electromagnetic wave and a general bianisotropic medium is far from being completely solved.. Nevertheless, significant theoretical advances have brought better understanding and descriptions in specific cases.

3 Measurements and inverse problems

3.1 MATERIAL MANUFACTURING

According to the very few publications reporting experimental results, it is obvious that manufacturing chiral media not very simple. Kuehl *et al.* [86] present a process suitable for mass production of composites including wire based helical springs. Samples have been realised for free-space measurements, with random or oriented distribution. Had and Tunnicliffe [90] addressed some of the issues involved in the fabrication of biisotropic and bianisotropic samples, for circular waveguide or free space characterisation.

3.2 MEASUREMENTS

It is naturally of fundamental interest to predict the material properties, but it is also of great importance to be able to measure the constitutive parameters. Different approaches are present in the literature, the most commons being the waveguide and the free-space techniques.

3.2.1 Waveguide

The specific case of Ω media has been treated by Norgren and He [13], [28], using a rectangular waveguide. The chirality parameter of a lossy chiral slab was extracted by Reinert *et al.* [87] from the scattering data for two different slab length, in a circular waveguide operating in the fundamental mode. Theron and Cloete [102] carried out some experiment on a 2m long chiral crystal, showing that not only dipoles but also quadrupoles should be included when modelling these media. A more general approach, based on the perturbation theory, was developed by Tretyakov and Viitanen [29]. It has been applied to arbitrary shaped waveguide and to resonators.

3.2.2 Free space

To extract the effective chirality parameter of a bi-isotropic slab, Luebbers *et al.* [11] combine the Method of Moments applied to a single scatterer and effective medium theory. Whites [89] presents a comparison between computed and measured optical rotation on chiral materials in the X frequency band. Smith *et al.*[88] interested themselves in determining the accuracy of a free-space set-up for measuring the microwave chiral parameter.

4 Waveguides and transmission lines

4.1 CIRCULAR WAVEGUIDES

The propagation in bianisotropic waveguides is analysed by Wu [54], using a linear-operator formalism, and introducing the concept of adjoint waveguides. A coaxial waveguide, with an inner inhomogeneous medium composed with cylindrical layers, and an outer specific conductor, was analysed by Jakoby and De Zutter [42]. A more conventional one filled with bianisotropic material was studied by Graglia *et al.* [74]. Xu [56] presents the characteristic equation of a circular waveguide filled with chiroferrite, which is a specific case of bianisotropic medium. Studying the propagation in circular chirowaveguides, Busse and Jacob [60] showed the existence of complex power and mode coupling.

4.2 ARBITRARY SHAPED WAVEGUIDES

For arbitrary shaped waveguides, one can also find different approaches. A numerical formulation has been developed by Valor and Zapata [32], based on the Finite Element Method, to treat lossy inhomogeneous bianisotropic waveguides. A generalisation of the Telegraphist's Equation has been fulfilled by Pereda *et al.* [52].
After analysing the field distribution in the waveguide, Kamenetskii formulated some mode orthogonality relations, including complex modes.

4.3 PLANAR TRANSMISSION LINES

The most common planar structures, like the patch studied by Toscano and Vegni [58], [78], are derived from the microstrip line. An approximate formulation based on a quasi-static image method has been carried out by Koivisto and Sten [35], while Hanson [44] developed numerically the dyadic Green's functions. Kyriacou and Sahalos [15] studied a multilayer microstrip line, illuminated by an oblique incident TEM wave.

4.4 OPEN WAVEGUIDES

The modal equation corresponding to a circular or planar open waveguide has been written by Mahmoud [37], while Qiu and Lu [67] developed the concept of chiral step-index optical fibber.

5 Reflection, diffraction and radiation problems

One of the most important field of applications of chiral media in the microwave range is the stealth technology and more especially absorptive coatings. But among their potential applications one finds components life filters and polarisation transformers, antennas. The corresponding problems can be expressed in terms of propagation, scattering and radiation.

5.1 SPHERICAL, CYLINDRICAL AND ELLIPSOIDAL STRUCTURES

Karonis *et al.* [27] examined the scattering of a plane wave by a chiral ellipsoid, while Kaklamani and Uzunoglu were interested in the radiation of a sphere containing an electric dipole.
Li *et al.* [10], [47], and Yin *et al.* [48] carried out the dyadic Green's functions for multilayer structures.

5.2 PLANAR, SINGLE- OR MULTILAYERED STRUCTURES

The propagation of light through a uniaxial gyrotropic film illuminated with an oblique incidence has been investigated by Konstantinova *et al.* [105]. The diffraction and/or reflection at the surface of an isotropic chiral medium has been presented by Georgieva [66], while Matsumoto *et al.* [26], Friden [23] and He [53] analysed the diffraction due to gratings on a chiral-dielectric planar slab.
More specifically investigating absorbing coatings, Tretyakov *et al.* [4], [95] analysed the influence of the shape and the chirality of the inclusion on the reflective properties of planar composite layers. Single and multi-layer screens have been described by Ge *et al.* [55]. Mariotte *et al.* [121] discussed on the application of such media to Radar cross-section management.
Concerning the filtering function, Chuprin *et al.* [119] presented a concept of semi-transparent screen using a linear array of circular helically conductive cylinders. According to Lindell and Sihvola [7], it is possible to achieve any polarisation transformation using two independent chiral planar slabs.

6 Conclusions

This review showed the variety of problems encountered when dealing with complex media like chiral, or bianisotropic materials. Significant advances have been made since the beginning of 1995 on modelling. The better understanding of the physical phenomenon, or the use of purely numerical approach helped to carry out equivalent asymptotic models for the chiral inclusions, leading to more accurate representations of the constitutive parameters, even if Tellegen media are still the subject of hard discussions!
The field of application of bianisotropic or chiral media remains very wide, from microwaves to optics, though very few experiments have been performed.
In the future, the progress in material manufacturing will probably help to keep to complex media all their interest.

REFERENCES

1. H. CORY (1996) Technion-Israel Institute of Technology, Israël
« Chiral devices - an overview of canonical problems », *JEWA Vol.9, N° 5/6, June 1996*
General overview

2. D.I. KAKLAMANI, N.K. UZUNOGLU (1995) National Technical University of Athens, Greece
« Radiation properties of a biaxially anisotropic dielectric sphere excited by a dipole », *Electromagnetics Vol.15 N° 6, December*

Method of moments - Sphere excited in the centre - Applications in optical signal processing, radomes, substrates, absorbers,...- Green's functions - Reflected field inside described by Fourier series - Continuity integral equations

3. H.Y.D. YANG, P.L.E. USLENGHI (1995) University of Illinois, Chicago, « Radiation characteristics of microstrip antennas on cylindrical bianisotropic substrates » *Electromagnetics Vol.15 N° 5, October*
Constitutive equations - Lossless media -ϵ, μ : diagonal - Conducting cylinder with dielectric coating - Axial, array of axial, wraparound dipoles - Far field - more radiation in dark region, more directivity, high effect thin substrates

4. S.A. TRETYAKOV, A.A. SOCHAVA, C.R. SIMOVSKI (1996) St. Petersburg State Technical University « Influence of chiral shapes of individual inclusions on the absorption in chiral composite media » *Electromagnetics Vol.16 N° 2, April*
Planar layer - similar reflective properties for normal incidence, but not for oblique for different chiral or non-chiral shapes - Extended Maxwell-Garnet - Chirality asymmetric non-interacting inclusion of 2nd order effect on reflection - low density

5. E.K.N. YUNG, R.S.K. WONG (1995) City University of Hong Kong, « Analysis of a wire antenna of arbitrary shape » *JEWA Vol.9 N° 7/8, August*
In wire grid model, introduction of bent segments - helical antenna - Method of moments

6. W.S. WEIGLHOFER (1995) University of Glasgow, A. LAKHTAKIA Pennsylvania State University « A brief Review of a New Development for Constitutive Relations of Linear Bi-Anisotropic Media » *IEEE Antennas and Propagation Magazine Vol 37 N° 3, June*
Mathematical consistency of constitutive relations - Uniformity constraint - Homogeneous media (extension to heterogeneous) - bi-isotropic # reciprocal - General uniaxial bi-anisotropic # non-reciprocal - axially, transversally uni-axial bi-anisotropic

7. I.V. LINDELL, A.H. SIHVOLA (1995) Helsinki University of Technology « Plane wave Reflection from Uniaxial Chiral Interface and Its Application to Polarisation Transformation » *IEEE Antennas and Propagation Magazine Vol 43 N° 12, December*
Transverse uniaxial - Parallel springs in host medium -Polarisation transformation : axial ratio, handedness by rotation of slab - quarter wave slab -1 slab : linear to any, 2 slabs : any to any

8. F. OLYSLAGER (1995) University of Gent « Time-Harmonic Two- and Three-Dimensional Green's Dyadics for General Uniaxial Bianisotropic Media » *IEEE Antennas and Propagation Magazine Vol 43 N°4, April*
3D Green's dyadics derived by analytical inverse Fourier transformation in a special case

9. K.W. WHITES (1995) University of Kentucky « Full-Wave Computation of Constitutive Parameters for Lossless Composite Chiral Materials » *IEEE Antennas and Propagation Magazine Vol 43 N° 4, April*
Thin wire - Method of moments - Average scattering by random distribution using Monte-Carlo - Constitutive parameters computation - Far-scattered field comparison between chiral sphere and random helices

10. L.W. LI, P.S. KOOI, M.S. LEONG, T.S. YEO (1995) National University of Singapore « A General Expression of Dyadic Green's Functions in Radially Multilayered Chiral Media » *IEEE Antennas and Propagation Magazine Vol 43 N° 3, March*
Randomly distributed sources inside the sphere - multiple reflections - scattering superposition - boundary conditions - compact form of the dyadic Green's function

11. R. LUEBBERS, H.S. LANGDON, F. HUNSBERGER, C.F. BOHREN, S. YOSHIKAWA (1995) Pennsylvania State University « Calculation and Measurements of the Effective Chirality Parameter of a Composite Chiral Material Over a Wide Frequency Band » *IEEE Antennas and Propagation Magazine Vol 43 N° 2, February*
Scattering of helix in infinite lossy matrix using Method of moments - Tumble average - No coupling between helices - circularly polarised waves

12. A.N. SERDYUKOV , I.V. SEMCHENKO (1996) Gomel State University A.H. SIHVOLA Helsinki University of Technology S.A. TRETYAKOV St.Petersburg State Technical University « Duality in Electromagnetics : Application to Tellegen media » *Electromagnetics Vol.16 N° 1, February*
Non-reciprocal isotropic = Tellegen - Reciprocal isotropic chiral = Pasteur - Antisymmetric magnetoelectric coupling (chiral) not affected by duality transformation - Symmetric coupling (non-reciprocity) affected.

13. M. NORGREN, S. HE (1995) Royal Institute of Technology, Stockholm « Reconstruction of the constitutive parameters for an Omega material in a rectangular waveguide » *International Symposium on Electromagnetic Theory URSI St. Petersburg, May*
Permittivity and permeability diagonal tensors - Coupling parameter Omega - Measurement on TEm0 modes - ε_{zz} not extracted with these modes - Reflection and transmission coefficients - Numerical results

14. F.G. BASS (1995) Institute of Radiophysics and Electronics, Kharkov G.Y. SLEPYAN, S.A. MAKSIMENKO, A.V. GUREVICH Institute of Nuclear Problems, Minsk A. LAKHTAKIA Pennsylvania State University « Beltrami-Maxwell vector solitons in non-linear chiral media » *International Symposium on Electromagnetic Theory URSI, St. Petersburg, May*
Pulse propagation - Schrödinger equations

15. G.A. KYRIACOU (1995) Demokritos University of Thrace, Xanthi J.N. SAHALOS University of Thessaloniki, Thessaloniki « A Wiener-Hopf type analysis of substrate-superstrate uniaxially anisotropic microstrip structures » *International Symposium on Electromagnetic Theory URSI, St. Petersburg, May*
Incident wave on the edge of a semi-infinite plane conductor introduced between 2 uniaxial anisotropic layers - Space domain Fourier Transform - Surface wave propagation

16. N.P. ZHUCK, A.S. OMAR (1995) Technische Universität Hamburg-Harburg « Effective electromagnetic response of a bi-isotropic medium with random collection of spherical inclusions : a random medium approach » *International Symposium on Electromagnetic Theory URSI, St. Petersburg, May*
Electrically small inclusions - bi-isotropic sphere in bi-isotropic medium - random medium

17. V.V. FISANOV (1995) Siberian Physical & Technical Institute, Tomsk « Bi-isotropic constitutive relations and edge condition » *International Symposium on Electromagnetic Theory URSI, St. Petersburg, May*
Constitutive relations - edge diffraction - Post-Jaggard formalism preferable

18. V.V. FISANOV (1995) Sirerian Physical & Technical Institute, Tomsk S.G. VASHTALOV University of Tomsk « Plane wave diffraction by the wedge and the half-plane in chiral media » *International Symposium on Electromagnetic Theory URSI, St. Petersburg, May*
Circular polarisation - Integral transforms - Scattered field - Post-Jaggard constitutive relations

19. A.H. SIHVOLA (1995) Helsinki University of Technology « In defence of isotropic magnetoelectric media on the bi-anisotropic electromagnetic battlefield » *International Symposium on Electromagnetic Theory URSI, St. Petersburg, May*
Bi-anisotropic linear media - Non-Reciprocal Bi-Anisotropic (Tellegen) media - Bi-isotropic bi-isotropic media can exist

20. A.V. TYUKHTIN (1995) Institute of Physics, University of St.Petersburg « The average boundary conditions for a grid of parallel conductors in some anisotropic moving and dispersive media » *International Symposium on Electromagnetic Theory URSI, St. Petersburg, May*
Small distance between conductors - anisotropic non-dispersive motionless medium - moving non-dispersive medium - moving cold plasma - Hot motionless plasma

21. G. Y. SLEPYAN, A.V. GUREVICH, S.A. MAKSIMENKO (1995) Institute of Nuclear Problems, Minsk « Eigenwaves in chiral media with periodically inhomogeneous constitutive parameters » *International Symposium on Electromagnetic Theory URSI, St. Petersburg, May*
Floquet-Bloch equations - Bragg diffraction - two-wave approximation

22. F. OLYSLAGER, B. BAEKELANDT, J. UREEL (1995) University of Gent, Gent « Mode propagation in and reflection on general two-dimensional uniaxial bianisotropic periodic and non-periodic structures » *International Symposium on Electromagnetic Theory URSI, St. Petersburg, May*
Uniaxial bianisotropic media homogeneous in the propagation direction and inhomogeneous in the transverse plane - Galerkin's method - Dielectric waveguide

23. J. FRIDÉN (1995) Göteborg University and Chalmers University of Technology, Göteborg « Inverse scattering for a homogeneous, dispersive and bi-anisotropic slab » *International Symposium on Electromagnetic Theory URSI, St. Petersburg,*
Reflection and transmission data - Green's functions - Reconstruction of medium parameters - Medium identification

24. K.A. BARSUKOV, A.A. SMIRNOVA (1995) St.Petersburg Electrical Engineering University « The Doppler effect in chiral media » *International Symposium on Electromagnetic Theory URSI, St. Petersburg, May*
Electrical oscillator moving with uniform velocity in chiral medium

25. A.D. CHUPRIN, A.D. SHATROV, A.N. SIVOV (1995) Institute of Radio Engineering and Electronics of the Russian Academy of Science, Moscow reg. « Theory of short-period gratings composed of multifilar helices » *International Symposium on Electromagnetic Theory URSI, St. Petersburg, May*
Diffraction - Normal incidence - Elliptic polarisation - Fourier series - Integral-differential equation - Linear array of helically conductive cylinders - Polarisation filter (left-handed reflected , right-handed partially transmitted)

26. K. MATSUMOTO, K. ROKUSHIMA 1995) Osaka Sangyo University, Osaka J. YAMAKITA Okayama Prefectural University, Okayama « Three-dimensional rigorous analysis of chiral dielectric grating diffraction » *Symposium on Electromagnetic Theory URSI, St. Petersburg, May*
Scattering - Isotropic chiral grating - Space harmonics - Oblique incidence

27. G.J. KARONIS, D.I. KAKLAMANI, P.V. FRANGOS, N.K. UZUNOGLU (1995) National Technical University of Athens « Scattering of a plane wave by a chiral ellipsoid » *Symposium on Electromagnetic Theory URSI, St. Petersburg, May*
Linear polarisation - isotropic chiral lossless ellipsoid - Fourier transform - Phase space - Integral equation solved by a Galerkin type Method of Moments - Comparison with published data on spheroidal and ellipsoidal raindrops

28. M. NORGREN, S. HE (1995) Royal Institute of Technology, Stockholm « Reconstruction of the Constitutive Parameters for an Omega Material in a Rectangular Waveguide » *IEEE-MTT Vol.43, N°6, June*
Permittivity and permeability diagonal tensors - Coupling parameter Omega - Measurement on TEm0 modes - ε_{zz} not extracted with these modes - Reflection and transmission coefficients - Numerical results

29. S.A. TRETYAKOV (1995) St.Petersburg State Technical University A.J. VIITANEN Helsinki University of Technology « Waveguide and Resonator Perturbation Techniques Measuring Chirality and Nonreciprocity Parameters of Biisotropic Materials » *IEEE-MTT Vol.43, N°1, January*
Measurement of phase constant or frequency shift - Degenerate modes required to have a first order effect due to chirality or nonreciprocity - Small biisotropic inclusions

30. I.V. LINDELL, A.H. SIHVOLA (1995) Helsinki University of Technology K. SUCHY University of Düsseldorf « Six-vector formalism in electromagnetics of bi-anisotropic media » *JEWA Vol.9 N° 7/8, August*
Six-vectors, six-dyadics - Reduction of vector equations in half and dyadic equations by four -Review of six-vector algebra

31. S.A. TRETYAKOV (1995) St.Petersburg State Technical University F. MARIOTTE French Atomic Energy Commission « Maxwell Garnett modelling of uniaxial chiral composites with bianisotropic inclusions » *JEWA Vol.9 N° 7/8, August*
Small bianisotropic inclusions in 3D array - Spherical shell - Averaged dyadic constitutive parameters - Analytical antenna model - Reflection of metal-backed screens - Influence of shape and orientation

32. L. VALOR, J. ZAPATA (1996) Universidad Politécnica de Madrid « An Efficient Finite Element Formulation to Analyse Waveguides with Lossy Inhomogeneous Bi-Anisotropic Materials » *IEEE-MTT Vol.44, N°2, February*
Magnetic field - Full tensors - Arbitrary cross section - Analysis of complex modes - Generalised eigenvalue problem solved by a method based on the subspace iteration algorithm

33. F. MARIOTTE , A. BOURGEADE (1995) French Atomic Energy Commission CESTA P. BANNELIER, F. GUÉRIN Thomson CSF, Orsay, Elancourt, IRCOM, Limoges « Numerical computations of the electromagnetic field scattered by complex chiral bodies » *JEWA Vol.9 N° 11/12, November*
Linear polarisation of incident wave - Scattering and diffraction - Surface integral equation using bi-dimensional finite elements - single or 3 turns helices, Omegas, non-chiral shapes - Theoretical and experimental comparison

34. S. TRETYAKOV, C. SIMOVSKI, T. KHARINA (1995) St. Petersburg State Technical University F. MARIOTTE French Atomic Energy Commission, CESTA S. BOLIOLI National Agency for Research on Aerospace, Research Centre of Toulouse « Scattering by Omega-shaped conductive particles and Omega composite modelling » *Symposium on Electromagnetic Theory URSI, St. Petersburg, May*
Analytical approach - Comparison with numerical approach - Antenna theory to single element - Generalised Maxwell Garnett

35. P.K. KOIVISTO, C.E. STEN (1995) Helsinki University of Technology « Quasi-static Image Method Applied to Bi-Isotropic Microstrip Geometry » *IEEE-MTT Vol.43, N°1, January*
Image method - Bi-isotropic medium - Green's function - Asymptotic series expansion for fields in terms of frequency - Infinitely thin conductor

36. F. MESA, M. HORNO 1995) University of Seville « Computation of Proper and Improper Modes in Multilayered Bianisotropic Waveguides » *IEEE-MTT Vol.43, N°1, January*
Grounded planar multilayer waveguide - Interfaces between layers simulated by impedance dyadics

37. S.F. MAHMOUD (1995) Kuwait University, Safat « Guided Modes on Open Chirowaveguides » *IEEE-MTT Vol.43, N°1, January*
Chiral rod and planar slab - Modal equation - Achiral case splits into RCP and LCP in chiral case near cut-off, and RCP (for right handed medium) far from cut-off - Surface wave and guided modes

38. A.H. SIHVOLA (1995) Helsinki University of Technology « Are Nonreciprocal Bi-Isotropic Media Forbidden Indeed ? » *IEEE-MTT Vol.43, N°9, September*
Tracelessness property of the nonreciprocity dyadic - Chromium oxide case - Covariance property in the form of Post not universal

39. E.O. KAMENETSKII (1996) Tel-Aviv University « Mode Orthogonality Relations and Field Structure in Chirowaveguides » *IEEE-MTT Vol.44, N°3, March*
Six component vector formulation from Maxwell's equations leads to power orthogonality - Complex modes - Scalar formulation using propagation equation -

40. G. OBERSCHMIDT, A.F. JACOB (1996) Technische Universität Braunschweig « Averaging Rules for the Scattering by Randomly Oriented Chiral Particles » *IEEE-MTT Vol.44, N°3, March*
Orientational averaging - Vector radiative transfer equation - Lorenz-Lorentz mixing formulas - Polarisabilities

41. M.E. PETERS, E.H. NEWMAN (1995) Ohio State University, Columbus « Method of Moments Analysis of Anisotropic Artificial Media Composed of Dielectric Wire Objects » *IEEE-MTT Vol.43, N°9, September*
Infinite 3D periodic array - Complex wavenumber(s) and eigenfunction currents and fields - Plane wave

42. B. JAKOBY, D. DE ZUTTER (1996) University of Gent « Analysis of Guided Waves in Inhomogeneous Bianisotropic Cylindrical Waveguides » *IEEE-MTT Vol.44, N°2, February*
Propagation of surface waves, fundamental and high order modes - Cylindrical microstrip line - Tellegen constitutive relations - Fourier series - Propagator matrix

43. G.G. GOSHIN, N.E. LUGINA (1995) Siberian Physical & Technical Institute, Tomsk « Boundary-value problems associated with the excitation of wire system on a chiral substrate » *Symposium on Electromagnetic Theory URSI, St. Petersburg, May*
Straight line wire - Logarithmic spiral wire - Fourier or Fourier-Bessel integral transform - Eigen value wave spectrum - Radiation pattern

44. G.W. HANSON (1996) University of Wisconsin-Milwaukee « A Numerical Formulation of Dyadic Green's Functions for Planar Bianisotropic Media with Application to Printed Transmission Lines » *IEEE-MTT Vol.44, N°1, January*
Two coupled 1D integral equations - Linear combinations of Green's functions for an isotropic half-space - Chiroferrite media - Microstrip transmission lines

45. W.S. WEIGLHOFER (1995) University of Glasgow « One-dimensional dyadic Green function for general linear bianisotropic media » *JEWA Vol.9 N° 11/12, November*
General, homogeneous bianisotropic medium - field dependence on one spatial coordinate - Explicit dyadic Green's functions

46. B. JAKOBY, F. OLYSLAGER (1996) University of Gent « Asymptotic expansions for Green's dyadics in bianisotropic media (Summary) » *JEWA Vol.10 N° 1, January (Summary of PIER publication)*
Spectral domain technique, 3D Fourier transform - Asymptotic expressions for Green's functions

47. L.W. LI, P.S. KOOI, M.S. LEONG, T.S. YEO (1995) National University of Singapore « Analytical representation of scattering dyadic Green's functions coefficients for cylindrically multilayered chiral media » *JEWA Vol.9 N° 9, September*
Randomly distributed sources inside a cylinder - multiple reflections - scattering superposition - boundary conditions - compact form of the dyadic Green's function

48. W. YIN, W. WANG, P. LI (1995) Northwestern Polytechnical University, Xi'an Jiaotong University, People's Republic of China « Dyadic Green's functions in arbitrary multilayered spherical chiral media and its applications » *JEWA Vol.9 N° 1/2, January*
Scattering superposition - Spherical inhomogeneous chirolens - Impedance matching - Geometrical optics

49. I.V. LINDELL (1995) Helsinki University of Technology F. OLYSLAGER University of Gent « Duality transformations, Green dyadics and plane-wave solutions for a class on bianisotropic media » *JEWA Vol.9 N° 1/2, January*
Duality transformation - Self-duality - Bianisotropic medium equivalent to two anisotropic media of affine-isotropic type

50. I.V. LINDELL (1996) Helsinki University of Technology F. OLYSLAGER University of Gent « Decomposition of electromagnetic sources in axially chiral uniaxial anisotropic media » *JEWA Vol.10 N° 1, January*
Generalisation of the TE/TM decomposition - Uniaxial affine transformation - Affine-isotropic media - Axial impedances

51. A. CHEREPANOV (1996) St. Petersburg State Technical University A.H. SIHVOLA Helsinki University of Technology « Internal electric field of anisotropic and bianisotropic spheres » *JEWA Vol.10 N° 1, January*
Electrostatic field - Uniform field inside a bianisotropic sphere - Extension to quasi-static electric field

52. J.A. PEREDA, A. VEGAS, A. PRIETO (1996) Universidad de Cantabria, Santander « On the Derivation of the Generalised Telegraphist's Equations for Full-Wave Analysis of Chirowaveguides » *IEEE Microwave and guided wave letters Vol.6, N°.4, April*
Parallel-plate chirowaveguide - Boundary condition on H leading to a formulation non valid if the medium in contact with the perfectly electrically conductive wall is anisotropic and/or chiral - Galerkin method - Expansion of E, B fields instead of E, H

53. S. HE (1996) Royal Institute of Technology, Stockholm « A uniform approximation to the scattering and propagation problem for a stratified bi-anisotropic slab » *International Journal of Infrared and Millimeter Waves Vol.17, N°2 February*

Approximate solutions - Frequency dependent parameters - Left and right moving waves

54. X. WU (1995) University of Science and Technology of China « A linear-operator formalism for bianisotropic waveguides » *International Journal of Infrared and Millimeter Waves Vol.16, N°, February*
Linear operator formalism - Adjoint waveguides - Variational approach - Bi-orthogonality relations - Waveguide discontinuity

55. F. GE, L. CHEN, J. ZHU (1996) Central Iron and Steel Research Institute, Beijing « Reflection characteristics of chiral microwave absorbing coatings » *International Journal of Infrared and Millimeter Waves Vol.17, N°1, January*
Metal-backed , one- or multi-layered absorbing coating -Reflection coefficient - Complex chiral admittance - Normal incidence - Plane wave

56. J.P. XU (1996) Ninjing University of Aeronautics and Astronautics « Propagation characteristics of a circular waveguide filled with a chiroferrite medium » *International Journal of Infrared and Millimeter Waves Vol.17, N°1, January*
Chiral objects in magnetically biased ferrite - Decoupling equations on electric and magnetic fields - Transverse components expressed in terms of longitudinal ones - Analytical characteristic equation

57. K. HAYATA, M. KOSHIBA (1995) Hokkaido University « Chirosolitons: Unique Spatial Solitons in Chiral Media » *IEEE-MTT Vol.43, N°8, August*
Kerr-type non-linear media - Liquid crystals - Superluminous solitons and two branches in the dark soliton solution - Isotropic chiral medium - Combination of chirality and non-linearity of electromagnetic medium

58. L. VEGNI, A. TOSCANO (1995) Third University of Rome « Input impedance of a chirostrip antenna » *25th European Microwave Conference, August*
Substrate above an infinite ground plane - Multilayer structure including isotropic and/or biisotropic - Spectral domain Green's function - Method of moments - Analytical evaluation which simplifies numerical computation

59. F. OLYSLAGER, E. LAERMANS, D. DE ZUTTER (1995) University of Gent « Rigorous Quasi-TEM Analysis of Multiconductor Transmission Lines in Bi-Isotropic Media - Part I : Theoretical Analysis for General Inhomogeneous Media and Generalisation to Bianisotropic Media - Part II: Numerical Solution for Layered Media » *IEEE-MTT Vol.43, N°7, July*
Bitransmission lines - General-shaped conductors - Lossless media - Circuit representation - Properties of circuit matrices (C, L, X, Z) - Spectral Green's function -Integral equation - Method of moments - Point-matching

60. G. BUSSE, A.F. JACOB (1995) Technische Universität Braunschweig « Complex Power and Mode Coupling in Circular Chirowaveguides » *IEEE-MTT Vol.43, N°5, May*
Lossless circular chirowaveguides - Complex power - Drude-Born-Fedorov constitutive relations - No orthogonality in the power sense

61. Y. XU, R.G. BOSISIO (1995) Ecole Polytechnique de Montréal « An Efficient Method for Study of General Bi-Anisotropic Waveguides » *IEEE-MTT Vol.43, N°5, April*
Coupled-mode equation - Circular, rectangular, Faraday chiral waveguides

62. D. CHENG (1996) Fudan University, Shanghai « Green dyadic and dipole radiation in triaxial omega medium » *Electronic Letters Vol.32, N°6, March*
Green's functions - Auxiliary fields - Coordinate transformation - Complex exponential-like attenuation of the wave

63. F. GE, J. ZHU, L. CHEN (1996) Central Iron and Steel Research Institute, Beijing « Effective electromagnetic and chiral parameters of chiral composite material » *International Journal of Infrared and Millimeter Waves Vol.17, N°2, February*
Randomly oriented non-interacting helices embedded in a non-chiral host medium - Permittivity and permeability independent on handedness, chirality dependent - Maximum chirality for R/P=0.23

64. S.A. TRETYAKOV, A.A. SOCHAVA, S.I. MASLOVSKI (1996) St. Petersburg State University « Plane wave propagation in a class of knotted media » *Electromagnetics, June*
Complex polarisability properties - Anisotropy, no magneto-electric coupling - Frequency dispersion model - Plane wave equation

65. K.M. FLOOD, D.L. JAGGARD (1996) University of Pensylvania, Philadelphia « Effective charge densities and current densities in isotropic chiral media » *Journal of the Optical Society of America A. Vol.12, N°1, January*
Moving observer's reference frame in a stationary chiral medium

66. E. GEORGIEVA (1996) University of Sofia « Reflection and refraction at the surface of an isotropic chiral medium : eigenvalue-eigenvector solution using a 4x4 matrix method » *Journal of the Optical Society of America A. Vol.12, N°10, October*
Drude-Condon constitutive relations - Circular incident polarisation - Achiral-chiral interface - Arbitrary angle of incidence

67. R.C. QIU, I.T. LU (1994) Weber Research Institute, Polytechnic University, Farmingdale « Guided waves in chiral optical fibbers » *Journal of the Optical Society of America A. Vol.11, N°12, December*
Step-index fibber - chiral core - Effects of chirality on modal propagation constants, group delay, waveguide dispersion

68. J.M. HICKS (1995) Georgetown University, Washington « Second order non-linear optical spectroscopy of chiral surfaces » *PIERS, Seattle,July*
Detection of dichroic and rotational behaviour in Second Harmonic Generation - SHG scattering efficiency - Wavelength dependence

69. A. LAKHTAKIA (1995) Pennsylvania State University « The Lorentz force vetoes the existence of Tellegen media » *PIERS, Seattle, July*
Non-existence of Tellegen media

70. C.M. KROWNE (1995) Naval Research Laboratory, Washington « On the forbidden nature of nonreciprocity in linear biisotropic media » *PIERS, Seattle, July*
Non-reciprocal to reciprocal coefficient ratio

71. W.S. WEIGLHOFER (1995) University of Glasgow « Closed-form expressions for the one-dimensional Dyadic Green function of general homogeneous and some inhomogeneous linear bianisotropic media » *PIERS, Seattle, July*
General, homogeneous bianisotropic medium - field dependence on one spatial coordinate - Explicit dyadic Green's functions - Helicoidal bianisotropic media

72. L.R. ARNAUT, L.E. DAVIS (1995) The Universitry of Manchester Institute of Science and Technology « On planar chiral structures » *PIERS, Seattle, July*
New planar elements - No chirality at grazing angles - Reconsideration of the notions of chirality and wave propagation in a multi-dimensional context - Generalised swastika

73. A. MALINOVSKI, S.V. POPOV, Y.P. SVIRKO, N.I. ZHELUDEV (1995) University of Southampton « Nonreciprocal interaction of light with Zinc-Blende semiconductors » *PIERS, Seattle, July*
Macroscopic Onsager symmetry - Time-non-reversibility - Ultra-short pulses - nonreciprocal rotation

74. R.D. GRAGLIA, M.S. SARTO, P.L.E. USLENGHI (1995) Politecnico di Torino, Università di Roma "La Sapienza" « Theory of coaxial cable filled with bianisotropic material » *PIERS, Seattle, July*
Transverse components expressed in terms of longitudinal ones - Coupled second-order differential equations - Decoupling conditions

75. W.S. WEIGLHOFER, S.F. NAGLE, A. LAKHTAKIA (1995) University of Glasgow, Pennsylvania State University « Axial propagation in helicoidal bianisotropic media : Dependence of mode handedness on pitch » *PIERS, Seattle, July*
Unidirectionnaly inhomogeneous - Helicoidal variations of the constitutive tensors along this direction - Spatial periodicity - Liquid crystals - Thin film technology - Polarisation-sensitive filters

76. H. CORY, I. ROSENHOUSE (1995) Technion-Israel Institute of Technology « Chiral filters synthesis » *PIERS, Seattle, July*
Normal incidence - Analogy to dielectric filter - Multilayered structure - Dependence of the performances on the incidence angle

77. C. MANOLATOU, J.T. JOHNSON, J.A. KONG, R.T. SHIN (1995) Massachusetts Institute of Technology, Cambridge « Mixing laws for helix loaded composite media » *PIERS, Seattle, July*
Effective isotropic medium model - Derivation of the chiral Maxwell Garnett mixing law - Helices arrays - Periodic Green's function

78. A. TOSCANO, L. VEGNI (1995) Third University of Rome « Spatial fields in planar integrated structures with chiral substrates » *PIERS, Seattle, July*
Lossless or lossy grounded biisotropic chiral slabs - Fourier spectral domain - Spectral microstrip dyadic Green's function for a mono-layered biisotropic structure

79. R. VACEK (1995) Charles University, Faculty of Mathematics and Physics, Prague « Chirality in non-linear media » *PIERS, Seattle, July*
Space-time Fourier transform - Taylor series expansion for permittivity and permeability - New constitutive relations - Non-linear Helmholtz-like wave equation

80. S.F. MAHMOUD (1995) Kuwait University, Safat « Electromagnetic fields of a current loop in the vicinity of a chiral slab » *PIERS, Seattle, July*
Rigorous solutions - Approximate closed forms for low and high frequency limits - Co- and crosspolarisation, near and far zone Fresnel reflection coefficients

81. B. MICHEL (1995) Max-Planck-Geselschaft A. LAKHTAKIA Pennsylvania State University « Effective medium theory for homogenising chiral particulate composites with strong property fluctuations » *PIERS, Seattle,July*
Spatial correlations between particles - Electromagnetic length scale - Generalisation of the Strong Permittivity Fluctuations Theory - Dyson equation

82. F.G. BASS (1995) Institute of Radiophysics and Electronics, Kharkov G.Y. SLEPYAN, S.A. MAKSIMENKO Institute of Nuclear Problems, Minsk « Self-focusing of electromagnetic waves in non-linear chiral media » *PIERS, Seattle, July*
Phenomenological non-linear optics extended to chiral media - Beltrami-Maxwell formalism - Generalisation of the Bespalov-Talanov criteria of the stability of non-linear plane waves

83. W.S. WEIGLHOFER (1995) University of Glasgow A. LAKHTAKIA Pennsylvania State University « Linear constitutive relations and Maxwell's equations : the case for mathematical consistency » *PIERS, Seattle, July*
Mathematical uniformity constraint for homogeneous bianisotropic media - Extension to inhomogeneous media

84. B. JAKOBY, F. OLYSLAGER (1995) University of Gent « Recent developments in dyadic Green's functions for bianisotropic media » *PIERS, Seattle, July*
Spectral domain technique - Asymptotic forms for Green's dyadics - Application to a class of diagonally or coaxially bianisotropic media

85. A.K. BHATTACHARYYA, R.P. JEDLICKA, S. SUBRAMANIUM, P.G. FRANCO (1995) Lincoln University, Jefferson City & New Mexico State University, Las Cruces « Theory and characterisation of chiral material in rectangular/square waveguide » *PIERS, Seattle, July*
Exact solution for a rectangular waveguide fully filled with chiral material - Analytical expressions for the crosspolarisation component intensity and the rotation of the plane of polarisation - Experimental results

86. S.A. KUEHL, S.S. GROVÉ, A.G. SMITH, I.P. THERON (1995) University of Stellenbosch, South Africa « Manufacture of microwave chiral materials and their electromagnetic properties » *Chiral'95, October*

NiCr helical lines - Lossless and lossy matrices - Isotropic and anisotropic samples - Experimental results - Rotation and ellipticity - Ku band

87. J. REINERT, G. BUSSE, A.F. JACOB (1995) Technische Universität Braunschweig « A procedure to extract the chirality parameter from waveguide measurements » *Chiral'95, October*
Lossy chiral slab - Circular waveguide - Mainly analytical parameters reconstruction using two different slab lengths - Fundamental mode

88. A.G. SMITH, J.H. CLOETE, I.P. THERON (1995) University of Stellenbosch, South Africa « The accuracy of a free-space system for measuring the electromagnetic properties of microwave chiral materials » *Chiral'95, October*
Ku band horn lens antennas - Inversion of experimental results - Accuracy of measurements

89. K.W. WHITES (1995) University of Kentucky « Comparison of computed and measured scattering by artificial chiral material » *Chiral'95, October*
Effective material simulation algorithm - Free-space measurements - Lens horn antennas - X band - Randomly distributed 3-turn metallic helices - Multilayer samples

90. S. HAQ, D.L. TUNNICLIFFE (1995) British Aerospace, Bristol « Fabrication of chiral composites for waveguide and free-space measurements » *Chiral'95, October*
Free-space and waveguide measurements - Sample manufacturing techniques - Bi-anisotropic samples using aligned helices

91. J.P. HÉLIOT, J. GAY, F. MARIOTTE, B. SAUVIAC (1995) French Atomic Energy Commission, Le Barp « A general thin wire approximation for dielectric or magnetic scatterers : validation in the case of chiral objects » *Chiral'95, October*
ARLENE code - Fourier series - Comparison between 1D wire approximation and 2D surface mesh

92. B. SAUVIAC, F. MARIOTTE, J.P. HÉLIOT, A. RIBOULET (1995) French Atomic Energy Commission, Le Barp « Numerical modelling of heterogeneous materials with dielectric or magnetic chiral inclusions » *Chiral'95, October*
ARLENE code - Randomly distributed helices in a dielectric host - Maxwell-Garnet mixing rules - Comparison between theoretical and experimental results

93. V.V. SHEVCHENKO (1995) Institute of Radioengineering and Electronics, R.A.S., Moscow « A novel model of a chiral electromagnetic particle » *Chiral'95, October*
Helically anisotropic conducting sphere - Circular polarised wave - Spherical harmonics

94. R.T. APPARAO (1995) Argonne National Laboratory V.V. VARADAN, V.K. VARADAN The Pennsylvania State University « On the preference for the chiral inclusion phase in chiral composite media and study of effective medium properties » *Chiral'95, October*
Vector multiple scattering technique - Randomly distributed inclusions - Quasi-crystalline approximation Dispersion equation - Circular polarisation

95. S.A. TRETYAKOV, C.R. SIMOVSKI, A.A. SOCHAVA (1995) St.Petersburg State Technical University « On the influence of inclusion chirality on reflective properties of planar layers » *Chiral'95, October*
Planar layer - similar reflective properties for normal incidence, but not for oblique for different chiral or non-chiral shapes - Extended Maxwell-Garnet - Analytical antenna model

96. A.H. SIHVOLA (1995) Helsinki University of Technology « Towards a theory for chiral and bi-anisotropic mixtures » *Chiral'95, October*
Spherical inclusions - Six-vector analysis - Combination between six-dyadic Maxwell-Garnett expression and the general family of mixing rules

97. M.V. KOSTIN, V.V. SHEVCHENKO (1995) Institute of Radioengineering and Electronics, R.A.S., Moscow « A novel model of an artificial chiral electromagnetic medium » *Chiral'95, October*
Helically conductive spheres - Space cubical lattice - Circular polarisation

98. R.E. RAAB (1995) University of Natal, Scottsville, South Africa A.H. SIHVOLA Helsinki University of Technology « On the existence of NRBI medium ? » *Chiral'95, October*
Static weak fields - Polarisability tensors - NRBI effects possible in some classes of cubic crystals

99. A.P. VINOGRADOV, S.N. STAROSTENKO (1995) SCAPE R.A.S. Moscow T.G. KHARINA, A.A. SOCHAVA St.Petersburg State Technical University « An experimentally simple method for determination ultra-high frequency constitutive parameters and their dispersion characteristics in composite materials » *Chiral'95, October*
Effective media parameters - Omega composites - Frequency behaviour - Generalisation of the Cole-Cole formula - Waveguide reflection and transmission measurements

100. I.P. THERON, J.H. CLOETE (1995) University of Stellenbosch, South Africa « Microwave optical activity of an artificial non-magnetic uniaxial chiral crystal » *Chiral'95, October*
Long wavelength regime - Magnetic dipole, electric dipole and quadrupole - Constitutive equations - Chiral hook - Regular array

101. L.R. ARNAUT, L.E. DAVIS (1995) The Universitry of Manchester Institute of Science and Technology « Effect of mutual coupling between aligned helices on the constitutive dyadics of high density bianisotropic composites » *Chiral'95, October*
Regular arrays of identical helices with random or aligned orientations - Body-centred cubic configuration - Multiple scattering - De-embedding procedure

102. I.P. THERON, J.H. CLOETE (1995) University of Stellenbosch, South Africa « Waveguide measurement of an artificial uniaxial chiral crystal in the long wavelength regime » *Chiral'95, October*
Circular waveguide - Magnetic dipole, electric dipole and quadrupole - Comparison between theoretical and experimental results

103. C.Y. CHUNG, K.W. WHITES (1995) University of Kentucky « Numerical extraction of the effective constitutive parameters for uniaxial bianisotropic chiral medium » *Chiral'95, October*
Uniaxial bianisotropic chiral - Oriented helices randomly distributed - Numerical (MOM) and analytical (Effective medium theory) approach - Constitutive parameters determined by equating the reflection coefficient obtained with both methods

104. L.R. ARNAUT (1995) The Universitry of Manchester Institute of Science and Technology « Axial representation of dipolarisability dyadics for chiral structures » *Chiral'95, October*
Single helix - Variation of eigenvalues and eigenvectors of dipolarisability dyadics with frequency, pitch/length ratio

105. A.F. KONSTANTINOVA, B.V. NABATOV, T.A. YU (1995) Institute of Crystallography, R.A.S., Moscow « Reflection and propagation of light through an uniaxial optically active slab » *Chiral'95, October*
Uniaxial gyrotropic film - Oblique incidence

106. I. LINDELL, A.H. SIHVOLA, P. PUSKA, L. RUOTANEN (1995) Helsinki University of Technology « Inequalities for medium parameters of idealised bi-anisotropic media based on the energy condition » *Chiral'95, October*
Non dispersive and lossless media - Six-vector formalism - Application to bi-isotropic, to uniaxial, to Omega uniaxial, to Omega nonuniaxial medium

107. V.V. FISANOV (1995) Siberian Physical & Technical Institute, Tomsk « Singularity exponent of the electromagnetic field near the chiroplasma edge » *Chiral'95, October*
Combination of magnetoactive plasma and chiral medium - Near field singular behaviour -Matrix formalism

108. P. HILLION (1995) Le Vesinet (France) « Harmonic pulse in a chirotropic medium » *Chiral'95, October*
Constitutive equations in the time domain

109. I.V. SEMCHENKO, A.I. SAMUSENKO (1995) Gomel State University « Boundary value problem for particular waves in chiral media with spiral structure » *Chiral'95, October*

Magneto-regulated crystals - Plasma into magnetic fields - Heterogeneous medium along the propagation direction - Reflection and transmission coefficients

110. G.W. HANSON, B. RAY (1995) University of Wisconsin-Milwaukee « Propagation characteristics of guided surface-wave modes on grounded slabs with anisotropic chirality » *Chiral'95, October*
Planar, inhomogeneous, linear bianisotropic, non dispersive medium - Volume equivalence principle - Coupled integral equations - Green's functions

111. A.L. TOPA, C.R. PAIVA, A.M. BARBOSA (1995) Instituto Superior Técnico, Lisboa « Rigorous analysis of a step discontinuity in a planar dielectric chiral slab » *Chiral'95, October*
Hybrid modes - Incident transverse electric and magnetic modes (ITE, ITM) - Mode matching - Least squares boundary residual method - Power balance criterion

112. G. BUSSE, M. HOFFMEISTER, M. KUNZE, A.F. JACOB (1995) Technische Universität Braunschweig « Properties of chiral filled circular cylindrical resonators » *Chiral'95, October*
Drude-Born-Fedorov constitutive equations - Resonant frequency and quality factor - High order modes

113. S.A. KHAKHOMOV, I.V. SEMCHENKO (1995) Gomel State University « The amplification of ultrasound in semiconductors by the electron drift in condition of forming of chiral properties by the rotating electric field » *Chiral'95, October*
Ferroelectric ceramics - Piezoelectric effects - Acoustic waves - Amplification - Energy exchange between acoustic Wave and drifting electrons in piezoelectric semiconductors

114. A.P. VINOGRADOV, V.E. ROMANENKO (1995) SCAPE R.A.S., Moscow « Artificial magnetics based on racemic helix inclusions » *Chiral'95, October*
Length of inclusions comparable to wavelength - Paramagnetic and diamagnetic effects - Thin wire helices - Approximation for Green's function

115. G.S. MITYURICH, V.V. SVIRIDOVA, V.P. ZELENYI, A.N. SERDYUKOV (1995) Gomel State University « Photoacoustic transformation under the interaction of opposing light waves in magnetically active double layer » *Chiral'95, October*
Tunnel interference of electromagnetic waves - Elliptical polarisations - Thermal activities - Control of the amplitude of photoacoustic signal by modifying the interference energy dissipation

116. G.S. MITYURICH, P.V. ASTAKHOV (1995) Gomel State University « Laser photodeflection spectroscopy of the dichroic ferromagnetic colloids » *Chiral'95, October*
Relative linear dichroïsm of diluted or concentrated magnetic fluids - Short pulses - Green's function for heat conduction equation

117. E.G. STARODUBTSEV, G.S. MITYURICH (1995) Gomel State University « Power transformation of electromagnetic radiation by gyrotropic supergratings formed by crystals of rhombic singony » *Chiral'95, October*
Non-magnetic absorbing gyrotropic crystals - Short period crystal structures-superlattices

118. Y.N. KAZANTSEV, G.A. KRAFTMAKER (1995) Institute of Radioengineering and Electronics, R.A.S., Moscow « "Chiral medium-ferrite" and "electric dipole-ferrite" structures : chiral-ferromagnetic, dipole+ferromagnetic and ferromagnetic resonances » *Chiral'95, October*
Waveguide - Reflection coefficients - Travelling waves or standing wave measurements - Frequency dependence - Resonances

119. A.D. CHUPRIN, A.D. SHATROV, A.N. SIVOV (1995) Institute of Radioengineering and Electronics, R.A.S., Moscow « Chiral semitransparent screen for circularly polarised wave » *Chiral'95, October*
Linear array of circular helically conductive cylinders - Circular polarisation - Rotation of the polarisation for a linearly polarised plane wave

120. L.R. ARNAUT (1995) The Universitry of Manchester Institute of Science and Technology « Computation of dipolarisability dyadics for helix-based bianisotropic media » *Proc. Journées Maxwell Int. Workshop, Bordeaux June*

Dipolarisability dyadics - Single thin-wire helix - Counterpropagating waves with averaged direction of propagation - Dependence on frequency, pitch/length ratio - Quasi-stationary or quasi-resonant excitation

121. F. MARIOTTE, B. SAUVIAC, D. GOGNY (1995) French Atomic Energy Commission, Le Barp « Application of heterogeneous chiral materials to the design of radar absorbing materials » *Proc. Journées Maxwell Int. Workshop, Bordeaux June*
Isotropic composites - Metallic chiral inclusions - Computation of constitutive parameters - Experimental results - low reflective shields

122. L.R. ARNAUT, L.E. DAVIS (1995) The Universitry of Manchester Institute of Science and Technology « Novel 8-port polarimetric waveguide junction for characterisation of synthetic chiral media » *25th European Microwave Conference, August*
X-band frequency - scattering matrix - Chain algorithm - Reflection, transmission, absorption, depolarisation - numerical extraction of material parameters

123. E.O. KAMENETSKII (1995) Tel-Aviv University, Israël « Theory of excitation of chirowaveguides » *IEE Proceeding.-Microwaves Antennas Propagation, Vol. 142, N°3 June*

124. E.O. KAMENETSKII (1996) Tel-Aviv University, Israël « Mode interactions in chirowaveguides » *Journal of Applied Physics 79 (12), 16 June*

125. E.O. KAMENETSKII (1996) Tel-Aviv University, Israël « On the technology of making chiral and bianisotropic waveguides for microwave propagation » *Microwaves and Optical Technology Letters / Vol. 11, N°2 February*

126. G. BUSSE, A.F. JACOB (1996) Technische Universität Braunschweig « Lossy chiral slab in circular waveguides » *IEE Proceeding.-Microwaves Antennas Propagation, Vol. 143, N°1 February*

127. G.Y. SLEPYAN, A.V. GUREVICH, S.A. MAKSIMENKO (1995) Institute of Nuclear Problems, Minsk « Floquet-Bloch waves in periodic chiral media » *Physical Review E (The American Physical Society) Vol. 51, N°3 March*

128. G.Y. SLEPYAN, S.A. MAKSIMENKO (1995) Institute of Nuclear Problems, Minsk F.G. BASS Ukraine Academy of Science, Kharkov A. LAKHTAKIA The Pennsylvania State University « Non-linear electromagnetics in chiral media : Self-action waves » *Physical Review E (The American Physical Society) Vol. 52, N°1 July*

129. F.G. BASS (1995) Ukraine Academy of Science, Kharkov G.Y. SLEPYAN, S.A. MAKSIMENKO Institute of Nuclear Problems, Minsk A. LAKHTAKIA The Pennsylvania State University « Time-Domain Beltrami-Maxwell solitons in certain non-linear chiral media » *Microwaves and Optical Technology Letters / Vol. 9, N°4 July*

130. K.E. OUGHSTUN (1995) « Transients in chiral media with single-resonance dispersion : comment » *JOSA A, Vol. 12, N°3, March 1995, p 626-628*

131. A. LAKHTAKIA (1996) The Pennsylvania State University W.S. WEIGLHOFER University of Glasgow « Dilute non-linear particular composites composed by linear bianisotropic inclusions in non-linear dielectric host media » *Opt. Eng. Vol.35, N°4 April 1996, pp 1112-1118*

132. F. GUÉRIN, P. BANNELIER, M. LABEYRIE (1995) Thomson CSF, Orsay, Elancourt IRCOM, Limoges « Scattering of electromagnetic waves by helices and applications to the modelling of chiral composites I : simple effective-medium theories, II : Maxwell Garnett treatment » *Journal of Physics D : Applied Physics Vol.28 1995, pp 623-656*

133. T.EGE (1995) University of Gaziantep, Turkey « Scattering by a two dimensional periodic array of conducting rings on a chiral slab. » *IEEE/AP-S Symposium, pp 1667-70, Newport Beach California*

134. A.O.KOCA AND T.EGE (1995) University of Gaziantep, Turkey « A novel frequency selective surface; Chiro-FSS. » *Microwave and Optical Technology Letters. Vol.10 pp 157-160*

135. A.O.KOCA AND T.EGE (1996) University of Gaziantep, Turkey « The reflection and transmission characteristics of a FSS comprising of crossed dipoles on a chiral slab » *8th Mediterranean Electrotechnical Conference, MELECON'96, pp 531-533. 13-16 May 1996 Bari, Italy.*

136. T.EGE AND A.O.KOCA (1996) University of Gaziantep, Turkey « Effect of Chiral Loading on frequency selective surfaces comprising of arrays of rings » *Progress in Electromagnetics Research Symposium, PIERS'96, 8-12 July 1996, Innsbruck, Austria.*

II.

Fundamental Issues of Bi-Anisotropic Electromagnetics

COVARIANT MULTIPOLE D AND H FIELDS FOR REFLECTION FROM A MAGNETIC ANISOTROPIC CHIRAL MEDIUM

E.B. GRAHAM AND R.E. RAAB
Physics Department, University of Natal, P/Bag X01, Scottsville, Pietermaritzburg 3209, South Africa.

Abstract. Difficulties have been noted in theories of reflection from chiral media where standard Maxwell boundary conditions are applied to the **D** and **H** fields. Such difficulties may be avoided by adopting covariant forms for **D** and **H**. The appropriate forms for a magnetic anisotropic chiral medium are derived in the electric quadrupole-magnetic dipole approximation in terms of the macroscopic property tensors of the medium.

1. Introduction

The use of standard Maxwell boundary conditions [1] on the **D** and **H** fields in theories of specular reflection from both isotropic [2, 3, 4] and anisotropic [5, 6] chiral media has been found to yield conflicting results which depend on the forms for **D** and **H**. For example, the differing theories for an isotropic chiral medium raise the fundamental question whether the optical activity of the medium is manifest in reflection for normal incidence at a vacuum-crystal interface. Analytical [3] and experimental [4] procedures have shown, in agreement with macroscopic space-time symmetry predictions [7], that an isotropic chiral medium in a vacuum does not display its preferential handedness in reflection at normal incidence. However, it may do so when the ambient achiral media are birefringent [8].

An important consideration in any reflection theory is whether the 2×2 reflection matrix exhibits the form that is allowed by symmetry. The constraints imposed by space-time symmetry on the reflection matrices for various time-even and time-odd systems have been identified [7, 9], and these provide a necessary means for assessing the correctness of a theory

A. Priou et al. (eds.), Advances in Complex Electromagnetic Materials, 55–66.

and also for distinguishing which forms of the constitutive relations for the **D** and **H** fields are suitable for use in the Maxwell boundary conditions.

We have found when considering reflection from several different systems, namely noncentrosymmetric uniaxial crystals [6], a fluid in a uniform electric field [10], and a single-domain crystal of antiferromagnetic chromium oxide [11], that reflection matrices with the correct symmetry properties are obtained from the standard Maxwell boundary conditions only when the **D** and **H** fields are taken in covariant form. Such forms are derived in the following section for a magnetic anisotropic chiral medium.

2. Covariant Multipole D and H Fields for a Magnetic Chiral Medium

When electromagnetic radiation interacts with matter the multipole contributions to a physical effect are ordered as follows [12]:

$$\textit{electric dipole} \gg \left\{ \begin{array}{l} \textit{electric quadrupole} \\ \textit{magnetic dipole} \end{array} \right. \gg \left\{ \begin{array}{l} \textit{electric octopole} \\ \textit{magnetic quadrupole} \end{array} \right. \gg \cdots \qquad (1)$$

We consider a source-free homogeneous chiral medium which may be both anisotropic and magnetic, and work to the order of electric quadrupoles and magnetic dipoles, since the behaviour of a chiral medium is in general described within this approximation [12, 13, 14, 15, 16].

The theory is based on Maxwell's equations for a source-free medium in the form [17]

$$\nabla \times \mathbf{E} = -\dot{\mathbf{B}} \qquad (2)$$
$$\nabla \cdot \mathbf{E} = \epsilon_0^{-1} \rho_b \qquad (3)$$
$$\nabla \times \mathbf{B} = \mu_0(\epsilon_0 \dot{\mathbf{E}} + \mathbf{J_b}) \qquad (4)$$
$$\nabla \cdot \mathbf{B} = 0\ . \qquad (5)$$

In (2)-(5), **E** and **B** are, respectively, the macroscopic electric and magnetic fields, an overdot denotes partial differentiation with respect to time, and ρ_b and $\mathbf{J_b}$ are, respectively, densities of bound charge and bound current which arise in the presence of matter.

It is common practice to rewrite (3) and (4) in the form

$$\nabla \cdot \mathbf{D} = 0 \qquad (6)$$
$$\nabla \times \mathbf{H} = \dot{\mathbf{D}} \qquad (7)$$

where the **D** and **H** fields, when expressed in terms of the properties of the medium, are collectively called the constitutive relations.

If the wavelength of the light wave is very much larger than the dimensions of a macroscopic volume element of material, ρ_b and $\mathbf{J_b}$ may be expressed in terms of multipole moment densities. To the order of electric quadrupoles and magnetic dipoles the appropriate forms are [18, 19, 20]

$$\rho_b = -\nabla_\alpha P_\alpha + \tfrac{1}{2}\nabla_\alpha \nabla_\beta Q_{\alpha\beta} - \dots \tag{8}$$

$$J_{b_\alpha} = \dot{P}_\alpha - \tfrac{1}{2}\nabla_\beta \dot{Q}_{\alpha\beta} + \epsilon_{\alpha\beta\gamma}\nabla_\beta M_\gamma + \dots \ . \tag{9}$$

In (8) and (9), a Greek subscript denotes a Cartesian component of a tensor, $\epsilon_{\alpha\beta\gamma}$ is the Levi-Civita (alternating) tensor, and P_α, $Q_{\alpha\beta}$ and M_α are, respectively, the electric dipole, electric quadrupole and magnetic dipole moments per unit macroscopic volume. These polarization densities are induced in a macroscopic volume element of the material by the light wave fields and their space and time derivatives. A time-dependent perturbation treatment, in which use is made of the Hamiltonian describing the interaction of a macroscopic volume element of the medium with a low intensity light wave of angular frequency ω, yields the following forms for the induced polarization densities in the electric quadrupole-magnetic dipole approximation [15, 21, 22]

$$\begin{aligned} P_\alpha &= \alpha_{\alpha\beta}E_\beta + \omega^{-1}\alpha'_{\alpha\beta}\dot{E}_\beta + \tfrac{1}{2}a_{\alpha\beta\gamma}\nabla_\gamma E_\beta + \tfrac{1}{2}\omega^{-1}a'_{\alpha\beta\gamma}\nabla_\gamma \dot{E}_\beta \\ &\quad + G_{\alpha\beta}B_\beta + \omega^{-1}G'_{\alpha\beta}\dot{B}_\beta + \dots \end{aligned} \tag{10}$$

$$Q_{\alpha\beta} = \mathfrak{a}_{\alpha\beta\gamma}E_\gamma + \omega^{-1}\mathfrak{a}'_{\alpha\beta\gamma}\dot{E}_\gamma + \dots \tag{11}$$

$$M_\alpha = \mathcal{G}_{\alpha\beta}E_\beta + \omega^{-1}\mathcal{G}'_{\alpha\beta}\dot{E}_\beta + \dots . \tag{12}$$

Because (10)-(12) are frequency-domain representations of the multipole moment densities the symbols $\dot{E}_\beta$, $\dot{B}_\beta$ and $\nabla_\gamma \dot{E}_\beta$ are equivalent, respectively, to $(-i\omega E_\beta)$, $(-i\omega B_\beta)$ and $(-i\omega\nabla_\gamma E_\beta)$ [15, 16, 23]. However, the use of an overdot is preferred here to show explicitly which contributions to the induced moments are due to the time variation of the light wave fields and field gradient.

The polarizability tensors on the right-hand sides of (10)–(12) are frequency-dependent volume properties of the medium and are in general complex to allow for absorption. The dispersive and absorptive parts of the complex polarizability tensors are related by the Kramers-Kronig relations [15]. Quantum-mechanical expressions have previously been derived for these polarizabilities, and are given in Appendix A. It follows from the hermiticity of the operators in (33)-(40) that the polarizability tensors in these equations are not all independent and satisfy the following relationships:

$$\mathfrak{a}_{\alpha\beta\gamma} = a_{\gamma\alpha\beta} \ , \quad \mathfrak{a}'_{\alpha\beta\gamma} = -a'_{\gamma\alpha\beta} \ , \quad \mathcal{G}_{\alpha\beta} = G_{\beta\alpha} \ , \quad \mathcal{G}'_{\alpha\beta} = -G'_{\beta\alpha} \ . \tag{13}$$

The behaviour of the polarizability tensors under time reversal is readily identified by inspection [22]. Of the time-even tensors in (10)-(12) $\alpha_{\alpha\beta}$

is well-known and accounts for electric dipole effects in matter such as the conventional linear birefringence in uniaxial and biaxial crystals, while $G'_{\alpha\beta}$, $\mathcal{G}'_{\alpha\beta}$, $a_{\alpha\beta\gamma}$ and $\mathrm{a}_{\alpha\beta\gamma}$ are the optical activity tensors [12, 14, 15, 16]. The remaining polarizability tensors are time-odd and are properties only of magnetic systems [24]. Of these $\alpha'_{\alpha\beta}$ is associated with the intrinsic Faraday effect in ferromagnets [25] and $G_{\alpha\beta}$, $\mathcal{G}_{\alpha\beta}$, $a'_{\alpha\beta\gamma}$ and $\mathrm{a}'_{\alpha\beta\gamma}$ describe nonreciprocal and magnetoelectric phenomena in magnetic crystals [22].

With the aid of (8)-(13), the Maxwell equations (3) and (4) may be expressed in terms of the independent property tensors of the medium. Thereafter, when use is made of (2) and of the symmetry properties (see Appendix A)

$$a_{\alpha\beta\gamma} = a_{\alpha\gamma\beta} \,, \qquad a'_{\alpha\beta\gamma} = a'_{\alpha\gamma\beta} \tag{14}$$

(3) and (4) can be manipulated into the forms in (6) and (7) with **D** and **H** in covariant form. For a light wave with time-dependence $\exp\{-i\omega t\}$, the covariant expressions for **D** and **H** in the electric quadrupole-magnetic dipole approximation are

$$D_\alpha = A_{\alpha\beta}E_\beta + T_{\alpha\beta}B_\beta \tag{15}$$
$$H_\alpha = U_{\alpha\beta}E_\beta + X_{\alpha\beta}B_\beta \tag{16}$$

where, in terms of the independent polarizability tensors in (10)-(12),

$$A_{\alpha\beta} = \epsilon_0\delta_{\alpha\beta} + \alpha_{\alpha\beta} - i[\alpha'_{\alpha\beta} + \tfrac{1}{3}(a'_{\alpha\beta\gamma} + a'_{\beta\alpha\gamma} + a'_{\gamma\alpha\beta})\nabla_\gamma] \tag{17}$$
$$T_{\alpha\beta} = [G_{\alpha\beta} - \tfrac{1}{3}\delta_{\alpha\beta}G_{\gamma\gamma} - \tfrac{1}{6}\omega\epsilon_{\beta\gamma\delta}a'_{\gamma\delta\alpha}] - i(G'_{\alpha\beta} - \tfrac{1}{2}\omega\epsilon_{\beta\gamma\delta}a_{\gamma\delta\alpha}) \tag{18}$$
$$U_{\alpha\beta} = -[G_{\beta\alpha} - \tfrac{1}{3}\delta_{\alpha\beta}G_{\gamma\gamma} - \tfrac{1}{6}\omega\epsilon_{\alpha\gamma\delta}a'_{\gamma\delta\beta}] - i(G'_{\beta\alpha} - \tfrac{1}{2}\omega\epsilon_{\alpha\gamma\delta}a_{\gamma\delta\beta}) \tag{19}$$
$$X_{\alpha\beta} = \mu_0^{-1}\delta_{\alpha\beta} \,. \tag{20}$$

In (17)-(20) the time-odd tensors that apply only to magnetic media have been placed in square brackets. For a lossless medium the polarizability tensors on the right-hand sides of (17)-(20) are real and we then find, in agreement with Post [26], that $A_{\alpha\beta}$ and $X_{\alpha\beta}$ are Hermitian (that is $A_{\alpha\beta} = A^*_{\beta\alpha}$, $X_{\alpha\beta} = X^*_{\beta\alpha}$) and $U_{\alpha\beta} = -T^*_{\beta\alpha}$, where an asterisk denotes complex conjugation.

Because (17)-(19) contain tensors which depend on the choice of origin used for the multipole moment operators in their quantum-mechanical expressions, it is necessary to check that **D** and **H** are themselves origin independent. With the aid of the origin shifts in (44)-(52), of the symmetry properties of $\alpha_{\alpha\beta}$ and $\alpha'_{\alpha\beta}$ in (31) and (32), respectively, and also of the identity

$$\delta_{\alpha\beta}G_{\gamma\gamma} = \epsilon_{\beta\gamma\delta}\epsilon_{\alpha\gamma\epsilon}G_{\delta\epsilon} + G_{\alpha\beta} \tag{21}$$

it can be shown that the time-even and time-odd contributions to $A_{\alpha\beta}$, $T_{\alpha\beta}$ and $U_{\alpha\beta}$ in (17)-(19) are each separately origin independent which, in turn, establishes the translational invariance of the covariant forms of **D** and **H** in (15) and (16).

3. Application to Macroscopic Electrodynamic Problems

To describe wave propagation in any medium it is necessary only to make use of Maxwell's equations augmented with suitable constitutive relations. It therefore follows that any pair of **D** and **H** fields which is consistent with the multipole forms of ρ_b and $\mathbf{J_b}$ in (8) and (9), respectively, will provide the same description of the propagating waves in the medium. The traditional multipole forms for **D** and **H**, namely

$$D_\alpha = \epsilon_0 E_\alpha + P_\alpha - \tfrac{1}{2}\nabla_\beta Q_{\alpha\beta} + \ldots \quad (22)$$

$$H_\alpha = \mu_0^{-1} B_\alpha - M_\alpha + \ldots \quad (23)$$

which appear in a number of standard electrodynamic texts [1, 18, 19], are an obvious rearrangement of (3), (4), (8) and (9). These **D** and **H** fields have been used in theories of chiral and other effects associated with wave propagation in gases [14, 15, 16, 27], and also in non-magnetic [12, 16] and magnetic [22, 25, 28] crystals and chiral composites [29]. Other forms of the constitutive relations, in which the **B** and **H** fields satisfy a scalar relationship, have also been used to describe wave propagation in non-magnetic and magnetic media [30, 31, 32]. This latitude in the choice of constitutive relations has led to the assumption that all forms may be regarded as equivalent in the treatment of other electromagnetic phenomena.

When considering reflection and transmission at an interface it is necessary to ensure that the propagating waves in each of the adjoining media are properly matched at the boundary. This constraint imposes conditions on the electromagnetic fields at the interface. Standard Maxwell boundary conditions require continuity across a source-free interface of the normal components of **D** and **B** and the tangential components of **E** and **H**. Because these conditions are derived from the integral forms of the field equations without consideration of constitutive relations, there is no apparent restriction on the forms for the **D** and **H** fields to be used in them.

It has been shown, however, that although the **D** and **H** fields in (22) and (23) ensure the translational invariance of Maxwell's equations, they do not themselves possess this property [20]. They are therefore unsuitable for use in theories of reflection because they do not preserve the translational invariance of the equations specifying continuity of the fields at an interface.

Other forms of **D** and **H** which are translationally invariant, but not covariant, may be identified from alternative arrangements of (8)-(12). However, when these forms are used in the standard Maxwell boundary conditions they yield reflection matrices that are at variance with the dictates of Reciprocity, that is time-inversion (see Appendix B).

We have found when considering reflection from both nonmagnetic and magnetic systems [6, 10, 11] that only the covariant forms of **D** and **H** in (15)-(20) yield reflection matrices with the correct spatial and temporal symmetry properties. Although our reflection matrices satisfy the requirements of space-time symmetry, it is nevertheless interesting to compare the properties predicted by our reflection theories with those observed experimentally, where data is available.

3.1. REFLECTION FROM CADMIUM SULFIDE (SYMMETRY 6*MM*)

Transmission and reflection effects in crystals belonging to the symmetry classes $6mm$, $4mm$ and $3m$ are described by a single origin-independent property [6, 12], namely

$$A = G'_{12} + \tfrac{1}{2}\omega(a_{113} - a_{311}) \quad . \tag{24}$$

When the optic axis is perpendicular to the plane of incidence, our reflection theory [6] yields the following expressions for the elements of the reflection matrix R defined in (54) (see Appendix B):

$$r_{pp} = \frac{n_o^2\cos\phi - (n_o^2 - \sin^2\phi)^{\frac{1}{2}}}{n_o^2\cos\phi + (n_o^2 - \sin^2\phi)^{\frac{1}{2}}} \tag{25}$$

$$r_{ss} = \frac{\cos\phi - (n_e^2 - \sin^2\phi)^{\frac{1}{2}}}{\cos\phi + (n_e^2 - \sin^2\phi)^{\frac{1}{2}}} \tag{26}$$

$$r_{ps} = r_{sp} = \frac{i\mu_0 cA\sin 2\phi}{[n_o^2\cos\phi + (n_o^2 - \sin^2\phi)^{\frac{1}{2}}][\cos\phi + (n_e^2 - \sin^2\phi)^{\frac{1}{2}}]} \tag{27}$$

where n_o and n_e are, respectively, the ordinary and extraordinary refractive indices and ϕ is the angle of incidence. Similar forms have previously been obtained by Ivchenko and Selkin [33] with the aid of additional boundary conditions.

It follows from (54) and (25)-(27) that for oblique incidence of either p or s linearly polarized beams the reflected light is elliptically polarized. Also, (27) shows that the reflectivities $R_{ps} = r_{ps}r_{ps}^*$ and $R_{sp} = r_{sp}r_{sp}^*$ are equal. It is evident from (27) that effects which depend on r_{ps} and r_{sp} vanish for normal incidence ($\phi = 0$) and have their maximum values when $\phi \simeq 45°$. These properties have all been observed for CdS crystals (class $6mm$) in the vicinity of the $B(n = 1)$ exciton resonance [34].

3.2. REFLECTION AT NORMAL INCIDENCE FROM A SINGLE DOMAIN CRYSTAL OF ANTIFERROMAGNETIC CHROMIUM OXIDE

Chromium oxide (Cr_2O_3) is a magnetoelectric crystal which has magnetic point group symmetry $\underline{\overline{3}\,m}$ below the Neel temperature $T_N = 307K$ [35]. Tables [24] show that for this crystal symmetry the property tensors $\alpha'_{\alpha\beta}$, $G'_{\alpha\beta}$ and $a_{\alpha\beta\gamma}$ in (17)-(19) are zero. The components of the remaining nonvanishing property tensors in (17)-(19) are specified relative to Cartesian crystallographic axes x, y, z where, following Birss [24], we have taken $C_3 \parallel z$ and $C_2 \parallel x$. We then find from the covariant **D** and **H** fields in (15)-(20) the following reflection matrices for normal incidence at a vacuum-crystal interface [11]:

$$R(z) = \begin{bmatrix} (1-n)/(1+n) & 2B/(1+n)^2 \\ -2B/(1+n)^2 & (1-n)/(1+n) \end{bmatrix} \tag{28}$$

when propagation is along the C_3 axis (optic axis), and

$$R(x) = \begin{bmatrix} (1-n)/(1+n) & -B/(1+n)^2 \\ B/(1+n)^2 & (1-n)/(1+n) \end{bmatrix} \tag{29}$$

when propagation is along the C_2 axis. In (28) and (29),

$$B = \tfrac{1}{3}\mu_0 c(G_{xx} - G_{zz} + \tfrac{1}{2}\omega a'_{xyz}) \tag{30}$$

and we have also set $n_x = n_z = n$. (This is a reasonable approximation since the measured birefringence [35] is $n_z - n_x = 5.8 \times 10^{-2}$ and also since this factor does not enter our results in (28) and (29).)

It follows from the form of the reflection matrices in (28) and (29) that the real and imaginary parts of B describe, respectively, rotation and circular dichroism in reflection. Inspection shows that, for an incident linearly polarized beam of azimuth 0° or 90°, the rotation for propagation along the C_3 axis is twice that for propagation along the C_2 axis and of opposite sign. Because B is related to time-odd property tensors, as (30) shows, the rotation and circular dichroism change sign when the crystal time-form is reversed. These predictions are in accord with experiment [35].

4. Conclusion

In order to obtain a correct description of the physical effects in reflection from any system, it is essential that the appropriate reflection matrix has the form that is required by space-time symmetry. The burden of this paper is that reflection matrices which satisfy the constraints of Reciprocity (time-inversion), and which also yield expressions for observables that are independent of the arbitrary origin to which the multipole moments are referred, can only be obtained from the standard Maxwell boundary conditions when use is made of the covariant forms for **D** and **H** in (15)-(20).

A. Appendix

The quantum-mechanical expressions for the polarizability tensors in (10)-(12) may be determined by using first-order perturbation theory and by working in the Barron-Gray gauge [36, 37]. Errors have, however, been noted in the forms that were previously reported [22]. The correct results for a macroscopic volume element Δv in the unperturbed quantum state n with energy E_n and for negligible absorption are

$$\alpha_{\alpha\beta} = 2\hbar^{-1}\Delta v \sum_j Z_{jn}\omega_{jn}\Re\langle n \mid P_\alpha \mid j\rangle\langle j \mid P_\beta \mid n\rangle = \alpha_{\beta\alpha} \quad (31)$$

$$\alpha'_{\alpha\beta} = -2\hbar^{-1}\Delta v \sum_j Z_{jn}\omega\Im\langle n \mid P_\alpha \mid j\rangle\langle j \mid P_\beta \mid n\rangle = -\alpha'_{\beta\alpha} \quad (32)$$

$$a_{\alpha\beta\gamma} = 2\hbar^{-1}\Delta v \sum_j Z_{jn}\omega_{jn}\Re\langle n \mid P_\alpha \mid j\rangle\langle j \mid Q_{\beta\gamma} \mid n\rangle = a_{\alpha\gamma\beta} \quad (33)$$

$$a'_{\alpha\beta\gamma} = -2\hbar^{-1}\Delta v \sum_j Z_{jn}\omega\Im\langle n \mid P_\alpha \mid j\rangle\langle j \mid Q_{\beta\gamma} \mid n\rangle = a'_{\alpha\gamma\beta} \quad (34)$$

$$\mathfrak{a}_{\alpha\beta\gamma} = 2\hbar^{-1}\Delta v \sum_j Z_{jn}\omega_{jn}\Re\langle n \mid Q_{\alpha\beta} \mid j\rangle\langle j \mid P_\gamma \mid n\rangle = \mathfrak{a}_{\beta\alpha\gamma} \quad (35)$$

$$\mathfrak{a}'_{\alpha\beta\gamma} = -2\hbar^{-1}\Delta v \sum_j Z_{jn}\omega\Im\langle n \mid Q_{\alpha\beta} \mid j\rangle\langle j \mid P_\gamma \mid n\rangle = \mathfrak{a}'_{\beta\alpha\gamma} \quad (36)$$

$$G_{\alpha\beta} = 2\hbar^{-1}\Delta v \sum_j Z_{jn}\omega_{jn}\Re\langle n \mid P_\alpha \mid j\rangle\langle j \mid M_\beta \mid n\rangle \quad (37)$$

$$G'_{\alpha\beta} = -2\hbar^{-1}\Delta v \sum_j Z_{jn}\omega\Im\langle n \mid P_\alpha \mid j\rangle\langle j \mid M_\beta \mid n\rangle \quad (38)$$

$$\mathcal{G}_{\alpha\beta} = 2\hbar^{-1}\Delta v \sum_j Z_{jn}\omega_{jn}\Re\langle n \mid M_\alpha \mid j\rangle\langle j \mid P_\beta \mid n\rangle \quad (39)$$

$$\mathcal{G}'_{\alpha\beta} = -2\hbar^{-1}\Delta v \sum_j Z_{jn}\omega\Im\langle n \mid M_\alpha \mid j\rangle\langle j \mid P_\beta \mid n\rangle \ . \quad (40)$$

In (31)-(40), $\omega_{jn} = \hbar^{-1}(E_j - E_n)$, $Z_{jn} = (\omega_{jn}^2 - \omega^2)^{-1}$ and $\Re$ and $\Im$ denote, respectively, the real and imaginary parts. The symmetry properties in (31) and (32) follow from the hermiticity of the operators, while those in (33)-(36) are a consequence of the intrinsic symmetry of the electric quadrupole moment operator $Q_{\alpha\beta}$, as is evident from (42).

The multipole moment operators in (31)-(40) are based on the following definitions for a distribution of discrete spinless charges [37]:

$$\textit{electric dipole} \quad : \quad \mu_\alpha = \sum q r_\alpha \tag{41}$$

$$\textit{electric quadrupole} \quad : \quad q_{\alpha\beta} = \sum q r_\alpha r_\beta \tag{42}$$

$$\textit{magnetic dipole} \quad : \quad m_\alpha = \sum (q/2m)\ell_\alpha \tag{43}$$

in which $\mathbf{r}$ specifies the position relative to an arbitrary origin 0 of charge q with mass m and angular momentum $\boldsymbol{\ell} = \mathbf{r} \times \mathbf{p}$. A displacement $\mathbf{R}$ of origin from the point 0 produces the following changes in these operators and in the independent polarizability tensors in (17)-(19):

$$\begin{aligned}
\Delta P_\alpha &= -R_\alpha(\Delta v)^{-1}\sum q && (44)\\
\Delta Q_{\alpha\beta} &= -R_\alpha P_\beta - R_\beta P_\alpha + R_\alpha R_\beta(\Delta v)^{-1}\sum q && (45)\\
\Delta M_\alpha &= -\epsilon_{\alpha\beta\gamma} R_\beta(\Delta v)^{-1}\sum (q/2m)p_\gamma && (46)\\
\Delta \alpha_{\alpha\beta} &= 0 && (47)\\
\Delta \alpha'_{\alpha\beta} &= 0 && (48)\\
\Delta a_{\alpha\beta\gamma} &= -R_\gamma \alpha_{\alpha\beta} - R_\beta \alpha_{\alpha\gamma} && (49)\\
\Delta a'_{\alpha\beta\gamma} &= -R_\gamma \alpha'_{\alpha\beta} - R_\beta \alpha'_{\alpha\gamma} && (50)\\
\Delta G_{\alpha\beta} &= -\tfrac{1}{2}\omega\epsilon_{\beta\gamma\delta} R_\gamma \alpha'_{\alpha\delta} && (51)\\
\Delta G'_{\alpha\beta} &= \tfrac{1}{2}\omega\epsilon_{\beta\gamma\delta} R_\gamma \alpha_{\alpha\delta} \ . && (52)
\end{aligned}$$

Equations (47)-(52) provide the basis for testing the origin independence of an expression for an observable which is derived within the electric quadrupole-magnetic dipole approximation.

B. Appendix

The Principle of Reciprocity is a form of time-inversion that may be properly applied to light scattering [38, 39, 40], including reflection (coherent light scattering) [7, 9], where it has recently been used to determine the constraints imposed by time-reversal symmetry on the reflection matrices for a system in two time-conjugated equilibrium states. A brief description of the role of Reciprocity in reflection is given here.

Consider the experiment, shown in Figure 1(a), in which an incident light beam from an unpolarized source S, after passage through a polarizer P, is reflected from a physical system in a state (t). A detector D records the intensity $I(t)$ of the reflected light that is transmitted by the analyzer A. Figure 1(b) depicts the time-reversed experiment in which the source and the detector are interchanged and the physical system, including any

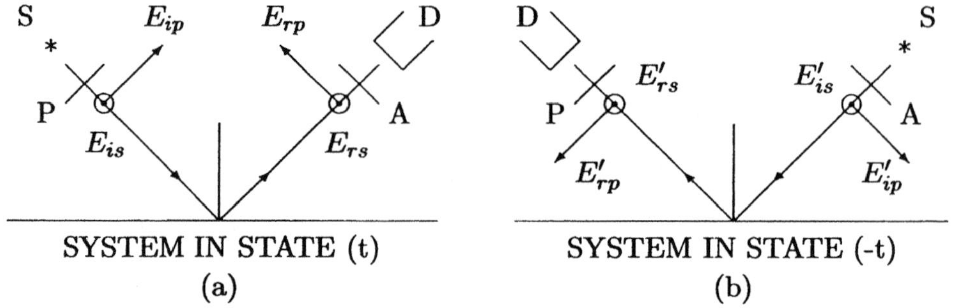

Figure 1. Reflection from a physical system in two time-conjugated equilibrium states: (a) time-forward experiment, (b) time-reversed experiment. S-unpolarized light-source, P-polarizer, A-analyzer, D-detector.

applied fields, is time-reversed. If the intensity recorded by the detector in the time-reversed experiment is $I(-t)$, then the Principle of Reciprocity states that

$$I(t) = I(-t) \ . \tag{53}$$

In the time-forward experiment (see Figure 1(a)), the components of the electric field of the incident light wave, E_{ip} and E_{is}, that are, respectively, parallel and perpendicular to the plane of incidence, may be related to those of the reflected wave, E_{rp} and E_{rs}, by the matrix equation [41]

$$\begin{pmatrix} E_{rp} \\ E_{rs} \end{pmatrix} = \begin{bmatrix} r_{pp} & r_{ps} \\ r_{sp} & r_{ss} \end{bmatrix} \begin{pmatrix} E_{ip} \\ E_{is} \end{pmatrix} = R \begin{pmatrix} E_{ip} \\ E_{is} \end{pmatrix} \tag{54}$$

where R is the 2×2 reflection matrix.

If R' is the reflection matrix for the time-reversed experiment in Figure 1(b), and is defined in a similar manner to R in (54), then the Reciprocity condition in (53) requires that the elements of R and R' satisfy the relationship [6]

$$\begin{bmatrix} r'_{pp} & r'_{ps} \\ r'_{sp} & r'_{ss} \end{bmatrix} = \begin{bmatrix} r_{pp} & -r_{sp} \\ -r_{ps} & r_{ss} \end{bmatrix} \ . \tag{55}$$

The relationship between R and R' in (55) places constraints on the elements of the reflection matrices for different physical systems. For example, if a system is nonmagnetic its properties are unchanged under time reversal. If, in addition, the system is isotropic, it is also insensitive to incidence from the left or the right. It then follows for a nonmagnetic isotropic system that

$$R = R' \tag{56}$$

and hence that

$$r_{ps} = -r_{sp} \quad . \tag{57}$$

Theories of reflection from a nonmagnetic isotropic chiral medium in which the elements of the reflection matrix do not satisfy the requirement in (57) are unable to describe correctly the observed physical effects [2, 3, 4].

References

1. Jackson, J.D. (1975) *Classical Electrodynamics*, 2nd edn, Wiley, New York, pp. 17-20, 232.
2. Schlagheck, U. (1973) A symmetry property in the theory of optical activity, *Z. Physik* **258**, 223-230.
3. Silverman, M.P. (1986) Reflection and refraction at the surface of a chiral medium: comparison of gyrotropic constitutive relations, invariant or noninvariant under a duality transformation, *J. Opt. Soc. Am.* A **3**, 830-837.
4. Luk'yanov, A.Yu. and Novikov, M.A. (1990) Reflection of light from the boundary of chiral gyrotropic medium, *JETP Lett.* **51**, 673-675.
5. Lalov, I.J. and Miteva, A.I. (1986) Reflection optical activity of uniaxial media, *J. Chem. Phys.* **85**, 5505-5511.
6. Graham, E.B. and Raab, R.E. (1996) Reflection from noncentrosymmetric uniaxial crystals: a multipole approach, *J. Opt. Soc. Am.* A **13**, 1239-1248.
7. Shelankov, A.L. and Pikus, G.E. (1992) Reciprocity in reflection and transmission of light, *Phys. Rev.* B **46**, 3326-3336.
8. Silverman, M.P. and Badoz, J. (1994) Interferometric enhancement of chiral asymmetries: ellipsometry with an optically active Fabry-Perot interferometer, *J. Opt. Soc. Am.* A **11**, 1894-1917.
9. Canright, G.S. and Rojo, A.G. (1992) Ellipsometry and broken time-reversal symmetry in the high-temperature superconductors, *Phys. Rev.* B **46**, 14078-14088.
10. Graham, E.B. and Raab, R.E. (1996) Reflection from a fluid in a uniform electric field, *Molec. Phys.* **88**, 1011-1023.
11. Graham, E.B. and Raab, R.E. Macroscopic theory of reflection from antiferromagnetic Cr_2O_3, *J. Phys. C : Condens. Matter* (submitted).
12. Graham, E.B. and Raab, R.E. (1990) Light propagation in cubic and other anisotropic crystals, *Proc. R. Soc. London* A **430**, 593-614.
13. Nakano, H. and Kimura, H. (1969) Quantum statistical-mechanical theory of optical activity, *J. Phys. Soc. Japan* **27**, 519-535.
14. Buckingham, A.D. and Dunn, M.B. (1971) Optical activity of oriented molecules, *J. Chem. Soc.* A, 1988-1991.
15. Barron, L.D. (1982) *Molecular Light Scattering and Optical Activity*, Cambridge University Press, Cambridge, pp. 74-80, 84-88, 243-246.
16. Raab, R.E. and Cloete, J.H. (1994) An eigenvalue theory of circular birefringence and dichroism in a non-magnetic chiral medium, *J. Electromagn. Waves Appl.* **8**, 1073-1089.
17. Purcell, E.M. (1985) *Electricity and Magnetism*, McGraw-Hill, New York, pp. 380, 432.
18. Rosenfeld, L. (1951) *Theory of Electrons*, North-Holland, Amsterdam, ch 2.
19. Robinson, F.N.H. (1973) *Macroscopic Electromagnetism*, Pergamon Press, Oxford, pp. 56-62.
20. Graham, E.B., Pierrus, J. and Raab, R.E. (1992) Multipole moments and Maxwell's equations, *J. Phys. B: At. Mol. Opt. Phys.* **25**, 4673-4684.
21. Buckingham, A.D. (1967) Permanent and induced molecular moments and long-range intermolecular forces, *Adv. Chem. Phys.* **12**, 107-142.

22. Graham, E.B. and Raab, R.E. (1992) Magnetic effects in antiferromagnetic crystals in the electric quadrupole-magnetic dipole approximation, *Phil. Mag.* B **66**, 269-284.
23. Kong, J.A. *Electromagnetic Wave Theory*, Wiley, New York, pp. 44-45.
24. Birss, R.R. (1963) Macroscopic symmetry in space-time, *Rep. Prog. Phys.* **26**, 307-360.
25. Graham, E.B. and Raab, R.E. (1991) Electric dipole effects in magnetic crystals, *J. Appl. Phys.* **69**, 2549-2551.
26. Post, E.J. (1962) *Formal Structure of Electromagnetics*, North Holland, Amsterdam, p. 132.
27. Imrie, D.A. and Raab, R.E. (1991) A new molecular theory of field gradient induced birefringence used for measuring electric quadrupole moments, *Molec. Phys.* **74**, 833-842.
28. Graham, E.B. and Raab, R.E. (1991) Non-reciprocal optical rotation in cubic antiferromagnets, *Phil. Mag.* B **64**, 267-274.
29. Theron, I.P. and Cloete, J.H. (1996) The optical activity of an artificial non-magnetic uniaxial chiral crystal at microwave frequencies, *J. Electromagn. Waves Appl.* **10**, 539-561.
30. Landau, L.D. and Lifshitz, E.M. (1960) *Electrodynamics of Continuous Media*, Pergamon Press, Oxford, p. 338.
31. Agranovich, V.M. and Ginzburg, V.L. (1984) *Crystal Optics with Spatial Dispersion, and Excitons*, 2nd edn, Springer-Verlag, Berlin, pp. 171-173.
32. Hornreich, R.M. and Shtrikman, S. (1968) Theory of gyrotropic birefringence, *Phys. Rev.* **171**, 1065-1074.
33. Ivchenko, E.L. and Selkin, A.V. (1982) Corrections to Maxwell boundary conditions in gyrotropic crystals, *Opt. Spectrosc.* (USA) **53**, 58-63.
34. Ivchenko, E.L. (1987) Spatial dispersion effects in the exciton resonance region, in I.E. Rashba and M.D. Sturge (eds.), *Excitons*, North-Holland, Amsterdam, pp. 167-170.
35. Krichevtsov, B.B., Pavlov, V.V., Pisarev, R.V., and Gridnev, V.N. (1993) Spontaneous non-reciprocal reflection of light from antiferromagnetic Cr_2O_3, *J. Phys. C: Condens. Matter* **5**, 8233-8244.
36. Barron, L.D. and Gray, C.G. (1973) The multipole interaction Hamiltonian for time dependent fields, *J. Phys. A : Math. Gen.* **6**, 59-61.
37. Raab, R.E. (1975) Magnetic multipole moments, *Molec. Phys.* **29**, 1323-1331.
38. Perrin, F. (1942) Polarization of light scattered by isotropic opalescent media, *J. Chem. Phys.* **10**, 415-427.
39. de Figueiredo, I.M.B. and Raab, R.E. (1980) A pictorial approach to macroscopic space-time symmetry, with particular reference to light scattering, *Proc. R. Soc. London* A **369**, 501-516.
40. Graham, C. (1980) Symmetry indications of the polarization state of light scattered by fluids in electric and magnetic fields, *Proc. R. Soc. London* A **369**, 517-535.
41. Azzam, R.M.A. and Bashara, N.M. (1979) *Ellipsometry and Polarized Light*, North-Holland, Amsterdam, p. 353.

JONES TRANSMISSION AND REFLECTION MATRICES FOR LOW-SYMMETRIC ABSORBING GYROTROPIC CRYSTALS

E.A. EVDISCHENKO, A.F. KONSTANTINOVA
Institute of Crystallography, Moscow, Russia

Abstract. Transmission and reflection matrices are obtained for low-symmetric absorbing gyrotropic crystal plates where eigenwave ellipticities can be different. Multiple reflections are taken into account. The eigenwave irreversibility is properly taken into account. The matrices are represented as the sums of four matrices, and each of them has its own physical meaning.

1. Introduction

Taking into account multiple reflections of eigenwaves in slabs demands corrections during the analysis of amplitudes and polarization measurement data. Unfortunately, up to now the exact solution to this problem for an anisotropic plate made of a low-symmetric absorbing crystal is not known in the literature. The intent of the article is to obtain analytical expressions for this case.

In [1, 2] the Jones matrices in general form were derived with allowance for multiple reflections from the boundaries of the plate:

$$T = T_2 T_c T_1 + T_2 Q \left(1 - Q\right)^{-1} T_c T_1 \tag{1}$$

$$R = R_1 + T_2 T_c R_2 \left(1 - Q\right)^{-1} T_c T_1 \tag{2}$$

where $Q = T_c R_1 T_c R_2$. Also, T is the transmission matrix, and R is the reflection matrix. The Jones matrices are T_j $(j = 1, 2, c)$ and R_j $(j = 1, 2)$. Indices $1, 2, c$ correspond to the crystal boundaries and to the crystal bulk, respectively.

In the general form T and R are suitable for computer calculations only. By taking advantage of (1) and (2) we can obtain the matrices T and R in the component form for low-symmetric absorbing gyrotropic crystals.

A. Priou et al. (eds.), Advances in Complex Electromagnetic Materials, 67–70.

2. Theory

T_j (or R_j) for an anisotropic plate can be written in terms of complex polarization parameters ξ_1, ξ_2 of the eigenwave of this plate:

$$T_j = \frac{1}{\xi_1 - \xi_2} \begin{bmatrix} \xi_1 t_2^{(j)} - \xi_2 t_1^{(j)} & t_1^{(j)} - t_2^{(j)} \\ \xi_1 \xi_2 \left(t_2^{(j)} - t_1^{(j)} \right) & \xi_1 t_2^{(j)} - \xi_2 t_1^{(j)} \end{bmatrix} \tag{3}$$

where $t_1^{(j)}$, $t_2^{(j)}$ (or $r_1^{(j)}$, $r_2^{(j)}$) are the transmission (or reflection) coefficients for the eigenwaves at the crystal boundaries ($j = 1, 2$) and in the crystal bulk ($j = c$):

$$t_m^{(c)} = \exp\left(i2\pi d N_m / \lambda \right), \quad N_m = n_m^{'} + i n_m^{''} \tag{4}$$

Here, $n_m^{'}$ and $n_m^{''}$ are the refractive index and the absorption coefficient of the m-th eigenwave, respectively. The question about values for $t_1^{(j)}, t_2^{(j)}$ (and $r_1^{(j)}$, $r_2^{(j)}$) for $j = 1, 2$ is considered later.

Complex polarization parameters of eigenwaves take the form

$$\xi_1 = i k_1, \quad \xi_2 = -i/k_2 \tag{5}$$

where k_1 and k_2 are the ellipticities of the orthogonal polarization ellipses of eigenwaves. The values ξ_1, ξ_2 change with the wave reflection from the second (exit) boundary of the plate, i.e., in the reverse path wave [3, 4]. Let us denote by $\xi_1^{\leftarrow}$, $\xi_2^{\leftarrow}$ the complex polarization parameters for the reverse path eigenwaves. The values ξ_1, ξ_2 and $\xi_1^{\leftarrow}$, $\xi_2^{\leftarrow}$ are connected with the relations [4]:

$$\xi_1 \xi_2^{\leftarrow} = \xi_1^{\leftarrow} \xi_2 = -1 \tag{6}$$

Therefore, we can write

$$\xi_1^{\leftarrow} = -1/\xi_2 = -i k_2; \quad \xi_2^{\leftarrow} = -1/\xi_1 = i/k_1 \tag{7}$$

If we substitute (3), (5), and (7) into (1) and (2), we obtain the expressions for T and D. To clarify the physical sense of the obtained bulky expressions for T and R we present them as the sums of matrices, so that each matrix would have its own physical sense. As the result we derive the matrices T and R in the following form:

$$\begin{aligned} T &= T_0 + A_1 T_1 + A_2 T_2 + A_3 T_3 + A_4 T_4 \\ &= \frac{1}{1 + k_1 k_2} \begin{bmatrix} h_1 + k_1 k_2 h_2 & -i k_2 (h_1 - h_2) \\ i k_1 (h_1 - h_2) & k_1 k_2 h_1 + h_2 \end{bmatrix} \\ &\quad + A_1 \begin{bmatrix} 1 & -i k_2 \\ i k_1 & k_1 k_2 \end{bmatrix} + A_2 \begin{bmatrix} k_1 k_2 & i k_2 \\ -i k_1 & 1 \end{bmatrix} \\ &\quad + A_3 \begin{bmatrix} k_1 & i \\ i k_1^2 & -k_1 \end{bmatrix} + A_4 \begin{bmatrix} k_2 & -i k_2^2 \\ -i & -k_2 \end{bmatrix} \end{aligned} \tag{8}$$

$$\begin{aligned} R &= R_0 + B_1R_1 + B_2R_2 + B_3R_3 + B_4R_4 \\ &= \frac{1}{1+k_1k_2}\begin{bmatrix} r_1 + k_1k_2r_2 & -ik_2\left(r_1 - r_2\right) \\ ik_1\left(r_1 - r_2\right) & k_1k_2r_1 + r_2 \end{bmatrix} \\ &+B_1\begin{bmatrix} 1 & -ik_2 \\ -ik_2 & -k_2^2 \end{bmatrix} + B_2\begin{bmatrix} -k_1^2 & -ik_1 \\ -ik_1 & 1 \end{bmatrix} \\ &+B_3\begin{bmatrix} k_1 & i \\ -ik_1k_2 & k_2 \end{bmatrix} + B_4\begin{bmatrix} k_1 & -ik_1k_2 \\ i & k_2 \end{bmatrix} \end{aligned} \tag{9}$$

where $h_m = t_m^{(1)} t_m^{(c)} t_m^{(2)}$ (m is the eigenwave number).

T_0 is the transmission matrix which describes a single reflection at the first boundary and a single reflection at the second boundary of the plane-parallel plate. Matrices $T_1 \cdots T_4$ describe the contribution of multiple reflection into the transmitted wave polarization. In this connection T_1, T_2 correspond to the transmission of eigenwaves which are generated on the first boundary. T_3, T_4 describe eigenwaves which are the results of the eigenwave irreversibility by reflections.

R_0 describes the incidence wave reflection from the first boundary. Matrices $R_1 \cdots R_4$ describe the additional contribution of waves which come out from the first boundary after multiple reflections inside the plate.

For example, multiplying T and R by the Jones vector of one of the plate's eigenwaves with complex polarization parameter $\xi_1 = ik_1$ we obtain the following expression for the Jones vectors of the transmitted E_t and reflected E_r waves:

$$E_t = T\begin{bmatrix} 1 \\ ik_1 \end{bmatrix} = (1 + k_1k_2)\left\{A_1\begin{bmatrix} 1 \\ ik_1 \end{bmatrix} + A_4\begin{bmatrix} 1 \\ -i/k_2 \end{bmatrix}\right\} \tag{10}$$

$$E_r = R\begin{bmatrix} 1 \\ ik_1 \end{bmatrix} = (1 + k_1k_2)\left\{r_1\begin{bmatrix} 1 \\ ik_1 \end{bmatrix} - B_1\begin{bmatrix} 1 \\ -ik_2 \end{bmatrix} - B_4\begin{bmatrix} 1 \\ i/k_1 \end{bmatrix}\right\} \tag{11}$$

From these expression we can see that the transmitted wave is built up of the waves with the polarizations of the first (incident) eigenwave $\xi_1 = ik_1$ and the second eigenwave $\xi_2 = -1/k_2$ with small amplitude A_4. The reflected wave is built up of three waves. The first of them has the polarization of the incident eigenwave $\xi_1 = ik_1$. Two additional waves have the polarizations of the eigenwaves in the crystal plate in the reverse path $\xi_1^{\leftarrow} = -ik_2$, $\xi_2^{\leftarrow} = i/k_1$ and the small amplitudes B_1, B_4.

Amplitudes A_j, B_j are contained in the expressions for T and R. Each of them depends on the ellipticities k_1, k_2 of both the eigenwaves and the values $t_1^{(j)}, t_2^{(j)}$ $(j = 1, 2, c)$, $r_1^{(j)}, r_2^{(j)}$ $(j = 1, 2)$. A_3, A_4, B_3, B_4 are proportional to k_1, k_2 and therefore become zero for the nongyrotropic crystals.

We find that the transmission and reflection coefficients of eigenwaves on the plate boundaries are equal to the Fresnel coefficients for isotropic media. This equality is true with an accuracy of the product of the birefringence and the ellipticity for each of the eigenwaves. It should be remembered that each eigenwave has its own refractive index.

For the transmission matrix T the amplitudes A_j take the form:

$$A_1 = h_1\eta_1; \quad A_2 = h_2\eta_2; \quad A_3 = \alpha_1\beta_2\eta_3; \quad A_4 = \alpha_2\beta_1\eta_3 \qquad (12)$$

$$\begin{aligned}
\eta_1 &= \Delta^{-1}\left\{p_1^2\left(1-k_1^2\right)\left(1-k_2^2\right)+p_1p_2\left(k_1+k_2\right)^2-p_1^2p_2^2\left(1+k_1k_2\right)^2\right\}\\
\eta_2 &= \Delta^{-1}\left\{p_2^2\left(1-k_1^2\right)\left(1-k_2^2\right)+p_1p_2\left(k_1+k_2\right)^2-p_1^2p_2^2\left(1+k_1k_2\right)^2\right\}\\
\eta_3 &= \Delta^{-1}\left(k_1+k_2\right)\left(p_1-p_2\right)
\end{aligned}$$

For the reflection matrix R the amplitudes B_j take the form:

$$B_1 = \alpha_1\beta_1\mu_1; \quad B_2 = \alpha_2\beta_2\mu_2; \quad B_3 = \alpha_1\beta_2\mu_3; \quad B_4 = \alpha_2\beta_1\mu_3 \qquad (13)$$

$$\begin{aligned}
\mu_1 &= \Delta^{-1}\left(1-k_1^2\right)\left[1-\left(1+k_1k_2\right)p_2^2\right]\\
\mu_2 &= \Delta^{-1}\left(1-k_2^2\right)\left[1-\left(1+k_1k_2\right)p_1^2\right]\\
\mu_3 &= \Delta^{-1}\left(k_1+k_2\right)\left[1-\left(1+k_1k_2\right)p_1p_2\right]
\end{aligned}$$

The following notations have been introduced in (12), (13):

$$\begin{aligned}
\alpha_j &= r_jt_j^{(c)}t_j^{(2)}; \quad \beta_j = t_j^{(1)}t_j^{(c)}; \quad p = r_jt_j^{(c)}\\
\Delta &= \left(1+k_1k_2\right)\left[\left(1+k_1k_2\right)^2\left(1-p_1^2\right)\left(1-p_2^2\right)+\left(k_1+k_2\right)^2\left(p_1-p_2\right)^2\right]
\end{aligned}$$

3. Conclusion

The derived transmission and reflection matrices permit to solve the problem of light propagation through a crystal plate with any collection of optical properties. Besides, it is possible to properly take into account multiple reflections and the eigenwave irreversibility.

References

1. Filippov, V.V. (1983) Method of obtaining dispersion equation for normal and surface waves in layered anisotropic systems, *Sov. Phys. Crystallogr.* **28(2)**, 136-139.
2. Shamburov, V.A. and E.A.Evdischenko, E.A. (1991) Exact Jones matrix for a plate of a natural gyrotropic nonmagnetic crystal, *Sov. Phys. Crystallogr.* **36(4)**, 473-475.
3. Fedorov, F.I. (1976) *Theory of gyrotropy* (in Russian), Nauka i Tekhnika, Minsk.
4. Shamburov, V.A. (1989) The phenomenon of natural waves parameters change by the reversal of the waves normal in the gyrotropic and absorbing crystals with low symmetry, *Collected Abstracts XII European Crystallographic Meeting*, p. 418.

INVERSE PROBLEM OF REFLECTION AND TRANSMISSION FOR A BIANISOTROPIC MEDIUM

G.N. BORZDOV
Department of Theoretical Physics
Byelorussian State University
Fr. Skaryny avenue, 4
220050 Minsk, Belarus

Abstract. On the basis of the surface impedance and characteristic matrix methods, the exact solutions of the inverse problems of reflection and transmission for motionless and uniformly moving bianisotropic media with temporal and spatial dispersion are obtained.

1. Introduction

Recently, considerable attention has been focused on the properties of chiral, biisotropic and bianisotropic media, especially, layered media and composite materials. Within the last few decades, several sets of constitutive equations for such media have been proposed, and their merits and shortcomings as well as the number of the independent material parameters and their properties have been intensively discussed in literature [1]–[8]. The progress in constructing artificial chiral media provides new impetus for theoretical studies concerning predictions of the effective media properties of composite materials and development of new techniques for measuring electromagnetic parameters of biisotropic and bianisotropic media [9]–[13].

Bianisotropic media are characterized by the constitutive equations [1, 3, 4, 5]

$$\boldsymbol{D} = \varepsilon \boldsymbol{E} + \alpha \boldsymbol{B}, \quad \boldsymbol{H} = \beta \boldsymbol{E} + \mu^{-1} \boldsymbol{B} \tag{1}$$

where ε and μ are the permittivity and permeability tensors, α and β are the pseudotensors of gyrotropy. Using the four-dimensional tensors

A. Priou et al. (eds.), Advances in Complex Electromagnetic Materials, 71–84.

$$F = \sum_{1\le i<j\le 4} F_{ij}\vartheta^i\wedge\vartheta^j$$
$$= B_3\vartheta^1\wedge\vartheta^2 - B_2\vartheta^1\wedge\vartheta^3 + B_1\vartheta^2\wedge\vartheta^3 + (E_1\vartheta^1 + E_2\vartheta^2 + E_3\vartheta^3)\wedge\vartheta^4 \tag{2}$$

$$G = \sum_{1\le i<j\le 4} G_{ij}\vartheta^i\wedge\vartheta^j$$
$$= H_3\vartheta^1\wedge\vartheta^2 - H_2\vartheta^1\wedge\vartheta^3 + H_1\vartheta^2\wedge\vartheta^3 + (D_1\vartheta^1 + D_2\vartheta^2 + D_3\vartheta^3)\wedge\vartheta^4 \tag{3}$$

one can replace Eqs. (1) by the Lorentz-covariant constitutive equation

$$G = MF = \sum_{1\le i<j\le 4,\, 1\le k<l\le 4} M_{ij}{}^{kl} F_{kl}\vartheta^i\wedge\vartheta^j \tag{4}$$

where

$$M = \sum_{1\le i<j\le 4,\, 1\le k<l\le 4} M_{ij}{}^{kl}\vartheta^i\wedge\vartheta^j\otimes e_k\wedge e_l \tag{5}$$

is the four-dimensional material tensor. Here, (e_i) and (ϑ^i) are the dual orthonormal bases in the Minkowski vector space V and its dual V^* (the space of 1-forms or, in other words, the space of covariant vectors), $e_i\rfloor\vartheta^j = e_i\lfloor\vartheta^j = \delta_i^j$, $i,j = 1,2,3,4$, δ_i^j is the Kronecker delta function, $\otimes$, $\wedge$ and $\rfloor$, $\lfloor$ are the tensor, exterior and interior products [11, 12]. For instance, $e_i\wedge e_j = e_i\otimes e_j - e_j\otimes e_i$, $e_1\rfloor(\vartheta^1\wedge\vartheta^2) = \vartheta^2$, $(e_1\wedge e_2)\lfloor\vartheta^1 = e_2$.

In [11, 12], by making use of the exterior algebra and the generalized impedance and characteristic matrix methods, the Lorentz-covariant solutions of direct and inverse problems for dispersionless linear media are obtained. In particular, one can use these solutions to find all 36 material parameters $M_{ij}{}^{kl}$ by measuring reflection and transmission coefficients.

In a homogeneous linear dispersive medium, G and F are related by the integral transformation [5, 14]

$$G(\boldsymbol{x}) = \int \mathcal{M}(\boldsymbol{y})F(\boldsymbol{x}-\boldsymbol{y})\,d^4y \tag{6}$$

where d^4y is the infinitesimal element of the space-time volume, and $\mathcal{M}$ is a tensor function of type (2,2). For a harmonic partial wave, Eq. (6) reduces to Eqs. (1) and (4), but ε, μ, α, β, and M now depend on the wave vector $\mathbf{k}$ as well as on the frequency ω (the four-dimensional wave vector $\boldsymbol{K} = \boldsymbol{k} + (\omega/c)e_4$) [5, 14]. Instead of using a separate set of such tensors for each partial wave, it is advantageous to extend the material equation (4) to some superpositions of waves by introducing the generalized material tensor defined on a set of evolution operators ($F(\boldsymbol{x}+\boldsymbol{y}) = \mathcal{F}(\boldsymbol{y})F(\boldsymbol{x})$) by [12, 13, 15, 16]

$$M(\mathcal{F}) = \int \mathcal{M}(\boldsymbol{y})\mathcal{F}(-\boldsymbol{y})\,d^4y \tag{7}$$

In this paper, we present the relations which make it possible to find the three-dimensional and four-dimensional generalized material parameters as well as the properties of eigenwaves in a dispersive bianisotropic medium, provided its reflection and transmission coefficients are given (measured).

2. Three-dimensional Material Parameters

2.1. TENSORS OF SURFACE IMPEDANCE γ AND NORMAL REFRACTION N_H

The three-dimensional tensors of surface impedance γ and normal refraction N_H [16, 17] are defined by

$$\boldsymbol{q}\times\boldsymbol{E} = \gamma\boldsymbol{H}_t, \quad \boldsymbol{q}\gamma = \gamma\boldsymbol{q} = 0 \tag{8}$$

$$\boldsymbol{H}_t(z) = \exp(ik_0 z N_H)\boldsymbol{H}_t(0), \quad \boldsymbol{q}N_H = N_H\boldsymbol{q} = 0 \tag{9}$$

where $\boldsymbol{q}$ is the unit vector normal to the interface, $\boldsymbol{H}_t = I\boldsymbol{H}$ is the tangential component of $\boldsymbol{H}$, $I = \mathbf{1} - \boldsymbol{q}\otimes\boldsymbol{q}$ is the projection operator of the interface, $\mathbf{1}$ is the unit tensor, $z = \boldsymbol{q}\cdot\boldsymbol{r}$, $k_0 = \omega/c$, c is the velocity of light in vacuum. In the general case, γ and N_H are complex nonsymmetric tensors. A nondegenerate operator N_H, i.e. N_H with different eigenvalues of multiplicity 1, describes the superpositions of two refracted eigenwaves excited in a homogeneous bianisotropic medium by an incident wave (see Fig. 1)

$$\boldsymbol{H}_t^{in}(\boldsymbol{r},t) = \boldsymbol{H}_t^{in}(0)\exp[i(k_0\boldsymbol{m}_{in}\cdot\boldsymbol{r} - \omega t)] \tag{10}$$

The eigenvectors $\boldsymbol{u}$ and the corresponding eigenvalues η of N_H $(N_H\boldsymbol{u} = \eta\boldsymbol{u}$, $\boldsymbol{q}\cdot\boldsymbol{u} = 0)$ specify the polarizations (in terms of $\boldsymbol{H}_t$) and the normal components of refraction vectors $\boldsymbol{m}$ $(\boldsymbol{m} = \boldsymbol{b}+\eta\boldsymbol{q}$, $\boldsymbol{b} = I\boldsymbol{m} = I\boldsymbol{m}_{in}$, $\boldsymbol{b}\cdot\boldsymbol{q} = 0$, $\eta = \boldsymbol{m}\cdot\boldsymbol{q})$ of these refracted eigenwaves (see Fig. 1). Here we use Fedorov's term "refraction vector" [1]. The latter is the wave vector $\boldsymbol{k}$ divided by the wave number k_0 in vacuum $(\boldsymbol{m} = \boldsymbol{k}/k_0)$, i.e. the product of the refractive index n and the unit wave normal $\boldsymbol{n}$ $(\boldsymbol{m} = n\boldsymbol{n})$. It should be also noted that, if N_H has a double eigenvalue η with a one-dimensional eigensubspace, Eqs. (8) and (9) describe the Fedorov–Petrov wave with linear dependence of amplitude on z [1, 15, 16, 18, 19].

Now, let us consider the plane wave approximation of a collimated incident beam, such as a laser beam. Let us assume, that the thickness of the slab under study and the angle of incidence have sufficiently large values so that the multiple reflected beames are separated and, therefore, they make no contributions to the reflection r and transmission d tensors (see Fig. 1). Making use of the definitions (8) and (9), we obtain

$$\gamma = \gamma_0(I - r)(I + r)^- \tag{11}$$

$$N_H = \frac{1}{ik_0}(I+r)d^- \frac{\partial d}{\partial z}(I+r)^- \tag{12}$$

where $d^- = (d_t I - d)/\overline{d}_t$ is the pseudoinverse tensor [16, 17] (since $\boldsymbol{q}\boldsymbol{d} = \boldsymbol{d}\boldsymbol{q} = \boldsymbol{q}\boldsymbol{r} = \boldsymbol{r}\boldsymbol{q} = 0$, r and $\boldsymbol{d}$ do not have the inverse tensors), d_t is the trace of d, $\overline{d}_t$ is the trace of the adjoint tensor $\overline{d}$ (if $\boldsymbol{q} = \boldsymbol{e}_3$, $d_t = d_{11} + d_{22}$, $\overline{d}_t = d_{11}d_{22} - d_{12}d_{21}$). We assume here, that the incident wave propagates in an isotropic medium with the permittivity ε_0 and permeability μ_0, so that its impedance tensor is defined by

$$\gamma_0 = \frac{\mu_0}{\eta_0}\boldsymbol{e}_1 \otimes \boldsymbol{e}_1 + \frac{\eta_0}{\varepsilon_0}\boldsymbol{e}_2 \otimes \boldsymbol{e}_2 \tag{13}$$

where $\boldsymbol{q} = \boldsymbol{e}_3$, $\boldsymbol{m}_{in} \cdot \boldsymbol{e}_2 = 0$, $\eta_0 = \sqrt{\varepsilon_0\mu_0 - \boldsymbol{b}^2}$ and $\boldsymbol{b}$ are the normal and tangential components of $\boldsymbol{m}_{in}$.

Thus, to determine γ and N_H it is necessary to measure the reflection tensor r and the so-called multiplicative derivative [20] $D_z d = d^- \partial d/\partial z$ of the transmission tensor $\boldsymbol{d}$. The latter can be measured in experiments with two similarly cut layers with slightly different values of thickness z. It should be mentioned that all numerical calculations with the tensors γ, r, d and so on, including the pseudoinversion, can be reduced to the algebraic operations with 2×2-matrices.

2.2. CHARACTERISTIC MATRIX $\mathcal{C}$ AND DIFFERENTIAL MATRIX OF PROPAGATION $\mathcal{A}$

The characteristic matrix $\mathcal{C}$ of a bianisotropic slab relates boundary values of $\boldsymbol{H}_t$ and $\boldsymbol{q}\times\boldsymbol{E}$ as follows [16, 17]

$$\begin{pmatrix} \boldsymbol{H}_t(z) \\ \boldsymbol{q}\times\boldsymbol{E}(z) \end{pmatrix} = \mathcal{C}(z) \begin{pmatrix} \boldsymbol{H}_t(0) \\ \boldsymbol{q}\times\boldsymbol{E}(0) \end{pmatrix} \tag{14}$$

The characteristic matrix $\mathcal{C}$ and the differential matrix of propagation $\mathcal{A}$ ($\partial\mathcal{C}/\partial z = i\mathcal{A}\mathcal{C}$) give the exact and exhaustive description of the field arising in the slab due to the multibeam interference.

Let us assume, that the waves with the same value of $\boldsymbol{b}$ incident on the slab from the both sides. Let us denote by γ_1, r_1 and d_1 (γ_2, r_2 and d_2) the surface impedance, reflection and transmission tensors of the wave incident onto the first (second) boundary, respectively. Let γ_3 and γ_4 be the impedances of the waves leaving the slab from the first and the second boundaries, respectively.

Let the reflection and transmission tensors r_1, r_2 and d_1, d_2 be measured. Then, the characteristic matrix $\mathcal{C}$ and the differential matrix of propagation $\mathcal{A}$ can be uniquely determined as follows:

$$\mathcal{C} = \begin{pmatrix} C_1 & C_2 \\ C_3 & C_4 \end{pmatrix}, \quad \mathcal{A} = \begin{pmatrix} A_1 & A_2 \\ A_3 & A_4 \end{pmatrix} \tag{15}$$

where

$$C_1 = \left[d_1 - (r_2 + I)d_2^-(r_1 + \gamma_3^-\gamma_1)\right](I - \gamma_3^-\gamma_1)^- \tag{16}$$

$$C_2 = \left[d_1 - (r_2 + I)d_2^-(r_1 + I)\right](\gamma_1 - \gamma_3)^- \tag{17}$$

$$C_3 = \left[\gamma_4 d_1 - (\gamma_4 r_2 + \gamma_2)d_2^-(r_1 + \gamma_3^-\gamma_1)\right](I - \gamma_3^-\gamma_1)^- \tag{18}$$

$$C_4 = \left[\gamma_4 d_1 - (\gamma_4 r_2 + \gamma_2)d_2^-(r_1 + I)\right](\gamma_1 - \gamma_3)^- \tag{19}$$

$$\begin{aligned} A_1 &= i\left[\frac{\partial d_1}{\partial z}d_1^-(r_2 + \gamma_4^-\gamma_2) + (r_2 + I)d_2^-\frac{\partial d_2}{\partial z}\right. \\ &- \left.(r_2 + I)d_2^-\frac{\partial r_1}{\partial z}d_1^-(r_2 + \gamma_4^-\gamma_2) - \frac{\partial r_2}{\partial z}\right](I - \gamma_4^-\gamma_2)^- \end{aligned} \tag{20}$$

$$\begin{aligned} A_2 &= i\left[\frac{\partial d_1}{\partial z}d_1^-(r_2 + I) + (r_2 + I)d_2^-\frac{\partial d_2}{\partial z}\right. \\ &- \left.(r_2 + I)d_2^-\frac{\partial r_1}{\partial z}d_1^-(r_2 + I) - \frac{\partial r_2}{\partial z}\right](\gamma_2 - \gamma_4)^- \end{aligned} \tag{21}$$

$$\begin{aligned} A_3 &= i\left[\gamma_4\frac{\partial d_1}{\partial z}d_1^-(r_2 + \gamma_4^-\gamma_2) + (\gamma_4 r_2 + \gamma_2)d_2^-\frac{\partial d_2}{\partial z}\right. \\ &- \left.(\gamma_4 r_2 + \gamma_2)d_2^-\frac{\partial r_1}{\partial z}d_1^-(r_2 + \gamma_4^-\gamma_2) - \gamma_4\frac{\partial r_2}{\partial z}\right](I - \gamma_4^-\gamma_2)^- \end{aligned} \tag{22}$$

$$\begin{aligned} A_4 &= i\left[\gamma_4\frac{\partial d_1}{\partial z}d_1^-(r_2 + I) + (\gamma_4 r_2 + \gamma_2)d_2^-\frac{\partial d_2}{\partial z}\right. \\ &- \left.(\gamma_4 r_2 + \gamma_2)d_2^-\frac{\partial r_1}{\partial z}d_1^-(r_2 + I) - \gamma_4\frac{\partial r_2}{\partial z}\right](\gamma_2 - \gamma_4)^- \end{aligned} \tag{23}$$

In an important particular case when the slab is surrounded by the isotropic medium, these expressions simplify, since $\gamma_1 = \gamma_4 = -\gamma_2 = -\gamma_3 = \gamma_0$ (13) and $\gamma_3^-\gamma_1 = -\gamma_4^-\gamma_2 = -I$.

In the general case, when $\mathcal{C}$ is a nondegenerate matrix, the field in a homogeneous bianisotropic slab consists of four eigenwaves, two of which propagate in the positive z-direction (see Fig. 1). The superposition of these two waves is described by the operators γ (8) and N_H (9). The values of $\boldsymbol{H}_t = \boldsymbol{u}$, $\boldsymbol{q} \times \boldsymbol{E} = \gamma\boldsymbol{u}$, and $\eta = \boldsymbol{m} \cdot \boldsymbol{q}$ for the eigenwave under study ($N_H\boldsymbol{u} = \eta\boldsymbol{u}$) can also be found from the equations

$$\mathcal{A}\begin{pmatrix} \boldsymbol{u} \\ \gamma\boldsymbol{u} \end{pmatrix} = k_0\eta\begin{pmatrix} \boldsymbol{u} \\ \gamma\boldsymbol{u} \end{pmatrix}, \quad \mathcal{C}\begin{pmatrix} \boldsymbol{u} \\ \gamma\boldsymbol{u} \end{pmatrix} = \exp(ik_0z\eta)\begin{pmatrix} \boldsymbol{u} \\ \gamma\boldsymbol{u} \end{pmatrix} \tag{24}$$

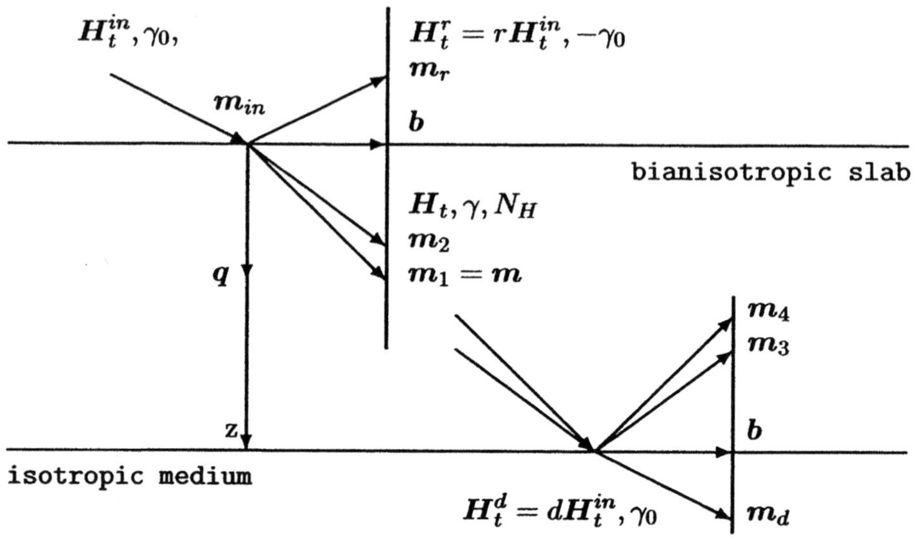

Figure 1. Refraction vectors of eigenwaves in a bianisotropic slab.

2.3. FIELD VECTORS OF EIGENWAVES

Let slabs I and II be differently cut off from the same bianisotropic material. Let $\boldsymbol{q}$ and $\boldsymbol{q}'$ be the corresponding unit interface normals (see Fig. 2). The relations obtained in the previous sections make it possible to find the vectors $\boldsymbol{H}_t \equiv \boldsymbol{u}$, $\boldsymbol{E}_t = (\gamma \boldsymbol{u}) \times \boldsymbol{q} \equiv \boldsymbol{w}$ and $\boldsymbol{m} = \boldsymbol{b} + \eta \boldsymbol{q}$ of an eigenwave in slab I as well as γ (8) and N_H (9). The same eigenwave, i.e., the wave with the same refraction vector $\boldsymbol{m} = \boldsymbol{b} + \eta \boldsymbol{q} = \boldsymbol{b}' + \eta' \boldsymbol{q}'$ and the same polarization, can be excited in slab II by the incident wave with the refraction vector

$$\boldsymbol{m}'_{in} = \boldsymbol{b}' + \boldsymbol{q}' \sqrt{\varepsilon_0 \mu_0 - \boldsymbol{b}'^2} \tag{25}$$

where $\boldsymbol{b}' = \boldsymbol{m} - \eta' \boldsymbol{q}'$, $\eta' = \boldsymbol{m} \cdot \boldsymbol{q}'$. By making a similar set of measurements and calculations for slab II, we can determine the parameters γ', N'_H, $\boldsymbol{u}'$ and $\boldsymbol{w}'$. Then, the field vectors $\boldsymbol{E}$, $\boldsymbol{B}$, $\boldsymbol{H}$ and $\boldsymbol{D}$ of the eigenwave under study can be found as follows:

$$\boldsymbol{E} = \boldsymbol{w}' + \boldsymbol{q}' \frac{(\boldsymbol{w}' \times \boldsymbol{w}) \cdot \boldsymbol{q}}{(\boldsymbol{q}' \times \boldsymbol{q}) \cdot \boldsymbol{w}}, \qquad \boldsymbol{B} = \boldsymbol{m} \times \boldsymbol{E} \tag{26}$$

$$H = u' + q' \frac{(u' \times u) \cdot q}{(q' \times q) \cdot u}, \qquad D = -m \times H \tag{27}$$

where

$$N_H u = \eta u, \quad w = (\gamma u) \times q, \quad N_H \neq \eta I \tag{28}$$

$$N_H' u' = \eta' u', \quad w' = (\gamma' u') \times q' \tag{29}$$

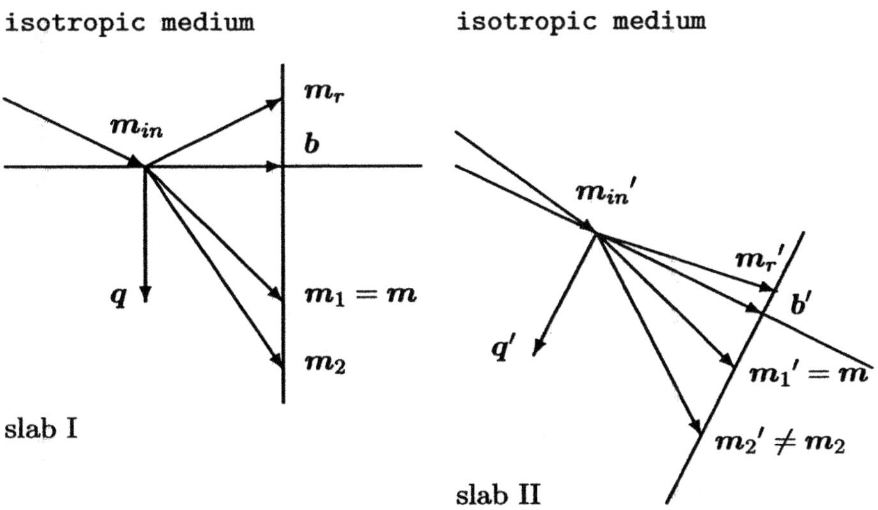

Figure 2. Refraction vectors of eigenwaves in slabs I and II.

If the eigenwave propagates along an isotropic axis ($N_H = \eta I$),

$$E = w' + \frac{q'}{w_1 \cdot a} \left[(w' - C w_2) \times w_1 \right] \cdot q, \qquad B = m \times E \tag{30}$$

$$H = u' + \frac{q'}{u_1 \cdot a} \left[(u' - C u_2) \times u_1 \right] \cdot q, \qquad D = -m \times H \tag{31}$$

where

$$C = \frac{(a \cdot u_1)(a \cdot w') - (a \cdot w_1)(a \cdot u')}{(a \cdot u_1)(a \cdot w_2) - (a \cdot w_1)(a \cdot u_2)}, \qquad a = q' \times q \tag{32}$$

Here, u_i and u' are arbitrary vectors satisfying the conditions $q \cdot u_i = q \cdot u' = 0$; $w_i = (\gamma u_i) \times q$, $i = 1, 2$; $w' = (\gamma' u') \times q'$. The expressions (30) and (31) describe the wave with the polarization parameter u' restricted only by the condition $q \cdot u' = 0$.

2.4. GENERALIZED MATERIAL PARAMETERS

Let the refraction vectors $\boldsymbol{m}_k$ and amplitudes $\boldsymbol{E}_k$, $\boldsymbol{H}_k$, $\boldsymbol{D}_k$ and $\boldsymbol{B}_k$ of arbitrary six eigenwaves with linearly independent six-vectors $(\boldsymbol{E}_k,\ \boldsymbol{H}_k)$ be found as outlined in sections 2.1–2.3. Then, the set $((\boldsymbol{E}_k,\ \boldsymbol{H}_k), k = 1, 2, \ldots, 6)$ is a basis in the six-dimensional $(\boldsymbol{E}, \boldsymbol{H})$–space, and any linear superposition of these six waves is described by the evolution operator

$$\mathcal{P}(\boldsymbol{r}, t) = \sum_{k=1}^{6} P_k \exp[i\omega_k(\boldsymbol{m}_k \cdot \boldsymbol{r}/c - t)] \tag{33}$$

and the constitutive equation

$$\begin{pmatrix} \boldsymbol{D}(\boldsymbol{r}, t) \\ \boldsymbol{B}(\boldsymbol{r}, t) \end{pmatrix} = M(\mathcal{P}) \begin{pmatrix} \boldsymbol{E}(\boldsymbol{r}, t) \\ \boldsymbol{H}(\boldsymbol{r}, t) \end{pmatrix}, \quad M(\mathcal{P}) = \begin{pmatrix} \varepsilon & \alpha \\ \beta & \mu \end{pmatrix} \tag{34}$$

where

$$M(\mathcal{P}) = \sum_{k=1}^{6} \begin{pmatrix} \boldsymbol{D}_k \otimes \boldsymbol{a}_k & \boldsymbol{D}_k \otimes \boldsymbol{b}_k \\ \boldsymbol{B}_k \otimes \boldsymbol{a}_k & \boldsymbol{B}_k \otimes \boldsymbol{b}_k \end{pmatrix} = \sum_{k=1}^{6} \begin{pmatrix} \boldsymbol{0} & -\boldsymbol{m}_k^{\times} \\ \boldsymbol{m}_k^{\times} & \boldsymbol{0} \end{pmatrix} P_k \tag{35}$$

is the matrix of generalized material parameters. Here, $\boldsymbol{m}^{\times}$ is the antisymmetric tensor dual to $\boldsymbol{m}$ $(\boldsymbol{m}^{\times}\boldsymbol{a} = \boldsymbol{m}\times\boldsymbol{a})$,

$$P_k = \begin{pmatrix} \boldsymbol{E}_k \otimes \boldsymbol{a}_k & \boldsymbol{E}_k \otimes \boldsymbol{b}_k \\ \boldsymbol{H}_k \otimes \boldsymbol{a}_k & \boldsymbol{H}_k \otimes \boldsymbol{b}_k \end{pmatrix} \tag{36}$$

is a projection operator ($P_k^2 = P_k$, $P_k P_n = 0$, $k \neq n$), $\boldsymbol{a}_k$ and $\boldsymbol{b}_k$ are defined by $\boldsymbol{a}_k \cdot \boldsymbol{E}_n + \boldsymbol{b}_k \cdot \boldsymbol{H}_n = \delta_{kn}$, i.e. the components of $\boldsymbol{a}_k$ and $\boldsymbol{b}_k$ can be found by the inversion of 6×6-matrix:

$$\begin{pmatrix} \boldsymbol{a}_1 & \boldsymbol{b}_1 \\ & \vdots & \\ \boldsymbol{a}_6 & \boldsymbol{b}_6 \end{pmatrix} = \begin{pmatrix} \boldsymbol{E}_1 & & \boldsymbol{E}_6 \\ & \cdots & \\ \boldsymbol{H}_1 & & \boldsymbol{H}_6 \end{pmatrix}^{-1} \tag{37}$$

In a bianisotropic medium with frequency dispersion, the constitutive equation (34) is valid for any monochromatic field. To find the corresponding material tensors $\varepsilon(\omega)$, $\mu(\omega)$, $\alpha(\omega)$ and $\beta(\omega)$, it is sufficient to calculate $M(\mathcal{P})$ for arbitrary six eigenwaves with the same frequency $\omega_k = \omega$, $k = 1, 2, \ldots, 6$. In this case, $M(\mathcal{P}) = M(\omega)$ depends only on ω.

3. Four-Dimensional Material Tensor

3.1. LORENTZ-COVARIANT SURFACE IMPEDANCE AND CHARACTERISTIC MATRIX METHODS

Lorentz-covariant surface impedance and characteristic matrix methods are developed in [11, 12, 13, 21]. The four-dimensional tensors of surface impedance γ and normal refraction $\mathcal{N}$, the Lorentz-covariant characteristic matrix $\mathcal{C}$ and the differential matrix of propagation $\mathcal{A}$ are defined by [11, 12, 13, 21]

$$h = \gamma\varphi, \quad \boldsymbol{Q}\gamma = \boldsymbol{\tau}\gamma = 0, \quad \gamma Q = \gamma\tau = 0 \tag{38}$$

$$\varphi(\boldsymbol{x}) = \exp(i\boldsymbol{x}\rfloor\tau)\exp(i\zeta\mathcal{N})\varphi(0) \tag{39}$$

$$\begin{pmatrix} \varphi \\ h \end{pmatrix}_{\boldsymbol{x}_2} = \mathcal{C}\begin{pmatrix} \varphi \\ h \end{pmatrix}_{\boldsymbol{x}_1}, \quad \frac{\partial \mathcal{C}}{\partial \zeta} = i\mathcal{A}\mathcal{C} \tag{40}$$

where

$$\varphi = (\boldsymbol{u}\wedge\boldsymbol{v})\rfloor(Q\wedge F) \in V^*, \qquad h = \boldsymbol{Q}\rfloor G \in V^* \tag{41}$$

$\boldsymbol{Q} \in V$ and $Q \in V^*$ are the contravariant and covariant four-dimensional vectors of normal to the interface ($Q_i = Q^i$, $i = 1,2,3$; $Q_4 = -Q^4$), $\zeta = \boldsymbol{x}\rfloor Q$, $\boldsymbol{\tau} \in V$ is the tangential component of the four-dimensional wave vector $\boldsymbol{K}_{in} = \boldsymbol{\tau} + \xi_0\boldsymbol{Q}$ of the incident wave, $\boldsymbol{u}$ and $\boldsymbol{v}$ are auxiliary vector parameters related with $\boldsymbol{Q}$ and $\boldsymbol{\tau}$ by the conditions $\boldsymbol{u}\rfloor Q = \boldsymbol{v}\rfloor\tau = 1$, $\boldsymbol{v}\rfloor Q = \boldsymbol{u}\rfloor\tau = 0$. The 1-forms φ and h (41) are continuous at the moving interface. Eqs. (39)-(41) are the Lorentz-covariant generalizations of Eqs. (8), (9), and (14). In particular, the eigenvectors φ and the corresponding eigenvalues ξ of $\mathcal{N}$ ($\mathcal{N}\varphi = \xi\varphi$) specify the polarizations and the normal components of four-dimensional wave vectors $\boldsymbol{K} = \boldsymbol{\tau} + \xi\boldsymbol{Q}$ of refracted eigenwaves. The measurements procedure for γ and $\mathcal{N}$ (39) is described in detail in [11, 12]. To express $\mathcal{C}$ and $\mathcal{A}$ (41) in terms of the reflection and transmission tensors (the measurement of the latter reduces [11] to the Jones matrix measurements), it is sufficient to replace all three-dimensional tensors in Eqs. (15)-(23) by four-dimensional ones, and z by $\zeta = \boldsymbol{x}\rfloor Q$. The impedance tensors γ_i in the explicit form and the pseudoinversion formula are presented in [11, 12].

3.2. FIELD AND INDUCTION TENSORS OF EIGENWAVES

By calculating eigenvalues and eigenvectors of $\mathcal{N}$, $\mathcal{A}$ and $\mathcal{C}$ at some fixed Q, τ, $\boldsymbol{u}$, and $\boldsymbol{v}$, one can find the parameters φ, $h = \gamma\varphi$ and $K = \tau + \xi Q$ of a refracted eigenwave. In addition to φ, h, and K, to find the tensors F (2) and G (3) of this wave it is necessary to excite it under a different set of conditions ($Q \mapsto Q'$, $\tau \mapsto \tau' = K - Q'(\boldsymbol{u}'\rfloor K)$, $\boldsymbol{u}'\rfloor Q' = 1$, $v' = v$) and to determine the corresponding polarization 1-forms φ' and h'. Thereafter, F and G can be restored as follows:

$$F = K\wedge f, \qquad G = *(L\wedge K) \tag{42}$$

where

$$f = \varphi' + f_u Q', \qquad L = \beta' + L_u Q' \tag{43}$$

$$f_u = a(\varphi',\ \varphi)/b(\varphi) \tag{44}$$

$$L_u = a(\varphi,\ \beta b(\varphi')/b(\varphi) - \beta')/b(\varphi) \tag{45}$$

$$\beta = (\boldsymbol{v} \wedge \boldsymbol{u})\rfloor{*h}, \qquad \beta' = (\boldsymbol{v}\wedge\boldsymbol{u}')\rfloor{*h'} \tag{46}$$

$$a(\varphi', \varphi) = *(\varphi' \wedge \varphi \wedge Q \wedge \tau), \tag{47}$$

$$b(\varphi) = a(\varphi, Q') = *(\varphi \wedge Q' \wedge Q \wedge \tau) \tag{48}$$

and the star operator $*$ is defined by the relation [11, 12] $*\omega = \boldsymbol{\omega}\rfloor(\vartheta^1 \wedge \vartheta^2 \wedge \vartheta^3 \wedge \vartheta^4)$, where ω and $\boldsymbol{\omega}$ (bold omega) are the s-form and the s-vector related by the operations of lowering and raising of the indices $s = 1, 2, 3, 4$.

If the eigenwave under study propagates along an isotropic axis, (i.e. $\mathcal{N}$ has a double eigenvalue ξ with two-dimensional eigensubspace), F and G are described by expressions (42) and (43), where

$$f_u = a(\varphi^1,\ D_0\varphi^2 - \varphi')/b(\varphi^1) \tag{49}$$

$$L_u = a(\beta^1,\ D_0\beta^2 - \beta')/b(\beta^1) \tag{50}$$

$$D_0 = \frac{b(\varphi^1)b(\beta') - b(\beta^1)b(\varphi')}{b(\varphi^1)b(\beta^2) - b(\beta^1)b(\varphi^2)} \tag{51}$$

and φ', φ^i are arbitrary 1-forms satisfying the conditions $\boldsymbol{u}'\rfloor\varphi' = \boldsymbol{v}\rfloor\varphi' = 0$, $\boldsymbol{u}\rfloor\varphi^i = \boldsymbol{v}\rfloor\varphi^i = 0$; $\beta^i = (\boldsymbol{v} \wedge \boldsymbol{u})\rfloor * (\gamma\varphi^i)$, $i = 1, 2$; $\beta' = (\boldsymbol{v} \wedge \boldsymbol{u}')\rfloor * (\gamma'\varphi')$. The polarization of this wave is specified by φ'.

3.3. GENERALIZED MATERIAL TENSOR

Let the wave vectors K^i and the amplitudes F^i and G^i of arbitrary six eigenwaves be measured as described above. Assume that the amplitudes $(F^i;\ i = 1, \ldots, 6)$ are linearly independent, so that they form a basis in the six-dimensional space of 2-forms. The basis $(\boldsymbol{s}_i)$ dual to (F^i) is defined by

$$\boldsymbol{s}_i = \sum_{1 \le j < k \le 4} s_i^{jk} \boldsymbol{e}_j \wedge \boldsymbol{e}_k \tag{52}$$

where

$$\begin{pmatrix} [\boldsymbol{s}_1] \\ \vdots \\ [\boldsymbol{s}_6] \end{pmatrix} = \begin{pmatrix} (F^1) & \cdots & (F^6) \end{pmatrix}^{-1} \tag{53}$$

and $[s_i]$ is the row matrix of components s_i^{jk}, (F^i) is the column matrix of components F_{jk}^i, $j < k$. Since $s_i \rfloor F^n = \delta_i^n$, $i, n = 1, 2, \ldots, 6$, superpositions of the above-mentioned six waves are described by the evolution operator

$$\mathcal{F}(x) = \sum_{n=1}^{6} F^n \otimes s_n \exp(ix \rfloor K^n) \tag{54}$$

and the generalized material tensor

$$M(\mathcal{F}) = \sum_{n=1}^{6} G^n \otimes s_n \tag{55}$$

The relations presented in this paper make it possible to uniquely determine $M(\mathcal{F})$ by the reflection and transmission coefficients of the medium. Naturally, these Lorentz-covariant relations can also be applied to a dispersive bianisotropic medium at rest. As with the matrix $M(\mathcal{P})$ (35), in a medium with frequency dispersion, $M(\mathcal{F})$ (55) reduces to $M(\omega)$ for arbitrary six eigenwaves which have the same frequency ω in the reference frame related with the medium.

Thus, to find the three-dimensional or four-dimensional material tensors of a bianisotropic medium, it is necessary:

1. to measure the reflection and transmission tensors of a bianisotropic slab and to calculate the tensors of normal refraction and surface impedance or the characteristic matrix and the differential matrix of propagation;
2. to calculate the polarization parameters $\boldsymbol{u}, \boldsymbol{w}$ or $\varphi, \boldsymbol{h}$ and the wave vectors of eigenwaves excited in the slab;
3. to excite one of these eigenwaves in slab II and to determine similarly its polarization parameters $\boldsymbol{u}', \boldsymbol{w}'$ or $\varphi', \boldsymbol{h}'$;
4. to calculate the complex amplitudes of the eigenwave under study;
5. to find similarly the amplitudes of any other five eigenwaves so that to obtain a basis of the six-dimensional amplitude space;
6. to compute the three-dimensional or four-dimensional material tensors from the formula (35) or (55).

4. Conclusion

The obtained exact solutions of the inverse boundary value problems for dispersive bianisotropic media make it possible to find the three-dimensional and four-dimensional tensors of surface impedance and normal refraction, the characteristic matrix and the differential matrix of propagation, the field vectors $\boldsymbol{E}$, $\boldsymbol{H}$, $\boldsymbol{D}$ and $\boldsymbol{B}$ and the field and induction tensors F

and G of eigenwaves. Besides, it is possible to find the three-dimensional $\varepsilon(\omega)$, $\mu(\omega)$, $\alpha(\omega)$, $\beta(\omega)$ and four-dimensional $M(\omega)$ material tensors of a spatially dispersionless bianisotropic medium and the generalized material tensors of a bianisotropic medium with time and spatial dispersion, provided the reflection and transmission coefficients of the medium under study are given (measured).

References

1. Fedorov, F.I. (1976) *Theory of Gyrotropy*. Nauka i Tekhnika, Minsk.
2. Landau, L.D., and Lifshitz, E.M. (1982) *Electrodynamics of Continuous media*. Nauka, Moscow.
3. Kong, J.A. and Cheng, D.K. (1968) Wave behavior at an interface of a semi-infinite moving anisotropic medium, *J. Appl. Phys.* **39**, 2282-2286.
4. Kong, J.A. (1972) Theorems of bianisotropic media, *Proc. IEEE* **60**, 1036-1046.
5. Serdyukov, A.N. (1976) On the phenomenological theory of optical and acoustical properties of crystals, *Kristallografiya* **21**, 275-282.
6. Lakhtakia, A., and Weiglhofer, W.S. (1994) Are linear, nonreciprocal, biisotropic media forbidden? *IEEE Trans. Microwave Theory Tech.* **42**, 1715-1716.
7. Sihvola, A.H. (1995) Are nonreciprocal bi-isotropic media forbidden indeed? *IEEE Trans. Microwave Theory Tech.* **43**, 2160-2162.
8. Serdyukov, A.N., Sihvola, A.H., Tretyakov, S.A., and Semchenko, I.V. (1996) Duality in electromagnetics: Application to Tellegen media, *Electromagnetics* **16**, 51-61.
9. Kampia, R.D., and Lakhtakia, A. (1992) Bruggeman model for chiral particulate composites, *J. Phys.* **D25**, 1390-1394.
10. Tretyakov, S.A., and Haliullin, D.Ya. (1993) Free-space techniques for biisotropic media parameters measurements, *Microw. and Optic. Technol. Lett.* **6**, 512-515.
11. Borzdov, G.N. (1992) Lorentz-covariant surface impedance and characteristic matrix methods with applications to measurements of material parameters of linear media, *Opt. Commun.* **94**, 159-173.
12. Borzdov, G.N. (1993) An intrinsic tensor technique in Minkowski space with applications to boundary value problems, *J. Math. Phys.* **34**, 3162-3196.
13. Borzdov, G.N. (1993) Lorentz-covariant solution of the inverse problem of reflection and transmission for a dispersive bianisotropic medium, in A. Sihvola, S. Tretyakov and I. Semchenko (eds.), *Proc. Bianisotropics'93*, Gomel, Belarus, pp. 60-62.
14. Polevoi, V.G. and Rytov, S.M. (1978) On the four-dimensional group velocity, *Usp. Fiz. Nauk* **125**, 549-565.
15. Barkovskii, L.M., Borzdov, G.N., and Fedorov, F.I. (1990) Evolution operators in electrodynamics of spatially dispersive media, *J. Mod. Opt.* **37**, 85-97.
16. Borzdov, G.N. (1990) Evolution operators of electromagnetic waves in crystals, *Sov. Phys. Cryst.* **35**, 313-326.
17. Barkovskii, L.M., Borzdov, G.N., and Lavrinenko, A.V. (1987) Fresnel's reflection and transmission operators for stratified gyroanisotropic media, *J. Phys.* **A20**, 1095-1106.
18. Borzdov, G.N. (1994) Fedorov–Petrov waves in bi-isotropic and bi-anisotropic media, in F. Mariotte and J.-P. Parneix (eds.), *Proc. CHIRAL'94*, Périgueux, France, pp. 323-328.
19. Borzdov, G.N. (1996) Waves with linear, quadratic and cubic coordinate dependence of amplitude in crystals, *Pramana—journal of physics* **46**, 245-257.
20. Gantmacher, F.R. (1988) *Theory of Matrices*. Nauka, Moscow.
21. Borzdov, G.N. (1994) Four-dimensional impedance tensor of a bi-anisotropic medium, in F. Mariotte and J.-P. Parneix (eds.), *Proc. CHIRAL'94*, Périgueux, France, pp. 329-334.

Two Comments by the Editors

A The "local" equations (4) as a model of reciprocal media take into account spatial dispersion effects, such as the effects of chirality. There are no other known physical effects which cause magnetoelectric coupling in reciprocal media but dispersion effects. Eq. (4) can be rewritten in terms of spatial derivatives of the fields in the right-hand side. In that sense, (4) are already "non-local" relations since they model non-local effects. The generalization in the form of Eq. (6) is necessary for considering moving media, however, in that case the kernels in the cross-terms are not independent, but proportional to that in the co-terms. In the general case, introducing non-local dependence of polarizations on the spatial derivatives of the fields, along with the non-local dependence on the fields themselves, might possibly lead to double inclusion of the same effects in the model.

B In spatially-dispersive media, there can exist more than two eigenwaves propagating along a given direction at a fixed frequency (*e.g.*, in "zebra" composites, see the paper by Lagarkov and Vinogradov in this volume, and in ferrite layers). One can see this, for example, when modelling the dispersion by considering the permittivity as a function of the wave vector. In the simplest case, writing the dispersion relation for plane waves of as $k = \omega\sqrt{\epsilon(k)}\sqrt{\epsilon_0\mu_0}$, it becomes obvious that this equation can have more than one solution. In this paper, such cases are excluded from the analysis.

Reply by G.N. Borzdov

Concerning A The monochromatic Maxwell equations in conjunction with the local equations (1) yield the equivalent non-local constitutive eguations

$$\boldsymbol{D} = \varepsilon'\boldsymbol{E} + \alpha'\mathrm{rot}\boldsymbol{E}, \quad \boldsymbol{B} = \mu'\boldsymbol{H} + \beta'\mathrm{rot}\boldsymbol{H}. \qquad (1')$$

It seems likely that this or similar transformation is mentioned in the comment. Clearly, there is as little sense in calling Eqs. (1) "non-local in some sense" with Eqs. (1′) in mind as in calling Eqs. (1′) "local" with Eqs. (1) in mind. Moreover, such three-dimensional transformations are inapplicable to the Lorentz-covariant constitutive equations (4) and (6), which should be treated in conjunction with the four-dimensional Maxwell equations $dF = 0$ and $d{*}G = 0$ (see, *e.g.*, [12-14,21]).

Eq. (6) describes a linear homogeneous motionless or uniformly moving dispersive medium with non-instantaneous and (or) non-local response. The claim of the Editors that the kernels in the cross-terms are always proportional to that in the co-terms is unjustified. In particular, being written in the four-dimensional form, the constitutive equations for the homogeneous bianisotropic medium with local (in the rest frame) non-instantaneous response, which are discussed in the paper by Lakhtakia and Weiglhofer (*IEEE Trans. Microwave Theory Tech.* **43**(1995)2722–2723), take the form of Eq. (6), where the kernels in the cross-terms (the magnetoelectric kernels) are independent of the kernels in the co-terms. Even more complex linear media with non-instantaneous and non-local response and independent kernels in the constitutive equations are treated in the articles by Serdyukov [5], and Polevoi and Rytov [14].

Eq. (6) is in complete conformity with both the definitions and physical meaning of the field and induction vectors (tensors) in electrodynamics of dispersive continuous media. A correct use of this equation together with Maxwell's equations $dF = 0$ and $d{*}G = 0$ can not lead to any "double inclusion of the same effects in the model" and, of course, Eq. (6) does not contain any spatial derivatives.

It is essential that I do not discuss in the paper any properties of the kernel in Eq. (6) and do not use any speculations about these properties in my solutions. On the contrary, I consider as one of main advantages of my approach the fact that first the complex amplitudes of eigenwaves (which are dictated by the medium itself and, hence, do not depend on the way one understands or describes dispersion) and then, by using these amplitudes, the generalized material tensor $\mathcal{M}(\mathcal{F})$ (7) itself can be expressed through the measurable values (reflection and transmission coefficients) without use of any preliminary speculations about the properties of the kernel.

Concerning B This comment may possibly bring one to a wrong conclusion that a dependence of material parameters on the wave vector almost without exception results in additional waves, whereas more likely such dependence will result in a "deformation" of the wave vector surface without any additional splitting of it and, hence, without any additional waves. The formula presented provides a vivid example. The solutions of the dispersion equation $k = \omega\sqrt{\epsilon(k)}\sqrt{\epsilon_0\mu_0}$ are specified by intersection points of the straight line $y = y_1(k) = k$ and the curve $y = y_2(k) = \omega\sqrt{\epsilon(k)}\sqrt{\epsilon_0\mu_0}$ in the (k, y)-plane (at fixed ω). In most cases the effects of spatial dispersion are small, so that in such cases the curve $y_2(k)$ differs very little from a straight line parallel to the k-axis, hence, it will most likely have only one intersection point with the straight line $y_1(k)$. In fact the function $y_2(k)$, i.e. $\epsilon(k)$, must have a rather specific form to provide the existence of additional intersection points. In any case, until $\epsilon(k)$ is not specified in the explicit form any statement about the obvious existence of additional eigenwaves is unjustified.

The last sentence of the comment B has a rather mysterious sense. By way of illustration, let us consider the three-dimensional problem for a motionless slab in vacuum. The problem under analysis in the paper is the reflection and transmission of a plane harmonic wave at oblique incidence onto a linear homogeneous slab. In this case, the boundary vales of the tangential components of $\boldsymbol{E}$ and $\boldsymbol{H}$ at the interfaces of the slab are linearly related by the 4×4 characteristic matrix C, and its eigenvalues specify the phase factors $\exp(ik_{jz}z)$ of all eigenwaves excited in the slab (z is the slab thickness, k_{jz} is the normal component of the wave vector $\boldsymbol{k}_j, j = 1, 2, 3, 4$). Either the last sentence of B makes the sense that the above statement is incorrect and should be rejected to find the place for additional waves in the slab or none of the homogeneous dispersive media is excluded from the analysis.

In summary, the solutions obtained in the paper can not lead to any "double inclusion of the same effects", and they are applicable to any linear homogeneous medium, no matter what the form of the dispersion equation in this medium.

CHIRAL EFFECTS AND EIGENWAVES IN BI-ANISOTROPIC OMEGA STRUCTURES

A.A. SOCHAVA, C.R. SIMOVSKI, AND S.A. TRETYAKOV
St. Petersburg State Technical University
195251 St. Petersburg, Russia

Abstract. Coupling effects in reciprocal bi-anisotropic composite materials are classified and analysed in this paper. In particular, it is shown that some composites with omega-shaped conductive inclusions embedded in an isotropic dielectric matrix can be considered as media with equivalent phantom chiral particles. Special cases of such media: the so-called pseudochiral composite and omega composite with all the particles located in parallel planes (planar omega composite) are studied. The eigenwave problem is solved for these cases. Analytical solutions are given for plane eigenwave polarizations and wavenumbers.

1. Introduction

In the last several years, complex artificial composite materials for microwave applications were studied very intensively. An important special type of the bi-anisotropic composites is the isotropic chiral medium. Microwave chiral materials can be realized as microstructures comprising small metal helices randomly distributed in isotropic matrices. When the helices are arranged in a certain order, we obtain bi-anisotropic chiral media (corresponding examples are studied in [1]).

Along with the well-known isotropic chiral composites, some novel microstructures were suggested. One of them is so-called *omega* composite which is formed by embedding small Ω-shaped particles in an isotropic matrix [2, 3]. An evident advantage of omega composites for microwave applications compared to chiral media is the possibility to fabricate them with the use of the planar technology. In the past, attention was mostly concentrated on the uniaxial omega media, which are formed of two or-

A. Priou et al. (eds.), Advances in Complex Electromagnetic Materials, 85–102.

thogonal sets of omega inclusions [4, 5]. In these media the magnetoelectric interaction does not produce any rotation of wave polarization.

In this paper we classify general bi-anisotropic media on the basis of the dyadic coupling parameter structure. We demonstrate that certain structures of omega particles possess properties of bi-anisotropic chiral composites and even can not be distinguished from them from electromagnetic measurements. Then, we consider two examples of such pseudochiral omega composites. Analysis of these omega structures and calculation of the parameters of eigenwaves allows us to conclude that pseudochiral omega structure can be used instead of chiral materials in some applications. For example, they can be used as rather effective polarization transformers.

2. General Relations and Classification of Bi-anisotropic Composite Media

Let us consider an arbitrary reciprocal bi-anisotropic medium. Such medium is governed by the corresponding general bi-anisotropic constitutive relations [4, 6]:

$$\mathbf{D} = \overline{\overline{\epsilon}} \cdot \mathbf{E} + j\sqrt{\epsilon_0\mu_0}\overline{\overline{L}} \cdot \mathbf{H} \qquad \mathbf{B} = \overline{\overline{\mu}} \cdot \mathbf{H} - j\sqrt{\epsilon_0\mu_0}\overline{\overline{L}}^T \cdot \mathbf{E} \tag{1}$$

where $\overline{\overline{L}}$ is the dimensionless coupling dyadic and $\overline{\overline{L}}^T$ is the transposed one. The time dependence in this paper is taken to be $\exp(j\omega t)$. The actual forms of $\overline{\overline{L}}$, which models the magnetoelectric coupling ([7], Appendix A) depend on the shape, size, and positions of bi-anisotropic inclusions. In the model of isotropic chiral media the dyadics in (1) are multiples of the unit dyadic and therefore they can be expressed in terms of scalar and pseudoscalar coefficients [7]:

$$\mathbf{D} = \epsilon\mathbf{E} + j\kappa\sqrt{\epsilon_0\mu_0}\,\mathbf{H} \qquad \mathbf{B} = \mu\mathbf{H} - j\kappa\sqrt{\epsilon_0\mu_0}\,\mathbf{E} \tag{2}$$

In [6], coupling dyadic $\overline{\overline{L}}$ in (1) was split into two parts: symmetric and antisymmetric. The first part is considered as that responsible for chiral effects, and the second part describes magnetoelectric effects in nonchiral uniaxial media. In [8], this splitting was carried out for trace-free dyadics. It means that the composites which do not contain chiral particles but are described by general trace-free coupling dyadics, can be considered as media with phantom chiral inclusions.

Let us now present the coupling dyadic in (1) as

$$\overline{\overline{L}} = \kappa\overline{\overline{I}} + \overline{\overline{M}} \tag{3}$$

where $\overline{\overline{I}}$ is the unit dyadic and the trace of $\overline{\overline{M}}$ is zero. For bi-anisotropic media with mirror symmetry $\kappa = 0$. We further decompose $\overline{\overline{M}}$ into symmetric and antisymmetric parts:

$$\overline{\overline{M}} = f\overline{\overline{N}} + K\overline{\overline{J}} \tag{4}$$

where $\overline{\overline{J}}$ can be presented in a certain coordinate system (x_{10}, y_{10}, z_{10}) as $\overline{\overline{J}} = \mathbf{z}_{10} \times \overline{\overline{I}}_t$, where $\overline{\overline{I}}_t$ is the two-dimensional unit dyadic in the plane $(x_{10} - y_{10})$, $\overline{\overline{I}}_t = \mathbf{x}_{10}\mathbf{x}_{10} + \mathbf{y}_{10}\mathbf{y}_{10}$. The dyadic $\overline{\overline{N}}$ can be diagonalized. It is physically clear that the diagonal terms in $\overline{\overline{N}}$ correspond to presence of virtual chiral inclusions. There must be at least two sets of inclusions of opposite handednesses, because the trace of $\overline{\overline{N}}$ is zero. The second term in (4) describes a uniaxial omega medium. So, every omega composite which is not a uniaxial omega medium can be considered as a combination of a bi-anisotropic chiral medium and a uniaxial omega one.

Thus, we suggest the following classification of reciprocal bi-anisotropic materials.

TABLE 1. Classification of reciprocal bi-anisotropic media

Coupling parameters	Class
$\kappa \neq 0,\ f = 0,\ K = 0$	Isotropic chiral medium
$\kappa \neq 0,\ f \neq 0,\ K = 0$	Anisotropic chiral medium
$\kappa = 0,\ f \neq 0,\ K = 0$	Pseudochiral medium
$\kappa = 0,\ f = 0,\ K \neq 0$	Uniaxial omega medium
$\kappa \neq 0,\ f = 0,\ K \neq 0$	Uniaxial chiral omega medium
$\kappa = 0,\ f \neq 0,\ K \neq 0$	Pseudochiral omega medium
$\kappa \neq 0,\ f \neq 0,\ K \neq 0$	General bi-anisotropic medium

3. Examples of Pseudochiral Omega Media

Now we study the omega composites without chiral inclusions, for which the magnetoelectric dyadic is trace-free: $\overline{\overline{L}} = \overline{\overline{M}}$. To understand the geometry of this pseudochiral medium consider at first the uniaxial omega structure [3, 4, 9, 10]. Let us position Ω particles of the first set so that their stems stretch along the x-axis, and the loops are in the $(x-z)$-plane. For particles of the second ensemble the stems go along the y-axis and the loops belong to the $(y - z)$ plane (see Figure 1a). Corresponding constitutive relations

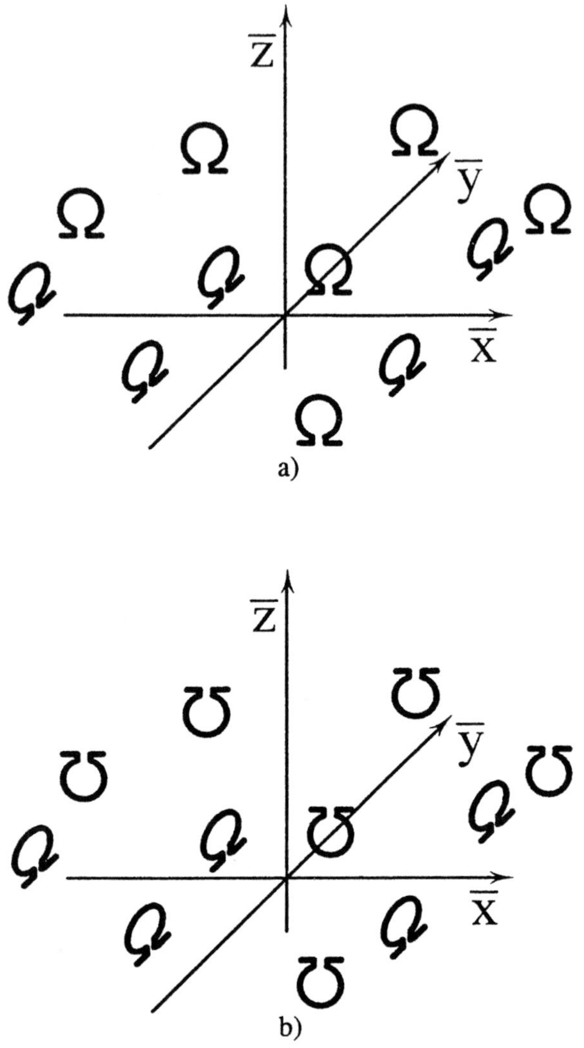

Figure 1. Geometry of the uniaxial and pseudochiral omega media

are given, for example, in [3]. It is known that the eigenpolarizations in such a medium are linear for all possible directions of propagation [4, 10].

Let us consider a "modified" omega medium, in which loops of the second ensemble particles are situated in the $(y-z)$ plane on the opposite side of the stems (Figure 1b). Finally, for the most general case we suppose that the omega particles in the two orthogonal sets have different dimensions or (and) volume concentrations. Then, the constitutive equations can be written as

$$\mathbf{D} = \overline{\overline{\epsilon}} \cdot \mathbf{E} + j\sqrt{\epsilon_0\mu_0}\,\overline{\overline{M}} \cdot \mathbf{H} \qquad \mathbf{B} = \overline{\overline{\mu}} \cdot \mathbf{H} - j\sqrt{\epsilon_0\mu_0}\,\overline{\overline{M}} \cdot \mathbf{E} \tag{5}$$

where

$$\overline{\overline{M}} = K_x \mathbf{x}_0\mathbf{y}_0 + K_y \mathbf{y}_0\mathbf{x}_0$$

The permittivity and permeability dyadics have the diagonal forms:

$$\overline{\overline{\epsilon}} = \epsilon_0(\epsilon_x \mathbf{x}_0\mathbf{x}_0 + \epsilon_y \mathbf{y}_0\mathbf{y}_0 + \epsilon_n \mathbf{z}_0\mathbf{z}_0) \tag{6}$$

(similar for $\overline{\overline{\mu}}$). The host medium is isotropic, and the difference between ϵ_n and μ_n and $\epsilon_{x,y}$ and $\mu_{x,y}$ is due to the particles influence.

Splitting all the constitutive dyadics into symmetric and antisymmetric parts and diagonalizing the symmetric part of the magnetoelectric dyadic we easily find that

$$\overline{\overline{M}} = \frac{1}{2}(K_x + K_y)(\mathbf{x}_0'\mathbf{x}_0' - \mathbf{y}_0'\mathbf{y}_0') - \frac{1}{2}(K_x - K_y)\overline{\overline{J}}$$

$$\frac{\overline{\overline{\epsilon}}}{\epsilon_0} = \frac{1}{2}(\epsilon_x + \epsilon_y)(\mathbf{x}_0'\mathbf{x}_0' + \mathbf{y}_0'\mathbf{y}_0') + \frac{1}{2}(\epsilon_x - \epsilon_y)(\mathbf{x}_0'\mathbf{y}_0' + \mathbf{y}_0'\mathbf{x}_0') + \epsilon_n \mathbf{z}_0\mathbf{z}_0$$

$$\frac{\overline{\overline{\mu}}}{\mu_0} = \frac{1}{2}(\mu_x + \mu_y)(\mathbf{x}_0'\mathbf{x}_0' + \mathbf{y}_0'\mathbf{y}_0') + \frac{1}{2}(\mu_x - \mu_y)(\mathbf{x}_0'\mathbf{y}_0' + \mathbf{y}_0'\mathbf{x}_0') + \mu_n \mathbf{z}_0\mathbf{z}_0$$

where $\overline{\overline{J}} = \mathbf{y}_0'\mathbf{x}_0' - \mathbf{x}_0'\mathbf{y}_0'$. Here the axes x' and y' are the bisectors between the x- and y-axes. In the special case when $\epsilon_x = \epsilon_y = \epsilon_t$, $\mu_x = \mu_y = \mu_t$, $K_x = -K_y = K$ we have the uniaxial omega medium. In the case when $\epsilon_x = \epsilon_y = \epsilon_t$, $\mu_x = \mu_y = \mu_t$, $K_y = K_x = K$ we obtain the medium which can be called *pseudochiral* medium, where $f = K$ and $\overline{\overline{N}} = \mathbf{x}_0'\mathbf{x}_0' - \mathbf{y}_0'\mathbf{y}_0'$ (see Eq. (4)). In that case the constitutive relations

$$\mathbf{D} = \epsilon_0(\epsilon_t \overline{\overline{I}}_t + \epsilon_n \mathbf{z}_0\mathbf{z}_0) \cdot \mathbf{E} + j\sqrt{\epsilon_0\mu_0}K(\mathbf{x}_0'\mathbf{x}_0' - \mathbf{y}_0'\mathbf{y}_0') \cdot \mathbf{H}$$

$$\mathbf{B} = \mu_0(\mu_t \overline{\overline{I}}_t + \mu_n \mathbf{z}_0\mathbf{z}_0) \cdot \mathbf{H} - j\sqrt{\epsilon_0\mu_0}K(\mathbf{x}_0'\mathbf{x}_0' - \mathbf{y}_0'\mathbf{y}_0') \cdot \mathbf{E} \tag{7}$$

are the same as if the composite were formed by two orthogonal sets of uniaxial chiral particles oriented along x' and y'. The particles (helices) in these phantom chiral sets are left- and right-handed, respectively. Here $\overline{\overline{I}}_t$ is the unit dyadic in the $(x' - y')$-plane: $\overline{\overline{I}}_t = \mathbf{x}_0'\mathbf{x}_0' + \mathbf{y}_0'\mathbf{y}_0' = \mathbf{x}_0\mathbf{x}_0 + \mathbf{y}_0\mathbf{y}_0$.

Finally, in the case when $K_y = 0$ we have only one set of parallel particles, i.e., a planar omega structure. The constitutive relations are the following:

$$\mathbf{D} = \overline{\overline{\epsilon}} \cdot \mathbf{E} + j\sqrt{\epsilon_0\mu_0}K_x(\mathbf{x}_0\mathbf{y}_0) \cdot \mathbf{H} \qquad \mathbf{B} = \overline{\overline{\mu}} \cdot \mathbf{H} - j\sqrt{\epsilon_0\mu_0}K_x(\mathbf{y}_0\mathbf{x}_0) \cdot \mathbf{E} \tag{8}$$

These equations we present in the form:

$$\mathbf{D} = \overline{\overline{\epsilon}} \cdot \mathbf{E} + \frac{j}{2}\sqrt{\epsilon_0\mu_0}K(\overline{\overline{N}} - \overline{\overline{J}}) \cdot \mathbf{H} \tag{9}$$

$$\mathbf{B} = \overline{\overline{\mu}} \cdot \mathbf{H} - \frac{j}{2}\sqrt{\epsilon_0\mu_0}K(\overline{\overline{N}} + \overline{\overline{J}}) \cdot \mathbf{E} \tag{10}$$

where the permittivity and permeability dyadics are given in (6), and the symmetric part of magnetoelectric dyadic is diagonalized as in (7). Note, that the constitutive equations for the planar omega composite in the form (10) were introduced in [8].

Thus, the planar omega medium can be presented as a combination of a composite with chiral particles and a uniaxial omega composite. The equivalent chiral particles form, again, two orthogonal ensembles of helices oriented along the $x'-$ and $y'-$axes as well as the stems of equivalent omega-particles. Therefore, the omega-medium considered above includes as particular cases the well-studied uniaxial one, the medium which we call pseudochiral composite, and the planar omega composite. It is important to note that the planar omega medium can consist not only of one set of parallel particles but, more generally, of two sets with orthogonal stems.

Next, we consider omega structures formed by two orthogonal sets of particles lying in planes parallel to the $(x - y)$-plane (Figure 2) and write the constitutive relations in the form

$$\mathbf{D} = \overline{\overline{\epsilon}} \cdot \mathbf{E} + j\sqrt{\epsilon_0\mu_0}\,(K_x\mathbf{x}_0\mathbf{z}_0 + K_y\mathbf{y}_0\mathbf{z}_0) \cdot \mathbf{H}$$

$$\mathbf{B} = \overline{\overline{\mu}} \cdot \mathbf{H} - j\sqrt{\epsilon_0\mu_0}\,(K_x\mathbf{z}_0\mathbf{x}_0 + K_y\mathbf{z}_0\mathbf{y}_0) \cdot \mathbf{E} \tag{11}$$

Here, the axes x and y are directed along the stems of the first and second sets of omega particles, and the z-axis is orthogonal to the particles plane. We present the permittivity and permeability as $\overline{\overline{\epsilon}} = \epsilon_0(\overline{\overline{\epsilon}}_t + \epsilon_n\,\mathbf{z}_0\mathbf{z}_0)$, $\overline{\overline{\mu}} = \mu_0(\overline{\overline{\mu}}_t + \mu_n\,\mathbf{z}_0\mathbf{z}_0)$.

In the $(x - y)$ plane we can define a vector $\mathbf{r}_0$

$$K\,\mathbf{r}_0 = K_x\mathbf{x}_0 + K_y\mathbf{y}_0$$

$$K = \sqrt{K_x^2 + K_y^2} \qquad \mathbf{r}_0 = \frac{K_x\mathbf{x}_0 + K_y\mathbf{y}_0}{\sqrt{K_x^2 + K_y^2}} \tag{12}$$

such that $K_x\mathbf{x}_0\mathbf{z}_0 + K_y\mathbf{y}_0\mathbf{z}_0 = K\mathbf{r}_0\mathbf{z}_0$, $K_x\mathbf{z}_0\mathbf{x}_0 + K_y\mathbf{z}_0\mathbf{y}_0 = K\mathbf{z}_0\mathbf{r}_0$. The vectors $\mathbf{z}_0$ and $\mathbf{r}_0$ define a coordinate system which is most appropriate to the present geometry. In that system, dielectric and magnetic properties are also anisotropic.

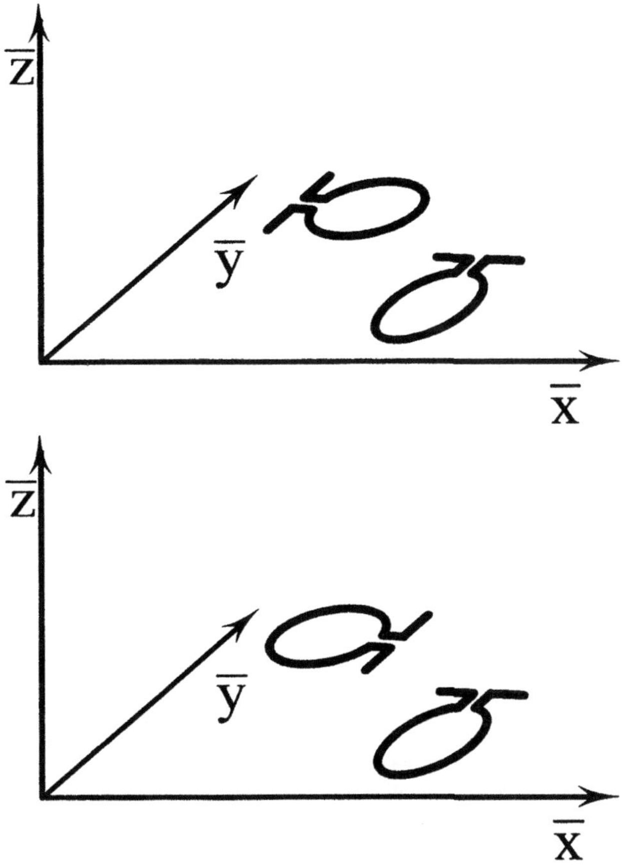

Figure 2. **Geometry of the planar omega media**

From (11) we have

$$\mathbf{D}=\overline{\overline{\epsilon}}\cdot\mathbf{E}+j\sqrt{\epsilon_0\mu_0}K(\mathbf{r}_0\mathbf{z}_0)\cdot\mathbf{H}\quad \mathbf{B}=\overline{\overline{\mu}}\cdot\mathbf{H}-j\sqrt{\epsilon_0\mu_0}K(\mathbf{z}_0\mathbf{r}_0)\cdot\mathbf{E} \tag{13}$$

These equations have the same form as (8) and can be also rewritten as

$$\mathbf{D}=\overline{\overline{\epsilon}}\cdot\mathbf{E}+\frac{j}{2}\sqrt{\epsilon_0\mu_0}K(\overline{\overline{N}}'+\overline{\overline{J}}')\cdot\mathbf{H} \tag{14}$$

$$\mathbf{B}=\overline{\overline{\mu}}\cdot\mathbf{H}-\frac{j}{2}\sqrt{\epsilon_0\mu_0}K(\overline{\overline{F}}'-\overline{\overline{J}}')\cdot\mathbf{E} \tag{15}$$

where $\overline{\overline{N}}'=\mathbf{r}_0\mathbf{z}_0+\mathbf{z}_0\mathbf{r}_0$, $\overline{\overline{J}}'=\mathbf{r}_0\mathbf{z}_0-\mathbf{z}_0\mathbf{r}_0$. The axes u,v,w of the coordinate system in which the dyadic $\overline{\overline{N}}'$ is diagonal are inclined by $\pi/4$ with respect to the axes x, y, z. The axes u, v correspond to the relation $\overline{\overline{N}}' = \mathbf{v}_0\mathbf{v}_0 - \mathbf{u}_0\mathbf{u}_0$. They lie in the $(x' - z)$-plane and are bisectors for x'- and z-axes.

3.1. MICROSCOPIC MEANING OF CHIRAL EFFECT IN OMEGA MEDIUM

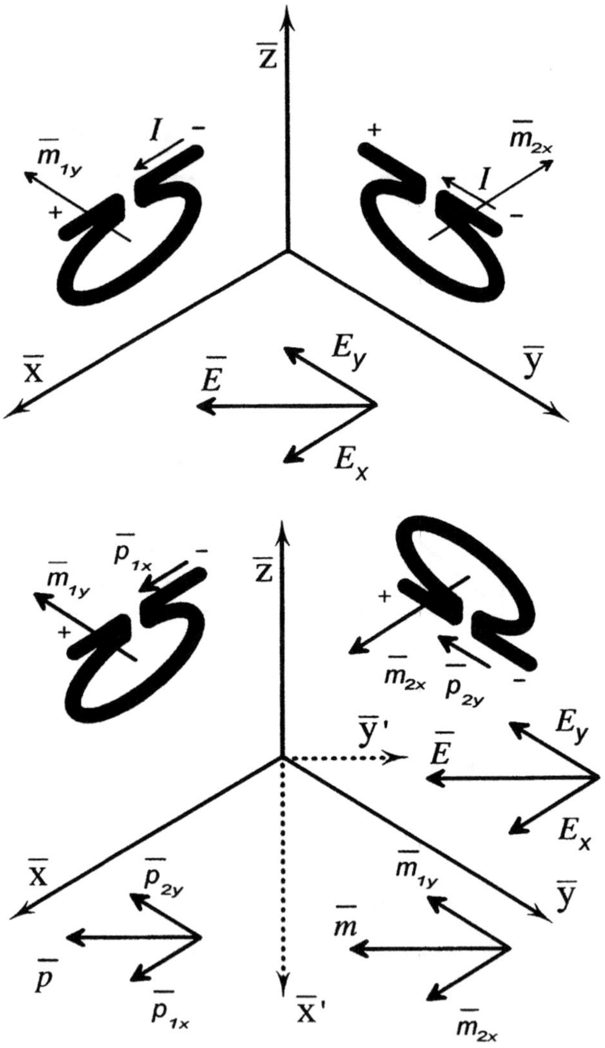

Figure 3. **Pertinent to the explanation of the microscopic chiral effect**

Let us consider at first a pseudochiral omega medium. The physical meaning of the fact that circular polarization of eigenwaves appears in such a medium can be easily clarified when we consider the electric and magnetic dipoles induced by an incident wave in any couple of orthogonal omega particles. For such a pair in the uniaxial composite the external electric field $\mathbf{E}^i$ polarized in the $(x-y)$-plane (see Figure 3) produces in both the particles electric dipole moments $\mathbf{p}_1$ and $\mathbf{p}_2$ oriented along their stems, and it is evident that the total dipole moment $\mathbf{p}$ is parallel to $\mathbf{E}^i$. The magnetic moments $\mathbf{m}_1$ and $\mathbf{m}_2$ induced due to the magnetoelectric coupling (i.e. the common induced current in the stem and in the loop portions of an omega particle) are oriented so that their vector sum is orthogonal to $\mathbf{p}$. Thus, $\mathbf{m}$ lies in the plane of the vectors $\mathbf{k}$ and $\mathbf{H}^i$. It means that the electric (magnetic) moment of the medium in this case has no component which would be parallel to the magnetic (electric) field vector. A similar conclusion can be easily reached for induced electric and magnetic moments due to external magnetic fields.

When particles magnetic moment induced by incident electric field ("coupling portion") is directed along the electric field (and *vice versa*) such a particle is equivalent to a chiral one. In the same way (see Figure 3) we easily find that both the total electric moment and total "coupling" magnetic moment of a pair of different omega particles in a pseudochiral composite are always oriented along the same axis. It means that a pair of omega particles in this medium can be presented as an equivalent chiral particle for a given direction of propagation.

To clarify the physical meaning of chiral properties of planar omega structure we can consider plane wave propagation in the case when the magnetic field is parallel to the particle plane and does not produce any effect. The induced magnetic moment is due to the electric field normal component, and its vector is orthogonal to the particle plane. This vector contains two components. The first one is oriented along the propagation direction, and the second component is parallel to $\mathbf{E}^i$ and therefore points along the corresponding component of $\mathbf{p}$.

Planar omega structures were studied earlier, for example, in [11]. However, the influence of electric and magnetic properties of omega particles on the composite permittivity and permeability were neglected in numerical calculations (in [11] $\epsilon_t = \epsilon_n$ and $\mu_t = \mu_n$, which contradicts to the suggestion of isotropy of the host dielectric matrix), and the propagation constants and polarization were studied for a particular case of identical sets of particles. Chiral effects were not discovered in [11]. In the following, we consider two examples of pseudochiral omega structures in more detail.

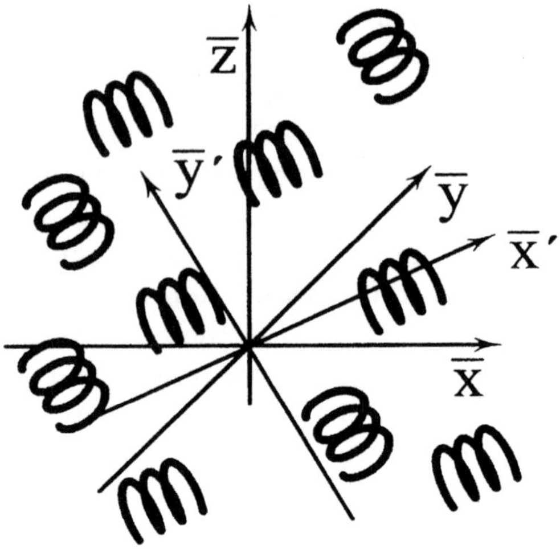

Figure 4. Equivalent structure of psedochiral omega composite

4. Eigenwaves in Bi-anisotropic Omega Structures

4.1. PSEUDOCHIRAL OMEGA MEDIUM

To study electromagnetic waves in pseudochiral "modified" omega media we solve the problem of eigenwaves in the coordinate system (x', y', z). It practically means that we consider the equivalent medium with chiral inclusions, whose structure is shown on Fig. 4. Substituting (7) into the Maxwell equations and applying the two-dimensional Fourier transform in the $(x' - y')$-plane we easily obtain the relations

$$-j\mathbf{k}_t \times \mathbf{E} + \frac{\partial}{\partial z}(\mathbf{z}_0 \times \mathbf{E}_t) = -j\omega\overline{\overline{\mu}} \cdot \mathbf{H} - k_0 K(\mathbf{x}'_0\mathbf{x}'_0 - \mathbf{y}_0\mathbf{y}_0) \cdot \mathbf{E}$$

$$-j\mathbf{k}_t \times \mathbf{H} + \frac{\partial}{\partial z}(\mathbf{z}_0 \times \mathbf{H}_t) = j\omega\overline{\overline{\epsilon}} \cdot \mathbf{E} - k_0 K(\mathbf{x}'_0\mathbf{x}'_0 - \mathbf{y}_0\mathbf{y}_0) \cdot \mathbf{H} \qquad (16)$$

We can eliminate the normal (n) field components, expressing them in terms of the transversal (t) components:

$$j\mathbf{k}_t \times \mathbf{E}_t = j\,\omega\mu_0\mu_n\, H_n\mathbf{z}_0 \qquad j\mathbf{k}_t \times \mathbf{H}_t = -j\,\omega\epsilon_0\epsilon_n\, E_n\mathbf{z}_0 \qquad (17)$$

Scalar multiplying (17) by $\mathbf{z}_0$ from the left side and substituting the result into (16) we obtain in the coordinate form four scalar equations:

$$A_{y'}\, H_{x'} - B\, H_{y'} = \omega\epsilon_0\epsilon_n(\beta E_{y'} + jcE_{x'}) \qquad (18)$$

$$A_{x'}\, H_{y'} - B\, H_{x'} = -\omega\epsilon_0\epsilon_n(\beta E_{x'} + jcE_{y'}) \qquad (19)$$

$$A_{y'}{}' E_{x'} - B\, E_{y'} = -\omega\mu_0\mu_n(\beta H_{y'} + jcH_{x'}) \tag{20}$$

$$A_{x'}{}' E_{y'} - B\, E_{x'} = \omega\mu_0\mu_n(\beta H_{x'} + jcH_{y'}) \tag{21}$$

where β is the propagation constant along the z–axis ($\partial/\partial z \Rightarrow -j\beta$), $A_{x,y} = -k_{x,y}^2 + k_0^2\mu_t\epsilon_n$, $A'_{x,y} = -k_{x,y}^2 + k_0^2\mu_n\epsilon_t$, $c = k_0 \cdot K$, $B = k_{x'}k_{y'}$. The first two equations in this system allow to find the wave admittance, defined as

$$-\mathbf{z}_0 \times \mathbf{H} = \overline{\overline{Y}} \cdot \mathbf{E}$$

In the coordinate form:

$$Y_{xx} = \frac{(jcB - \beta A_{y'})\omega\epsilon_0\epsilon_n}{\Delta} \qquad Y_{xy} = \frac{(\beta B - jcA_{y'})\omega\epsilon_0\epsilon_n}{\Delta}$$

$$Y_{yx} = \frac{(\beta B - jcA_{x'})\omega\epsilon_0\epsilon_n}{\Delta} \qquad Y_{yy} = \frac{(jcB - \beta A_{x'})\omega\epsilon_0\epsilon_n}{\Delta} \tag{22}$$

where $\Delta = k_0^2\epsilon_n\mu_t(k_0^2\epsilon_n\mu_t - k_t^2)$. We can also obtain the wave admittance in the coordinate-free form, if after substitution (17) into (16) we multiply these equations by $\mathbf{z}_0$ and use the relation

$$\overline{\overline{N}} \cdot (\mathbf{z}_0 \times \overline{\overline{I}}) = -\mathbf{z}_0 \times \overline{\overline{N}}$$

where $\overline{\overline{N}}$ corresponds to (7). We obtain:

$$(-j\beta\overline{\overline{I}} - c\mathbf{z}_0 \times \overline{\overline{N}}) \cdot \mathbf{E}_t = j\omega\mu_0(\mu_t\overline{\overline{I}}_t - \frac{k_t^2}{k_0^2\epsilon_n}\frac{\mathbf{k}_t\mathbf{k}_t}{k_t^2}) \cdot \mathbf{z}_0 \times \mathbf{H}_t$$

$$(-j\beta\overline{\overline{I}} + c\mathbf{z}_0 \times \overline{\overline{N}}) \cdot \mathbf{z}_0 \times \mathbf{H}_t = j\omega\epsilon_0(\epsilon_t\overline{\overline{I}}_t - \frac{k_t^2}{k_0^2\mu_n}\frac{\mathbf{z}_0 \times \mathbf{k}_t\mathbf{z}_0 \times \mathbf{k}_t}{k_t^2}) \cdot \mathbf{E}_t$$

After simple dyadic algebra we find the wave admittance

$$\overline{\overline{Y}} = \frac{1}{j\omega\mu_0}\left(\frac{k_0^2\epsilon_n}{k_0^2\epsilon_n\mu_t - k_t^2}\frac{\mathbf{k}_t\mathbf{k}_t}{k_t^2} + \frac{\mathbf{z}_0 \times \mathbf{k}_t\mathbf{z}_0 \times \mathbf{k}_t}{\mu_t k_t^2}\right) \cdot$$

$$(j\beta\overline{\overline{I}} + c\mathbf{z}_0 \times \overline{\overline{N}}) \tag{23}$$

The equivalence of (23) with the coordinate representation (22) can be easily checked.

Substituting the magnetic field components into (20) and (21) we obtain after some algebra the system of equations for $E_{x'}, E_{y'}$ which allows to find the dispersion equation and the polarization relations:

$$E_{x'}[(\beta A'_{y'} - jcB)\Delta + (\beta A_{y'} - jcB)XC^2] +$$

$$E_{y'}[(jcA'_{x'} - \beta B)\Delta + (jcA_{y'} - \beta B)XC] = 0 \tag{24}$$

$$E_{x'}[(cA'_{y'} + j\beta B)\Delta + (cA_{x'} + j\beta B)XC^2] -$$
$$E_{y'}[(cB_{x'} + j\beta A'_{x'})\Delta + (j\beta A_{x'} + cB)XC] = 0 \tag{25}$$

where $C = k_0^2 \epsilon_n \mu_n$, $X = \beta^2 + c^2$. In the dispersion equation all the terms with βc cancel out, which finally yields the equation for β in the form $(\beta^2 + c^2)(X^2 - fX + g) = 0$. The roots $\beta = \pm jc$ are spurious, and the coefficients in the bi-quadratic equation are

$$f = k_0^2 \mu_t \epsilon_t - k_t^2 \frac{\mu_t \epsilon_n + \mu_n \epsilon_t}{\mu_n \epsilon_n}$$

$$g = k_t^4 \frac{\mu_t \epsilon_t}{\mu_n \epsilon_n} + (k_0^2 \mu_t \epsilon_t)^2 - k_0^2 \mu_t \epsilon_t (\mu_n \epsilon_t + \mu_t \epsilon_n) \frac{k_t^2}{\mu_n \epsilon_n}$$

The result is the same as that for the uniaxial omega medium [4]:

$$\beta_1^2 = k_0^2 \mu_t \epsilon_t - k_t^2 \frac{\mu_t}{\mu_n} - (k_0 K)^2 \tag{26}$$

$$\beta_2^2 = k_0^2 \mu_t \epsilon_t - k_t^2 \frac{\epsilon_t}{\epsilon_n} - (k_0 K)^2 \tag{27}$$

Parameters of the polarization ellipse on the $(x - y)$-plane can be obtained from the relation which follows immediately from (24):

$$\frac{E_{x'}}{E_{y'}} = e_{xy} = \Big\{ [jk_0 K(k_0^2 \mu_n \epsilon_t - k_{x'}^2) - \beta k_{x'} k_{y'}](k_0^2 \mu_t \epsilon_n - k_t^2) k_0^2 \mu_t \epsilon_n -$$
$$[jk_0 K(k_0^2 \mu_t \epsilon_n - k_{y'}^2) - \beta k_{x'} k_{y'}](\beta^2 + k_0^2 K^2) k_0^2 \mu_n \epsilon_n \Big\} / G \tag{28}$$

where

$$G = [\beta(k_0^2 \mu_n \epsilon_t - k_{y'}^2) - jk_0 K k_{x'} k_{y'}](k_0^2 \mu_t \epsilon_n - k_t^2) k_0^2 \mu_t \epsilon_n -$$
$$[\beta(k_0^2 \mu_t \epsilon_n - k_{y'}^2) - jk_0 K k_{x'} k_{y'}](\beta^2 + k_0^2 K^2) k_0^2 \mu_n \epsilon_n \tag{29}$$

As a check, consider a special case when $k_{x'} = \pm k_{y'}$. This relation means that the wave propagation vector lies in the plane in which one ensemble of real omega-particles is located. In this case the magnetoelectric interaction with one ensemble is absent, and one of the eigenwaves interacts practically with one set of particles so that the wave vector lies in the same plane as the particles (it corresponds to the transversal propagation in planar omega structure). Of course, the polarization must be linear for these cases.

Substituting $k_{x'} = k_{y'}$ and (26) into (28) and (29) we easily find after the simplest algebra $e_{xy} = 1$. Substituting again $k_{x'} = k_{y'}$ and (27) into the same relations we obtain $e_{xy} = -1$. For $k_{x'} = -k_{y'}$ the result is dual. It means that the eigenwaves are actually linearly polarized, β_1 corresponds to the TE-wave and β_2 corresponds to the TM-wave. This result can be considered as a check for both the dispersion and polarization equations.

Note, that though the vectors $\mathbf{E}$ and $\mathbf{H}$ are not orthogonal even in the case of the linear polarization (it takes place for the z–propagation, too) but it is easy to show that $\mathbf{E}\cdot\mathbf{B} = \mathbf{H}\cdot\mathbf{D} = 0$, which is well-known in the optics of uniaxial magnetic crystals.

4.2. PLANAR OMEGA MEDIUM

To consider an arbitrary propagation direction, let us present the electric and magnetic fields as sums of the transverse and normal components:

$$\mathbf{E} = E_n\mathbf{z}_0 + \mathbf{E}_t \qquad\qquad \mathbf{H} = H_n\mathbf{z}_0 + \mathbf{H}_t \tag{30}$$

and apply the two-dimensional Fourier transform in the transverse plane to the Maxwell equations with constitutive relations (11). It results in

$$\begin{aligned} -j\mathbf{k}_t\times\mathbf{E} + \frac{\partial}{\partial z}(\mathbf{z}_0\times\mathbf{E}_t) &= -j\omega\overline{\overline{\mu}}\cdot\mathbf{H} - k_0\left[K_x\mathbf{z}_0\mathbf{x}_0 + K_y\mathbf{z}_0\mathbf{y}_0\right]\cdot\mathbf{E} \\ -j\mathbf{k}_t\times\mathbf{H} + \frac{\partial}{\partial z}(\mathbf{z}_0\times\mathbf{H}_t) &= j\omega\overline{\overline{\epsilon}}\cdot\mathbf{E} - k_0\left[K_x\mathbf{x}_0\mathbf{z}_0 + K_y\mathbf{y}_0\mathbf{z}_0\right]\cdot\mathbf{H} \end{aligned} \tag{31}$$

Consider the term inside the square brackets. It can be rearranged as

$$\left[K_x\mathbf{z}_0\mathbf{x}_0 + K_y\mathbf{z}_0\mathbf{y}_0\right]\cdot\mathbf{E} = \mathbf{z}_0\left(K_x E_x + K_y E_y\right)$$

$$\left[K_x\mathbf{x}_0\mathbf{z}_0 + K_y\mathbf{y}_0\mathbf{z}_0\right]\cdot\mathbf{H} = \left(K_x\mathbf{x}_0 + K_y\mathbf{y}_0\right)H_z$$

The field equations can be written separately for the normal (n) and transverse (t) field components:

$$\begin{aligned} j\mathbf{k}_t\times\mathbf{E}_t &= j\,\omega\mu_0\mu_n H_n\mathbf{z}_0 + k_0K\mathbf{z}_0\mathbf{r}_0\cdot\mathbf{E}_t \\ j\mathbf{k}_t\times\mathbf{H}_t &= -j\,\omega\epsilon_0\epsilon_n E_n\mathbf{z}_0 \end{aligned} \tag{32}$$

$$\begin{aligned} j\mathbf{k}_t\times\mathbf{z}_0E_n - \frac{\partial}{\partial z}(\mathbf{z}_0\times\mathbf{E}_t) &= j\omega\mu_0\overline{\overline{\mu}}_t\cdot\mathbf{H}_t \\ j\mathbf{k}_t\times\mathbf{z}_0H_n - \frac{\partial}{\partial z}(\mathbf{z}_0\times\mathbf{H}_t) &= -j\omega\epsilon_0\overline{\overline{\epsilon}}_t\cdot\mathbf{E}_t + k_0KH_n\mathbf{r}_0 \end{aligned} \tag{33}$$

Eliminating the normal components, after some algebra we arrive to the following equations for the transverse fields:

$$\frac{\partial}{\partial z}(\mathbf{z}_0\times\mathbf{E}_t) = -j\omega\mu_0\left[\overline{\overline{\mu}}_t\cdot\mathbf{H}_t + \frac{1}{k_0^2\epsilon_n}\mathbf{k}_t\times(\mathbf{k}_t\times\mathbf{H}_t)\right],$$

$$\frac{\partial}{\partial z}(\mathbf{z}_0 \times \mathbf{H}_t) = j\frac{k_0}{\eta_0}\bar{\bar{\epsilon}}_t \cdot \mathbf{E}_t - \frac{j}{k_0\eta_0\mu_n}(\mathbf{z}_0 \times \mathbf{k}_t)(\mathbf{z}_0 \times \mathbf{k}_t) \cdot \mathbf{E}_t$$

$$+K\frac{(\mathbf{r}_0 \cdot \mathbf{k}_t)}{\eta_0\mu_n}\bar{\bar{J}} \cdot \mathbf{E}_t - \frac{jk_0}{\eta_0\mu_n}K^2\,\mathbf{r}_0\mathbf{r}_0 \cdot \mathbf{E}_t \tag{34}$$

In the vector form the wave equation following from (34) reads

$$\frac{\partial^2}{\partial z^2}\mathbf{E}_t + \left[k_0^2\bar{\bar{\mu}}_t' \cdot \bar{\bar{\epsilon}}_t - \frac{\mathbf{k}_t\mathbf{k}_t \cdot \bar{\bar{\epsilon}}_t}{\epsilon_n} - \frac{\bar{\bar{\mu}}_t' \cdot (\mathbf{z}_0 \times \mathbf{k}_t)(\mathbf{z}_0 \times \mathbf{k}_t)}{\mu_n}\right.$$

$$-\frac{j\,k_0K\,(\mathbf{r}_0 \cdot \mathbf{k}_t)}{\mu_n}\bar{\bar{\mu}}_n' \cdot \bar{\bar{J}} - \frac{k_o^2K^2}{\mu_n}\bar{\bar{\mu}}_t' \cdot \mathbf{r}_0\mathbf{r}_0 - j\frac{K(\mathbf{r}_0 \cdot \mathbf{k}_t)}{k_0\epsilon_n\mu_n}\mathbf{k}_t(\mathbf{z}_0 \times \mathbf{k}_t)$$

$$\left.+\frac{K^2\,(\mathbf{r}_0 \cdot \mathbf{k}_t)}{\epsilon_n\mu_n}\mathbf{k}_t\mathbf{r}_0\right] \cdot \mathbf{E}_t = 0, \tag{35}$$

where $\bar{\bar{\mu}}_t' = \mu_y\mathbf{x}_0\mathbf{x}_0 + \mu_x\mathbf{y}_0\mathbf{y}_0$.

Detailed study of the wave equation in the vector form becomes involved. In cartesian components, substituting a plane-wave solution in the form $\exp(\pm j\beta z)$, we obtain two equations for the transverse electric field components:

$$\left[-\beta^2 + k_0^2\mu_y\epsilon_x - \frac{k_x^2\epsilon_x}{\epsilon_n} - \frac{k_y^2\mu_y}{\mu_n} - \frac{k_0^2K_x^2\mu_y}{\mu_n} + j\frac{K(\mathbf{r}_0 \cdot \mathbf{k}_t)k_xk_y}{k_0\epsilon_n\mu_n}\right.$$

$$\left.+\frac{KK_x(\mathbf{r}_0 \cdot \mathbf{k}_t)k_x}{\epsilon_n\mu_n}\right]E_x + \left[k_xk_y\left(\frac{\mu_y}{\mu_n} - \frac{\epsilon_y}{\epsilon_n}\right) + j\frac{K(\mathbf{r}_0 \cdot \mathbf{k}_t)k_0\mu_y}{\mu_n}\right.$$

$$\left.-\frac{k_0^2K_xK_y\mu_y}{\mu_n} - j\frac{K(\mathbf{r}_0 \cdot \mathbf{k}_t)k_x^2}{k_0\epsilon_n\mu_n} + \frac{KK_y(\mathbf{r}_0 \cdot \mathbf{k}_t)k_x}{\epsilon_n\mu_n}\right]E_y = 0 \tag{36}$$

$$\left[k_xk_y\left(\frac{\mu_x}{\mu_n} - \frac{\epsilon_x}{\epsilon_n}\right) - j\frac{K(\mathbf{r}_0 \cdot \mathbf{k}_t)k_0\mu_x}{\mu_n} - \frac{k_0^2K_xK_y\mu_x}{\mu_n} + j\frac{K(\mathbf{r}_0 \cdot \mathbf{k}_t)k_y^2}{k_0\epsilon_n\mu_n}\right.$$

$$\left.+\frac{KK_x(\mathbf{r}_0 \cdot \mathbf{k}_t)k_y}{\epsilon_n\mu_n}\right]E_x + \left[-\beta^2 + k_0^2\mu_x\epsilon_y - \frac{k_y^2\epsilon_y}{\epsilon_n} - \frac{k_x^2\mu_x}{\mu_n} - \frac{k_0^2K_y^2\mu_x}{\mu_n}\right.$$

$$\left.-j\frac{K(\mathbf{r}_0 \cdot \mathbf{k}_t)k_xk_y}{k_0\epsilon_n\mu_n} + \frac{KK_y(\mathbf{r}_0 \cdot \mathbf{k}_t)k_y}{\epsilon_n\mu_n}\right]E_y = 0 \tag{37}$$

Here k_x and k_y stand for the cartesian components of the transverse propagation factor k_t. To solve for the propagation factor β we equate the determinant of (36) and (37) to zero. That leads to a biquadratic equation whose four solutions give propagation factors of two eigenwaves. The solutions can be found in the analytic form:

$$\beta_{1,2}^2 = \frac{1}{2}\left[k_0^2(\mu_x\epsilon_y + \mu_y\epsilon_x) - k_x^2\left(\frac{\epsilon_x}{\epsilon_n} + \frac{\mu_x}{\mu_n}\right) - k_y^2\left(\frac{\epsilon_y}{\epsilon_n} + \frac{\mu_y}{\mu_n}\right)\right.$$
$$\left. - \frac{k_0^2}{\mu_n}(K_x^2\mu_y + K_y^2\mu_x) + \frac{K^2\,(\mathbf{r}_0\cdot\mathbf{k}_t)^2}{\epsilon_n\mu_n} \pm \sqrt{D}\right] \tag{38}$$

$$D = k_0^4(\mu_x\epsilon_y - \mu_y\epsilon_x)^2 + 2k_0^2(\mu_x\epsilon_y - \mu_y\epsilon_x)\left[k_x^2\left(\frac{\epsilon_x}{\epsilon_n} - \frac{\mu_x}{\mu_n}\right) - k_y^2\left(\frac{\epsilon_y}{\epsilon_n} - \frac{\mu_y}{\mu_n}\right)\right]$$
$$+\left[k_x^2\left(\frac{\epsilon_x}{\epsilon_n} - \frac{\mu_x}{\mu_n}\right) + k_y^2\left(\frac{\epsilon_y}{\epsilon_n} - \frac{\mu_y}{\mu_n}\right)\right]^2 + \left[\frac{K^2\,(\mathbf{r}_0\cdot\mathbf{k}_t)^2}{\epsilon_n\mu_n}\right.$$
$$\left. - \frac{k_0^2}{\mu_n}(K_x^2\mu_y + K_y^2\mu_x)\right]^2 - 2\frac{K^2\,(\mathbf{r}_0\cdot\mathbf{k}_t)^2}{\epsilon_n\mu_n}\left[k_x^2\left(\frac{\epsilon_x}{\epsilon_n} + \frac{\mu_x}{\mu_n}\right) + k_y^2\left(\frac{\epsilon_y}{\epsilon_n} + \frac{\mu_y}{\mu_n}\right)\right]$$
$$+\frac{2k_0^2}{\mu_n^2}(k_x^2\mu_x + k_y^2\mu_y)(K_x^2\mu_y + K_y^2\mu_x) + \frac{4k_0^2 K_x K_y k_x k_y(\mu_x\epsilon_y + \mu_y\epsilon_x)}{\epsilon_n\mu_n}$$
$$+\frac{2k_0^2}{\epsilon_n\mu_n}(\mu_x\epsilon_y - \mu_y\epsilon_x)(K_y^2k_y^2 - K_x^2k_x^2) + \frac{2k_0^2}{\epsilon_n\mu_n}(K_x^2\mu_y - K_y^2\mu_x)(k_x^2\epsilon_x - k_y^2\epsilon_y)$$
$$+\frac{2k_0^4}{\mu_n}(K_x^2\mu_y - K_y^2\mu_x)(\mu_x\epsilon_y - \mu_y\epsilon_x) \tag{39}$$

In a special case when the two sets of omega particles are identical and the matrix is isotropic ($\epsilon_x = \epsilon_y = \epsilon_t$, $\mu_x = \mu_y = \mu_t$), the above result simplifies to that obtained in [11].

5. Numerical Results for Planar Omega Structure

Let us consider the eigenwave polarizations. We can use equations (36) and (37) to determine the relation between $\mathbf{E}_x$ and $\mathbf{E}_y$. It defines the projection of the polarization ellipse on the $(x-y)$-plane. For an arbitrary wave propagation direction the polarization ellipse can be easily reconstructed from its projection on that plane. The ellipticity coefficient e in the $(x-y)$-plane is expressed in form of the absolute value of the polarization vector $\mathbf{P}$ [12]:

$$\mathbf{P} = P\mathbf{z}_0 = \mathbf{E}_t \times \mathbf{E}_t^*/j\,\mathbf{E}_t\cdot\mathbf{E}_t^* \qquad P = 2e/(e^2+1) \tag{40}$$

Here $P = -2\mathrm{Im}(A)/(1+|A|^2), A = a_{12}/a_{11}$, or $A = a_{22}/a_{21}$ when $a_{11} = 0$ and the coefficients at E_x, E_y in the system (36), (37) are denoted as $a_{11},\ a_{12},\ a_{21},\ a_{22}$ in the evident way. The value of P ranges from -1 to 1.

Obviously, when $P = 0$, the field is linearly polarized, $P = 1$ corresponds to the right-handed circular polarization, and the field is left-hand circularly polarized when $P = -1$.

In order to see how different material parameters and geometry affect the polarization properties of bi-anisotropic omega structures we calculate polarization vector magnitudes and consider properties of the planar omega structure.

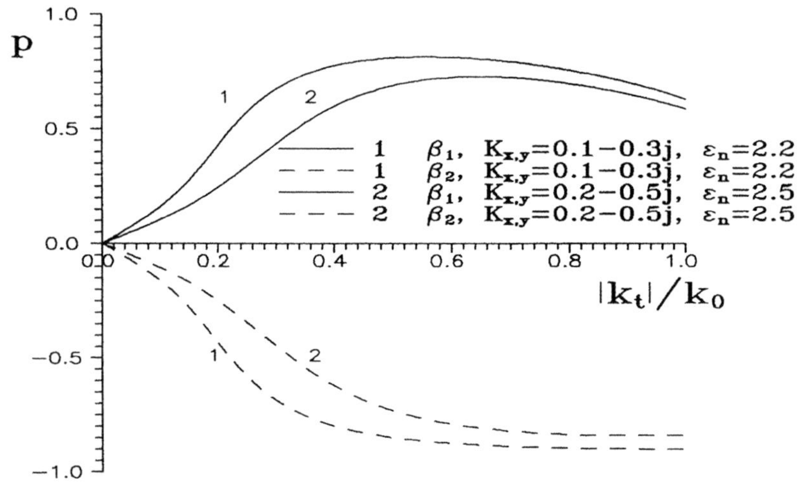

Figure 5. Polarization vector magnitude P as a function of the normalized transverse propagation factor $|k_t|/k_0$. Values of material parameters correspond to the resonance frequency area

Dependence of the polarization vector P on the normalized transverse propagation factor $|k_t|/k_0$ and corresponding dependence for the normal propagation factors $\beta_{1,2}$ are demonstrated in Figures 5 and 6. Curves (1) on Figure 5 correspond to two eigenpolarizations of waves for the following material parameters: the permeabilities $\mu_n = 1.2 - 0.2j$, $\mu_{x,y} = 1.0$, permittivities $\epsilon_n = 2.2$, $\epsilon_{x,y} = 2. - 0.5j$ and coupling parameters $K_{x,y} = 0.1 - 0.3j$. The second pair of curves is calculated for higher values of coupling parameters $K_{x,y} = 0.2 - 0.5j$ and permittivities $\epsilon_n = 2.5$, $\epsilon_{x,y} = 3. - 0.7j$. The permeabilities are $\mu_n = 1.4 - 0.4j$, $\mu_{x,y} = 1.0$. All the parameters were chosen near the resonance for material parameters and were calculated with the use of the model [9].

It is easy to see that for propagation along the z-axis and for arbitrary orientation of the vector $\mathbf{k}_t$ the eigenwaves are linearly polarized and eigenvectors are oriented along characteristic directions. For oblique propagation two main directions of the transverse propagation vector $\mathbf{k}_t$ are interesting.

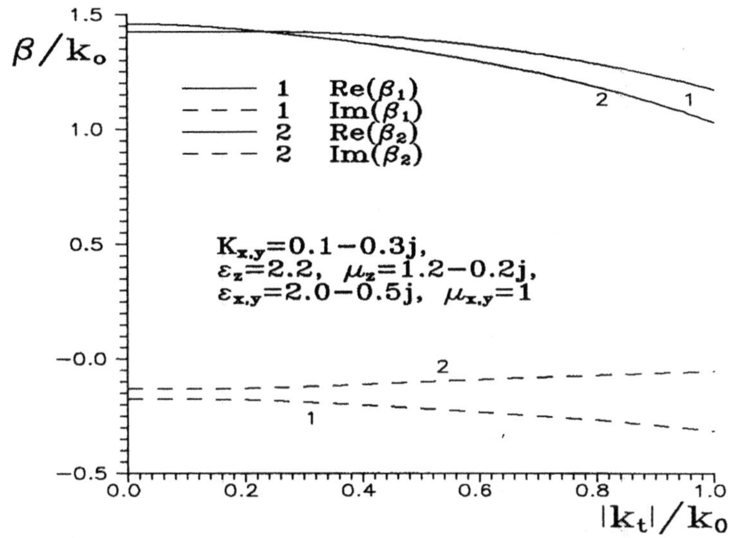

Figure 6. Normalized propagation factors in the z-direction $\beta_{1,2}$ as a function of the normalized transverse propagation factor $|k_t|/k_0$. Values of material parameters are the same as for previous figure

The first direction coincides with the vector $\mathbf{r}_0$ (the bisector of the angle between $\mathbf{x}_0$ and $\mathbf{y}_0$) and the second direction is orthogonal to the first one in the $(x-y)$-plane. For the first case the magnetoelectric effects due to two sets of omega-particles are compensated. The eigenpolarizations are linear for various angles between $\mathbf{k} = \beta\mathbf{z}_0 + \mathbf{k}_t$ and $\mathbf{z}_0$. For the second case with increasing angles both the eigenpolarizations become elliptical and in the limit they are nearly circular with the opposite directions of rotation.

Let us consider a special case of the normal incidence on a planar omega slab. For the incident wave with the component $\mathbf{E}_t$ orthogonal to $\mathbf{r}_0$ the Maxwell equations take the form

$$\frac{\partial}{\partial z}E_t = jk_0\eta_0\mu_y H_t$$

$$\frac{\partial}{\partial z}H_t = j\frac{k_0}{\eta_0}\epsilon_t E_t - j\frac{k_0}{\eta_0\mu_z}K^2 E_t \tag{41}$$

For the axial propagation ($\mathbf{k}_t = 0$), and for this orientation of $\mathbf{E}_t$ the normal propagation factor simplifies to

$$\beta_2^2 = k_0^2\mu_t\left(\epsilon_t - \frac{K^2}{\mu_n}\right)$$

Defining a renormalized material parameter as $\epsilon_x' = \epsilon_x - K^2/\mu_z$ we can reformulate the Maxwell equations in terms of the equations for an equivalent simple dielectric.

6. Conclusion

Basic properties of reciprocal bi-anisotropic composite materials were considered and classified in this paper. We studied the chiral effects in composites with omega-shaped conductive inclusions. The physical meaning of chiral effects was clarified. In particular, we considered properties of two examples of so-called pseudochiral omega structures. Analysis of these omega structures suggest that pseudochiral omega structure and planar omega composites can be used instead of chiral materials in some applications. Along with the propagation factors, eigenpolarization were studied in detail for both composites. The theory of reflection and transmission in slabs and design of wave polarization transformers we plan to report in the near future.

References

1. Lindell, I.V. and Sihvola, A.H. (1995) Plane wave reflection from uniaxial chiral interface and its application to polarization transformation, *IEEE Transactions on Antennas Propagat.*, **43**, 1397-1405.
2. Saadoun, M.M.I. and Engheta, N. (1992) A reciprocal phase shifter using novel pseudochiral or Ω medium, *Microw. and Opt. Technol. Lett.* **5**, 184-188.
3. Tretyakov, S.A. and Sochava, A.A. (1993) Proposed composite material for nonreflecting shields and antenna radomes, *Electron. Lett.* **29**, 1048-1049.
4. Tretyakov, S.A. and Sochava, A.A. (1994) Novel uniaxial bianisotropic materials: reflection and transmission in planar structures, *Progress In Electromagnetic Research,* **9**, 157-179.
5. Tretyakov, S.A. and Sochava, A.A. (1994) Reflection and transmission of plane electromagnetic waves in uniaxial bianisotropic materials, *J. of Infrared and Millimeter Waves*, **15**, 829-855.
6. Sihvola, A.H. and Lindell, I.V. (1995) Material effects in bi-anisotropic electromagnetics", *IEICE Transactions on Electronics (Japan)*, **E78-C**, 1383-1390.
7. Lindell, I.V., Sihvola, A.H., Tretyakov, S.A. and Viitanen, A.J. (1994) *Electromagnetic waves in chiral and bi-isotropic media,* Artech House, Boston and London.
8. Sihvola, A.H. and Lindell, I.V. (1995) Magnetoelectric properties of reciprocal bianisotropic materials with focus on the polarizability of chiro-omega spheres, *Int. J. of Applied Electromagnetics and Mechanics*, **6**, 113-130.
9. Tretyakov, S.A., Simovski, C.R., Mariotte, F., Bolioli, S. and Kharina T. (1995) Scattering by Ω-shaped conductive particles and omega composite modelling, *URSI International Electromagnetic Theory Symposium, S.-Petersburg*, 679-681.
10. Tretyakov, S.A. and Sochava, A.A. (1993) Eigenwaves in uniaxial chiral omega media, *Microw. and Optical Technol. Lett.*, **6**, 701-705.
11. Norgren M. and He S. (1993) Electromagnetic reflection and transmission for a dielectric-Ω interface and a Ω slab, *Department of Electromagnetic Theory, Royal Institute of Technology, Report* **93-14**.
12. Lindell, I.V. (1992) *Methods for Electromagnetic Field Analysis*, Clarendon Press, Oxford.

PULSE DISTORTION BY A LOSSY, RESONANT CHIRAL MEDIUM

S.A. MAKSIMENKO AND G.YA. SLEPYAN
Institute of Nuclear Problems, Belarus State University,
11 Bobruiskaya Str., Minsk, 220050, Belarus

AND

A. LAKHTAKIA
CATMAS – Computational and Theoretical Materials Sciences Group, Department of Engineering Science and Mechanics,
The Pennsylvania State University,
University Park, PA 16802-1401, USA

Abstract. Pulse propagation in a linear, lossy chiral medium with single–resonance dispersion is considered. A relevant time–domain Green function is derived, after assuming the medium is weakly dispersive. The distortions of a transform-limited Gaussian pulse are analyzed in the slow–amplitude approximation. The role of chirality and material dumping in the propagating pulse shaping are discussed.

1. Introduction

Time–domain fields in a linear chiral medium have attracted recent attention [1]–[8]. A general Green's function approach to the description of the temporal response of chiral media has been formulated by Gvozdev and Serdyukov [1] and Kristensson and Rikte [2, 3]. This approach and a wave–splitting one were used in Refs. [2, 3] to describe transient signal wavefront propagation. Correct to the first–order saddlepoint approximation, Engheta and Zablocky [4, 5] examined time–domain fields at length in a lossless chiral medium with single–resonance dispersion. Hillion [7, 8] studied the distortion of a pulse through a chiral slab, but he made the constitutive parameters of the chiral medium frequency–independent. Because the optical rotatory power as the circular dichroism spectra of a single–

A. Priou et al. (eds.), Advances in Complex Electromagnetic Materials, 103–113.

resonance chiral medium have remarkable features in the same frequency range [9], ignoring the dispersive characteristics of a chiral medium is not advisable. Furthermore, Oughstun [6] pointed out the inapplicability of the results obtained in [5] because damping had been ignored. Indeed, recent investigations [10, 11] have shown that even small damping can play an important role. In this paper, we describe the problem of pulse distortions in a chiral medium with certain reasonable restrictions imposed on the constitutive parameters and on the initial pulse duration.

2. Basic Relations

Let us suppose that all space $-\infty < z < \infty$ is occupied by a linear, homogeneous chiral medium, and electromagnetic pulses are launched into the halfspaces $z \geq 0$ from a source wholly specified on the plane $z = 0+$ for all time t. The frequency–domain constitutive relations of the chiral medium are given in the Boys–Post notation as [12]

$$\begin{aligned} \mathbf{D}_\omega(\mathbf{r}) &= \epsilon(\omega)\mathbf{E}_\omega(\mathbf{r}) + i\xi(\omega)\mathbf{B}_\omega(\mathbf{r}), \\ \mathbf{H}_\omega(\mathbf{r}) &= i\xi(\omega)\mathbf{E}_\omega(\mathbf{r}) + \frac{1}{\mu(\omega)}\mathbf{B}_\omega(\mathbf{r}), \end{aligned} \tag{1}$$

where $\mathbf{r}$ is the position vector and the subscript ω on the field quantities indicates that an $\exp(-i\omega t)$ time–dependence is implicit. The quantities $\epsilon(\omega)$, $\mu(\omega)$ and $\xi(\omega)$ are the three constitutive parameters of the chiral medium.

In a linear chiral medium, time–harmonic electromagnetic fields are economically represented by two frequency–domain Beltrami fields [12],

$$\mathbf{Q}^{\pm}_\omega(\mathbf{r}) = (1/2)[\mathbf{E}_\omega(\mathbf{r}) \pm i\eta(\omega)\mathbf{H}_\omega(\mathbf{r})]\,, \tag{2}$$

of different helicities. Superscripts $\pm$ correspond to the left–handed and right–handed types, respectively [12]. The quantity

$$\eta(\omega) = \sqrt{\frac{\mu(\omega)}{\epsilon(\omega) + \mu(\omega)\xi^2(\omega)}} \tag{3}$$

is the chiral medium's intrinsic impedance at angular frequency ω.

On taking the inverse Fourier transform of Eq. (2), time–domain Beltrami fields can be defined as

$$\mathbf{Q}^{\pm}(\mathbf{r},t) = \frac{1}{2}[\mathbf{E}(\mathbf{r},t) \pm i\int_0^\infty \eta(t')\mathbf{H}(\mathbf{r},t-t')\,dt'], \tag{4}$$

where $\eta(t)$ is the inverse Fourier transform of $\eta(\omega)$. The wavenumbers corresponding to the frequency–domain Beltrami fields are given by

$$k_\pm(\omega) = \omega[\pm\xi(\omega)\mu(\omega) + \sqrt{\xi^2(\omega)\mu^2(\omega) + \epsilon(\omega)\mu(\omega)}\,]\,. \tag{5}$$

Thus, the source–free time–domain Maxwell curl postulates can be specialized for a chiral medium as

$$\nabla \times \mathbf{Q}^{\pm}(\mathbf{r},t) = \pm \int_0^{\infty} k_{\pm}(t')\mathbf{Q}^{\pm}(\mathbf{r},t-t')dt' \,, \tag{6}$$

with $k_{\pm}(t)$ defined as the inverse Fourier transform of $k_{\pm}(\omega)$.

Let the pulse fields depend only on spatial variable z. From [14], we know that any time–harmonic source confined to the plane $z = 0$ can be configured to radiate either exclusively left– or exclusively right–handed frequency–domain fields into both halfspaces. Hence, the boundary values

$$\mathbf{Q}^{\pm}(z = 0+, t) = A_0^{\pm}(t)e^{-i\omega_c t}\mathbf{e}^{\pm} \tag{7}$$

as well as the initial condition

$$\lim_{t \to -\infty} \mathbf{Q}^{\pm}(z,t) = \mathbf{0}, \qquad z > 0 \,, \tag{8}$$

are appropriate for investigating the evolution of Beltrami pulses with infinitely wide spectra (for example, Gaussian or soliton–like pulses) launched into the halfspace $z \geq 0$. Here, $\mathbf{e}^{\pm} = \{1, \pm i, 0\}/\sqrt{2}$, ω_c is the carrier frequency.

To prescribe the electromagnetic response of the chosen chiral medium we adopt single–resonance Lorentz models for the permittivity $\epsilon(\omega)$ [15] and the permeability $\mu(\omega)$ [16], i.e.,

$$\epsilon(\omega) = \epsilon_0 \left(1 - \frac{\omega_p^2}{\omega^2 - \omega_0^2 + 2i\delta\omega}\right), \quad \mu(\omega) = \mu_0 \left(1 - \frac{\omega_m^2}{\omega^2 - \omega_0^2 + 2i\delta\omega}\right). \tag{9}$$

The quantities ϵ_0 and μ_0 are the permittivity and the permeability of free space; ω_0 and δ are the frequency and linewidth of the resonance; while ω_p and ω_m characterize the oscillator strengths of the respective transitions. Here we assume that ω_m does not exceed ω_p (in dielectric media $\omega_m \to 0$). To achieve maximal dispersion of the propagating pulse, we next impose the condition

$$|\omega_c - \omega_0| \ll \omega_0, \tag{10}$$

For the chirality we apply the well–known Condon model of optical activity [17]:

$$\xi(\omega) = \frac{ac\eta_0^{-1}\omega}{\omega^2 - \omega_0^2 + 2i\delta\omega}, \tag{11}$$

where a is the rotatory strength of the resonance, $\eta_0 = \sqrt{\mu_0/\epsilon_0}$ and $c = 1/\sqrt{\epsilon_0\mu_0}$. Let us now impose a set of restrictions on the constitutive as well as the pulse parameters. First, we suppose that the criterion

$$\frac{\omega_p^2}{\omega_0^2} \ll \frac{\delta}{\omega_0} \ll 1, \tag{12}$$

is satisfied. This condition may serve very well for weakly dispersive and weakly absorbing media at optical and microwave frequencies [18]. Using this condition and assuming the chirality to be sufficiently weak so that the inequality

$$\frac{(ac)^2}{\omega_0^2} / \frac{2\delta}{\omega_0} \ll \frac{\omega_p^2}{\omega_0^2}, \tag{13}$$

holds true, we can get rid of the square root in Eq. (5) and reduce it to a more tractable form as follows:

$$k_{\pm}(\omega) = \frac{\omega}{c}\left(1 - \frac{\tilde{\omega}_p^2 \pm ac\omega}{\omega^2 - \omega_0^2 + 2i\delta\omega}\right), \tag{14}$$

where $\tilde{\omega}_p^2 = (\omega_p^2 + \omega_m^2)/2$. Restriction (13) has allowed us to neglect the term $\xi^2\mu^2$ in Eq. (5).

Let us note that $Im[k_{\pm}(\omega)] > 0$ for a lossy passive chiral medium; otherwise, gain shall occur in the Beltrami pulses of one handedness. Assuming $|\omega - \omega_c| < \Delta\omega$ and using Eq. (10), we obtain the following restriction on the spectral bandwidth $\Delta\omega$ of the boundary value of the pulse:

$$\frac{\Delta\omega}{\omega_0} < \frac{\tilde{\omega}_p^2}{\omega_0^2} / \frac{ac}{\omega_0} - 1 \equiv q - 1. \tag{15}$$

This limitation becomes very significant for highly chiral media; then, $q - 1 \to 0+$ and, hence, $\Delta\omega$ must be made to go to zero for our approximate approach to remain applicable.

3. Evaluation of Time–Domain Green Functions

Let us consider the temporal response of the chiral medium (with dispersion defined by Eq. (14)) to a left–handed Beltrami pulse (i.e., $A_0^-(t) = 0$). The evolution of the chosen Beltrami pulse in time and space is determined by the function $\mathbf{Q}^+(z,t) = A^+(z,t)\exp\{-i\omega_c\tau\}\mathbf{e}^+$, where $\tau = t - z/c$ is the retarded time. The amplitude $A^+(z,t)$ is defined by the convolution integral

$$A^+(z,t) = \int_{-\infty}^{\infty} e^{-i\omega_c t'} G^+(z, t - t') A_0^+(t')dt', \tag{16}$$

where the temporal Green function is given by

$$G^+(z,t) = \frac{1}{2\pi}\int_{-\infty}^{\infty} e^{-i\omega t + ik_+(\omega)z} d\omega. \tag{17}$$

The substitution $\omega = i(p - \delta)$ reduces Eq. (17) to the form of the inverse Laplace transform:

$$G^+(z,t) = e^{-iaz-\delta\tau}\frac{1}{2\pi i}\int_{\delta-i\infty}^{\delta+i\infty} dp\, e^{p\tau} \exp\left\{-\frac{c_1^+ z}{p + i\Omega_0} - \frac{c_2^+ z}{p - i\Omega_0}\right\} \tag{18}$$

where $\Omega_0^2 = \omega_0^2 - \delta^2$ and

$$c_{1,2}^{+} = \frac{\omega_0^2}{2c\Omega_0}\left(\frac{\tilde{\omega}_p^2 - 2i\delta ac}{\Omega_0 \pm i\delta} \pm ac\right). \tag{19}$$

As the next step, we expand the integrand on the right side of Eq. (18) as the sum

$$\exp\left\{-\frac{c_1^+ z}{p + i\Omega_0} - \frac{c_2^+ z}{p - i\Omega_0}\right\} = 1 + B_1 + B_2 + B_1 B_2, \tag{20}$$

where

$$B_{1,2} = \exp\left\{-\frac{c_{1,2}^+ z}{p \pm i\Omega_0}\right\} - 1. \tag{21}$$

Using [19], we can then present $G^+(z,t)$ by

$$G^+(z,t) = e^{-iaz}\left\{\delta(\tau) + G_1^+(z,\tau) + G_2^+(z,\tau) + G_3^+(z,\tau)\right\}, \tag{22}$$

where

$$G_{1,2}^+(z,\tau) = -e^{\mp i\Omega_0\tau - \delta\tau}\sqrt{\frac{c_{1,2}^+ z}{\tau}}\, J_1(2\sqrt{c_{1,2}^+ z\tau})\,\Theta(\tau), \tag{23}$$

$$G_3^+(z,\tau) = \int_0^\tau G_1^+(z,\tau')G_2^+(z,\tau-\tau')d\tau'\,\Theta(\tau). \tag{24}$$

In these equations, $\Theta(\cdot)$ is the unit step function and $J_1(\cdot)$ is the Bessel function. An expression for the $G^-(z,t)$ can be easily obtained on replacing a by $-a$.

4. Evolution of Beltrami–Gaussian Pulses

Let us suppose the boundary value (at $z = 0+$) of the pulse has the Gaussian envelope

$$A_0^+(t) = \exp\left\{-t^2/2t_c^2\right\}, \tag{25}$$

with t_c referred to as the pulse duration. The substitution of Eqs. (22) and (25) into Eq. (16) leads us to the integrals to be solved

$$A_j^+(z,\tau) = \int_{-\infty}^{\infty} G_j^+(z,\tau - t')e^{-i\omega_c t' - t'^2/2t_c^2}dt', \quad j = 1,2,3. \tag{26}$$

Using Eq. (23) in Eq. (26) and then setting $x = (\tau - t')/t_c$, we get for $A_{1,2}^+(z,\tau)$:

$$A_{1,2}^+(z,\tau) = -\sqrt{\zeta_{1,2}^+}\,e^{-i\omega_c t}\int_0^\infty e^{i\lambda_{1,2}x - \Gamma x}e^{-(x-T)^2/2}J_1(2\sqrt{\zeta_{1,2}^+ x})\frac{dx}{\sqrt{x}}, \tag{27}$$

where the dimensionless parameters are given as follows:

$$\zeta^+_{1,2} = c^+_{1,2} z t_c, \quad \lambda_{1,2} = (\omega_c \mp \Omega_0) t_c, \quad \Gamma = \delta t_c, \quad T = \tau / t_c. \tag{28}$$

Let us now suppose that the pulse duration is long enough so that $t_c \gg 1/\omega_c$. As can be established from Eqs. (10) and (28), the inequality $\lambda_2 \gg 1$ holds true, while $\lambda_1 \sim 1$ if $\delta/\omega_0 \ll 1$ also. Then, for estimating $A^+_2(z,\tau)$, we can apply the stationary phase method to Eq. (27) and get $A^+_2(z,\tau) = O(1/\lambda_2)$. Certain manipulations of Eq. (24) yield the same estimation of the third integral in Eq. (26): $A^+_3(z,\tau) = O(1/\lambda_2)$. Thus, condition $t_c \gg 1/\omega_c$ allows us to present a reasonably accurate final result as follows:

$$A^+(z,t) \approx e^{-iaz} \left\{ A^+_0(\tau) + A^+_1(z,\tau) \right\}, \tag{29}$$

with $A^+_0(\tau)$ and $A^+_1(z,\tau)$ given by Eqs. (25) and (27), respectively.

Equation (29) for the left–handed Beltrami–Gaussian pulse, and the analogous one for the right–handed (i.e., $\mathbf{e}^-$–polarized) Beltrami–Gaussian pulse, enable us to examine the effects of dispersion and chirality on the pulse shapes. Certain computed profiles of the pulse intensities $|A^{\pm}(z,t)|^2$ are shown in Figs. 1–4. In all plots, the launched pulse shapes on the boundary are shown as dashed curves.

Fig. 1 illustrates the modification of the temporal profiles by the chirality of the medium. According to Eq. (28), only coefficients $\zeta^{\pm}_1 = c^{\pm}_1 z t_c$ (which can be interpreted as *effective propagation lengths*) depend on a and, consequently, are responsible for the differences exhibited by Beltrami–Gaussian pulses of opposite helicities. In the zeroth–order approximation with respect to δ/ω_0, $\zeta^{\pm}_1 \propto z(1 \pm 1/q)$ with q determined by Eq. (15). All other quantities being fixed, the product aq is constant. As a increases, q decreases so that the effective propagation length of the $\mathbf{e}^-$–polarized pulse tends to zero while that of $\mathbf{e}^+$–polarized pulse remains fixed. As a result, the $\mathbf{e}^-$–polarized pulse approaches its initial shape (see the curves labelled 1 in Figs. 1a–1c) while the $\mathbf{e}^+$–polarized pulse is only weakly affected (curves labelled 2 in Figs. 1a–1c) when q decreases. In the achiral limit (i.e., $1/q \to 0$) both Beltrami–Gaussian pulses take the same shape, shown as curve 3 in Fig. 1a.

Fig. 2 shows how the damping δ affects the shape of propagating pulses. As the effective propagation length of the $\mathbf{e}^+$–polarized pulse is thrice that of the $\mathbf{e}^-$–polarized pulse, pulse distortions due to dispersion must be more significant for the former pulse than for the latter one. In the limiting case of zero damping (see the curves labelled 4 in Figs. 2a and 2b), the $\mathbf{e}^+$–polarized pulse displays a characteristic double–peak shape with the main part of transmitted energy concentrated in the delayed peak, but the delayed peak for the $\mathbf{e}^-$–polarized pulse is hardly prominent.

The effect of damping is strongly dependent on the relation between the parameters $\Delta\omega$ and δ. When $\Delta\omega \gg \delta$, damping may not be significant. For

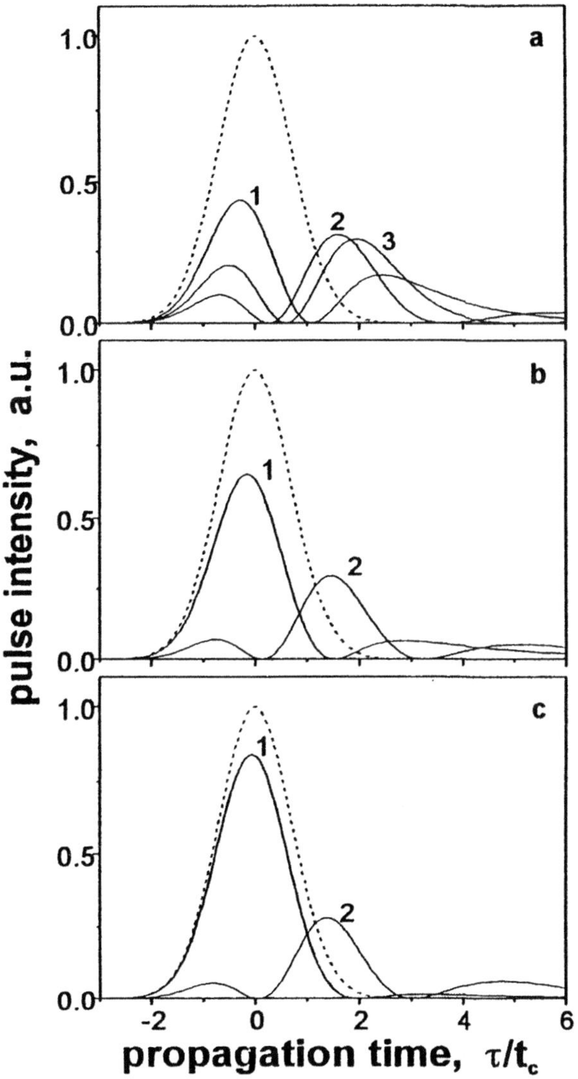

Figure 1. The effect of chirality on the shapes of Beltrami–Gaussian pulses. The curves labelled 1 (e^-–polarized pulse) and 2 (e^+–polarized pulse) were obtained with $\delta/\omega_0 = 10^{-3}$, $\omega_c t_c = 100$, $z\omega_c/c = 150$, $\tilde{\omega}_p/\omega_0 = 10^{-2}$ and $\omega_c = \omega_0$. Fig 1a: $1/q = 0.5$; Fig. 1b: $1/q = 0.75$; Fig. 1c: $1/q = 0.90$.

instance, the curves labelled 1 in Figs. 2a and 2b were obtained at $\delta/\Delta\omega \sim \delta t_c = 0.1$. But even then, the delayed peaks are suppressed although the initial peaks are not much affected by absorption. The curves labelled 2 and 3 in Fig. 2 show that the delayed peaks in propagating pulses turn out

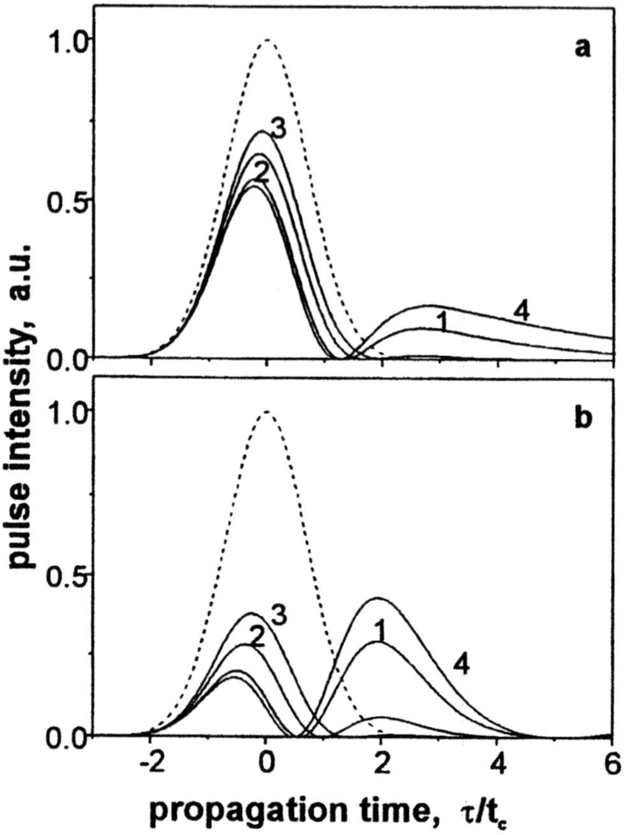

Figure 2. The effect of material damping δ on the shapes of propagating pulses. For all curves, $\omega_c t_c = z\omega_c/c = 100$, $ac/\omega_0 = 5 \cdot 10^{-4}$, $\bar{\omega}_p/\omega_0 = 10^{-2}$ and $\omega_c = \omega_0$. The curves are labelled $1(\delta/\omega_0 = 10^{-3})$, $2(\delta/\omega_0 = 5 \cdot 10^{-2})$, $3(\delta/\omega_0 = 10^{-2})$ and $4(\delta/\omega_0 = 0)$. Fig. 2a: e^-–polarized pulse; Fig. 2b: e^+–polarized pulse.

to be completely suppressed when δ either approaches or exceeds $\Delta\omega$.

Fig. 3 gives an indication of the variation of pulse shape with the normalized propagation length $z\omega_c/c$. As this length increases, the initial pulse shape is lost while the pulse is transformed into a set of low–intensity temporal oscillations.

Fig. 4 presents the temporal profiles of propagating pulses at different levels of frequency detuning quantitated as $(\omega_c - \omega_0)/\omega_0$. Since all calculations were made with $1/q = 0.75$, the effective propagating length for the e^-–polarized pulse is relatively small and the pulse shape remains approximately Gaussian at any detuning. As for the e^+–polarized pulse, its shape is noticeably distorted when the detuning is small (curves labelled 1 and 2 in Fig. 4b), but approaches its initial Gaussian shape as the detuning

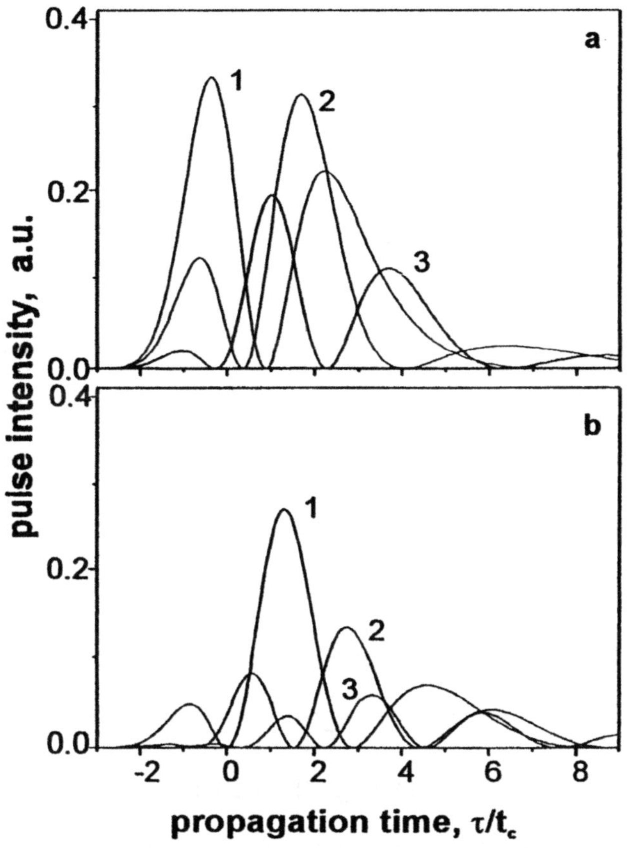

Figure 3. The modification of pulse shape with increasing normalized propagation length $z\omega_c/c$. For all curves, $\omega_c t_c = 100$, $\delta/\omega_0 = 10^{-3}$, $ac/\omega_0 = 5 \cdot 10^{-4}$, $\tilde{\omega}_p/\omega_0 = 10^{-2}$ and $\omega_c = \omega_0$. The curves are labelled $1(z\omega_c/c = 200)$, $2(z\omega_c/c = 400)$ and $3(z\omega_c/c = 800)$. Fig. 3a: e^-–polarized pulse; Fig. 3b: e^+–polarized pulse.

becomes more pronounced (curves labelled 3 and 4 in Fig. 4b).

All four figures demonstrate a certain analogy with the ones obtained for a transform–limited Gaussian pulse diffracted by a volume grating [20]–[22]. That analogy can be understood by referring to the fact that, in spite of completely different dispersion laws used for the two problems, almost all of the chosen pulse spectrum lies beyond the group velocity dispersion (GVD) band defined by the function $d^2k(\omega)/d\omega^2$. But GVD (together with other higher–order derivatives of the wavenumbers) is responsible for the pulse shape distortions [23].

Acknowledgements: S. A. Maksimenko and G. Ya. Slepyan acknowledge partial support from the International Science Foundation under Contract

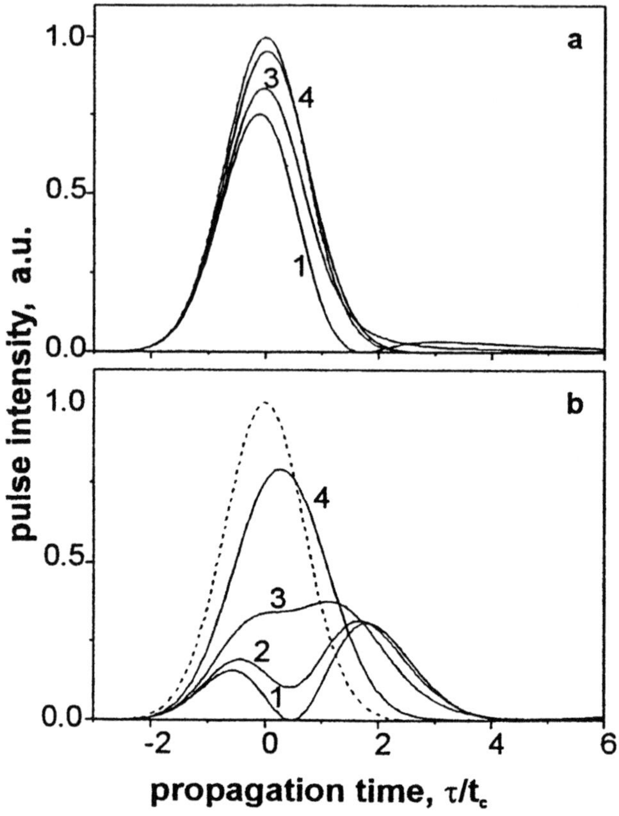

Figure 4. Temporal profiles of propagating Beltrami-Gaussian pulses at different levels of frequency detuning. For all curves, $\omega_c t_c = z\omega_c/c = 100$, $\delta/\omega_0 = 10^{-3}$, $ac/\omega_0 = 7.5 \cdot 10^{-4}$ and $\tilde{\omega}_p/\omega_0 = 10^{-2}$. The curves are labelled $1((\omega_c - \omega_0)/\omega_0 = 0)$, $2((\omega_c - \omega_0)/\omega_0 = 0.005)$ $3((\omega_c - \omega_0)/\omega_0 = 0.10)$ and $4((\omega_c - \omega_0)/\omega_0 = 0.20)$. Fig. 4a: e^-–polarized pulse; Fig. 4b: e^+–polarized pulse.

No. U9Y000 and Belarus Foundation for Fundamental Research. This work was done in part while A. Lakhtakia was visiting the University of Glasgow; he thanks the Board of Trustees, Pennsylvania State University for granting him a sabbatical leave of absence as well as the Engineering and Physical Sciences Research Council (UK) for Grant No. GR/K76979.

References

1. Gvozdev, V. V. and Serdyukov, A. N. (1981) Radiation of electromagnetic waves in a dispersive gyrotropic medium, *Optics and Spectroscopy* **50**, 187-190.
2. Kristensson, G. and Rikte, S. (1992) Transient electromagnetic waves in reciprocal bi-isotropic media, *J. Electromag. Waves Appl.* **6**, 1517–1535.
3. Kristensson, G. and Rikte, S. (1993) Transient wave propagation in reciprocal bi-

isotropic media at oblique incidence, *J. Math. Phys.* **34**, 1339–1359.
4. Engheta, N. and Zablocky, P. G. (1991) Effect of chirality on the transient signal wave front, *Opt. Lett.* **16**, 1924–1926.
5. Zablocky, P. G. and Engheta, N. (1993) Transients in chiral media with single-resonance dispersion, *J. Opt. Soc. Am. A* **10**, 740–758.
6. Oughstun, K. E. (1995) Transients in chiral media with single-resonance dispersion: comment, *J. Opt. Soc. Am. A* **12**, 626–628.
7. Hillion, P. (1994) Time-dependent electromagnetic pulse on a chiral slab I, *Int. J. Appl. Electromag. Mater.* **4**, 221–230.
8. Hillion, P. (1994) Time-dependent electromagnetic pulse on a chiral slab II, *Int. J. Appl. Electromag. Mater.* **4**, 231–238.
9. Lakhtakia, A. (1992) Comments on "Accelerated particle radiation in chiral media," *J. Appl. Phys.* **71**, 3059–3060.
10. Oughstun, K. E. and Sherman, G. C. (1988) Propagation of electromagnetic pulses in a linear dispersive medium with absorption (the Lorentz medium), *J. Opt. Soc. Am. B* **5**, 817–849.
11. Balictsis, C. M. and Oughstun, K. E. (1993) Uniform asymptotic description of ultrashort Gaussian-pulse propagation in a causal, dispersive dielectric, *Phys. Rev. E* **47**, 3645–3669.
12. Lakhtakia, A. (1994) *Beltrami Fields in Chiral Media*, World Scientific, Singapore.
13. Post, E. J. (1962) *Formal Structure of Electromagnetics,* North-Holland, Amsterdam.
14. Lakhtakia, A. (1992) Dyadic Green's functions for an isotropic chiral half-space bounded by an anisotropic impedance plane, *Int. J. Electron.* **72**, 493–497.
15. Kittel, C. (1974) *Introduction to Solid State Physics,* Wiley Eastern, New Delhi, p. 472.
16. Robin, T., Souillard, B., Accolas, R., Theret, P. and Bolioli, S. (1994) Modelling of electromagnetic properties of helix-loaded chiral media, in *Proc. Chiral'94: 3rd International Workshop on Chiral, Bi-isotropic and Bi-anisotropic Media,* Périgueux, France, pp. 103–108.
17. Charney, E. (1985) *The Molecular Basis of Optical Activity*, Krieger, Malabar, FL, USA, Sec. 2.4.
18. Grosse, P. (1979) *Freie Electronen in Festkörpern*, Springer–Verlag, Berlin.
19. Bateman, H. and Erdelyi, A. (1954) *Tables of Integral Transforms, I*, McGraw–Hill, New York.
20. Baryshevsky, V. G. and Maksimenko, S. A. (1992) Light pulse dispersion under Laue diffraction from a spatial holographic grating, *Opt. Commun.* **94**, 379–388.
21. Baryshevsky, V. G. and Maksimenko, S. A. (1994) Chirped pulse distortion in a volume reflection grating, *Opt. Commun.* **110**, 401–409.
22. Maksimenko, S. A. (1994) Group velocity dispersion in an all-pass Bragg filter, *Opt. Lett.* **19**, 1783–1785.
23. Akhmanov, S. A., Vysloukh, V. A. and Chirkin, A.S. (1992) *Optics of Femtosecond Laser Pulses,* AIP, New York.

III.

Continuum Modelling of Complex Media

NON-LOCAL RESPONSE OF COMPOSITE MATERIALS IN MICROWAVE RANGE

A. N. LAGARKOV AND A. P. VINOGRADOV
Scientific Center for Applied Problems in Electrodynamics,
Russian Academy of Sciences,
Izhorskaja 13/19, Moscow, 127412 IVTAN, Russia

1. Introduction

Composite materials have recently found wide industrial use. This in turn demands development of more sophisticated descriptions of an interaction of electromagnetic waves with such inhomogeneous systems. Commonly, for this aim the macroscopic Maxwell equations are employed. These equations are obtained by homogenization (averaging) of the microscopic Lorentz equations and contain additional unknown fields: magnetic field $\vec{H}$ and electrical induction $\vec{D}$. To complete the system, one needs to introduce, so-called, constitutive relations connecting these new fields with the averaged values of the electric $\vec{E}$ and magnetic $\vec{B}$ fields. Although there exists a well developed theory of constitutive relations for materials without dispersion and with temporal dispersion, we still have run into obstacles while describing system with spatial dispersion. This communication is devoted to some issues of spatial dispersion problem.

2. Local Constitutive Relations (Weak Spatial Dispersion)

The object of the study is a composite material that is a dielectric matrix loaded by different inclusions. The size of a separate inclusion is characterized by the length a_{incl} whereas inhomogeneity of the matrix could be scaled with a_{mat} which is the mean distance between inclusions. Illumination of such a composite by electromagnetic wave brings into the problem another set of scales which are λ_0 (the free space wavelength), λ_{mat} (the wavelength inside the matrix), and λ_{incl} (the wavelength inside an inclusion).

Treating an inhomogeneous system as homogeneous means that some kind of averaging (homogenization) over a "physically small" scale L is

A. Priou et al. (eds.), Advances in Complex Electromagnetic Materials, 117–130.

performed. The scale L is not an inherent scale of the electromagnetic problem. We have some freedom while choosing it. Obviously, the L should be greater than the scale $a = \max(a_{\text{incl}}, a_{\text{mat}})$ of an inhomogeneity.

If $L = 0$ we deal with non-averaged fields. It is natural to anticipate that, increasing L we sooner or later come to a certain limit where the fields cease depending upon spatial variables on the scale L. It means that the values of the averaged fields in neighboring L-domains differ in amount of the order of $L/\lambda_0 \ll 1$, only. The procedure of homogenization could be considered as successful if we find such a compromise value of L that smooths microscopic fluctuations of the fields without touching the wave nature of the fields.

In the static case, since the behavior of systems is governed by "scale-less" (Laplace) equation, the increasing of L can be unlimited. Moreover, there is no difference if we increase L or reduce a. The relevant parameter is the ratio (L/a). There are some well-known algorithms for homogenization [1, 2, 3]. For periodic structures the algorithm is rigorously justified (see, e.g., [4, 5]). This algorithm implies passage to the limit $a \to 0$.

For time-varying fields, in the long-wavelength limit

$$\tau = \max\{a_{\text{incl}}, a_{\text{mat}}\}/\min\{\lambda_{\text{incl}}, \lambda_{\text{mat}}\} \ll 1,$$

we can still use the aforementioned static ($\tau = 0$) algorithms substituting proper local values of permittivity $\varepsilon(\omega)$. The strong inequality $a \ll \lambda_{\text{int}} = \min(\lambda_{\text{mat}}, \lambda_{\text{incl}})$ allows us to choose the averaging scale L so that

$$\max(a_{\text{incl}}, a_{\text{mat}}) < L < \min(\lambda_{\text{mat}}, \lambda_{\text{incl}}).$$

On the scales smaller than L, the static approximation is justifiable. Nevertheless, we lose the mathematical strictness. Even the algorithms rigorous for static fields become approximate ones.

From the mathematical point of view, insolvency of the static methods lies in impossibility either to infinitely reduce a or to increase L. Indeed, as going to the limit $(L/a) \to \infty$ when $a \to 0$ the quantities $\tau_{\text{mat}} = a_{\text{mat}}/\lambda_{\text{mat}}$, $\tau_{\text{incl}} = a_{\text{incl}}/\lambda_{\text{incl}}$, and $\kappa = \varepsilon_{\text{mat}}/\varepsilon_{\text{incl}}$ should be kept constant to preserve the physical pattern of the phenomena. The first condition demands renormalization of frequency by passing from a to $a' < a$: $\omega' = \left(\frac{a}{a'}\sqrt{\frac{\varepsilon_{\text{mat}}(\omega)}{\varepsilon_{\text{mat}}(\omega')}}\right)\omega$. The other conditions would be satisfied if and only if $\kappa = \varepsilon_{\text{mat}}/\varepsilon_{\text{incl}}$ is independent of frequency.[1] Although an increase of L does not influence physical pattern until L is limited by the value of $\min(\lambda_0, \lambda_{\text{int}})$, we arrive at a problem with several scales and can not apply the mathematics developed for

[1]Moreover, in the limit $a \to 0$ one can not properly treat a bounded system because in this limit the effects of any order in $a\sqrt{\varepsilon}/d$ are lost.

the "scale-less" problem. If we must, for some reasons, overcome the limitation $L < \lambda_{\text{int}}$ we have to be careful, since it is possible to come to a physically unsound description.

From the physical point of view, the troubles with alternating fields are due to the loss of uniqueness of the homogenization. To obtain the complete physical description one should average both the constitutive equation

$$< D >= \varepsilon^{(1)}_{\text{eff}} < E >$$

and the expression for energy

$$< DE >=< \varepsilon E^2 >= \varepsilon^{(2)}_{\text{eff}} < E >^2$$

For curl-free fields, in both cases we obtain the same result for effective permittivity: $\varepsilon_{\text{eff}} = \varepsilon^{(1)} = \varepsilon^{(2)}$. Indeed, in this case, the correlation $< DE >$ splits into the product $< D >< E >$ [6] and we arrive at the following chain of equalities proving our statement:

$$\varepsilon^{(2)}_{\text{eff}} < E >^2 =< \varepsilon E^2 >=< DE >=< D >< E >= \varepsilon^{(1)}_{\text{eff}} < E >^2$$

It is not the case for solenoidal fields. For example, direct averaging of the Maxwell equations and the wave equation deduced from them might give different results (see [7]).

Fortunately, in case of a widespread situation: $| \varepsilon_{\text{incl}} | \gg | \varepsilon_{\text{mat}} | > 1$ we can overcome the limitation $L < \lambda_{\text{int}}$ and obtain a unique result. In this case, there is a frequency range where $\lambda_{\text{incl}} \ll \lambda_{\text{mat}} < \lambda_0$. The averaging scale L can be chosen so that $\lambda_{\text{incl}} \leq a_{\text{incl}}, a_{\text{mat}} < L < \lambda_{\text{mat}} < \lambda_0$. One has to deal with solenoidal fields only inside the region confined to the inclusion volume. Outside the inclusions (in matrix) the fields may be considered as curl-free. At low concentration of inclusions[2] the averaging procedure can be completely carried out, resulting in local constitutive equations. In this case, we can no longer confine ourselves only to effective permittivity. Depending upon the structure of the material, additional constitutive parameters should be introduced (see [19]). It is the eddy current permeability [8, 9, 10] for composites loaded by highly conducting inclusions, chirality for composites loaded by inclusions without central symmetry [11, 12, 13], or effective magnetic permeability for composites loaded by dielectric inclusions if a TE mode is excited in separate inclusion [14].

The situation can be significantly complicated if there are several lengths a_{mat} involved in the problem. As examples we can refer to percolation system [2], to media with specially arranged omega [13] or chiral [13] inclusions, artificial magnet [23] media, etc.

[2]There should be neither percolation channel through the system nor large clusters comparable with the wavelength inside the matrix.

The general features of the problem are displayed while examining permeability. Effective permeability has its origin in circulating character of currents on the scale of inhomogeneity a. The electromagnetic field by itself is naturally circulating on the scale λ and, as a consequence, current is circulating on this scale. In other words, we should distinguish circulation of currents due to inhomogeneity of the medium and due to original properties of electromagnetic fields. L.D. Landau [15] stated this is possible until one can neglect polarization current $j_{\text{pol}} = \partial P/\partial t = -\left[i\omega(\varepsilon - 1)/4\pi\right]E$ with respect to conducting one.[3] On the other hand, in composite media the conducting currents $j_{\text{cond}} = \sigma E$ flowing in metal inclusion are continuated by the Maxwell displacement (polarization) currents $j_{\text{pol}} = \partial P/\partial t = -\left[i\omega(\hat{\varepsilon} - 1)/4\pi\right]E$ outside the inclusion. Hence, the role of polarization currents is always important, especially in case of inclusions with high permittivity where $j_{\text{pol}} \gg j_{\text{cond}}$.

The Landau restriction $j_{\text{pol}} \ll j_{\text{cond}}$ follows from the following definition of the "macroscopic" magnetic M and electric P polarizations, connecting them with currents inside the material [15]:

$$\int_V \vec{j}dv = \int_V \frac{\partial \vec{P}}{\partial t}dv \tag{1}$$

and

$$\int \vec{M}^{(\text{Landau})}dv = \frac{1}{2c}\int \left[\vec{r} \times \vec{j}\right] dv \tag{2}$$

Equality (2) is valid if

$$\frac{\partial \vec{P}}{\partial t} \ll c \,\text{curl}\vec{M} \tag{3}$$

It seems that this requirement is too strong. To introduce the magnetic polarization we need unique representation of an averaged current as

$$<\vec{j}>= c\,\text{curl}\vec{M} + \frac{\partial \vec{P}}{\partial t} \tag{4}$$

Certainly, we have to change the definition of $\vec{P}$ and $\vec{M}$. It is easy to calculate the electric and magnetic dipole moments of individual inclusion (see [15]) separating this way ($c\,\text{curl}\,\vec{M}$) and $\partial\vec{P}/\partial t$ in (4). One might expect that this is a solution to the problem. However, if we consider a group of neighboring inclusions, we can see that there exists a magnetic moment due to the currents flowing through the group, whereas on the scale of inclusion these currents contribute to the electric moment according to (1). Thus,

[3]The Gaussian system of units is followed.

we can see that there is an open question how to divide the current into electric and magnetic parts (see [15, 16, 17])[4].

To clear up the problem let us imagine a composite sample[5] as a sum of cells whose characteristic size is l_1. We express the total current $\vec{j} = [\sigma(\vec{r}) - (i\omega(\varepsilon(\vec{r}) - 1)/4\pi))\vec{E}$ as a sum of the average part $\langle\vec{j}\rangle^{(l_1)}$ and the fluctuating one $\delta\vec{j}^{(l_1)}$: $\vec{j} = \langle\vec{j}_i\rangle^{(l_1)} + \delta\vec{j}^{(l_1)}$. The brackets $\langle\rangle^{(l_1)}$ denote averaging over a cell volume. The $\langle j_i\rangle^{(l_1)}$ can be considered as $\partial P^{(l_1)}/\partial t$. It varies only on scales greater than l_1. At the scale l_1

$$\mathrm{div}\vec{j} = \mathrm{div}(\langle\vec{j}\rangle^{(l_1)} + \delta\vec{j}^{(l_1)}) = \mathrm{div}\delta\vec{j}^{(l_1)} = 0$$

and the fluctuating part can be represented as a *curl* of some vector $\vec{M}^{(l_1)}$. The vector $\vec{M}^{(l_1)}$ represents a magnetic moment of the current $\delta\vec{j}^{(l_1)}$ if all current lines of $\delta\vec{j}^{(l_1)}$ are confined to the volume of the cell. In this case we can continue $\delta\vec{j}^{(l_1)}$ outside the cell with zero-valued function. The vector $\vec{M}^{(l_1)}$ outside the cell is equal to zero, too. Consideration of a volume of integration being slightly greater than the cell permits to neglect of the surface integral in the following line of expressions:

$$\begin{aligned}\frac{1}{2c}\int\left[\vec{r}\times\delta\vec{j}^{(l_1)}\right] &= \frac{1}{2}\int\left[\vec{r}\times\mathrm{curl}\vec{M}^{(l_1)}\right]\\ &= \frac{1}{2}\oint\left[\vec{r}\times\left[d\vec{s}\times\vec{M}^{(l_1)}\right]\right]\\ &- \frac{1}{2}\int\left[\left[\vec{M}^{(l_1)}\times\mathrm{curl}\right]\times\vec{r}\right]dv = \int\vec{M}^{(l_1)}dv\end{aligned}$$

A part of current lines may undoubtely come out from the cell. Nevertheless, as $\langle\delta\vec{j}^{(l_1)}\rangle = 0$, each current line going out of the cell has a conjugate one oppositely directed. As the contribution of the lines into the surface integral is small and random for each cell, we shall neglect it.

[4]To avoid this non-uniqueness some authors refuse consideration of permeability [18] at all. Since the effective permeability is an effect of weak spatial dispersion they introduce only a permittivity tensor of the form (see [18, 19]):

$$\varepsilon_{ij}(\omega,\vec{k}) = \varepsilon^{tr}(\omega)\left(\delta_{ij} - \frac{k_ik_j}{k^2}\right) + \varepsilon^{l}(\omega)\frac{k_ik_j}{k^2} \tag{5}$$

In the recent review [17], avoiding the problem of the non-uniqueness, the authors showed that the use of permittivity and permeability or confining to permittivity tensor is a matter of taste because the latter case gives incomplete descriptions and needs the introduction of additional surface currents flowing on each surface. These surface currents are expressed through magnetization of the material. Hence consideration of bounded systems needs introduction of permeability and brings us back to the problem of separation of the total current into two parts (4).

[5]Further on, we shall consider "symmetrical" inclusions neglecting the first order effects in (ka), such as chirality (for microscopic deduction of the chirality factor see [16, 20]).

It is worth emphasizing that $\vec{M}^{(l_1)}$ is independent of the origin position. Introducing $\vec{x} = \vec{r} - \langle\vec{r}\rangle$, where $\langle\vec{r}\rangle$ is the position of the cell center relative to the origin we can rewrite the expression for the magnetic moment density $\vec{M}^{(l_1)}$ associated with the fluctuating on scale l_1 part of current as:

$$\begin{aligned}
M_i^{(l_1)} &= \frac{e_{ikl}}{2cV_{cell}}\int_{V_{cell}} r_k \delta j_l d^3r = \\
&= \frac{e_{ikl}}{2cV_{cell}}\int_{V_{cell}} (x_k + \langle r_k\rangle)(j_l - \langle j_l\rangle) d^3r = \\
&= \frac{e_{ikl}}{2cV_{cell}}\int_{V_{cell}} x_k j_l d^3r + \frac{e_{ikl}}{2cV_{cell}}\int_{V_{cell}} \langle r_k\rangle j_l d^3r - \\
&- \frac{e_{ikl}}{2cV_{cell}}\int_{V_{cell}} x_k \langle j_l\rangle d^3r - \frac{e_{ikl}}{2cV_{cell}}\int_{V_{cell}} \langle r_k\rangle\langle j_l\rangle d^3r = \\
&= \frac{e_{ikl}}{2cV_{cell}}\int_{V_{cell}} x_k j_l d^3r \qquad (6)
\end{aligned}$$

So defined, magnetic moment is equal to the "Landau" moment minus the moment associated with the whole cell considered as an infrangible, elementary volume. Indeed,

$$\begin{aligned}
M_i^{(\mathrm{Landau})} &= \frac{e_{ikl}}{2cV_{cell}}\int_{V_{cell}} r_k j_l d^3r = \\
&= \frac{e_{ikl}}{2cV_{cell}}\int_{V_{cell}} x_k \delta j_l d^3r + \frac{e_{ikl}}{2cV_{cell}}\int_{V_{cell}} \langle r_k\rangle \delta j_l d^3r - \\
&+ \frac{e_{ikl}}{2cV_{cell}}\int_{V_{cell}} x_k \langle j_l\rangle d^3r + \frac{e_{ikl}}{2cV_{cell}}\int_{V_{cell}} \langle r_k\rangle\langle j_l\rangle d^3r = \\
&= M_i^{l_1} + \frac{e_{ikl}}{2c}\langle r_k\rangle\langle j_l\rangle \qquad (7)
\end{aligned}$$

Since the fluctuating part $\delta\vec{j}^{(l_1)}$ of the current relates to the magnetic density in the standard way: $\delta\vec{j_l}^{(l_1)} = c\,\mathrm{curl}\vec{M}^{(l_1)}$, the current inside the cell is presented as the sum:

$$\vec{j} = \langle\vec{j}\rangle^{(l_1)} + c\,\mathrm{curl}\vec{M}^{(l_1)} \qquad (8)$$

The $\langle\vec{j}\rangle^{(l_1)}$ does not vary inside each cell but it fluctuates on scales $l_2 > l_1$. To take into account this possibility we proceed in spirit of renormalization group theory [21]. We divide the space into cells of size $l_2 = nl_1$ $(n > 1)$ and present the current $\langle\vec{j}\rangle^{(l_1)}$ as the sum of an average part $\left\langle\langle\vec{j}\rangle^{(l_1)}\right\rangle^{(l_2)}$ and a fluctuating one $\delta(\langle\vec{j}\rangle^{(l_1)})^{(l_2)})$. Then the current in l_2-cell can be rewritten as

$$\begin{aligned}
\vec{j} &= \langle\vec{j}\rangle^{l_1} + c\,\mathrm{curl}\vec{M}^{(l_1)} = \\
&= \left\langle\langle\vec{j}\rangle^{(l_1)})\right\rangle^{(l_2)}) + \delta(\langle\vec{j}\rangle^{(l_1)})^{(l_2)}) + c\,\mathrm{curl}\,\vec{M}^{(l_1)}
\end{aligned}$$

The new fluctuating part can be expressed in terms of magnetic density of the scale l_2:

$$M_i^{(l_2)} = \frac{1}{2cV_{cell}} e_{ikl} \int_{V_{l_2-cell}} r_k \delta j_l^{(l_2)} d^3r \tag{9}$$

After these steps we again come back to expression (4) with

$$\vec{M} = \vec{M}^{(l_1)} + \vec{M}^{(l_2)} \tag{10}$$

Repeating the previous procedure we eventually arrive at a scale L_* where the current is uniform (they change only on scale λ which is assumed to be substantially larger than L_*). It is L_* which plays the role of "physically small" length. On averaging over the scale L_* we arrive at (4).

Substituting the expression for the average current into Lorentz equation we obtain homogenized macroscopic Maxwell equation of the form

$$\begin{aligned} \operatorname{curl} \vec{H} &= \frac{4\pi}{c} \langle \vec{j} \rangle^{(L_*)} + \frac{1}{c} \frac{\partial \langle \vec{E} \rangle^{(L_*)}}{\partial t} = \\ &= \left(\frac{4\pi}{c} \sigma_{\text{eff}} - \frac{i\omega(\varepsilon'_{\text{eff}})}{c} \right) \langle \vec{E} \rangle^{(L_*)} = -\frac{i\omega\varepsilon_{\text{eff}}}{c} \langle \vec{E} \rangle^{(L_*)} \end{aligned}$$

where $\langle \vec{j} \rangle^{(L_*)} = \langle \ldots \langle \vec{j} \rangle^{l_1} \ldots \rangle^{(L_*)} = \varepsilon_{\text{eff}} \langle \vec{E} \rangle^{(L_*)}$ and $\vec{H} = \vec{B} - 4\pi \sum_i \vec{M}^{(l_i)}$. That completes our averaging procedure for alternating fields. Thus, we have seen how to introduce permeability for composite media for $\lambda_{\text{mat}} \gg L$.

3. Non-local Constitutive Relation (Strong Spatial Dispersion)

If $a < \lambda_{\text{int}} < \lambda_0$ the problem of homogenization is still reasonable but we have to homogenize the Maxwell equations instead of the Laplace one. The procedure should results in a non-local theory. The non-locality may be displayed as a spatial dispersion of effective permittivity [24], if we consider bulk properties, or as Higher Order Impedance Boundary Condition [25, 26], if we consider eigenvalue problem.

Of all possible effects of non-locality, we confine ourselves to the one which seems to be the most important in effect, namely, the effect of the second channel of energy transport. This possibility becomes evident if we consider bulk properties of an infinitely extended system. In isotropic case, the non-local constitutive relation for harmonically varying in time fields (see [31]) is of the form:

$$\vec{D}(\vec{r}, \omega) = \int \epsilon(\vec{r}, \vec{r}', \omega) \vec{E}(\vec{r}', \omega) d^3r' \tag{11}$$

Applying spatial Fourier transformation we arrive at Fourier transformation of (11) which looks like the usual one but with the permittivity depending upon the wave number. This is the reason why the phenomenon is called spatial dispersion. The corresponding dispersion equation of eigenwaves is

$$k^2 = \varepsilon(\omega, k)\frac{\omega^2}{c^2} \tag{12}$$

The dispersion equation may have more than one solution. It means that at a given frequency there exist more than one wave traveling in a given direction. This in turn causes a problem of additional boundary conditions. Indeed, as an incident wave of any polarization induces inside the material more than one refracted wave we feel a shortage of boundary conditions.

Let us consider physical sense of additional refracted waves. For this aim we reconsider the well-known model of regular inhomogeneous system ("zebra"-medium) that is a slab with periodically modulated permittivity. To simplify the mathematics we shall use two different models: in the "zebra-1" medium the permittivity is assumed to vary continuously: $\varepsilon = \varepsilon_0 + \varepsilon_1 \cos(k_\varepsilon x)$ (see [41, 42]). The "zebra-2" medium is a stratified system which is a sandwich (layers of thickness h_1 and h_2 with permittivity ε_1 and ε_2) periodically reproduced.

Consideration of bounded "zebra-1" medium will help us to understand the structure of additional waves. Plane eigenwaves in such a medium can propagate along the layers and perpendicular to them. For electrically y-polarized wave, traveling along the z axis, the Maxwell equations reduce to one equation for the y-component of the electric field:

$$\nabla^2 E + k_0^2 \varepsilon(x) E = 0 \tag{13}$$

The solution to this equation can be found (see [41, 42]) in terms of separable functions of x and z: $X(x)Z(z)$. This yields $Z(z) = A\exp(\alpha z) + A'\exp(-\alpha z)$ and the Mathieu equation for $X(x)$. The latter can be reduced to an eigenvalue problem (see [41, 42]) with eigenvalues equal to the wave numbers of eigenwaves:

$$D_{m,l}\beta_l = \alpha^2 \beta_m \tag{14}$$

with

$$\begin{aligned} D_{l,l} &= -\varepsilon_0 k_0^2 + k_\varepsilon^2 l^2, \\ D_{l,l+1} = D_{l,l-1} &= -\varepsilon k_\varepsilon^2/2 \\ D_{m,l} &= 0 \quad \text{otherwise} \end{aligned} \tag{15}$$

The eigenvector β_l presents the structure of the corresponding eigenwave. Each eigenwave consists of a set of oblique waves with lateral wave vector proportional to k_ε (see [41, 42]). Since we consider the situation $h < \lambda_0$ or $k_0 < k_\varepsilon$ all these oblique waves totally reflect on the surface.

Outside the slab, the solution includes incident, reflected, and a set of evanescent waves:

$$E(x,z) = \begin{cases} e^{(-ik_0 z)} + \sum_{l=-\infty}^{\infty} E_l e^{\left(ik_\varepsilon lx + iz\sqrt{k_0^2-(k_\varepsilon l)^2}\right)} & z > 0 \\ \sum_{l=-\infty}^{\infty} E_l' e^{(ik_\varepsilon lx + iz\sqrt{k_0^2-(k_\varepsilon l)^2})} & z < -d \end{cases} \quad (16)$$

Thus, we can see that additional waves not only propagate through the system but also exist as evanescent waves outside. This results in a non-local relation outside inhomogeneous medium (see also [27]).

Problem (15) has an infinite set of solutions which, in principal, should correspond to some non-local, "homogenized" permittivity operator. Reconstruction of the corresponding kernel is a difficult task and it is possible in rare cases (for example, for infinitely expanded "zebra-2"-system). For this medium, the dispersion equation has the following form (see [7]):

$$\sqrt{k_0^2\varepsilon_2 - k^2}\tan\left(h_2\sqrt{k_0^2\varepsilon_2 - k^2}\right) + \\ + \sqrt{k_0^2\varepsilon_1 - k^2}\tan\left(h_1\sqrt{k_0^2\varepsilon_1 - k^2}\right) = 0 \quad (17)$$

Comparing (17) and (12) we can guess the equation for $\varepsilon(k)$:

$$\sqrt{k_0^2\varepsilon_2 - k^2}\tan\left(h_2\sqrt{k_0^2\varepsilon_2 - k^2}\right) + \quad (18)$$

$$+ \sqrt{k_0^2\varepsilon_1 - k^2}\tan\left(h_1\sqrt{k_0^2\varepsilon_1 - k^2}\right) = (k_0^2\varepsilon - k^2)(h_1 + h_2) \quad (19)$$

that yields

$$\varepsilon(k) = \frac{k^2}{k_0^2} + \frac{\sqrt{k_0^2\varepsilon_2 - k^2}\tan\left(h_2\sqrt{k_0^2\varepsilon_2 - k^2}\right)}{(h_1 + h_2)k_0^2} + \\ + \frac{\sqrt{k_0^2\varepsilon_1 - k^2}\tan\left(h_1\sqrt{k_0^2\varepsilon_1 - k^2}\right)}{(h_1 + h_2)k_0^2}$$

In the long wave limit, we arrive at the well-known [7, 22] expression:

$$\varepsilon(k) \to \frac{h_2\varepsilon_2 + h_1\varepsilon_1}{(h_1 + h_2)} = \langle\varepsilon\rangle \text{ as } k \to \infty \quad (20)$$

We have seen that several independent waves may propagate in inhomogeneous media at a fixed frequency. Since obtaining an explicit form of the kernel for bounded system is a difficult mathematical problem just for infinite system, below we consider a model kernel that yields two eigenwaves. We assume that propagation of waves through our hypothetical medium with spatial dispersion (MWSD) is governed by the following generalization of the non-local wave equation [29-32]:

$$\nabla \times \left[\nabla \times \vec{E}(\vec{r},\omega)\right] - k_0^2\varepsilon_0\vec{E}(\vec{r},\omega) = \frac{k_0^2}{4\pi}\int_{V_{MWSD}} d_3r' G(\vec{r},\vec{r}')\vec{E}(\vec{r}',\omega) \quad (21)$$

where we put the integral over a domain with the MWSD. The function $\varepsilon(z)$ is equal to the local value of permittivity for usual media and it is constant inside our MWSD. In accord with (15)–(16) a non-local interaction would be expanded in the vicinity range of an inhomogeneous medium sample (see also [27]). We consider equation (21) to be valid both inside and outside the MWSD. This approach differs from the universally adopted one where the non-local form of Maxwell equations is used for domain enclosing the MWSD [28, 29, 31]. Such a generalization permits us to describe the additional waves (16) not only inside the MWSD but also outside it. The kernel of the permittivity operator should express the propagation of a waves from any point up to the point of observation, where these waves are summarized. Going back to the aforementioned problem of "zebra"-media slab located in stratified medium (one of the layer is mentioned as a perfectly conducting screen) with all the layer perpendicular to the z axis, we propose the kernel for the permittivity operator in the form of Green function to the following problem (see also [28, 29]):

$$\left(\partial_z^2 + p(z)^2\right) G(z,z') = \delta(z-z') \quad (22)$$

where $p(z)$ is a piecewise function. Inside j-th layer $p(z) = p_j = \sqrt{k_0^2\varepsilon_j - k_*^2}$, ε_j is a permittivity of the material in the j-th layer, k_* is a parameter having a sense of transverse component of the wavevector, with respect to the z direction. As we have seen the value of k_* should be considered as a constant independent of surrounding media.[6] This form of the kernel warrants an existence of two waves (see [28, 29]). The choice of p_j in such a form reflects the fact that the additional waves may be damped out or propagate depending on optical density of the material. They will damp or propagate depending on the relation between the local wavenumber and the value of h. For optically low-dense materials ($\varepsilon_j' < k_*^2/k_0^2$) side lobes exhibit

[6]For "zebra-1" medium k_* should be equal to k_ε.

total internal reflection and the waves outside the material are damped out (p_j is a purely imaginary quantity). For $\varepsilon_j' > k_*^2/k_0^2$ the waves can propagate through the material without attenuation.

Confining ourselves to the system of three layers (the central layer is considered to be made of the MWSD) placed on metal screen at $z = 0$ we put the following boundary conditions on the function $G(z)$: it should be continuous with its first derivative over the half space $z > 0$ except $z = z'$. Specifically, on the boundaries $z = a$ and $z = b$, G should be continuous[7]:

$$G(a+0,z') = G(a-0,z') \qquad G(b+0,z') = G(b-0,z') \tag{23}$$
$$\frac{dG(a+0,z')}{dz} = \frac{dG(a-0,z')}{dz} \qquad \frac{dG(b+0,z')}{dz} = \frac{dG(b-0,z')}{dz}$$

On the metal surface G vanishes[8]

$$G(0,z') = 0 \tag{24}$$

The solution for (21-23) has the form:

$$G(z,z') = \begin{cases} \frac{1}{2p_0}\begin{pmatrix} e^{-ip_0|z-z'|} & -e^{-ip_0(z-z')} + \\ & +\left(Ue^{ip_1 z'} - Ve^{ip_1 z'}\right)e^{ip_0 x} \end{pmatrix} & z > b \\ \frac{1}{2ip_1}\begin{pmatrix} e^{ip_1|z-z'|} & +\left(Me^{ip_1 z'} - Ne^{ip_1 z'}\right)e^{ip_1 z} + \\ & +\left(\tilde{M}e^{ip_1 z'} - \tilde{N}e^{ip_1 z'}\right)e^{-ip_1 z} \end{pmatrix} & a < z < b \\ \frac{1}{2ip_2}\begin{pmatrix} e^{ip_2|z-z'|} - e^{ip_2(z-z')} + \\ +\left(Qe^{ip_2 z'} - Pe^{ip_2 z'}\right)\left(e^{ip_2 z} - e^{-ip_2 z}\right) \end{pmatrix} & 0 < z < a \end{cases} \tag{25}$$

Using the technique of solution to the equation developed by Maradudin and Mills [28] we can obtain both the solution and additional boundary conditions [32, 33].

Considering a multi-layer system we can experimentally observe the existence of ARW (additional refracted waves). Indeed, the ARW produce additional resonance minimum of reflection coefficient since they have their own value of wavenumber corresponding to new value of permittivity. The minima were observed both in computer simulation [32, 33, 37] and in experiment [32] (see Figure 1).

The spatial resonance of the ARW causes significant deviation from Fresnel's formulae manifesting in a deep minimum of the reflection coefficient. The effect might be referred to as "hidden Wood effect". Anomalies in

[7] If Eq. 22 is written for magnetic field H, we should use the condition $(\mu_+/k_+)dG(z_{\text{interface}}+0,z')/dz = (\mu_-/k_-)dG(z_{\text{interface}}-0,z')/dz$ instead of continuity of derivative.

[8] Certainly, this condition should be changed to $\frac{\partial}{\partial z}G(0,z') = 0$ if we consider equation for magnetic field.

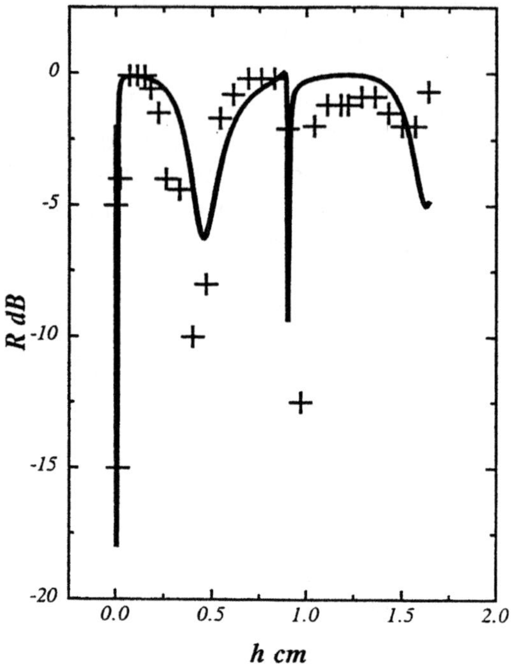

Figure 1. The dependence of the reflection coefficient from the "zebra-1" medium sample backed by metal screen against the distance d between the sample and screen.

reflection, which are well known as Wood's anomalies [34, 35, 36, 38, 39, 40], appear in high spectrum orders if $a < \lambda_0$. These anomalies are due to interaction of high spectrum orders with surface waves [34, 35, 36]. We think that in our case the anomalies in reflection are determined by interaction of internal waves with surface waves (hidden Wood effect) and have the same nature as anomalies observed in [37] in computer simulation.

We have seen that in the general case, inhomogeneity causes non-locality. If the scale of inhomogeneity is smaller than the free space wavelength and larger than the local wavelength, internal waves may be generated. The exterior problem can still be considered in terms of a macroscopically uniform medium.

We can imagine the following picture of the phenomenon. If $\lambda_{int} < a$ just inside the matrix material, strong scattering of an electromagnetic wave inside the host material by inclusions must produce side lobes. Due to total reflection these side lobes confine to the volume occupied by the composite material. The sum of these side lobes constitutes an additional refracted wave (ARW) which transfers into evanescent wave outside the MWSD. Thus, the rigorous description of the system as a uniform one demands introducing additional refracted waves not only inside the MWSD but also outside it, and a usual medium slab obtains features of spatial dispersive medium when located nearby a spatially dispersive material. Indeed, we can not still choose proper a scale for averaging in order to obtain local constitutive equations. The averaging region outside the system may include significant part of the system and, vice versa, the averaging region inside the planar system will include a part of surrounding medium. Although this fact is known for the optical literature [27], it does not play any significant role in optics, but it can not be ignored in microwave range.

Acknowledgement The authors wish to express their gratitude to Professor S.A. Tretyakov for useful discussion.

References

1. Landauer, R. (1978) Electrical conductivity in inhomogeneous media, in J.C. Garland and D.B. Tanner (Eds.), *Proc. of AIP Conference,* **40**, New York, American Institute of Physics, 2-45.
2. Bergman, D. J., Stroud, D. (1992) Physical properties of macroscopically inhomogeneous media, *Solid. St. Phys.*, **46**, 148-269.
3. Lagarkov, A. N., Sarychev, A. K., Smychkovich, Yu. R., and Vinogradov, A. P. (1992) Effective medium theory for microwave dielectric constant and magnetic permeability of conducting stick composites, *J. Electro. Waves Applic.* **6**, 1159-1176.
4. Sanchez-Palencia, E. (1980) *Non-Homogeneous Media and Vibration Theory*, Springer Verlag, New York.
5. Bakhvalov, N. S., Panasenko G. P. (1989) Homogenization. Averaging processes in periodic media, in *Mathematical Problems in the Mechanics of Composite Materials*, Kluwer Academic Publisher, Dordrecht.
6. Velikhov, E. P., Dykhne, A. M., (1963) in *Proc. VI Int. Symp. Ion. Phen. in Gases*, Paris, 511.
7. Brekhovskikh, I. M. (1960) *Waves in Layered Media* Academic Press, New York.
8. Vinogradov, A.P., Panina, L. V., Sarychev, A.K. (1989) Method for calculating the dielectric constant and magnetic permeability in percolating systems, *Sov. Phys. Dokl.* **34**, 530-532.
9. Schelkunoff, S. A. and Friis, H. T. (1952) *Antenna Theory and Practice*, Wiley, New York.
10. Lewin, L (1975) *Theory of Waveguides.*, Newnes-Butterworths, London.
11. Jaggard, D. L., Mickelson, A. R., and Papas, C. H. (1979) On electromagnetic waves in chiral media, *Appl. Phys.* **18**, 211-216.
12. Lakhtakia, A., Varadan, V. K., Varadan, V.V. (1989) *Time-harmonic Electromagnetic Fields in Chiral Media*, Lecture Notes in Physics **335**, Springer, New York.
13. Lindell, I. V., Sihvola, A.H., Tretyakov, S.A., Viitanen A.J. (1994) *Electromagnetic Waves in Chiral and Bi-Isotropic Media*, Artech House, Boston.
14. Kukolev, I. V., Lagarkov, A.N., Matitsin, S. M., Rosanov, K. N., and Vinogradov A. P. (1992) Investigation of effective magnetic permeability of inhomogeneous dielectric materials at microwaves, in *Proc. of the Material Research Society Spring Meeting*, San Francisco No. L6.14.
15. Landau, L.D., Lifshitz, E.M. (1982) *Electrodynamics of Continuous Media*, Pergamon Press, New York.
16. Born, M. (1933) *Optik*, Springer, Berlin (in German).
17. Golubkov, A.A., Makarov, V.A. (1995) Boundary conditions for electromagnetic field on the surface of media with weak spatial dispersion, *Uspekhi Fizicheskikh*

Nauk **165**, 339-346 (in Russian)
18. Silin, V.P., Rukhadze, A.A. (1961) *Electromagnetic Properties of Plasma and Plasma-like Media*, GosAtomIzdat, Moscow (in Russian).
19. Ryazanov, I.M. (1987) *Electrodynamics of Condensed Matter*, (in Russian).
20. Vinogradov, A.P. (1993) Microscopic properties of a chiral object, in Ari Sihvola, Sergei Tretyakov, Igor Semchenko (Eds.) *Proc. Int. Seminar on Electrodynamics of Chiral and Bianisotropic Media Bianisotropics'93*, Helsinki University of Technology, Electromagnetics Laboratory Report 159.
21. Stanley, H.E. (1971) *Introduction to Phase Transitions and Critical Phenomena*, Clarendon Press, Oxford.
22. Rytov, S.M. (1955) *J. Experimental and Theoretical Physics (JETF)*, **29**, 5 (in Russian)
23. Vinogradov, A.P., Romanenko, V.E. (1995) Artificial magnetics based on racemic helix inclusions, in A. Sihvola et al. (Eds.), *Proc. of 4th Intl. Conf. on Chiral, Bi-isotropic and Bi-anisotropic Media, CHIRAL'95*, The Pennsylvania State University, State College, pp. 143-148.
24. Ryghov, Yu.A., Tamoikin, V.V. (1970) Radiation and transmitance of electromagnetic waves in randomly inhomogeneous media, *Radifizika*, **13**, 356 (in Russian).
25. Hoppe, D.J., Rahmat-Samii, Y. (1995) *Impedance Boundary Conditions in Electromagnetics*, Taylor & Francis, Washington.
26. Senior, T. B. A., Volakis, J.L. (1995) *Approximate Boundary Conditions in Electromagnetics*, The IEE Press, London.
27. Agranovich, V.M., Yudson, V.I. (1973) Boundary conditions in media with spatial dispersion, *Opt. Com.* **7**, 121-124.
28. Maradudin, A.A., Mills, D.L. (1973) Effect of spatial dispersion on the properties of a semi-infinite dielectric, *Phys. Rev. B* **7**, 2787-2810.
29. Agarwal, G.S., Pattanayak, D.N., Wolf, E. (1974) *Phys. Rev. B* **8**, 1447-1463.
30. Birman, J.L. (1982) Chapter 2, in J.L. Birman (Ed.) *Excitons*, North Holland Co.
31. Agranovich, V.M., Ginzburg, V.L. (1966) *Spatial Dispersion in Crystal Optics and the Theory of Excitons*, Interscience, New York.
32. Busarov, I.G., Lagarkov, A.N., Sterlina, I.G., Vinogradov, A.P. (1992) Interaction of electromagnetic waves with heterogeneous systems. Non-Fresnels type of reflection, in *Proc. of the International Conference STATPHYS 18*, Berlin, pp. 269-280.
33. Vinogradov A.P., Lagarkov, A.N., Romanenko, V.E. (1996) Effects of spatial dispersion in composite materials at mw range, *Doclady Acdemii Nauk*, **349**, 182-184 (in Russian).
34. Stewart, J.E., Gallaway, W.S. (1962) Diffraction anomalies in grating spectrophotometers *Applied Optics* **1**, 421-427.
35. Hessel, A., Oliner, A.A. (1965) A new theory of Wood's anomalies on optical gratings, *Applied Optics* **4**, 1275-1281.
36. Palmer, C.H., Evering, F.C., and Nelson, F.M. (1965) Diffraction anomalies for gratings of rectengular profile, *Appl. Opt.* **4**, 1271-1278.
37. Botten, L.C., Craig, M.S., McPhedran, R.C., Adams, J. L., and Andrewartha, J. R. (1981) The finitely conducting lamellar diffraction grating, *Optica Acta* **28**, 1087-1101.
38. Botten, L.C., Craig, M.S., McPhedran, R. C. (1981) Highly conducting lamellar diffraction gratings, *Optica Acta* **8**, 1103-1109.
39. McPhedran, R.C., Waterworth, M.D. (1972) A theoretical demonstration of properties of grating anomalies (S-polarization), *Optica Acta* **19**, 877-891.
40. McPhedran, R.C., Botten, L.C., Craig, M.S., Neviere, M., and Maystre, D., (1982) Lossy lamellar gratings in the quasistatic limit, *Optica Acta* **29**, 289-294.
41. Burckhardt, C.B. (1966) Diffraction of a plane wave at a sinusoidally stratified dielectric grating, *J. Opt. Soc. of Am.* **56**, 1502-1511.
42. Kaspar, F.G. (1973) Diffraction by thick, periodically stratified grating with complex dielectric constant, *J. Opt. Soc. of Am.* **63**, 37-45.

EFFECTIVE-MEDIUM THEORIES FOR BI-ISOTROPIC MIXTURES

Six-vectors, Material Homogenization, and Racemization of Chiral Materials

A.H. SIHVOLA
Helsinki University of Technology, Electromagnetics Laboratory
Otakaari 5 A, FIN–02150 Espoo, Finland
and
Lund Institute of Technology, Electromagnetic Theory
Box 118, S–22100 Lund, Sweden

1. Introduction

It is perhaps impossible to count the number of classes into which electromagnetic materials can be grouped. Media can be insulating, semiconducting, superconducting, non-linear, reciprocal, chiral, non-homogeneous, dispersive, anisotropic, ferromagnetic, gyrotropic, etc. Within the chiral community of researchers, one quite often talks about complex or complicated materials, by which one may mean, as a matter of fact, any deviation from the isotropic behaviour.

In the following, the discussion is limited to linear bi-anisotropic non-homogeneous materials. The objective is to look for ways to solve the homogenization problem, in other words to present mixing formulas for heterogeneous non-isotropic materials to calculate the effective material parameters of the mixture. In the field of purely dielectric mixtures, an enormous literature exists on various mixing rules, starting from the venerable recipe by Maxwell Garnett from the year 1904. But not so for bi-isotropic (magnetoelectric but isotropic) and bi-anisotropic (magnetoelectric and anisotropic) materials. The message of the present article (that has also been preached elsewhere, see [1]) is that this shortcoming can be very elegantly remedied with the use of a six-vector formalism for bi-anisotropic materials. With six-vectors and six-dyadics, many of the existing dielectric mixing formulas find their natural generalization for mixtures composed of non-isotropic material components. Toward the end of the article, an example of the six-vector mechanism is presented where the optically active properties of a mixture of two otherwise identical but stereochemi-

A. Priou et al. (eds.), Advances in Complex Electromagnetic Materials, 131–144.

cally enantiometric components (the components are mirror images of each other) are calculated.

2. Dielectric Mixing Formulas

Let us look first at a two-component dielectric mixture. Be one of the phases the background material into which the other one is dispersed in the form of spheres. What is the macroscopic permittivity of this mixture on scales larger than the sphere radius? The family of mixing formulas according to [2]:

$$\frac{\epsilon_{\mathrm{eff}} - \epsilon_0}{\epsilon_{\mathrm{eff}} + 2\epsilon_0 + \nu(\epsilon_{\mathrm{eff}} - \epsilon_0)} = f \frac{\epsilon - \epsilon_0}{\epsilon + 2\epsilon_0 + \nu(\epsilon_{\mathrm{eff}} - \epsilon_0)} \tag{1}$$

predicts the effective permittivity ϵ_{eff} of this mixture where the inclusion permittivity is ϵ and that of the background is ϵ_0 (although often used in this vacuum–background form, this formula goes well to apply for background permittivities differing from the free-space permittivity). The volume fraction of the inclusions is f. This expression contains a number of important mixing rules that can be picked up with the dimensionless parameter ν.

Maxwell Garnett formula. For example, the choice $\nu = 0$ reproduces the very-well-known Maxwell Garnett formula [3]:

$$\epsilon_{\mathrm{eff}} = \epsilon_0 + 3f\epsilon_0 \frac{\epsilon - \epsilon_0}{\epsilon + 2\epsilon_0 - f(\epsilon - \epsilon_0)} \tag{2}$$

Sometimes this formula is termed as the ATA (average T-matrix approximation). But other integer values for ν in Equation (1) give other well-known mixing rules.

Coherent potential formula. The value $\nu = 3$ gives the so-called "Coherent Potential formula" [4] (CP). Quite often this formula is used in solid-state physics where it is sometimes known as the GKM (Gyorffy, Korringa, and Mills) rule. And sometimes the attribute "quasicrystalline approximation" (QCA) is added to coherent potential [5].

Bruggeman symmetric formula. If, in Equation (1), we choose $\nu = 2$, we are left with a formula that goes under various names. Sometimes it is called Böttcher mixing rule [6], sometimes — especially in remote sensing studies — as the Polder and Van Santen formula [7]. Often one sees it with the plain label EMT rule (effective medium theory) but perhaps even more often it is called the "Bruggeman formula," although this label is somewhat ambigious because Bruggeman in his large article [8] presented several mixing rules (one of which comes close to the rule of Equation (1) with the choice $\nu = 1$). Perhaps it may be more precise to call this $\nu = 2$ - formula with the name "Bruggeman symmetric rule," [9] which name would carry the indication of the fact that this mixing rule indeed treats both the background and the inclusion phases on a symmetrical basis. This property of the $\nu = 2$ -formula will be demonstrated by a numerical example later in the present article.

3. Bi-anisotropic Materials

3.1. CONSTITUTIVE RELATIONS

What, then, if the materials to be mixed have a more complicated electromagnetic response than purely dielectric and isotropic? To answer this question we need to describe the polarizability behavior of the materials in terms of constitutive equations. Linear bi-anisotropic media obey the following constitutive relations [10]

$$\mathbf{D} = \overline{\overline{\epsilon}} \cdot \mathbf{E} + \overline{\overline{\xi}} \cdot \mathbf{H} \tag{3}$$

$$\mathbf{B} = \overline{\overline{\zeta}} \cdot \mathbf{E} + \overline{\overline{\mu}} \cdot \mathbf{H} \tag{4}$$

Here the relation between the electric and magnetic fields ($\mathbf{E}$ and $\mathbf{H}$) and the electric and magnetic flux densities ($\mathbf{D}$ and $\mathbf{B}$) is contained in the permittivity $\overline{\overline{\epsilon}} = \epsilon_0 \overline{\overline{\epsilon}}_r$ and permeability $\overline{\overline{\mu}} = \mu_0 \overline{\overline{\mu}}_r$, and the magnetoelectric crosscoupling in $\overline{\overline{\xi}}$ and $\overline{\overline{\zeta}}$. The dyadic nature (double bar above a symbol) of these material quantities emphasises the anisotropy of the material.

3.2. SUBCLASSES OF BI-ANISOTROPIC MATERIALS

A useful and physically meaningful distinction in the magnetoelectric parameters is the separation of the dyadics $\overline{\overline{\xi}}$ and $\overline{\overline{\zeta}}$ into a reciprocal part $\overline{\overline{\kappa}}$ and a non-reciprocal part $\overline{\overline{\chi}}$ [11]:

$$\overline{\overline{\xi}} = \left(\overline{\overline{\chi}}^T - \mathrm{j}\overline{\overline{\kappa}}^T \right) \sqrt{\mu_0 \epsilon_0} \tag{5}$$

$$\overline{\overline{\zeta}} = \left(\overline{\overline{\chi}} + \mathrm{j}\overline{\overline{\kappa}} \right) \sqrt{\mu_0 \epsilon_0} \tag{6}$$

where the free-space parameters μ_0, ϵ_0 are included for nondimensionalization of the new dyadics $\overline{\overline{\chi}}$ and $\overline{\overline{\kappa}}$. The coefficient j, indicating time-harmonic notation $\exp(\mathrm{j}\omega t)$, is there to guarantee real values for the parameters in the case of lossless media. Losses make the parameters complex. The superscript T denotes the transpose operation on a dyadic.

The reciprocity condition [12]

$$\overline{\overline{\xi}} = -\overline{\overline{\zeta}}^T \tag{7}$$

is in harmony with the magnetoelectric splitting (5)–(6), and so also the losslessness (absence of dissipation) condition [13]

$$\overline{\overline{\xi}} = \overline{\overline{\zeta}}^\dagger \tag{8}$$

where † denotes conjugate transpose.

The bi-isotropic material is an important special case of bi-anisotropic materials [14]. The description of bi-isotropic media requires four material parameters: all dyadics are (due to isotropy) multiples of the unit dyadic: $\overline{\overline{\epsilon}} = \epsilon\overline{\overline{I}}$, $\overline{\overline{\mu}} = \mu\overline{\overline{I}}$, $\overline{\overline{\xi}} = (\chi - j\kappa)\sqrt{\mu_0\epsilon_0}\,\overline{\overline{I}}$, and $\overline{\overline{\zeta}} = (\chi + j\kappa)\sqrt{\mu_0\epsilon_0}\,\overline{\overline{I}}$. Here χ and κ are the dimensionless non-reciprocity and chirality parameters, respectively.

And finally, important special cases of the bi-isotropic medium are

- *Pasteur medium* Reciprocal chiral medium. $\kappa \neq 0$ and $\chi = 0$.
- *Tellegen medium* Non-reciprocal non-chiral medium. $\kappa = 0$ and $\chi \neq 0$.

3.3. SIX-VECTOR FORMALISM

Six-vector approach has been developed to account for bi-anisotropic effects in electromagnetics problems [15]. Six-vectors combine electric and magnetic quantities (that both have three components) into a single vector with six components and treat those as one unit. The electromagnetic six-vector field e and six-vector flux density d look like

$$\mathsf{e} = \begin{pmatrix} \mathbf{E} \\ \eta\mathbf{H} \end{pmatrix}, \qquad \mathsf{d} = \begin{pmatrix} c\eta\mathbf{D} \\ c\mathbf{B} \end{pmatrix} \tag{9}$$

where use is made of the idea by Rikte [16] to balance dimensionally the electric and magnetic field strenghts and flux densities. In the normalization, $c = 1/\sqrt{\mu_0\epsilon_0}$ and $\eta = \sqrt{\mu_0/\epsilon_0}$. Now the constitutive relations (3)–(4) can be written as a single equation:

$$\mathsf{d} = \mathsf{M} \cdot \mathsf{e} \tag{10}$$

where

$$\mathsf{M} = \begin{pmatrix} \overline{\overline{\epsilon}}_r & c\overline{\overline{\xi}} \\ c\overline{\overline{\zeta}} & \overline{\overline{\mu}}_r \end{pmatrix} \tag{11}$$

is the six-dyadic of the material parameters. It has a 6×6-element matrix representation and therefore the full medium description requires 36 parameters.

See also the works [17, 18, 19] where six-vector algebra has been very recently applied to electromagnetic wave reflection problems and bi-anisotropic superlattice analyses.

4. Bi-anisotropic Mixtures

Using six-vectors and six-dyadics, the dielectric mixing rule (1) can be generalized into mixtures with non-isotropic inclusions. First, define the effective material six-matrix as

$$\mathsf{M}_{\text{eff}} = \begin{pmatrix} \overline{\overline{\epsilon}}_{\text{eff,r}} & c\overline{\overline{\xi}}_{\text{eff}} \\ c\overline{\overline{\zeta}}_{\text{eff}} & \overline{\overline{\mu}}_{\text{eff,r}} \end{pmatrix} \tag{12}$$

In Reference [1], the conjecture is made that the effective material matrix obeys the direct generalization of (1) in the following way

$$\begin{aligned}(\mathsf{M}_{\text{eff}} - \mathsf{M}_0) \cdot [\mathsf{M}_{\text{eff}} + 2\mathsf{M}_0 + \nu(\mathsf{M}_{\text{eff}} - \mathsf{M}_0)]^{-1} \\ = f(\mathsf{M} - \mathsf{M}_0) \cdot [\mathsf{M} + 2\mathsf{M}_0 + \nu(\mathsf{M}_{\text{eff}} - \mathsf{M}_0)]^{-1}\end{aligned} \tag{13}$$

The inclusion six-dyadic M can be fully bi-anisotropic (36 parameters), but the medium to which the polarizabilities are refered to is allowed to be in the most general case bi-isotropic (4 parameters). This medium is, for the Maxwell Garnett case, the background medium M_0, and for the Bruggeman case, it is the effective medium M_{eff}. The formula (13) seems to work for all known Maxwell Garnett cases (isotropic, bi-isotropic, and bi-anisotropic in isotropic mixtures), and also for all previously known cases of the Bruggeman symmetric formula, isotropic and bi-isotropic mixtures, and, of course, for any value of ν in the scalar case (isotropic dielectric mixtures).

Note, however, that since the terms in Equation (13) are dyadic, the order of multiplication and inversions needs more attention than in the scalar isotropic case. Equation (13) is a second-order matrix equation for M_{eff} which is more difficult to solve than the pure scalar dielectric case of Equation (1). One approach to solve the unknown material parameters is an iterative approach which has been shown to be successful [1]. Another one would call for eigenvectors of the dyadics, which allows to decouple the 6×6 matrix into six different second-order equations that can be solved scalarically [20].

4.1. MAXWELL GARNETT FORMULA

Let us look more closely at the Maxwell Garnett formula which can be written from (13) with the choice $\nu = 0$. We have for the effective material dyadic:

$$\mathsf{M}_{\text{eff}} = \mathsf{M}_0 + 3f\mathsf{M}_0 \cdot [\mathsf{M} + 2\mathsf{M}_0 - f(\mathsf{M} - \mathsf{M}_0)]^{-1} \cdot (\mathsf{M} - \mathsf{M}_0) \qquad (14)$$

Perhaps it is important to make a reminder of the limitations of this formula:

- As all quasistatic mixing rules, this is also limited to low frequencies. It can be used for time-varying electromagnetic fields as long as the wavelength of the field is considerably larger than the size of the inclusion spheres. In that case the effective material parameters are frequency dependent through the frequency dependence of the components.
- The background material has to be isotropic or bi-isotropic. If it is anisotropic, the depolarizability dyadic in the polarization calculation is no longer 1/3 for spheres (as is assumed in the derivation of (13)) but will be transformed in a way determined by the anisotropy of the environment.
- The (bi)-anisotropic inclusion spheres have to be aligned so that their material dyadic has the same orientation in the global coordinate system. If there is an orientation distribution of the inclusion particles, the orientational averaging has to be included for the inclusion material matrix (which can be done, for that matter) and the formula (14) needs to be modified.

However, the formula (14) is considerably simpler than its earlierly known tree-dyadic counterpart, which can be calculated from (14) by dyadic rules. Even for the special case of isotropic background (μ_0, ϵ_0), the components read

$$\begin{aligned}\overline{\overline{\epsilon}}_{\text{eff}}/\epsilon_0 &= \overline{\overline{I}} + 3f\overline{\overline{D}}_1^{-1} \\ &\cdot\left(\overline{\overline{\epsilon}}_r - \overline{\overline{I}} - (1-f)\overline{\overline{\xi}} \cdot \left[\overline{\overline{\mu}}_r + 2\overline{\overline{I}} - f(\overline{\overline{\mu}}_r - \overline{\overline{I}})\right]^{-1} \cdot \overline{\overline{\zeta}}/(\mu_0\epsilon_0)\right) \qquad (15)\end{aligned}$$

$$\begin{aligned}\overline{\overline{\xi}}_{\text{eff}} &= 3f\overline{\overline{D}}_1^{-1} \cdot \overline{\overline{\xi}} \\ &\cdot\left(\overline{\overline{I}} - (1-f)\left[\overline{\overline{\mu}}_r + 2\overline{\overline{I}} - f(\overline{\overline{\mu}}_r - \overline{\overline{I}})\right]^{-1} \cdot (\overline{\overline{\mu}}_r - \overline{\overline{I}})\right) \qquad (16)\end{aligned}$$

$$\begin{aligned}\overline{\overline{\zeta}}_{\text{eff}} &= 3f\overline{\overline{D}}_2^{-1} \cdot \overline{\overline{\zeta}} \\ &\cdot\left(\overline{\overline{I}} - (1-f)\left[\overline{\overline{\epsilon}}_r + 2\overline{\overline{I}} - f(\overline{\overline{\epsilon}}_r - \overline{\overline{I}})\right]^{-1} \cdot (\overline{\overline{\epsilon}}_r - \overline{\overline{I}})\right) \qquad (17)\end{aligned}$$

$$\overline{\overline{\mu}}_{\text{eff}}/\mu_0 = \overline{\overline{I}} + 3f\overline{\overline{D}}_2^{-1} \cdot\left(\overline{\overline{\mu}}_r - \overline{\overline{I}} - (1-f)\overline{\overline{\zeta}}\cdot\left[\overline{\overline{\epsilon}}_r + 2\overline{\overline{I}} - f(\overline{\overline{\epsilon}}_r - \overline{\overline{I}})\right]^{-1}\cdot\overline{\overline{\xi}}/(\mu_0\epsilon_0)\right) \quad (18)$$

where

$$\begin{aligned}\overline{\overline{D}}_1 &= \overline{\overline{\epsilon}}_r + 2\overline{\overline{I}} - f(\overline{\overline{\epsilon}}_r - \overline{\overline{I}}) \\ &\quad -(1-f)^2\overline{\overline{\xi}}\cdot\left[\overline{\overline{\mu}}_r + 2\overline{\overline{I}} - f(\overline{\overline{\mu}}_r - \overline{\overline{I}})\right]^{-1}\cdot\overline{\overline{\zeta}}/(\mu_0\epsilon_0) \quad (19) \\ \overline{\overline{D}}_2 &= \overline{\overline{\mu}}_r + 2\overline{\overline{I}} - f(\overline{\overline{\mu}}_r - \overline{\overline{I}}) \\ &\quad -(1-f)^2\overline{\overline{\zeta}}\cdot\left[\overline{\overline{\epsilon}}_r + 2\overline{\overline{I}} - f(\overline{\overline{\epsilon}}_r - \overline{\overline{I}})\right]^{-1}\cdot\overline{\overline{\xi}}/(\mu_0\epsilon_0) \quad (20)\end{aligned}$$

4.2. CHIRAL–IN–CHIRAL MIXTURES

A very powerful example of the use of six-dyadic mixing rules is a calculation of the effective material parameters of bi-isotropic mixtures. Then the inclusion material dyadic in Equation (14) contains four unit dyadics, and the number of free parameters drops from 36 to 4. Nevertheless, no-one has been able to write down an explicit solution for that kind of mixture with three-vector formulation (without incorporating polarizability expressions into the formulas which make the appearance of the resulting parameters very clumsy). With six-dyadics it can be done easily. And numerically it is extremely straightforward to calculate the effective parameters with the formula (14).

The following three lines are sufficient to do the calculation in Mathematica software:

```
M    = {{Eps, Chi - I Kap}, {Chi + I Kap, Mu}}
MO   = {{Eps0, Chi0 - I Kap0}, {Chi0 + I Kap0, Mu0}}
Meff = MO + 3 f MO . Inverse[M + 2MO - f(M-MO)] . (M-MO)
```

This code is enough to enumerate the Maxwell Garnett prediction for any bi-isotropic–in–bi-isotropic mixture. One only needs to put in the eight material parameters forming the two bi-isotropic material matrices, and the volume fraction of the inclusions.

4.2.1. *Racemization example*

Let us make use of the bi-isotropic–in–bi-isotropic code. Take the reciprocal bi-isotropic example: isotropic chiral background and isotropic chiral inclusion phase. Let the background medium be right-handed ($\kappa_0 < 0$, see [21]) and the inclusions left-handed ($\kappa > 0$). Furthermore, let us choose

(maxwell) In[28]:=

```
Plot[Im[Meff[[2,1]]], {f,0,1}, PlotRange->All]
```

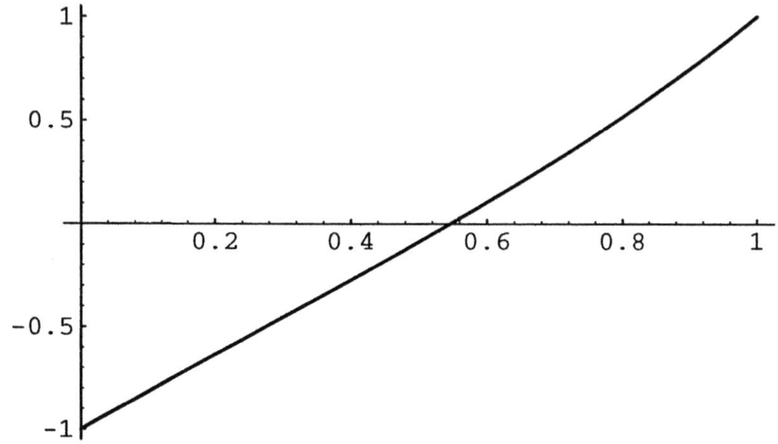

(maxwell) Out[28]=

```
-Graphics-
```

Figure 1. The effective chirality parameter of a left-handed–in–right-handed mixture. The two phases are mirror-image enantiomers, *i.e.*, their permittivities are the same ($\epsilon_r = 2$), the permeabilities are the same ($\mu_r = 1.5$), and their chirality parameters are equal in magnitude and opposite in sign: $\kappa = -\kappa_0 = 1$. Note the racemization density: about 55% in volume of the inclusion phase is needed to make the mixture non-chiral.

the two phases such that they are otherwise identical but they are mirror images of each other. Then their permittivities and permeabilities are the same for both ($\epsilon = \epsilon_0$ and $\mu = \mu_0$), and the chirality parameters are equal in magnitude but opposite in sign: $\kappa = -\kappa_0$.

What is the chirality parameter of the mixture? Obviously the mixture is predominantly right-handed for low volume fractions ($f \ll 1$) and on the average left-handed for high volume fractions ($f \approx 1$). But because the chirality parameter is an analytic function of the volume fraction, there has to be a zero-crossing point. In other words, the mixture looks racemic in the sense that it displays no optical activity. The plane of a linearly polarized electromagnetic wave does not suffer rotation when passing through this mixture.

Substituting the values for Mathematica:

```
M  = {{2, -I}, {I, 1.5}}
M0 = {{2, I}, {-I, 1.5}}
```

one can ask for the effective chirality parameter `Im[Meff[[2,1]]]` as a function of the volume fraction. The result is illustrated in Figure 1.

Figure 1 clearly shows the gradual transition from dextrorotatory to levorotatory macroscopic behavior of the mixture. An interesting detail revealed by the result is that the racemic point is not at a 50/50 mixture. Rather, a slightly higher concentration ($f \approx 0.55$) of the inclusion phase is needed to compensate for the rotatory power of the background material. This is, however, the prediction according to Maxwell Garnett. One could anticipate that the Bruggeman symmetric formula would demand the racemization mixing to happen exactly at the midpoint on the volume fraction axis.

This is indeed the case. The Bruggeman formula treats the background and the inclusion phases symmetrically, in other words, this mixing rule is invariant with respect to the following change of parameters: $\mathsf{M} \rightarrow \mathsf{M}_0$, $\mathsf{M}_0 \rightarrow \mathsf{M}$, and $f \rightarrow 1 - f$. Figure 2 shows the effective chirality parameter of the same mixture as Figure 1, calculated [20] according to four different members of the mixing rule family (13), including the symmetrical Bruggeman rule ($\nu = 2$). This curve intersects the racemic point at $f = 0.5$. For mixing rules where $\nu < 2$ (like, for example, Maxwell Garnett formula) the racemization happens when the inclusion phase volume exceeds that of the background. On the other hand, mixing formulae with $\nu > 2$ (like, for example, the Coherent Potential rule) predict that the racemic point is attained for mixtures which contain more background material than inclusion material.

5. Six-dyadic Generalization of the Relative Material Parameters

In studies of heterogeneous dielectric mixtures, the concept of the relative permittivity is rather uninteresting. The relative permittivity is calculated from the absolute permittivity by a plain division by the vacuum constant ϵ_0. But the situation is not as simple in the analysis of bi-anisotropic materials. Since the division corresponds to a more subtle operation of inversion in the six-dyadic regime, one may suggest that also the concept of relative material matrix or dyadic needs to be defined with care. To illustrate the aspects needed in the treatment of relative material parameters of non-isotropic media, let us consider in the following lossless bi-isotropic and chiral materials. Then all four material dyadics are unit dyadic multiples and the 6×6-matrix is effectively a 2×2-matrix.

For chiral and bi-isotropic materials, the six-dyadic matrix M is in the lossless case an Hermitian matrix:

$$\mathsf{M} = \begin{pmatrix} \epsilon_r & \chi - \mathrm{j}\kappa \\ \chi + \mathrm{j}\kappa & \mu_r \end{pmatrix} \tag{21}$$

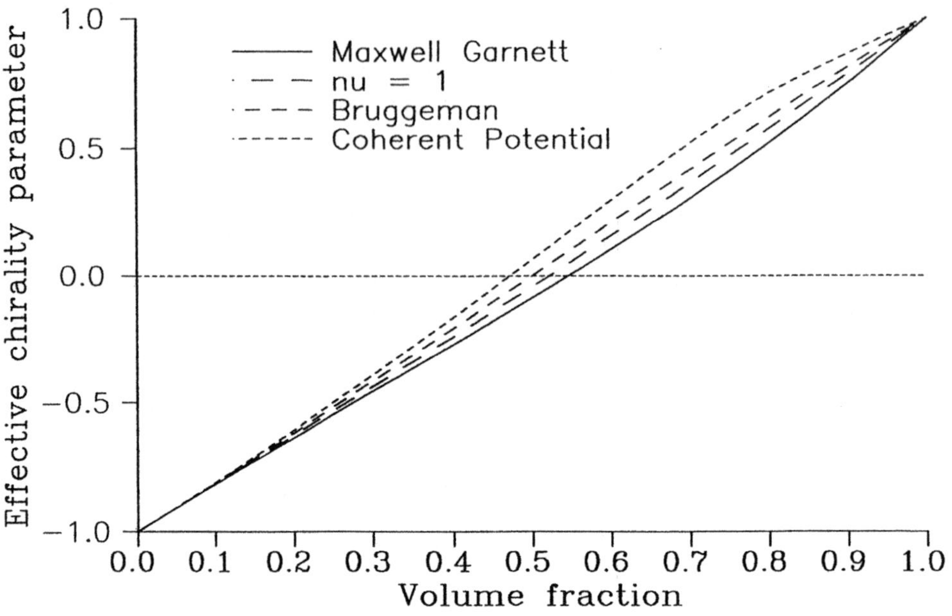

Figure 2. The effective chirality parameter of a left-handed–in–right-handed mixture. The parameters of the mixture are the same as in Figure 1. Four different mixing rules are illustrated: $\nu = 0, 1, 2, 3$. The curve $\nu = 2$ is symmetrical with respect to the racemic line $\kappa_{\text{eff}} = 0$, and corresponds to the Bruggeman formula but it is the only member of the mixing rule family which gives the racemication point at $f = 0.5$. The curves are calculated with the eigenvector solution for the six-vector mixing rule [20].

which means that the off-diagonal elements are complex conjugates of each other. Note that in spite of the explicit appearance of j in (21), all four material parameters are real [22].

5.1. STRAIGHTFORWARD NORMALIZATION

Let us first try to define the material matrix relative to a background material with matrix M_0 using the operation

$$\mathsf{M}_r = \mathsf{M}_0^{-1} \cdot \mathsf{M} \tag{22}$$

Now this the definition is symmetric in the sense that if we calculate the background matrix relative to M as

$$\mathsf{M}_{0,r} = \mathsf{M}^{-1} \cdot \mathsf{M}_0 \tag{23}$$

the following is easily seen to be valid:

$$\mathsf{M}_{0,r} = \mathsf{M}_r^{-1} \tag{24}$$

However, the definition (22) has the following disadvantage: a chiral material with an Hermitian matrix does not have an Hermitian relative material matrix, even if the background to which the normalization is made is nonchiral. To see this, look at the matrix of a reciprocal chiral material relative to a nonchiral dielectric–magnetic background medium with parameters $\epsilon_{0,r}$ and $\mu_{0,r}$ (where, of course, these are not the free-space permittivity and permeability), using the definition (22):

$$\mathsf{M}_r = \begin{pmatrix} \epsilon_{0,r} & 0 \\ 0 & \mu_{0,r} \end{pmatrix}^{-1} \cdot \begin{pmatrix} \epsilon_r & -\mathrm{j}\kappa \\ +\mathrm{j}\kappa & \mu_r \end{pmatrix} = \begin{pmatrix} \epsilon_r/\epsilon_{0,r} & -\mathrm{j}\kappa/\epsilon_{0,r} \\ +\mathrm{j}\kappa/\mu_{0,r} & \mu_r/\mu_{0,r} \end{pmatrix} \tag{25}$$

Now this matrix is not Hermitian because $\mathsf{M}_r^\dagger \neq \mathsf{M}_r$. In other words M_r does not represent an isotropic chiral material.

The same problem of non-Hermitian relative material matrices appears also if we try to represent the relative material matrix with the definition

$$\mathsf{M}_r = \mathsf{M} \cdot \mathsf{M}_0^{-1} \tag{26}$$

instead of (22).

5.2. BALANCED NORMALIZATION

To maintain the chiral character of the material matrix in the normalization operation, another definition for the relative material matrix has to be made. In [20], the following definition is introduced:

$$\mathsf{M}_r = \sqrt{\mathsf{M}_0}^{-1} \cdot \mathsf{M} \cdot \sqrt{\mathsf{M}_0}^{-1} \tag{27}$$

This definition retains the Hermitian character of a chiral material matrix for both dielectric–magnetic background and fully bi-isotropic background relativization. However, this relative definition (27) has another unexpected property: if M_r is diagonal, the complementary relative matrix $\mathsf{M}_{0,r}$ is not diagonal in general, where naturally the balanced definition

$$\mathsf{M}_{0,r} = \sqrt{\mathsf{M}}^{-1} \cdot \mathsf{M}_0 \cdot \sqrt{\mathsf{M}}^{-1} \tag{28}$$

is used. In other words, if a handed material is racemic or "non-handed" relative to another handed background material, the reverse is not the case. Instead, this background medium is "relatively chiral" to the inclusion material.

The condition for a material (with parameters $\epsilon_r, \mu_r, \kappa$ of the material matrix) to be "relatively racemic" with respect to the material $\epsilon_{0,r}, \mu_{0,r}, \kappa_0$ is the following:

$$\kappa/\kappa_0 = \frac{\epsilon_r\mu_{0,r} + \epsilon_{0,r}\mu_r + (\epsilon_r + \mu_r)\sqrt{\epsilon_{0,r}\mu_{0,r} - \kappa_0^2}}{2\epsilon_{0,r}\mu_{0,r} + (\epsilon_{0,r} + \mu_{0,r})\sqrt{\epsilon_{0,r}\mu_{0,r} - \kappa_0^2}} \tag{29}$$

This relation satisfied, the relative matrix M_r is diagonal. But it is easy to see that if this condition (29) holds, the reverse (with inclusion and background parameters interchanged) is not necessarily valid. In other words, $\mathsf{M}_{0,r}$ is not diagonal, although it is Hermitian.

The paradox of relative handednesses Another consequence of the relativization (27) is the following paradox: we may find special values for two handed materials in such a way that one material is left-handed with respect to the other but so is also the other relative to the first one! To take an example, consider the isotropically chiral left-handed inclusion matrix

$$\mathsf{M} = \begin{pmatrix} 3 & -j1.421 \\ +j1.421 & 2 \end{pmatrix} \tag{30}$$

and the left-handed background matrix

$$\mathsf{M}_0 = \begin{pmatrix} 2 & -j1 \\ +j1 & 1.5 \end{pmatrix} \tag{31}$$

By "left-handedness" it is meant that the imaginary part of the upper right off-diaginal element is negative, and the imaginary part of the lower left element positive (see also [21]).

Now the inclusion material relative to the background (using the relativization (27)) is

$$\mathsf{M}_r = \begin{pmatrix} 1.51663 & +j0.00091 \\ -j0.00091 & 1.31237 \end{pmatrix} \tag{32}$$

and the background material relative to the inclusion (with the square-root normalization principle (28)):

$$\mathsf{M}_{0,r} = \begin{pmatrix} 0.659361 & +j0.000716 \\ -j0.000716 & 0.761976 \end{pmatrix} \tag{33}$$

As can be seen, the two relative matrices are not inverses of each other: $\mathsf{M}_r \neq \mathsf{M}_{0,r}^{-1}$. Furthermore indeed, although the relative chirality parameters (the off-diagonal elements) are small, the inclusion material (M) is more right-handed than the background material (M_0) but on the other hand, the opposite holds, too: the background material is more right-handed than the inclusion material!

6. Conclusion

Hopefully the present article has demonstrated the power of six-vector formalism as it is wielded on problems of complex materials modeling. Six-matrices or six-dyadics, as they incorporate all dielectric, magnetic, and magnetoelectric linear constitutive parameters, reduce the complexity of the algebraic formalism that appears on the coupled three-vector formulation of bi-anisotropic materials. The coupled dyadic equations for the effective parameters of heterogeneous non-isotropic media are simplified to a form which appears as simple as in the description of the corresponding pure dielectric problem. Hence also the risks of making mistakes in algebra are much smaller when using six-vectors than on the three-vector level. When we write $\mathsf{d} = \mathsf{M} \cdot \mathsf{e}$ instead of $\mathbf{D} = \epsilon \mathbf{E}$ we only have to bear in mind the rules of six-dyadic manipulations: taking the inverses, for example, is a more complicated operation than a pure division by a scalar. The nontriviality of the relative material parameters was illustrated by the analysis for different normalization principles of bi-isotropic media, and was shown to result either into nonsymmetric and non-Hermitian chirality matrices or in a paradox with relative handednesses. The numerical example of the paper, the racemization analysis of a mixture that consist of enantiometrically paired materials, showed the manner how the optical activity of chiral–in–chiral mixtures can be calculated with six-vectors and how the different mixing rules differ in their predictions for the effective chirality parameter.

References

1. Sihvola A.H. and Pekonen, O.P.M. (1996) Effective medium formulas for bi-anisotropic mixtures, *J. Physics D: Appl. Phys.*, **29**, 514-521.
2. Sihvola, A. (1989) Self-consistency aspects of dielectric mixing theories, *IEEE Trans. Geosci. Remote Sensing*, **27(4)**, 403-415.
3. Garnett, J.C. Maxwell (1904) Colours in metal glasses and metal films, *Trans. of the Royal Society (London)*, **CCIII**, 385-420.
4. Kohler, W.E. and Papanicolaou, G.C. (1981) Some applications of the coherent potential approximation, in P.L. Chow, W.E. Kohler, and G.C. Papanicolaou (eds.), *Multiple Scattering and Waves in Random Media*, North-Holland Publishing Company, New York, pp. 199-223.
5. Tsang, L., Kong, J.A., and Shin, R.T. (1985) *Theory of Microwave Remote Sensing*, Wiley, New York.
6. Böttcher, C.J.F. (1952) *Theory of Electric Polarization*, Elsevier, Amsterdam.
7. Polder, D. and Van Santen, J.H. (1946) The effective permeability of mixtures of solids, *Physica*, **XII(5)**, 257-271.
8. Bruggeman, D.A.G. (1935) Berechnung verschiedener physikalischer Konstanten von heterogenen Substanzen, I. Dielektrizitätskonstanten und Leitfähigkeiten der Mischkörper aus isotropen Substanzen, *Annalen der Physik*, 5. Folge, Band 24, 636-664.

9. Landauer, R. (1978) Electrical conductivity of inhomogeneous media, in J.C. Garland and D.B. Tanner (eds.), *Electrical Transport and Optical Properties of Inhomogeneous Media*, (AIP Conference Proceedings, **40**), American Institute of Physics, New York, pp. 2-43.
10. Lindell, I.V. (1994) *Methods for Electromagnetic Field Analysis*, Clarendon Press, Oxford.
11. Sihvola, A.H. and Lindell, I.V. (1995) Material effects in bi-anisotropic electromagnetics, *IEICE Trans. Electronics (Japan)*, **E78-C(10)**, 1383-1390.
12. Kong, J.A. (1986) *Electromagnetic Wave Theory*, Wiley, New York, p. 403.
13. Reference [12], p. 51.
14. Lindell, I.V., Sihvola, A.H., Tretyakov, S.A., and Viitanen, A.J. (1994) *Electromagnetic Waves in Chiral and Bi-Isotropic Media,* Artech House, Boston and London.
15. Lindell, I.V., Sihvola, A.H., and Suchy, K. (1994) Six-vector formalism in electromagnetics of bianisotropic media, *J. of Electromagnetic Waves and Applications*, **9(7/8)**, 887-903.
16. Rikte, S. (1994) Reconstruction of bi-isotropic material parameters using transient electromagnetic fields, *Lund Institute of Technology*, Technical Report LUTEDX/(TEAT–7033)/1–22.
17. Borzdov, G.N. (1996) Inverse problem of reflection and transmission for a bianisotropic medium, *Proceedings of Chiral'96*, this volume.
18. Borzdov, A.N. (1996) Effective material parameters of plane stratified bi-anisotropic superlattices, *Proceedings of Chiral'96*, this volume.
19. Kamenetskii, E.O. (1996) Electrodynamics of chirowaveguides: mode orthogonality, mode excitation, mode interactions, *Proceedings of Chiral'96*, this volume.
20. Sihvola, A.H. and Olyslager, F. (1996) Eigenvector approach for solving bianisotropic mixing formulas, *Helsinki University of Technology, Electromagnetics Laboratory Rept. Series*, **218**. To appear in *Radio Science*.
21. The affiliation of handedness of the material with the sign of the chirality parameter κ is not unproblematic. In the present paper, the loose use of the handed character of a material follows the rationale that a positive κ for a homogeneous medium leads to the left-handed rotation of the polarization of a propagating wave (Reference [14], p. 35). Note, however, that for a given material, the rotation direction may be left-handed at a certain frequence range, and right-handed for other frequencies. This fact is known as the Cotton effect.
22. Reference [14], p. 50.

EFFECTIVE MATERIAL PARAMETERS OF PLANE STRATIFIED BIANISOTROPIC SUPERLATTICES

A.N. BORZDOV
Department of Theoretical Physics
Byelorussian State University
Fr. Skaryny avenue, 4
220050 Minsk, Belarus

Abstract. The superlattices formed by periodic sets of plane bianisotropic layers with the different thicknesses are theoretically studied in the long wavelength limit. The layers from the materials with the constitutive equations $\mathbf{D} = \varepsilon\mathbf{E} + \alpha\mathbf{H}$, $\mathbf{B} = \beta\mathbf{E} + \mu\mathbf{H}$ are considered, where dielectric permittivity ε, magnetic permeability μ and pseudotensors of gyrotropy α and β are assumed to be complex nonsymmetric tensors. The effective tensors of permittivity, permeability and two pseudotensors of gyrotropy characterizing the structure under consideration are found in compact coordinate-free form. Two particular structures are analyzed in detail, namely, the superlattices of fan type and folded type.

1. Introduction

At present, two different approaches are widely used to study the effective properties of plane stratified superlattices in the long wavelength limit.

The first one is based on the use of the characteristic matrix technique. The characteristic matrix [1] of a layer relates the tangential components of the field vectors at the layers boundaries. It has the form $P_n = \exp(ikl_nM_n)$ [2], where k denotes the free-space wave number, l_n is the layer thickness, and matrix M_n, depending on the tensors of permittivity, permeability, gyrotropy, angle of incidence, specifies the optical property of the n-th layer. The characteristic matrix of the unit cell, which is of the first importance when treating the periodical structures, is

$$P = \exp(ikl_NM_N)\dots\exp(ikl_2M_2)\exp(ikl_1M_1)$$

A. Priou et al. (eds.), Advances in Complex Electromagnetic Materials, 145–154.

(here N is the number of the layers in the unit cell). As matrices M_n do not commute, $P \neq \exp(l_1 M_1 + l_2 M_2 + \ldots + l_N M_N)$. But in some instances at appropriately long wavelength the characteristic matrix can be brought to the form $P = \exp(ikLM)$, where $L = l_1 + l_2 + \ldots + l_N$ is the superlattice period, while M depends on the angle of incidence and some "effective" tensors in the same way as in homogeneous medium. The above outlined method was used, for example, to find permittivity and permeability tensors of the medium equivalent in a long wavelength approximation to a periodic chiral arrangement of identical bianisotropic sheets [3]. In that work the permittivity and permeability tensors were assumed to be biaxial, while the gyrotropy tensors considered to be isotropic.

The other approach is based on the use of continuity at layers interfaces of some components of the field vectors. It is assumed that $L \ll \lambda$ and hence the field inside the layers can be considered to be constant. That makes it possible to define the averaged over the superlattice volume fields (Maxwell fields) $\mathbf{D}$, $\mathbf{E}$, $\mathbf{B}$, $\mathbf{H}$ and, finally, to obtain relating these vectors effective tensors of permittivity, permeability and the pseudotensor of gyrotropy. By this method the permeability and permittivity tensors of some superlattices were found [4, 5, 6]. Gyrotropic properties of the superlattices formed by two altering flat layers were investigated in [7, 8].

In this paper we make use of the latter approach to study the properties of the plane stratified bianisotropic superlattices in the long wavelength limit. We consider the superlattices of the most general type assuming the permittivity, permeability and tensors of gyrotropy of the layers to be complex nonsymmetric tensors. Therefore, the layers can be made from absorbing anisotropic and gyrotropic magnetic crystals, subject to the influence of external electric and magnetic fields and elastic deformations. The effective tensors of permittivity, permeability and two pseudotensors of gyrotropy characterizing the structure under consideration are found in compact coordinate-free form. Two particular structures, namely, the superlattices of fan type and folded type are analyzed in detail.

2. Basic equations

Let us consider the superlattice formed by the periodic set of plane bianisotropic layers with different thicknesses l_i ($i = 1, 2, \ldots N$, where N is number of the layers which constitute the unit cell). We assume for the layers materials the following constitutive equations

$$\mathbf{D}_i = \varepsilon_i \mathbf{E}_i + \alpha_i \mathbf{H}_i, \quad \mathbf{B}_i = \beta_i \mathbf{E}_i + \mu_i \mathbf{H}_i \tag{1}$$

which can be also written in the matrix form

$$\begin{pmatrix} \mathbf{D}_i \\ \mathbf{B}_i \end{pmatrix} = \mathcal{R}_i \begin{pmatrix} \mathbf{E}_i \\ \mathbf{H}_i \end{pmatrix}, \quad \mathcal{R}_i = \begin{pmatrix} \varepsilon_i & \alpha_i \\ \beta_i & \mu_i \end{pmatrix} \tag{2}$$

If ε_i, μ_i, α_i, β_i ($i = 1, 2, \ldots N$) are complex nonsymmetric tensors, then equations (1) or (2) describe an absorbing anisotropic and gyrotropic medium, subject to the influence of external electric and magnetic fields and elastic deformations.

Let $\mathbf{q}$ be the unit normal to the boundaries. Then the projection operators ρ onto the one-dimensional subspace given by $\mathbf{q}$, and I onto the two-dimensional subspace of the interfaces are

$$\rho = \mathbf{q} \otimes \mathbf{q}, \quad I = 1 - \rho \tag{3}$$

and have the properties

$$\rho + I = 1, \quad I\rho = \rho I = 0, \quad \rho_t = 1, \quad I_t = 2 \tag{4}$$

Here, 1 is the unit tensor of the three-dimensional space, subindex t denotes the trace of a tensor and $\otimes$ denotes the tensor product. The operators ρ and I allow us to find the continuous at the layers boundaries normal components $\rho\mathbf{D}_i$, $\rho\mathbf{B}_i$ and tangential components $I\mathbf{E}_i$, $I\mathbf{H}_i$ of the vectors $\mathbf{D}_i$, $\mathbf{B}_i$ and $\mathbf{E}_i$, $\mathbf{H}_i$.

In the same way, the operators

$$\mathcal{Q} = \begin{pmatrix} \rho & 0 \\ 0 & \rho \end{pmatrix}, \quad \mathcal{T} = \begin{pmatrix} I & 0 \\ 0 & I \end{pmatrix} \tag{5}$$

with the properties

$$\mathcal{Q} + \mathcal{T} = \mathcal{E} = \begin{pmatrix} 1 & 0 \\ 0 & 1 \end{pmatrix}, \quad \mathcal{T}\mathcal{Q} = \mathcal{Q}\mathcal{T} = 0 \tag{6}$$

$$\mathcal{Q}_t = 2, \quad \mathcal{T}_t = 4$$

enable us to find the normal and tangential components of the vectors $\begin{pmatrix} \mathbf{D}_i \\ \mathbf{B}_i \end{pmatrix}$ and $\begin{pmatrix} \mathbf{E}_i \\ \mathbf{H}_i \end{pmatrix}$.

In the case when $L/\lambda \ll 1$, the fields $\mathbf{D}_i$, $\mathbf{B}_i$, $\mathbf{E}_i$, $\mathbf{H}_i$ inside the layers can be considered to be constant. The fields averaged over the superlattice volume are determined by the well-known equations [6]

$$\mathbf{D} = \sum_{i=1}^{N} f_i \mathbf{D}_i, \quad \mathbf{B} = \sum_{i=1}^{N} f_i \mathbf{B}_i, \quad \mathbf{E} = \sum_{i=1}^{N} f_i \mathbf{E}_i, \quad \mathbf{H} = \sum_{i=1}^{N} f_i \mathbf{H}_i \tag{7}$$

where $f_i = l_i/L$ is the relative thickness of the i-th layer,

$$\sum_{i=1}^{N} f_i = 1 \tag{8}$$

The continuity conditions at the interfaces of layers with regard to equations (7)-(8) now can be written as

$$\mathcal{T}\begin{pmatrix} \mathbf{E}_i \\ \mathbf{H}_i \end{pmatrix} = \mathcal{T}\begin{pmatrix} \mathbf{E} \\ \mathbf{H} \end{pmatrix}, \quad \mathcal{Q}\begin{pmatrix} \mathbf{D}_i \\ \mathbf{B}_i \end{pmatrix} = \mathcal{Q}\begin{pmatrix} \mathbf{D} \\ \mathbf{B} \end{pmatrix} \tag{9}$$

The sought-for effective tensors ε, μ, α, β relate fields $\mathbf{D}$, $\mathbf{B}$, $\mathbf{E}$, $\mathbf{H}$ as follows:

$$\begin{pmatrix} \mathbf{D} \\ \mathbf{B} \end{pmatrix} = \mathcal{R}\begin{pmatrix} \mathbf{E} \\ \mathbf{H} \end{pmatrix}, \quad \mathcal{R} = \begin{pmatrix} \varepsilon & \alpha \\ \beta & \mu \end{pmatrix} \tag{10}$$

By averaging (2) with the help of (6), (9) we find

$$\begin{pmatrix} \mathbf{D} \\ \mathbf{B} \end{pmatrix} = \sum_{i=1}^{N} f_i \mathcal{R}_i \begin{pmatrix} \mathbf{E}_i \\ \mathbf{H}_i \end{pmatrix} = \sum_{i=1}^{N} f_i \mathcal{R}_i \mathcal{T}\begin{pmatrix} \mathbf{E} \\ \mathbf{H} \end{pmatrix} + \sum_{i=1}^{N} f_i \mathcal{R}_i \mathcal{Q}\begin{pmatrix} \mathbf{E}_i \\ \mathbf{H}_i \end{pmatrix} \tag{11}$$

$$\mathcal{Q}\begin{pmatrix} \mathbf{E}_i \\ \mathbf{H}_i \end{pmatrix} = \mathcal{R}_{iq}^{-}\left[\mathcal{Q}\begin{pmatrix} \mathbf{D} \\ \mathbf{B} \end{pmatrix} - \mathcal{Q}\mathcal{R}_i\mathcal{T}\begin{pmatrix} \mathbf{E} \\ \mathbf{H} \end{pmatrix}\right] \tag{12}$$

where the operator $\mathcal{R}_{iq}^{-}$ is the pseudoinverse [9, 2] to the operator

$$\mathcal{R}_{iq} = \mathcal{Q}\mathcal{R}_i\mathcal{Q} = \begin{pmatrix} \varepsilon_{iq}\mathbf{q}\otimes\mathbf{q} & \alpha_{iq}\mathbf{q}\otimes\mathbf{q} \\ \beta_{iq}\mathbf{q}\otimes\mathbf{q} & \mu_{iq}\mathbf{q}\otimes\mathbf{q} \end{pmatrix} \tag{13}$$

$$\varepsilon_{iq} = \mathbf{q}\varepsilon_i\mathbf{q}, \quad \mu_{iq} = \mathbf{q}\mu_i\mathbf{q}, \quad \alpha_{iq} = \mathbf{q}\alpha_i\mathbf{q}, \quad \beta_{iq} = \mathbf{q}\beta_i\mathbf{q} \tag{14}$$

The operator $\mathcal{R}_{iq}^{-}$ has the properties

$$\mathcal{R}_{iq}^{-}\mathcal{R}_{iq} = \mathcal{R}_{iq}\mathcal{R}_{iq}^{-} = \mathcal{Q} \tag{15}$$

and in explicit form it is defined as

$$\mathcal{R}_{iq}^{-} = \frac{1}{\delta_i}\begin{pmatrix} \mu_{iq}\mathbf{q}\otimes\mathbf{q} & -\alpha_{iq}\mathbf{q}\otimes\mathbf{q} \\ -\beta_{iq}\mathbf{q}\otimes\mathbf{q} & \varepsilon_{iq}\mathbf{q}\otimes\mathbf{q} \end{pmatrix} \tag{16}$$

where

$$\delta_i = \varepsilon_{iq}\mu_{iq} - \alpha_{iq}\beta_{iq}$$

By expressing $\mathcal{T}\begin{pmatrix} \mathbf{D} \\ \mathbf{B} \end{pmatrix}$ and $\mathcal{Q}\begin{pmatrix} \mathbf{D} \\ \mathbf{B} \end{pmatrix}$ in terms of $\begin{pmatrix} \mathbf{E} \\ \mathbf{H} \end{pmatrix}$ with regard to eqs. (11)-(12) we finally find

$$\mathcal{T}\begin{pmatrix} \mathbf{D} \\ \mathbf{B} \end{pmatrix} + \mathcal{Q}\begin{pmatrix} \mathbf{D} \\ \mathbf{B} \end{pmatrix} = \begin{pmatrix} \mathbf{D} \\ \mathbf{B} \end{pmatrix} = \mathcal{R}\begin{pmatrix} \mathbf{E} \\ \mathbf{H} \end{pmatrix} \tag{17}$$

where

$$\mathcal{R} = \begin{pmatrix} \varepsilon & \alpha \\ \beta & \mu \end{pmatrix} = \mathcal{N}_L \mathcal{R}_Q \mathcal{N}_R + \sum_{i=1}^{N} f_i \mathcal{R}_i \mathcal{T}_{Ri} \tag{18}$$

$$\mathcal{N}_L = \sum_{i=1}^{N} f_i \mathcal{N}_{Li}, \quad \mathcal{N}_R = \sum_{i=1}^{N} f_i \mathcal{N}_{Ri}, \quad \mathcal{R}_Q = \left(\sum_{i=1}^{N} f_i \mathcal{R}_{iq}^{-} \right)^{-} \tag{19}$$

$$\mathcal{N}_{Li} = \mathcal{R}_i \mathcal{Q} \mathcal{R}_{iq}^{-}, \quad \mathcal{N}_{Ri} = \mathcal{R}_{iq}^{-} \mathcal{Q} \mathcal{R}_i, \quad \mathcal{T}_{Ri} = \mathcal{E} - \mathcal{N}_{Ri} \tag{20}$$

Here, operators $\mathcal{N}_L$, $\mathcal{N}_R$, $\mathcal{N}_{Li}$, $\mathcal{N}_{Ri}$ and $\mathcal{T}_{Ri}$ are projection operators with the properties

$$\mathcal{N}_L^2 = \mathcal{N}_L, \ \mathcal{N}_R^2 = \mathcal{N}_R, \ \mathcal{N}_{Li}^2 = \mathcal{N}_{Li}, \ \mathcal{N}_{Ri}^2 = \mathcal{N}_{Ri}, \ \mathcal{T}_{Ri}^2 = \mathcal{T}_{Ri} \tag{21}$$

$$(\mathcal{N}_L)_t = (\mathcal{N}_R)_t = (\mathcal{N}_{Li})_t = (\mathcal{N}_{Ri})_t = 2, \ (\mathcal{T}_{Ri})_t = 4$$

It is essential that

$$\left(\sum_{i=1}^{N} f_i \mathcal{R}_i \mathcal{T}_{Ri} \right) \mathcal{Q} = \mathcal{Q} \left(\sum_{i=1}^{N} f_i \mathcal{R}_i \mathcal{T}_{Ri} \right) = 0 \tag{22}$$

$$\mathcal{Q}\mathcal{N}_L = \mathcal{Q}, \ \mathcal{N}_L \mathcal{Q} = \mathcal{N}_L, \ \mathcal{N}_R \mathcal{Q} = \mathcal{Q}, \ \mathcal{Q}\mathcal{N}_R = \mathcal{N}_R$$

The coordinate-free formulae (18)-(20) allow one to express the effective tensors of the superlattice in terms of the parameters of the layers. Unlike the cumbersome coordinate formulae, the derived compact expressions are convenient for the analysis of the superlattices of general type. In doing so the properties (6), (21-22) of the projection operators notably simplify the analysis. For example, using (21-22), (15) from Eq. (18) we immediately obtain

$$\mathcal{R}_Q = \mathcal{Q}\mathcal{R}\mathcal{Q} = \begin{pmatrix} \varepsilon_q \mathbf{q} \otimes \mathbf{q} & \alpha_q \mathbf{q} \otimes \mathbf{q} \\ \beta_q \mathbf{q} \otimes \mathbf{q} & \mu_q \mathbf{q} \otimes \mathbf{q} \end{pmatrix} \tag{23}$$

where

$$\varepsilon_q = \frac{1}{\delta} \sum_{i=1}^{N} f_i \frac{\varepsilon_{iq}}{\delta_i}, \quad \mu_q = \frac{1}{\delta} \sum_{i=1}^{N} f_i \frac{\mu_{iq}}{\delta_i} \tag{24}$$

$$\alpha_q = \frac{1}{\delta} \sum_{i=1}^{N} f_i \frac{\alpha_{iq}}{\delta_i}, \quad \beta_q = \frac{1}{\delta} \sum_{i=1}^{N} f_i \frac{\beta_{iq}}{\delta_i}$$

and

$$\delta = \left(\sum_{i=1}^{N} f_i \frac{\varepsilon_{iq}}{\delta_i}\right)\left(\sum_{i=1}^{N} f_i \frac{\mu_{iq}}{\delta_i}\right) - \left(\sum_{i=1}^{N} f_i \frac{\alpha_{iq}}{\delta_i}\right)\left(\sum_{i=1}^{N} f_i \frac{\beta_{iq}}{\delta_i}\right) \tag{25}$$

In some particular cases, the derived in this Section general formulae can be simplified still further.

3. Explicit expressions for effective material tensors in some particular cases

If $\mathbf{q}$ is the eigenvector of each of the tensors ε_i, μ_i, α_i, β_i $(i = 1, 2, \dots N)$ then $\mathcal{N}_L = \mathcal{N}_R = \mathcal{Q}$, $\mathcal{T}_{Ri} = \mathcal{T}$ and from the general formulae (18)-(25) we obtain explicit expressions for effective material tensors of superlattice

$$\varepsilon = \varepsilon_q \mathbf{q} \otimes \mathbf{q} + \varepsilon I, \qquad \varepsilon I = \sum_{i=1}^{N} f_i \varepsilon_i I \tag{26}$$

$$\mu = \mu_q \mathbf{q} \otimes \mathbf{q} + \mu I, \qquad \mu I = \sum_{i=1}^{N} f_i \mu_i I \tag{27}$$

$$\alpha = \alpha_q \mathbf{q} \otimes \mathbf{q} + \alpha I, \qquad \alpha I = \sum_{i=1}^{N} f_i \alpha_i I \tag{28}$$

$$\beta = \beta_q \mathbf{q} \otimes \mathbf{q} + \beta I, \qquad \beta I = \sum_{i=1}^{N} f_i \beta_i I \tag{29}$$

In general, the tensors ε, μ, α, β have different sets of eigenvectors and hence do not commute. For example, the biaxial crystals of the rhombic system and uniaxial crystals have mutually commutative tensors of permittivity, permeability and gyrotropy [10]. Nevertheless, the superlattice formed by these crystals is generally characterized by noncommutative effective tensors. On the other hand, the tensors ε, μ, α, β (26)-(29) have a common eigenvector $\mathbf{q}$ and, thus, the lowest possible symmetry of the effective medium under consideration is the symmetry of monoclinic system [10].

Since εI, μI, αI, βI are planar tensors [9, 2] ($\varepsilon I \mathbf{q} = \mathbf{q} \varepsilon I = 0$ and so on), then ε_q, μ_q, α_q, β_q are the eigenvalues of the tensors ε, μ, α, β, corresponding to their common eigenvector $\mathbf{q}$. Consequently, the values ε_q, μ_q, α_q, β_q are given by Eqs. (24),(25) and do not depend on the orientation of the preferred axes of the material tensors, provided $\mathbf{q}$ is their common eigenvector. On the contrary, two other eigenvalues of each of the tensors (26)-(29) depend on the mutual orientation of the crystals axes. These eigenvalues are the nonzero eigenvalues of the planar tensors εI, μI, αI, βI whose eigenvectors lie in the plane of the interfaces. All eigenvectors of the material tensors depend on the relative thicknesses of the layers.

We are not imposing any restrictions on the orientation of the preferred axes of the crystals except that of $\mathbf{q}$ to be an eigenvector of all material tensors. Therefore, one of the preferred axes can not be lying in the plane of the boundaries. For the sake of simplicity from here on we shall assume the material tensors of the layers to be real symmetric. In this case the tensors $\varepsilon_i I$, for example, can be written in the form [10]

$$\varepsilon_i I = \varepsilon_{ia}\mathbf{a}_{i\varepsilon} \otimes \mathbf{a}_{i\varepsilon} + \varepsilon_{ib}\mathbf{b}_{i\varepsilon} \otimes \mathbf{b}_{i\varepsilon} \tag{30}$$

$$\mathbf{a}_{i\varepsilon}^2 = \mathbf{a}_{i\varepsilon}^{*2} = \mathbf{b}_{i\varepsilon}^2 = \mathbf{b}_{i\varepsilon}^{*2} = 1, \ \mathbf{a}_{i\varepsilon} \cdot \mathbf{b}_i = 0, \ \mathbf{a}_{i\varepsilon} \cdot \mathbf{q} = \mathbf{q} \cdot \mathbf{b}_{i\varepsilon} = 0$$

Here, asterisk $*$ denotes the complex conjugation.

The planar tensor εI (26) have the same form as the tensors $\varepsilon_i I$:

$$\varepsilon I = \varepsilon_{I+}\tau_+ + \varepsilon_{I-}\tau_- = \varepsilon_{I+}\mathbf{a}_\varepsilon \otimes \mathbf{a}_\varepsilon + \varepsilon_{I-}\mathbf{b}_\varepsilon \otimes \mathbf{b}_\varepsilon \tag{31}$$

From Eqs. (26),(30) we obtain

$$\varepsilon_{I\pm} = \frac{1}{2}\left(\varepsilon_{ac} + \varepsilon_{bc} \pm \sqrt{(\varepsilon_{ac} - \varepsilon_{bc})^2 - 4\sum_{k=2}^{N} f_k \varepsilon'_k \sum_{j=1}^{k-1} f_j \varepsilon'_j \sin^2\phi_{jk}}\right) \tag{32}$$

where

$$\varepsilon_{ac} = \sum_{i=1}^{N} f_i \varepsilon_{ia}, \ \varepsilon_{bc} = \sum_{i=1}^{N} f_i \varepsilon_{ib}, \ \varepsilon'_k = \varepsilon_{ka} - \varepsilon_{kb} \tag{33}$$

and ϕ_{jk} is the angle between the vectors $\mathbf{a}_j$ and $\mathbf{a}_k$.

One can find dyads $\tau_+ = \mathbf{a}_\varepsilon \otimes \mathbf{a}_\varepsilon$ and $\tau_- = \mathbf{b}_\varepsilon \otimes \mathbf{b}_\varepsilon$ using the formula [2]:

$$\tau_\pm = \pm\frac{\varepsilon I - \varepsilon_{I\mp} I}{\varepsilon_{I+} - \varepsilon_{I-}} \tag{34}$$

Since structurally all the tensors (26)-(29) are the same, we have for μ, α, β expressions, similar to formulae (30)-(34).

Now let us consider in more detail two particular structures, which we shall refer to as the fan and folded superlattices by analogy with the fan and folded birefringent filters [11].

3.1. FAN SUPERLATTICE

This superlattice is formed by identical bianisotropic layers rotated at a uniform rate across the period. In this case we have $\varepsilon_{1a} = \varepsilon_{2a} = \ldots =$

$\varepsilon_{Na} \equiv \varepsilon_a$, $\varepsilon_{1b} = \varepsilon_{2b} = \ldots = \varepsilon_{Nb} \equiv \varepsilon_b$, $\mu_{1a} = \mu_{2a} = \ldots = \mu_{Na} \equiv \mu_a$, $\mu_{1b} = \mu_{2b} = \ldots = \mu_{Nb} \equiv \mu_b$, $f_1 = f_2 = \ldots = f_N = 1/N$, $\phi_{12} = \phi_{23} = \ldots = \phi_{N-1\ N} \equiv \phi$ and Eq. (32) reduces to

$$\varepsilon_{I\pm} = \frac{1}{2}\left(\varepsilon_a + \varepsilon_b \pm (\varepsilon_a - \varepsilon_b)\,\Phi_N\right) \tag{35}$$

where

$$\Phi_N = \frac{\sin(N\phi)}{N\sin\phi}$$

One can obtain identical formulae also for the eigenvalues of the tensors μI, αI and βI. The formulae (31),(34),(35) and similar ones for μI, αI, βI are generalizations of obtained in [3] expressions to the case when α, β are tensors. Besides, we do not demand the preferred axes of the tensors to be in the plane of the interfaces.

If $\phi = \pm(m\pi)/N$ $(m = 1, 2, \ldots)$, then $\varepsilon_{I+} = \varepsilon_{I-}$ and the superlattice in the long wavelength approximation is equivalent to the transversely isotropic (uniaxial) medium:

$$\varepsilon = \frac{1}{2}(\varepsilon_a + \varepsilon_b)\,I + \varepsilon_q \mathbf{q}\otimes\mathbf{q} \tag{36}$$

$$\mu = \frac{1}{2}(\mu_a + \mu_b)\,I + \mu_q \mathbf{q}\otimes\mathbf{q} \tag{37}$$

$$\alpha = \frac{1}{2}(\alpha_a + \alpha_b)\,I + \alpha_q \mathbf{q}\otimes\mathbf{q} \tag{38}$$

$$\beta = \frac{1}{2}(\beta_a + \beta_b)\,I + \beta_q \mathbf{q}\otimes\mathbf{q} \tag{39}$$

Otherwise, the equivalent medium is biaxial with the noncommuting in general material tensors. From Eqs. (31), (34), (35) it follows that the eigenvalues $\varepsilon_{I\pm}$ as well as the orientation of the vectors $\mathbf{a}_\varepsilon$ and $\mathbf{b}_\varepsilon$ vary with varying ϕ (of course, it is also true for the eigenvalues and eigenvectors of the tensors μI, αI, βI). It is extremely important that the eigenvectors of each of the tensors εI, μI, αI, βI are rotating through the same angle with varying ϕ. Let us find, for example, the angle Φ between the vectors $\mathbf{a}_{1\varepsilon}$ and $\mathbf{a}_\varepsilon$. From Eqs. (26), (31), (34), (35) we have

$$\cos^2\Phi = \frac{1}{2}\left(1 + \cos(N-1)\,\phi\right) = \cos^2\frac{(N-1)\,\phi}{2} \tag{40}$$

Therefore, Φ actually depends only on the angle ϕ, number of the layers N but does not depend on the tensor ε.

3.2. FOLDED SUPERLATTICE

The unit cell of this superlattice is formed by only two layers which in general have different thicknesses and different sets of the tensor constants $\varepsilon_i I$, $\mu_i I$, $\alpha_i I$, $\beta_i I$ $(i = 1, 2)$. Equations (31), (34) hold true while for the eigenvalues of the tensor εI we have

$$\varepsilon_{I\pm} = \frac{1}{2}\left(\varepsilon_{ac} + \varepsilon_{bc} \pm \sqrt{(\varepsilon_{ac} - \varepsilon_{bc})^2 - 4 f_1 f_2 \varepsilon_1' \varepsilon_2' \sin^2 \phi_{12}}\right) \tag{41}$$

One can obtain the similar formulae also for the eigenvalues of the tensors μI, αI and βI.

If $\varepsilon_{1a} = \varepsilon_{2a} \equiv \varepsilon_a$, $\varepsilon_{1b} = \varepsilon_{2b} \equiv \varepsilon_b$, but $f_1 \neq f_2$ then

$$\varepsilon_{I\pm} = \frac{1}{2}\left(\varepsilon_a + \varepsilon_b \pm (\varepsilon_a - \varepsilon_b)\sqrt{1 - 4(1 - f_1) f_1 \sin^2 \phi_{12}}\right) \tag{42}$$

The angle Φ between the vectors $\mathbf{a}_{1\varepsilon}$ and $\mathbf{a}_\varepsilon$ (between the vectors $\mathbf{a}_{1\mu}$ and $\mathbf{a}_\mu$ and so on) is given by

$$\cos^2 \Phi = \frac{1}{2}\left(1 + \frac{1 - 2(1 - f_1)\sin^2 \phi_{12}}{\sqrt{1 - 4(1 - f_1) f_1 \sin^2 \phi_{12}}}\right) \tag{43}$$

Thus, in this case, the mutual orientation of the corresponding eigenvectors of the tensors εI, μI, αI, βI is not affected by varying of ϕ_{12} and f_1.

It is interesting to compare the effective properties of the fan and folded superlattices. From Eq. (24) one can easily see that the eigenvalues ε_q, μ_q, α_q, β_q for both these superlattices coincide. The eigenvectors of each of the tensors εI, μI, αI, βI are mutually orthogonal under any conditions for both structures.

The eigenvalues $\varepsilon_{I\pm}$ of the fan (35) and folded (42) superlattices coincide provided

$$\sqrt{1 - 4(1 - f_1) f_1 \sin^2 \phi_{12}} = |\Phi_N| \tag{44}$$

That holds true also for the eigenvalues of the tensors μI, αI, βI. Both sides of Eq. (44) are varied within the same range. Because of that, Eq. (44) can be satisfied at any $|\Phi_N|$ if f_1 and (or) ϕ_{12} are properly chosen.

It is clear from the above discussion that both structures are effectively equivalent to the same homogeneous medium in the long wavelength limit if the condition (44) is met. It is possible, though, that one of these media is turned through some angle over $\mathbf{q}$ relative to the other one. The fan and folded superlattices are exactly equivalent to the same medium if $\cos^2 \Phi$ (40) is equal to $\cos^2 \Phi$ (43). Since Eq. (44) depends on the two values ϕ_{12}

and f_1, this latter condition and condition (44) can be met simultaneously. After simple algebra we find the values

$$f_1 = \frac{1 - \Phi_N^2}{2\,(1 - |\Phi_N| \cos{(N-1)\,\phi})} \tag{45}$$

$$\sin^2 \phi_{12} = \frac{1 - \Phi_N^2}{4 f_1\,(1 - f_1)} \tag{46}$$

providing complete equivalence of the fan and folded superlattices.

4. Conclusion

The general formulae derived in this paper agree with the expressions found previously by other authors for particular cases. The principal feature of obtained formulae for the effective tensors of the plane stratified bianisotropic superlattices is that they are given in compact coordinate-free form. The given examples show that it considerably facilitates the analysis of dependence of the superlattice effective parameters on the properties of the layers. However, only structures with real symmetric material tensors having the unit normal to the interfaces as an eigenvector, were studied in detail.

References

1. Born, M. and Wolf, E. (1968) *Principles of Optics*. Pergamon Press, Oxford.
2. Barkovskii, L.M., Borzdov, G.N., and Lavrinenko, A.V. (1987) Fresnel's reflection and transmission operators for stratified gyroanisotropic media, *J. Phys.* **A 20**, 1095-1106.
3. Reese, P.S., and Lakhtakia, A. (1990) A periodic chiral arrangement of thin identical bianisotropic sheets: effective properties, *Optik* **86**, 47-50.
4. Raj, N. and Tilley, D.R. (1987) Polariton and effective-medium theory of magnetic superlattices, *Phys. Rev.* **B 36**, 7003-7007.
5. Almeida, N.S., and Mills, D.L. (1988) Effective-medium theory of long-wavelength spin waves in magnetic superlattices, *Phys. Rev.* **B 38**, 6698-6710.
6. Agranovich V.M. (1991) Dielectric permeability and influence of external fields on optical properties of superlattices, *Solid State Communications* **78**, 747-750.
7. Semchenko, I.V. (1990) Gyrotropic properties of superlattices in the long wavelength approximation, *Crystallography* **35**, 1047-1050.
8. Semchenko, I.V. (1994) Gyrotropic properties of superlattices with magnetic structure in long-wave approximation, in F. Mariotte and J.-P. Parneix (eds.), *Proc. of CHIRAL'94*, Périgueux, France, pp. 175-178.
9. Gantmacher, F.R. (1988) *Theory of Matrices*. Nauka, Moscow.
10. Fedorov, F.I. (1976) *Theory of Gyrotropy*. Nauka i Tekhnika, Minsk.
11. Title, A.M., and Rosenberg, W.J. (1981) Tunable birefringent filters, *Optical Engineering* **20**, 815-823.

WEAKLY NONLINEAR CHIRAL COMPOSITES: THE BRUGGEMAN AND THE MAXWELL GARNETT MODELS

G.YA. SLEPYAN
Institute of Nuclear Problems, Belarus State University,
11 Bobruiskaya Str., Minsk, 220050, Belarus

A. LAKHTAKIA
CATMAS – Computational and Theoretical Materials Sciences Group, Department of Engineering Science and Mechanics,
The Pennsylvania State University,
University Park, PA 16802-1401, USA

AND

S.A. MAKSIMENKO
Institute of Nuclear Problems, Belarus State University,
11 Bobruiskaya Str., Minsk, 220050, Belarus

Abstract. The Bruggeman and the Maxwell Garnett models for the effective constitutive parameters (ECPs) of a chiral composite made of a linear chiral material and a cubic nonlinear dielectric material with weak nonlinearity are set up. All three ECPs are shown to be nonlinear in identical fashion.

1. Introduction

Many different types of applications in solid state physics, optics and radiophysics have stimulated studies of composite materials [1]–[5]. Qualitatively novel properties demonstrable in chiral composites have attracted particular interest in recent years [6]–[12]. These are linear composites—formed by dispersing chiral inclusions in some dielectric or chiral host medium—and their effective constitutive parameters (ECPs) are independent of the field strengths.

A. Priou et al. (eds.), Advances in Complex Electromagnetic Materials, 155–162.

As nonlinear dielectric composite materials have been quite heavily investigated during the last decade [13]–[17], we proposed the concept of a nonlinear chiral composite as a suspension of chiral inclusions in a nonlinear dielectric host medium recently [18]. Hayata and Koshiba [19] made an analogous proposal also, but did not critically inspect the ECPs. Lakhtakia and Weiglhofer [20] developed the Maxwell Garnett model for nonlinear dielectric inclusions immersed in a linear chiral host medium.

In this paper, we look at a weakly nonlinear chiral composite made of two component materials: one of the components is linear chiral, the other is nonlinear dielectric with intensity–dependent permittivity. In Section 2, we use previously derived results [11] to extract the scattering characteristics of an electrically small chiral sphere immersed in another chiral medium. The expressions derived in Section 2 are used to develop the Bruggeman model of the nonlinear chiral composite in Section 3, and the Maxwell Garnett model for a special case is sketched out in Section 4.

2. Scattering by an Electrically Small Chiral Sphere

Suppose an electrically small chiral sphere of volume V is embedded in another chiral medium. The Drude–Born–Fedorov constitutive relations [21, 22] of the external medium are given as

$$\mathbf{D} = \epsilon_a \left[\mathbf{E} + \beta_a \nabla\times\mathbf{E}\right], \quad \mathbf{B} = \mu_a \left[\mathbf{H} + \beta_a \nabla\times\mathbf{H}\right], \tag{1}$$

in which the constitutive parameters ϵ_a, μ_a and β_a are functions of the angular frequency ω. The constitutive relations of the internal medium are given by

$$\mathbf{D} = \epsilon_b \left[\mathbf{E} + \beta_b \nabla\times\mathbf{E}\right], \quad \mathbf{B} = \mu_b \left[\mathbf{H} + \beta_b \nabla\times\mathbf{H}\right], \tag{2}$$

Both media are assumed linear for now—but we shall later apply the expressions given in this section for weakly nonlinear media—and an $\exp(-i\omega t)$ time–dependence is implicit.

Suppose next that a certain time–harmonic field is incident on the sphere. The sphere is electrically small, so the incident field is approximately uniform over the region occupied by the sphere, the incident field amplitudes being denoted by $\{\mathbf{E}_a, \mathbf{H}_a\}$ in that region. The electromagnetic field induced inside the sphere is uniform too, with amplitudes denoted by $\{\mathbf{E}_b, \mathbf{H}_b\}$. Lakhtakia *et al.* [11] have shown that

$$\begin{pmatrix} \mathbf{E}_b \\ \mathbf{H}_b \end{pmatrix} = \begin{pmatrix} F_{11}(\xi_b;\xi_a) & F_{12}(\xi_b;\xi_a) \\ F_{21}(\xi_b;\xi_a) & F_{22}(\xi_b;\xi_a) \end{pmatrix} \begin{pmatrix} \mathbf{E}_a \\ \mathbf{H}_a \end{pmatrix} \tag{3}$$

in matrix notation, where the compact set $\xi = \{\epsilon, \mu, \beta\}$, and

$$
\begin{aligned}
F_{11}(\xi_b;\xi_a) &= 3\epsilon_a[\omega^2\epsilon_b\mu_a\mu_b\beta_b(\beta_a+2\beta_b)-(2\mu_a+\mu_b)]/\Delta(\xi_b;\xi_a),\\
F_{12}(\xi_b;\xi_a) &= -3i\omega\mu_a\mu_b(\beta_a\epsilon_a-\beta_b\epsilon_b)/\Delta(\xi_b;\xi_a),\\
F_{21}(\xi_b;\xi_a) &= 3i\omega\epsilon_a\epsilon_b(\beta_a\mu_a-\beta_b\mu_b)/\Delta(\xi_b;\xi_a),\\
F_{22}(\xi_b;\xi_a) &= 3\mu_a[\omega^2\epsilon_a\epsilon_b\mu_b\beta_b(\beta_a+2\beta_b)-(2\epsilon_a+\epsilon_b)]/\Delta(\xi_b;\xi_a),\\
\Delta(\xi_b;\xi_a) &= \omega^2\epsilon_a\epsilon_b\mu_a\mu_b(\beta_a+2\beta_b)^2-(2\mu_a+\mu_b)(2\epsilon_a+\epsilon_b).
\end{aligned}
\tag{4}
$$

As the sphere is electrically small, the fields scattered by it are the same as if radiated by an electric dipole moment $\mathbf{p}$ and a magnetic dipole moment $\mathbf{m}$ co–located with it [11]; thus,

$$
\begin{pmatrix} \mathbf{p} \\ \mathbf{m} \end{pmatrix} = \begin{pmatrix} a_{11}(\xi_b;\xi_a) & i a_{12}(\xi_b;\xi_a) \\ -i a_{21}(\xi_b;\xi_a) & a_{22}(\xi_b;\xi_a) \end{pmatrix} \begin{pmatrix} \mathbf{E}_a \\ \mathbf{H}_a \end{pmatrix}, \tag{5}
$$

where

$$
\begin{aligned}
a_{11}(\xi_b;\xi_a) &= 3\epsilon_a\left[\omega^2\epsilon_a\epsilon_b\mu_a\mu_b(\beta_a+2\beta_b)(\beta_a-\beta_b)\right.\\
&+ \left.(\epsilon_a-\epsilon_b)(2\mu_a+\mu_b)\right]/\Delta(\xi_b;\xi_a),\\
a_{12}(\xi_b;\xi_a) &= 9\omega\epsilon_a\mu_a\mu_b(\beta_a\epsilon_a-\beta_b\epsilon_b)/\Delta(\xi_b;\xi_a),\\
a_{21}(\xi_b;\xi_a) &= 9\omega\epsilon_a\epsilon_b\mu_a(\beta_a\mu_a-\beta_b\mu_b)/\Delta(\xi_b;\xi_a),\\
a_{22}(\xi_b;\xi_a) &= 3\mu_a\left[\omega^2\epsilon_a\epsilon_b\mu_a\mu_b(\beta_a+2\beta_b)(\beta_a-\beta_b)\right.\\
&+ \left.(\mu_a-\mu_b)(2\epsilon_a+\epsilon_b)\right]/\Delta(\xi_b;\xi_a).
\end{aligned}
\tag{6}
$$

3. Bruggeman Model

Let us now implement the Bruggeman approach for estimating the ECPs of the chosen nonlinear chiral composite. As stated earlier, one material component of this composite is linear chiral, and the other is nonlinear dielectric with intensity–dependent permittivity.

The volumetric proportion of the chiral material is denoted by f, that of the nonlinear dielectric material being $(1-f)$. Electrically small spheres of both component materials are supposed to be dispersed homogeneously and randomly in the homogenized composite. Let the fields induced inside the chiral component be denoted by $\{\mathbf{E}_{ch}, \mathbf{H}_{ch}\}$, and those inside the nonlinear dielectric medium as $\{\mathbf{E}_{nld}, \mathbf{H}_{nld}\}$. The constitutive relations of the two component materials are specified as

$$\mathbf{D} = \epsilon_{ch} \left[\mathbf{E}_{ch} + \beta_{ch} \nabla \times \mathbf{E}_{ch}\right], \quad \mathbf{B} = \mu_{ch} \left[\mathbf{H}_{ch} + \beta_{ch} \nabla \times \mathbf{H}_{ch}\right], \tag{7}$$

and

$$\mathbf{D} = (\epsilon_{nld0} + \epsilon_{nld1} |\mathbf{E}_{nld}|^2)\mathbf{E}_{nld}, \quad \mathbf{B} = \mu_0 \mathbf{H}_{nld}, \tag{8}$$

respectively, where μ_0 is the permeability of free space. While ξ_{ch} is independent of $\{\mathbf{E}_{ch}, \mathbf{H}_{ch}\}$, $\xi_{nld} = \{\epsilon_{nld0} + \epsilon_{nld1}|\mathbf{E}_{nld}|^2, \mu_0, 0\}$ displays weak nonlinearity.

The homogenized composite is assumed to be effectively isotropic and chiral, its constitutive relations being

$$\mathbf{D} = \epsilon_{Br} \left[\mathbf{E}_{eff} + \beta_{Br} \nabla \times \mathbf{E}_{eff}\right], \quad \mathbf{B} = \mu_{Br} \left[\mathbf{H}_{eff} + \beta_{Br} \nabla \times \mathbf{H}_{eff}\right], \tag{9}$$

where $\{\mathbf{E}_{eff}, \mathbf{H}_{eff}\}$ are the macroscopically averaged fields in the composite. Application of the Bruggeman approach and (5) yields the following three simultaneous, independent, nonlinear equations [12]:

$$\begin{aligned} f a_{11}(\xi_{ch}; \xi_{Br}) + (1-f) a_{11}(\xi_{nld}; \xi_{Br}) &= 0, \\ f a_{22}(\xi_{ch}; \xi_{Br}) + (1-f) a_{22}(\xi_{nld}; \xi_{Br}) &= 0, \\ f a_{12}(\xi_{ch}; \xi_{Br}) + (1-f) a_{12}(\xi_{nld}; \xi_{Br}) &= 0. \end{aligned} \tag{10}$$

From (3), it follows that

$$\begin{pmatrix} \mathbf{E}_{nld} \\ \mathbf{H}_{nld} \end{pmatrix} = \begin{pmatrix} F_{11}(\xi_{nld}; \xi_{Br}) & F_{12}(\xi_{nld}; \xi_{Br}) \\ F_{21}(\xi_{nld}; \xi_{Br}) & F_{22}(\xi_{nld}; \xi_{Br}) \end{pmatrix} \begin{pmatrix} \mathbf{E}_{eff} \\ \mathbf{H}_{eff} \end{pmatrix}, \tag{11}$$

and

$$\begin{pmatrix} \mathbf{E}_{ch} \\ \mathbf{H}_{ch} \end{pmatrix} = \begin{pmatrix} F_{11}(\xi_{ch}; \xi_{Br}) & F_{12}(\xi_{ch}; \xi_{Br}) \\ F_{21}(\xi_{ch}; \xi_{Br}) & F_{22}(\xi_{ch}; \xi_{Br}) \end{pmatrix} \begin{pmatrix} \mathbf{E}_{eff} \\ \mathbf{H}_{eff} \end{pmatrix}. \tag{12}$$

We know that naturally occurring linear chiral materials generally have permeabilities close to that of free space; and the nonlinear material component is purely dielectric. Nonlinearity is a small effect, the root cause of nonlinearity in the chosen chiral composite being purely a dielectric effect. Thus, we simply set

$$\begin{aligned} \epsilon_{Br} &= \epsilon_{Br0} + \epsilon_{Br1} \mathbf{E}_{eff} \cdot \mathbf{E}^*_{eff}, \\ \mu_{Br} &= \mu_{Br0} + \mu_{Br1} \mathbf{E}_{eff} \cdot \mathbf{E}^*_{eff}, \\ \beta_{Br} &= \beta_{Br0} + \beta_{Br1} \mathbf{E}_{eff} \cdot \mathbf{E}^*_{eff}. \end{aligned} \tag{13}$$

(Parenthetically, the first terms on the right sides of (13) are the linear terms considered by Kampia and Lakhtakia [12].)

Let us therefore replace ξ_{Br} in (10) with $\xi_{Br}^{(1)} = \{\epsilon_{Br0} + \epsilon_{Br1}|\mathbf{E}_{eff}|^2, \mu_{Br0} + \mu_{Br1}|\mathbf{E}_{eff}|^2, \beta_{Br0} + \beta_{Br1}|\mathbf{E}_{eff}|^2\}$; i.e.,

$$\begin{aligned} f a_{11}(\xi_{ch}; \xi_{Br}^{(1)}) + (1-f) a_{11}(\xi_{nld}; \xi_{Br}^{(1)}) &= 0, \\ f a_{22}(\xi_{ch}; \xi_{Br}^{(1)}) + (1-f) a_{22}(\xi_{nld}; \xi_{Br}^{(1)})) &= 0, \\ f a_{12}(\xi_{ch}; \xi_{Br}^{(1)}) + (1-f) a_{12}(\xi_{nld}; \xi_{Br}^{(1)})) &= 0. \end{aligned} \tag{14}$$

Furthermore, in keeping with (13) let us use [20, 23]

$$\mathbf{E}_{nld} \cong F_{11}(\xi_{nld}; \xi_{Br}^{(1)}) \mathbf{E}_{eff}, \tag{15}$$

thereby ignoring the contribution of $\mathbf{H}_{eff}$ in (11). Following [12], let us symbolically denote the system (14) as

$$h_j(u, v, w; \xi_{ch}; \xi_{nld}; f) = 0, \quad j = 1, 2, 3, \tag{16}$$

where u, v and w respectively denote ϵ_{Br}, μ_{Br} and β_{Br} given in (13). The way to finding an approximate solution of (16) is to expand the left sides as a Maclaurin series with respect to $|\mathbf{E}_{eff}|^2$ and then truncate each series after the terms containing the first derivative.

The zeroth order terms yield the following system of three transcendental equations:

$$h_j(\epsilon_{Br0}, \mu_{Br0}, \beta_{Br0}; \xi_{ch}; \xi_{nld0}; f) = 0, \quad j = 1, 2, 3, \tag{17}$$

where $\xi_{nld0} = \{\epsilon_{nld0}, \mu_0, 0\}$. The system (17) can be solved using a modified Newton–Raphson technique [12].

The first order terms yields the following system of equations for ϵ_{Br1}, μ_{Br1} and β_{Br1}:

$$\epsilon_{Br1} \frac{\partial h_j^{(0)}}{\partial u} + \mu_{Br1} \frac{\partial h_j^{(0)}}{\partial v} + \beta_{Br1} \frac{\partial h_j^{(0)}}{\partial w} + \epsilon_{nld1} \frac{d|\mathbf{E}_{nld}|^2}{d|\mathbf{E}_{eff}|^2} \frac{\partial h_j^{(0)}}{\partial \epsilon_{nld}} = 0, \tag{18}$$

where $j = 1, 2, 3$. The superscript (0) in these equations indicates that the derivatives are evaluated at $u = \epsilon_{Br0}, v = \mu_{Br0}$ and $w = \beta_{Br0}$ determined from (17) and ignoring ϵ_{nld1}. Next we use the relation (15) after neglecting the nonlinear terms therein. As a result, (18) can be rewritten in the following form:

$$\epsilon_{Br1} \frac{\partial h_j^{(0)}}{\partial u} + \mu_{Br1} \frac{\partial h_j^{(0)}}{\partial v} + \beta_{Br1} \frac{\partial h_j^{(0)}}{\partial w} = -\epsilon_{nld1} |F_{11}(\xi_{nld0}; \xi_{Br0})|^2 \frac{\partial h_j^{(0)}}{\partial \epsilon_{nld}}, \tag{19}$$

where $j = 1, 2, 3$. Equations (17) and (19) allow us to determine the ECPs of the chosen nonlinear chiral composite within the framework of the Bruggeman model.

4. Maxwell Garnett Model

Equations (17) and (19) admit only numerical solution. But in many instances, the Bruggeman model can be reduced to the simpler Maxwell Garnett model [24], which yields closed–form expressions for the ECPs.

Let $f \ll 1$. Within the framework of the Maxwell Garnett model, the specific chiral composite is described by constitutive relations (9) but with ϵ_{Br}, μ_{Br} and β_{Br} replaced by ϵ_{MG}, μ_{MG} and β_{MG}, respectively. Analogous to the relations given in [12] for linear chiral composites, we have

$$\begin{aligned}
\epsilon_{MG} &= \epsilon_{nld}\Delta_{MG}/\left\{[(2\epsilon_{nld}+\epsilon_{ch})+f(\epsilon_{nld}-\epsilon_{ch})][(2\mu_0+\mu_{ch}) \right. \\
&\quad \left. -2f(\mu_0-\mu_{ch})]+2\omega^2(f-1)(f+2)\epsilon_{nld}\mu_0\epsilon_{ch}\mu_{ch}\beta_{ch}^2\right\}, \\
\mu_{MG} &= \mu_0\Delta_{MG}/\left\{[(2\epsilon_{nld}+\epsilon_{ch})-2f(\epsilon_{nld}-\epsilon_{ch})][(2\mu_0+\mu_{ch}) \right. \\
&\quad \left. +f(\mu_0-\mu_{ch})]+2\omega^2(f-1)(f+2)\epsilon_{nld}\mu_0\epsilon_{ch}\mu_{ch}\beta_{ch}^2\right\}, \\
\beta_{MG} &= 9f\epsilon_{ch}\mu_{ch}\beta_{ch}/\Delta_{MG},
\end{aligned} \tag{20}$$

where

$$\begin{aligned}
\Delta_{MG} &= [(2\epsilon_{nld}+\epsilon_{ch})-2f(\epsilon_{nld}-\epsilon_{ch})][(2\mu_0+\mu_{ch})-2f(\mu_0-\mu_{ch})] \\
&\quad -4\omega^2(f-1)^2\epsilon_{nld}\mu_0\epsilon_{ch}\mu_{ch}\beta_{ch}^2,
\end{aligned} \tag{21}$$

and

$$\epsilon_{nld} = \epsilon_{nld0} + \epsilon_{nld1}|\mathbf{E}_{nld}|^2. \tag{22}$$

Let us suppose $\epsilon_{MG} = \epsilon_{MG0}+\epsilon_{MG1}|\mathbf{E}_{eff}|^2$, $\mu_{MG} = \mu_{MG0}+\mu_{MG1}|\mathbf{E}_{eff}|^2$, and $\beta_{MG} = \beta_{MG0} + \beta_{MG1}|\mathbf{E}_{eff}|^2$. The two–term Maclaurin expansions of the right sides of (20) with respect to $|\mathbf{E}_{nld}|^2$ then lead, in view of (15), to the following ECP estimates:

$$\begin{aligned}
\epsilon_{MG} &= \epsilon_{MG0} + \epsilon_{nld1}\frac{\partial\epsilon_{MG}^{(0)}}{\partial\epsilon_{nld}}|F_{11}(\xi_{nld0};\xi_{MG0})|^2|\mathbf{E}_{eff}|^2, \\
\mu_{MG} &= \mu_{MG0} + \epsilon_{nld1}\frac{\partial\mu_{MG}^{(0)}}{\partial\epsilon_{nld}}|F_{11}(\xi_{nld0};\xi_{MG0})|^2|\mathbf{E}_{eff}|^2, \\
\beta_{MG} &= \beta_{MG0} + \epsilon_{nld1}\frac{\partial\beta_{MG}^{(0)}}{\partial\epsilon_{nld}}|F_{11}(\xi_{nld0};\xi_{MG0})|^2|\mathbf{E}_{eff}|^2,
\end{aligned} \tag{23}$$

where $\xi_{MG0} = \{\epsilon_{MG0}, \mu_{MG0}, \beta_{MG0}\}$. The use of the right side of (15) in lieu of $|\mathbf{E}_{nld}|^2$ here lends (23) a Bruggeman*esque* flavor, which is fully justified from a local field perspective [24].

On replacing $\epsilon_{MG}, \mu_{MG}, \beta_{MG}$ and ϵ_{nld} by $\epsilon_{MG0}, \mu_{MG0}, \beta_{MG0}$ and ϵ_{nld0}, respectively, in (20), the linear quantities $\epsilon_{MG0}, \mu_{MG0}$ and β_{MG0} may be obtained. In (23), the superscript (0) indicates that the derivatives are to be evaluated after replacing ϵ_{nld} by ϵ_{nld0}. The final expressions for the nonlinear ECPs are cumbersome to write down, but they can be easily manipulated using symbolic algebra programs such as Mathematica and Macsyma.

5. Conclusion

In order to validate our results, we set $\beta_{ch} = 0$ and $\mu_{ch} = \mu_0$ in the equations obtained earlier. Then, it follows from (4) that $F_{11}(\xi_{nld}; \xi_{Br}^{(1)}) \rightarrow 1$ when $f \rightarrow 0$. Furthermore, (14) reduces to the formula given by Zeng *et al* [13] for nonlinear dielectric composites. Finally, the results of Section 3 reduce to those of Hui [25] for dielectric composites when $f \rightarrow 1$.

The main result of the presented work can be stated in the following manner: all three ECPs of a composite comprised by a linear chiral material and a cubic nonlinear dielectric material turn out to be nonlinear *in identical fashion.* This result holds for both the Bruggeman and the Maxwell Garnett models, but was not accounted for by Hayata and Koshiba [19], who considered only the permittivity to be nonlinear. Such an approximation corresponds to the particular case obtained by setting $\alpha_1 = \alpha_2 = \beta_1 = \beta_2$ in Eq. (11) of [18].

The approach developed in this paper provides the basis for the theoretical description of nonlinear chiral composites. Such artificial media seem to have an essential potentiality for applications owing to their unusual properties. Recent advances in the nano–technology afford a possibility for the design of nonlinear chiral composites (dielectric and magnetic) for the visible range. Using a semiconductor as a host medium, one can obtain an artificial chiral semiconductor for the microwave range.

Acknowledgments: S.A. Maksimenko and G.Ya. Slepyan are supported by the International Science Foundation under Contract No U9Y000.

References

1. Stroud, D. (1975) Generalized effective–medium approach to the conductivity of an inhomogeneous material, *Phys. Rev. B* **12**, 3368–3373.
2. Bergman, D. J. (1978) The dielectric constant of a composite material – a problem in classical physics, *Phys. Rep.* **43**, 377–407.

3. Ward, L. (1988) *The optical constants of bulk materials and films,* Adam Hilger, Bristol, UK.
4. Belevtsev, B. I. (1990) Superconductivity and electron localization in disordered two-dimensional metal systems, *Usp. Fiz. Nauk* **160**, 65-98.
5. A. Lakhtakia (ed) (1996) *Selected papers on linear optical composite materials,* SPIE Opt. Engg. Press, Bellingham, WA, USA.
6. Lakhtakia, A., Varadan, V.K., and Varadan, V.V. (1990) Dilute random distribution of small chiral spheres, *Appl. Opt.* **28**, 3627–3632.
7. Sihvola, A.H. and Lindell, I.V. (1992) Analysis of chiral mixtures, *J. Electromagn. Waves Appl.* **7**, 553–572.
8. Lakhtakia, A., Varadan, V.K., and Varadan, V.V. (1993) On the Maxwell–Garnett model for chiral composites, *J. Mater. Res.* **8**, 917–922.
9. Mariotte, F. (1994) *Heterogeneous chiral materials: modeling and applications,* D. Sc. Dissertation, University of Bordeaux I, Bordeaux, France.
10. Lakhtakia, A. and Shanker, B. (1993) Beltrami fields within continuous source regions, volume integral equations, scattering algorithms, and the extended Maxwell–Garnett model, *Int. J. Appl. Electromagn. Mater.* **4**, 65–82.
11. Lakhtakia, A., Varadan, V.K., and Varadan, V.V. (1991) Effective properties of a sparse random distribution of non–interacting small chiral spheres in a chiral host medium, *J. Phys. D: Appl. Phys.* **24**, 1–6.
12. Kampia, R. and Lakhtakia, A. (1992) Bruggeman model for chiral particulate composites, *J. Phys. D: Appl. Phys.* **25**, 1390–1394.
13. Zeng, X.C., Bergman, D.J., Hui, P.M., and Stroud, D. (1988) Effective–medium theory for weakly nonlinear composites, *Phys. Rev. B.* **38**, 10970–10973.
14. Levy O. and Bergman, D.J. (1992) Clausius–Mossotti approximation for a family of nonlinear composites, *Phys. Rev. B* **46**, 7189–7192.
15. Yu, K.W., Hui, P.M., and Stroud, D. (1993) Effective dielectric response of nonlinear composites, *Phys. Rev. B* **47**, 1450–1456.
16. Levy, O., Yagil, Y., and Bergman, D.J. (1994) Field–induced tuning of the optical properties of nonlinear composites near resonance, *J. Appl. Phys.* **76**, 1431–1435.
17. Levy, O. (1995) Nonlinear properties of partially resonant composites, *J. Appl. Phys.* **77**, 1696–1700.
18. Slepyan, G.Ya., Maksimenko, S.A., Bass, F.G., and Lakhtakia, A. (1995) Nonlinear electromagnetics in chiral media: self–action of waves, *Phys. Rev. E* **52**, 1049–1058.
19. Hayata, K. and Koshiba, M. (1995) Chirosolitons: unique spatial solitons in chiral media, *IEEE Trans. Microw. Theory Tech.* **43**, 1814–1818.
20. Lakhtakia, A. and Weiglhofer, W.S. (1996) Maxwell Garnett approach for nonlinear dilute particulate composites with bi–isotropic host media, *Int. J. Electron.* **80**, 665–676.
21. Fedorov, F.I. (1976) *Theory of gyrotropy,* Nauka i Tekhnika, Minsk, Belarus.
22. Lakhtakia, A. (1994) *Beltrami fields in chiral media,* World Scientific, Singapore.
23. Lakhtakia, A. and Weiglhofer, W.S. (1996) Dilute nonlinear particulate composites comprised by linear bianisotropic inclusions in nonlinear dielectric host media, *Opt. Engg.* **35**, 1112–1118.
24. Aspnes, D.E. (1982) Local–field effects and effective–medium theory: a microscopic perspective., *Am. J. Phys.* **50**, 704–709.
25. Hui, P.M. (1993) Higher order nonlinear response in dilute random composites, *J. Appl. Phys.* **73**, 4072–4073

OPTICALLY INDUCED ROTATING SPATIALLY UNIFORM STRUCTURE IN CHIRAL MEDIA

I.N. AKHRAMENKO AND I.V. SEMCHENKO
Department of Physics, Gomel State University,
Sovyetskaya Str. 104, 246699, Gomel, Belarus

1. Introduction

Recently, a great number of investigations have been done on the parametric interactions of electromagnetic waves in gyrotropic media. This is because of the possibilities of applying these effects in the design of devices controlling various characteristics of electromagnetic signals in SHF (super high frequency) and IR (infra-red) bands. In an earlier article [1] the interaction of electromagnetic waves in a crystal with spirally modulated transient permittivity was considered. The present paper is devoted to the study of the electromagnetic wave propagation in a naturally gyrotropic crystal in which the induced anisotropy is spatially uniform and rotates in time. This case is interesting because it allows an exact solution of Maxwell's equations for a non-stationary medium.

2. Creation of the Anisotropy in the Medium

In the following scheme, the idea is to make use of the Kerr effect to create forced anisotropy in a naturally gyrotropic crystal. The Kerr effect is a nonlinear optical phenomenon, and a sufficiently strong electric field is required. Let us consider two intensive circularly polarized electromagnetic waves

$$\tilde{\mathbf{E}}_{1,2} = \tilde{\mathrm{E}}_0 \mathbf{n}_{\pm} \exp\left[i\left(K_{1,2}z - \Omega_{1,2}t\right)\right] \tag{1}$$

These two waves have the opposite directions of rotation which is described by the unit circular polarization vectors

$$\mathbf{n}_{\pm} = \frac{\mathbf{u}_x \pm i\mathbf{u}_y}{\sqrt{2}}$$

A. Priou et al. (eds.), Advances in Complex Electromagnetic Materials, 163–168.

and different frequencies Ω_1 and Ω_2. This external influence can cause in a crystal a rotating anisotropy with the permittivity tensor of the following form:

$$\epsilon\,(z,t) = U(z,t)\epsilon U^{-1}(z,t) \tag{2}$$

Here, the rotation operator $U(z,t) = \exp(\varphi \mathbf{u}_z^{\times})$ has the effect of a turn by the angle φ in the xy-plane [2]. The uniaxial permittivity tensor ϵ is spatially and temporally constant in the rotating coordinate system and has the form

$$\epsilon = \epsilon_0 - 2\Delta\epsilon \mathbf{u}_x \cdot \mathbf{u}_x \tag{3}$$

ϵ_0 is the undisturbed permittivity tensor, $\Delta\epsilon$ describes the induced anisotropy. Here $\mathbf{u}_x \cdot \mathbf{u}_x$ is the dyadic product of the unit vectors $\mathbf{u}_x$ and $\mathbf{u}_x$, where the simple scalar ϵ_0 is understood as being multiplied by the unit dyadic.

There are conditions when the induced anisotropy becomes spatially uniform. This happens when the wave numbers are equal:

$$K_1(\Omega_1) = K_2(\Omega_2) \tag{4}$$

In this case the gyrotropy contribution is compensated by the difference of frequencies of the modulating waves [3]:

$$(\Omega_2 - \Omega_1)\sqrt{\epsilon_0} = \alpha(\Omega_1 + \Omega_2) \tag{5}$$

Here α is the gyrotropy parameter of the crystal. The induced optical axis rotates in time around the direction of the propagation of external waves (z-direction) with the frequency $\Delta\Omega = (\Omega_2 - \Omega_1)/2$. Usually, the rotation frequency is about 10^{11} Hz. This effect is illustrated in Figure 1.

3. Propagation of a Weak Electromagnetic Signal in the Medium with Induced Fast Rotating Uniaxial Anisotropy

Next, let us consider the propagation of a weak electromagnetic wave with the frequency $\omega_0 \ll \Omega_{1,2}$ in such a medium. The propagation is described by the wave equation

$$\nabla_z \nabla_z \mathbf{E} + \frac{1}{c^2}\frac{\partial^2}{\partial t^2}(\epsilon(t)\mathbf{E}) = 0 \tag{6}$$

Here c is the velocity of light in vacuum. Please note that the permittivity $\epsilon(t)$ is spatially uniform. Let us make the transformation to the rotating coordinate system ($z' = z$):

$$\begin{pmatrix} x' \\ y' \\ z' \end{pmatrix} = U(t) \begin{pmatrix} x \\ y \\ z \end{pmatrix} \tag{7}$$

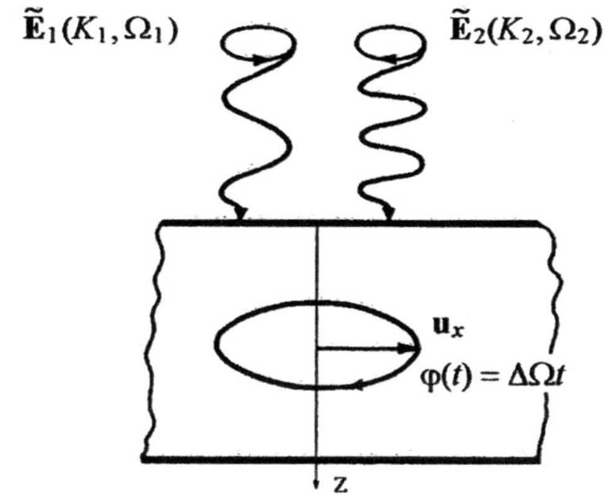

Figure 1. Creation of the rotating and spatially uniform anisotropy in naturally gyrotropic nonlinear crystal with the use of two circular light waves. Note that the circle lies in the xy-plane

This system is connected with the rotation of the induced axis $\mathbf{u}_x$, which allows us to determine the exact solution of Maxwell's equations [3]. Furthermore, one can find an exact solution for the dispersion equation which determines the wave numbers $k_1, \cdots k_4$ and ellipticities $\xi_1, ...\xi_4$ of the eigenmodes:

$$k_{1,2}(\omega) = \left\{ \frac{1}{2} \left[\frac{\omega_2^2}{c^2} \bar{\epsilon}(\omega_2) + \frac{\omega_1^2}{c^2} \bar{\epsilon}(\omega_1) \right] \pm \right.$$

$$\left. \left[\frac{1}{4} \left(\frac{\omega_2^2}{c^2} \bar{\epsilon}(\omega_2) - \frac{\omega_1^2}{c^2} \bar{\epsilon}(\omega_1) \right)^2 + \frac{\omega_1^2 \omega_2^2}{c^4} \Delta\epsilon(\omega_1) \Delta\epsilon(\omega_2) \right]^{1/2} \right\}^{1/2}$$

$$k_{3,4} = -k_{2,1} \tag{8}$$

$$\xi_i(\omega) = \frac{\Delta\epsilon(\omega_1)\omega_2^2}{\bar{\epsilon}(\omega_2)\omega_2^2 - k_i^2 c^2} \tag{9}$$

The parameters used in this expressions are: $\omega_{1,2} = \omega \pm \Delta\Omega$, $\bar{\epsilon} = \epsilon_0 - \Delta\epsilon$.

4. Reflection and Transmission

Let us consider the problem of reflection from and transmission through a slab of the medium described above. The result is [3] that the only one

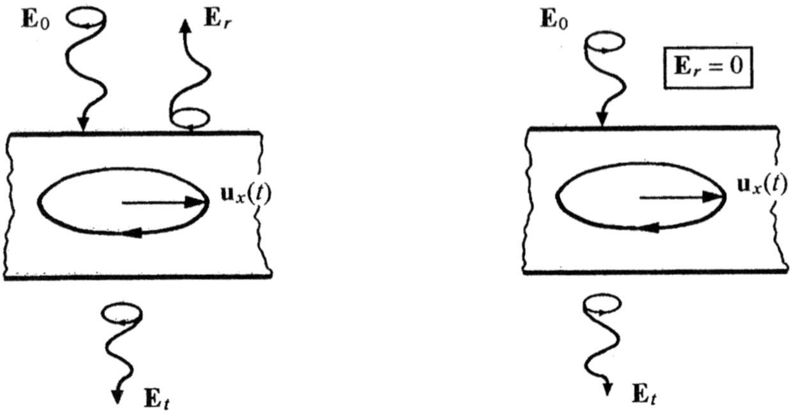

Figure 2. Selective reflection of a weak electromagnetic wave from a crystal with induced anisotropy. Note that the incoming wave $\mathbf{E}_0$ is much weaker than the external high frequency waves $\tilde{\mathbf{E}}_1$ and $\tilde{\mathbf{E}}_2$ responsible for the forced anisotropy

circularly polarized wave can be reflected from the slab. The only reflected wave has to have $\omega_0 \approx \Delta\Omega$ and direction of rotation of the vector $\mathbf{E}_0$ that coincides with that of the induced optical axis. This is illustrated in Figure 2.

The frequency dependence of the reflection (R) and transmission (T) coefficients obey the following expressions:

$$R = \frac{\Delta\epsilon^2 \Delta\Omega^2 \sin^2 [\Delta k(\beta) L/2]}{4\bar{\epsilon}\beta^2 + \Delta\epsilon^2 \Delta\Omega^2 \cos^2 [\Delta k(\beta) L/2]} \tag{10}$$

$$T = \frac{\Delta\epsilon^2 \Delta\Omega^2 + 4\bar{\epsilon}^2\beta^2}{4\bar{\epsilon}\beta^2 + \Delta\epsilon^2 \Delta\Omega^2 \cos^2 [\Delta k(\beta) L/2]} \tag{11}$$

where L is the crystal thickness, $\beta = \omega_0 - \Delta\Omega$, and $\Delta k = k_1 - k_2$.

Moreover, selective amplification of the weak electromagnetic wave has been demonstrated to be possible. The parameter $Q = R + T - 1$ can be introduced to describe the amplification of the test signal. The frequency dependencies of the reflection (R) and transmission (T) coefficients as well as of the parameter Q are represented on Figure 3.

5. Polarization Rotation

Finally, the rotation of the polarization plane of the wave propagating in the considered medium has been studied. The $\theta = \varphi/L$, where φ is the

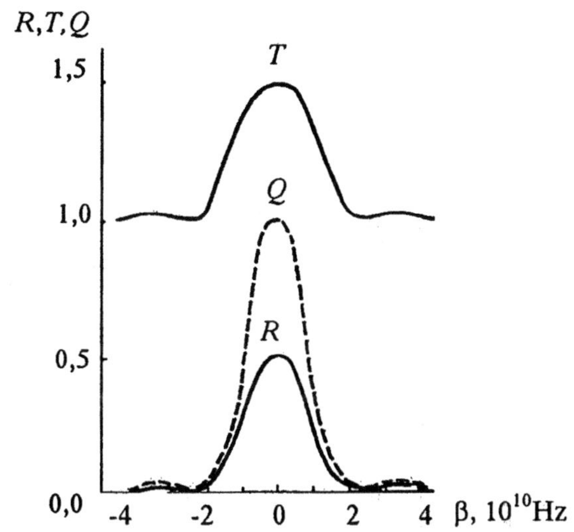

Figure 3. The frequency dependencies of the reflection (R) coefficient, transmission (T) coefficient and the parameter Q. Here $\beta = \omega_0 - \Delta\Omega$

rotation angle of the principal axis of polarization ellipse. The following formula has been obtained for θ:

$$\theta = [k_1(\omega_0 - \Delta\Omega) - k_2(\omega_0 + \Delta\Omega)]/2 = \frac{\omega_0^4 \Delta\epsilon^2}{8\bar{\epsilon}\sqrt{\bar{\epsilon}}c\Delta\Omega(\omega_0^2 - \Delta\Omega^2)} \tag{12}$$

This expression may be transformed into the well-known de Vries formula for the rotation ability of the cholesteric liquid crystals (CLC) [4]. This can be seen after the substitutions $\Delta\Omega\sqrt{\bar{\epsilon}}/c \to q$ and $q = 2\pi/p$, where p is the cholesteric spiral pitch. Figure 4 illustrates the frequency dependence of the rotation angle of the polarization ellipse principal axis.

6. Conclusion

It has been demonstrated that crystals with spatially uniform rotating anisotropy, induced by two intensive circularly polarized light waves, possess optical properties similar to those of cholesteric liquid crystals [5]. It is possible to achieve selective reflection of a weak electromagnetic signal. This requires that the frequency and the polarization rotation direction of the

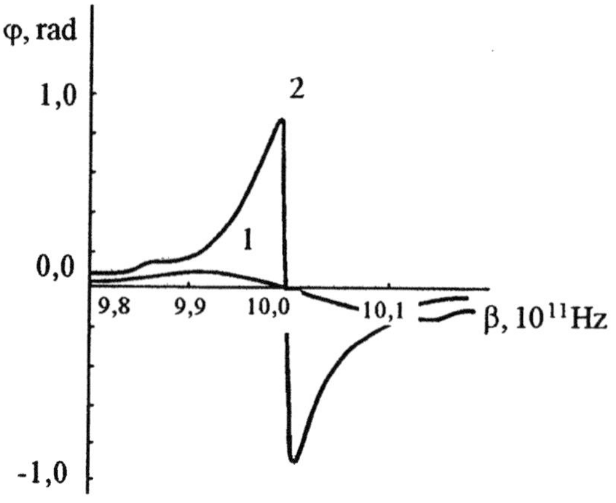

Figure 4. The frequency dependence of the rotation angle of the principal axis of the polarization ellipse. The parameters of the curves are (1): L=5 cm, (2): L=10 cm

signal coincide with the frequency and rotation direction of the induced uniaxial structure. Parametric amplification of the signal is also possible under the same conditions. The frequency of the signal is in the super high frequency (SHF) or infra-red (IR) band. Gyrotropic properties of the crystal with induced spatially uniform rotating anisotropy were indeed found to be similar to those of cholesteric liquid crystals [5].

References

1. Akhramenko, I.N. and Semchenko, I.V. (1994) Electromagnetic waves in chiral crystals with optically induced rotating helicoidal anisotropy. *"Chiral'94"*, Proceedings, Périgueux, France, (editors: Frederic Mariotte and Jean-Paul Parneix), pp. 439–443.
2. Fedorov, F.I. (1979) *Theory of gyrotropy*, Nauka i Tekhnica, Minsk, (in Russian).
3. Akhramenko, I.N., Semchenko, I.V. and Serdukov, A.N. (1988) Selective amplification of electromagnetic waves in the medium with rotating uni-axial structure, *Journal Prikladnoi Spectroscopii (Journal of Applied Spectroscopy)*, **49(2)**, 304–308.
4. P.G. de Gennes (1977) *The Physics of Liquid Crystals*, Mir, Moscow, (Russian translation of the French original (1974)).
5. Belyakov, V.A. and Sonin, A.S. (1982) *Cholesteric liquid crystals optics.* Nauka, Moscow, (in Russian)

FARADAY EFFECT AND MAGNETOGYRATION IN SUPERLATTICES IN THE LONG WAVELENGTH APPROXIMATION

E.G. STARODUBTSEV, I.V. SEMCHENKO, G.S. MITYURICH
Department of Physics, Gomel State University
Sovyetskaya Str. 104, 246699, Gomel, Belarus

1. Introduction

The period of superlattices in practise is often from 1 to 10 nm [1].[1] This makes it possible to use the long wavelength approximation in analysis of superlattices, in a wide spectral range of electromagnetic radiation. In this case a superlattice can be considered as a homogeneous medium characterized by a set of effective parameters: dielectric, magnetic, non-linear optic and other (see References, e.g., [2, 3]). Natural gyrotropic properties of a superlattice satisfying the requirement of the long wavelength approximation were determined in [2, 3]. As we believe, the only realizable superlattices formed by naturally optically active crystals are structures which include layers of SiO_2 [4, 5]. At the same time practically all experimentally realized superlattices can possess induced gyrotropy properties whose phenomenological theory was not developed for superlattices, as far as we know.

The aim of this paper is the calculation (within the frame of the long wavelength approximation) of tensors characterizing usual and induced [6] (in presence of governing electric field) Faraday effects and magnetogyration in superlattices. The main features of these effects exhibited in superlattices are also investigated.

2. Calculation of Effective Tensors

In the long wavelength approximation one can assume that electromagnetic field existing in a superlattice does not change at distances of the order of the superlattice period $D = d^{(1)} + d^{(2)}$, where $d^{(r)}$ (here and thereafter

[1]Superlattices are periodically layered crystal structures.

A. Priou et al. (eds.), Advances in Complex Electromagnetic Materials, 169–176.

$r = 1, 2$) are the thicknesses of the monocrystal layers constituting the superlattice. Let us choose an orthogonal system of coordinates with the axis Z perpendicular to the boundaries of the layers forming the superlattice. From the condition of additivity of the electric and magnetic moments in the volume of superlattice [2, 7] we have the relations connecting the normal components of the field strengths $E_3^{(e)}, H_3^{(e)}$ and tangential components of the inductions $D_r^{(e)}, B_r^{(e)}$ of electromagnetic field in the superlattice and in the layers:

$$A^{(e)} = xA^{(1)} + (1-x)A^{(2)} \tag{1}$$

where $A = E_3, H_3, D_r, B_r, \quad x = d^{(1)}/D$, and indices $e, 1, 2$ define the quantities characterizing the effective medium, the first and the second monocrystal layers, respectively.

When defining the effective superlattice characteristics, we will suppose that the phenomenological material equations [8]

$$\mathbf{D}^{(s)} = \varepsilon^{(s)}\mathbf{E}^{(s)} + i\alpha^{(s)}\mathbf{H}^{(s)}, \quad \mathbf{B}^{(s)} = \mu^{(s)}\mathbf{H}^{(s)} - i\alpha_t^{(s)}\mathbf{E}^{(s)} \tag{2}$$

hold both for the monocrystals forming the superlattice and for the effective medium: $s = 1, 2, e$. Here, the index t denotes the transpose tensor, and $i^2 = -1$. Eqs. (2) are defined by the complex nonsymmetric tensors $\varepsilon^{(s)}, \mu^{(s)}, \alpha^{(s)}$ which describe media with various kinds of anisotropy, gyrotropy, and absorption.

As usually, components of the tensors of the permittivity and permeability ε, μ are significantly larger than the components of the pseudotensor α, which describes the effects of optical activity [8]. All the calculations below are made in the linear (in components of α) approximation.

According to the methods proposed in [7], we have from Eqs. (2)

$$\begin{aligned}
E_3 = \frac{1}{\varepsilon_{33}} \Bigg[& D_3 - \varepsilon_{31}E_1 - \varepsilon_{32}E_2 - i\left(\alpha_{31} - \frac{\mu_{31}}{\mu_{33}}\alpha_{33}\right)H_1 \\
& - i\left(\alpha_{32} - \frac{\mu_{32}}{\mu_{33}}\alpha_{33}\right)H_2 - i\frac{\alpha_{33}}{\mu_{33}}B_3\Bigg] \\
H_3 = \frac{1}{\mu_{33}} \Bigg[& B_3 - \mu_{31}H_1 - \mu_{32}H_2 + i\left(\alpha_{13} - \frac{\varepsilon_{31}}{\varepsilon_{33}}\alpha_{33}\right)E_1 \\
& + i\left(\alpha_{23} - \frac{\varepsilon_{32}}{\varepsilon_{33}}\alpha_{33}\right)E_2 + i\frac{\alpha_{33}}{\varepsilon_{33}}D_3\Bigg]
\end{aligned} \tag{3}$$

Substitution of expressions (3) into (1) and taking into account continuity of the quantities E_r, H_r, D_3, B_3 on the layer boundaries, defines the expressions

$$\frac{1}{\varepsilon_{33}}, \quad \frac{\varepsilon_{3j}}{\varepsilon_{33}}, \quad \frac{\varepsilon_{r3}}{\varepsilon_{33}}, \quad \varepsilon_{rj} - \frac{\varepsilon_{r3}\varepsilon_{3j}}{\varepsilon_{33}} \tag{4}$$

$$\frac{1}{\mu_{33}}, \quad \frac{\mu_{3j}}{\mu_{33}}, \quad \frac{\mu_{r3}}{\mu_{33}}, \quad \mu_{rj} - \frac{\mu_{r3}\mu_{3j}}{\mu_{33}} \tag{5}$$

$$\begin{gathered}\frac{\alpha_{33}}{\mu_{33}\varepsilon_{33}}, \quad \frac{1}{\varepsilon_{33}}(\alpha_{3m} - \alpha_{33}\frac{\mu_{3m}}{\mu_{33}}), \quad \frac{1}{\mu_{33}}(\alpha_{m3} - \alpha_{33}\frac{\varepsilon_{m3}}{\varepsilon_{33}}) \\ \alpha_{mn} - \alpha_{3n}\frac{\varepsilon_{m3}}{\varepsilon_{33}} - (\alpha_{m3} - \alpha_{33}\frac{\varepsilon_{m3}}{\varepsilon_{33}})\frac{\mu_{3n}}{\mu_{33}}\end{gathered} \tag{6}$$

Here the indices r, j, m, n take the values 1 and 2. Quantities (4)–(6) (summation over repeating indices is not made) when substituted instead of the parameter A in equation (1) define all the components of the tensors $\varepsilon^{(e)}$, $\mu^{(e)}$, $\alpha^{(e)}$ for arbitrary crystallographic symmetry of the layers. Note that nonsymmetry of the tensors $\varepsilon^{(s)}$, $\mu^{(s)} (s = 1, 2, e)$ leads to differences of the expressions (4)–(6) from the analogous expressions for symmetric tensors $\varepsilon^{(s)}$, $\mu^{(s)}$ [3, 7].

The influence of low-frequency magnetic field $\mathbf{H}^0$ on the material tensors of the layers and the superlattice can be described by the relations [9]

$$\varepsilon^*_{lp} = \varepsilon_{lp} + i\gamma_{lpk}H^0_k, \quad \mu^*_{lp} = \mu_{lp} + i\xi_{lpk}H^0_k, \quad \alpha^*_{lp} = \alpha_{lp} + i\nu_{lpk}H^0_k \tag{7}$$

where $l, p, k = 1, 2, 3$, and the asterisk marks the components of the tensors disturbed by the field $\mathbf{H}^0$. $\varepsilon_{lp}, \mu_{lp}$ are the nondisturbed parts of the tensors of permittivity and permeability, γ_{lpk}, ξ_{lpk} and ν_{lpk} are the axial and polar third-rank tensors describing the Faraday effect and the influence of magnetic field on optical activity (linear magnetogyration), respectively [9, 10].

Expressions (4)–(6), and (1) are correct both for disturbed and undisturbed tensors $\varepsilon^{(e)}, \mu^{(e)}, \alpha^{(e)}$. At the same time, the expansion of the quantities (4)–(6) into the Taylor series with an accuracy to the first order in small parameters $\varepsilon^*_{lp} - \varepsilon_{lp}, \mu^*_{lp} - \mu_{lp}, \alpha^*_{lp} - \alpha_{lp}$ and the following averaging (1) leads to the quantities:

$$\begin{gathered}(\varepsilon_{33}\mu^0_{33})^{-1}\gamma_{3r3}, \quad \varepsilon^{-1}_{33}(\gamma_{3rj} - m^0_j\gamma_{3r3}) \\ (\mu^0_{33})^{-1}(\gamma_{rj3} - e_r\gamma_{3j3} - e_j\gamma_{r33}) \\ \gamma_{rjp} - e_r\gamma_{3jp} - e_j\gamma_{r3p} - m^0_p(\gamma_{rj3} - e_r\gamma_{3j3} - e_j\gamma_{r33})\end{gathered} \tag{8}$$

$$
\begin{aligned}
&(\varepsilon_{33}\mu_{33}\mu_{33}^0)^{-1}\nu_{333}, \quad (\varepsilon_{33}\mu_{33})^{-1}(\nu_{33r} - m_r^0\nu_{333}) \\
&(\varepsilon_{33}\mu_{33}^0)^{-1}(\nu_{3p3} - m_p\nu_{333}) \\
&(\varepsilon_{33})^{-1}[\nu_{3pj} - m_p\nu_{33j} - m_j^0(\nu_{3p3} - m_p\nu_{333})] \\
&(\mu_{33}\mu_{33}^0)^{-1}(\nu_{p33} - e_p\nu_{333}) \\
&(\mu_{33})^{-1}[\nu_{p3j} - e_p\nu_{33j} - m_j^0(\nu_{p33} - e_p\nu_{333})] \\
&(\mu_{33}^0)^{-1}(\nu_{pr3} + e_p m_r\nu_{333} - e_p\nu_{3r3} - m_r\nu_{p33}) \\
&\nu_{prj} - m_j^0\nu_{pr3} + e_p m_r(\nu_{33j} - m_j^0\nu_{333}) \\
&- e_p(\nu_{3rj} - m_j^0\nu_{3r3}) - m_r(\nu_{p3j} - m_j^0\nu_{p33})
\end{aligned} \tag{9}
$$

where $r, j, p = 1, 2,\ \ m_r = \mu_{r3}/\mu_{33},\ \ m_r^0 = \mu_{3r}^0/\mu_{33}^0,\ \ e_p = \varepsilon_{p3}/\varepsilon_{33}$ and summation over repeating indices is not made. For calculation of (8) and (9) we have taken into account the validity of the relations (1) for the field strength and induction of governing magnetic field, as well as the frequency dispersion: tensor μ corresponds to the frequency of the high-frequency wave, and tensor μ^0 corresponds to the frequency of the governing magnetic field (low-frequency wave). The expressions for the components of tensor $\xi^{(e)}$ follow from expressions (8) after substitutions $\varepsilon_{lp} \to \mu_{lp}, \varepsilon_r \to \mu_r, \gamma_{lpq} \to \xi_{lpq}$ (except the quantities m_j^0 which do not change). Quantities (8) and (9) define all the independent components of tensors $\gamma^{(e)}, \xi^{(e)}, \nu^{(e)}$ after appropriate averaging (1).

It follows from the relations derived above that symmetrical properties of the layers tensors: symmetry of tensors ε, μ and antisymmetry of γ, ξ in the first pair of indices [9] will take place also for the corresponding effective tensors of the superlattice. It makes it possible to express tensors γ, ξ through the second rank tensors (e_{lqp} is the Levi-Civita tensor)

$$
\gamma_{lpk} = e_{lqp}\eta_{qk}, \quad \xi_{lpk} = e_{lqp}\rho_{qk} \tag{10}
$$

describing the Faraday effect. Substitution of relations (10) in (8) gives

$$
\begin{aligned}
&(\varepsilon_{33}\mu_{33}^0)^{-1}\eta_{r3}, \quad \varepsilon_{33}^{-1}(\eta_{rj} - m_j^0\eta_{r3}) \\
&(\mu_{33}^0)^{-1}(\eta_{33} + e_1\eta_{13} + e_2\eta_{23}) \\
&\eta_{3p} + e_1\eta_{1p} + e_2\eta_{2p} - m_p^0(\eta_{33} + e_1\eta_{13} + e_2\eta_{23})
\end{aligned} \tag{11}
$$

where $r, j, p = 1, 2$. The expressions for the quantities $\rho_{lp}^{(e)}$ follow from (11) after the above indicated substitutions and the substitution $\eta_{lp} \to \rho_{lp}$.

From expressions (4) and (11) one can get the expressions defining the components of the effective vector of magnetic gyration of the superlattice $\mathbf{G}^{(e)} = \eta^{(e)}\mathbf{H}^0$ ($\varepsilon^* = \varepsilon + i\mathbf{G}^{\times}$, where "×" indicates a tensor dually conjugate to a vector,[2] $r = 1, 2$). After the averaging (1),

$$\varepsilon_{33}^{-1} G_r, \qquad G_3 + \varepsilon_{33}^{-1}(G_1\varepsilon_{13} + G_2\varepsilon_{23}) \tag{12}$$

Expressions (8), (9), (11), (12) characterize the linear Faraday effect and magnetogyration in a short-period superlattice subject to the long wavelength approximation.

For many important applications there is an interest in the research of possibility to govern the Faraday effect in crystal media with the help of external electric fields [6]. These tasks appear, in particular, when designing modulators of phase nonreciprocity for optical gyroscopes [11]. It was shown [6] that in crystal media the influence of electric field on the Faraday effect can be rather different. For example, there are conditions when the induced Faraday effect takes place but the usual one (without external electric field) is absent. Even more possibilities of governing the Faraday effect by electric field can be expected in superlattices.

When studying the Faraday effect in superlattices at the presence of constant (governing) electric field $\mathbf{E}^0$, we will proceed from the expansion of tensors of dielectric permittivity [6]

$$\varepsilon_{ij}^{(s)}(\mathbf{E}^0, \mathbf{H}^0) = \varepsilon_{ij}^{(s)} + \rho_{ijk}^{(s)} E_k^0 + \beta_{ijk}^{(s)} H_k^0 + \Delta_{ijkl}^{(s)} E_k^0 H_l^0 \tag{13}$$

where $s = 1, 2, e;\quad i, j, k = 1, 2, 3$. In Eq. (13) tensors $\rho_{ijk}^{(s)}$ and $\beta_{ijk}^{(s)}$ describe linear electrooptic effect and the usual Faraday effect. According to the Onsager relations, we have for the tensors $\Delta_{ijkl}^{(s)}$ characterizing the influence of constant electric field on the Faraday effect [6],

$$\Delta_{ijkl}^{(s)} = -\Delta_{jikl}^{(s)} = e_{ijm}\chi_{mkl}^{(s)} \tag{14}$$

where $m = 1, 2, 3,\quad e_{ijm}$ is the Levi-Civita tensor, $\chi_{mkl}^{(s)}$ is a polar tensor of the third rank describing the studied effect. Carrying out the same expansion of relations (4) in Taylor series as that described above, and taking into account Eqs. (13) and (14), we get the expressions defining the

[2]In dyadic notation, this antisymmetric tensor is written using the cross product of the vector with the unit dyadic. *–(Editors)*

tensor $\chi^{(e)}$ through the known tensors [6] of the layers $\chi^{(1)}, \chi^{(2)}$ (using the averaging rule as in (1)):

$$\begin{aligned}
&\varepsilon_{33}^{-1}\chi_{ijp}, \quad (\varepsilon_{33}\varepsilon_{33}^{0})^{-1}\chi_{i3j}\\
&(\varepsilon_{33}\mu_{33}^{0})^{-1}\chi_{ij3}, \quad (\varepsilon_{33}\varepsilon_{33}^{0}\mu_{33}^{0})^{-1}\chi_{i33}\\
&\chi_{3ij}, \quad (\varepsilon_{33}^{0})^{-1}\chi_{33j}\\
&(\mu_{33}^{0})^{-1}\chi_{3j3}, \quad (\varepsilon_{33}\mu_{33}^{0})^{-1}\chi_{333}
\end{aligned} \tag{15}$$

where $i, j, p = 1, 2$. Expressions (15) are obtained for the cases of the layers with diagonal tensors ε, μ (for optical frequencies), ε^0, μ^0 (for low frequencies, corresponding to the fields $\mathbf{E}^0$,$\mathbf{H}^0$).

3. Analysis

Let us examine the Faraday effect in superlattices formed by class $\overline{4}3m$ crystals to which most of the realized superlattices belong. Let us assume $\mu(H^0) = 1$, which corresponds to the case of non-magnetic media. Then relations (11) give the expressions for the following non-zero components of tensor $\eta^{(e)}$

$$\begin{aligned}
&\eta_{11}^{(e)} = \eta_{22}^{(e)} = <\eta/\varepsilon> / <1/\varepsilon>\\
&\eta_{33}^{(e)} = <\eta/\mu^0> / <1/\mu^0>
\end{aligned} \tag{16}$$

where the angular brackets denote the averaging in view of (1), e.g., $<1/\varepsilon>= x/\varepsilon^{(1)} + (1-x)/\varepsilon^{(2)}$. According to relations (16), the general form of the tensor $\eta^{(e)}$ in this case at any x, $0 < x < 1$, coincides with the form of similar tensor for uniaxial crystals, except of classes $3, \overline{3}, 4, \overline{4}, 6, \overline{6}$ [12]. At $x = 0, 1$ from Eqs. (16) we have: $\eta_{11}^{(e)} = \eta_{22}^{(e)} = \eta_{33}^{(e)}$, which corresponds to the limiting transition to cubic monocrystal.

According to formulas (15), in case of the induced Faraday effect in superlattice from class $\overline{4}3m$ crystals tensor $\chi^{(e)}$ has the following non-zero components:

$$\begin{aligned}
&\chi_{132}^{(e)} = \chi_{231}^{(e)} = \frac{<\chi/(\varepsilon\varepsilon^0)>}{<1/\varepsilon><1/\varepsilon^0>}\\
&\chi_{123}^{(e)} = \chi_{213}^{(e)} = \frac{<\chi/(\varepsilon\mu^0)>}{<1/\varepsilon><1/\mu^0>}
\end{aligned} \tag{17}$$

$$\chi_{312}^{(e)} = \chi_{321}^{(e)} = <\chi>$$

Taking into account the form of the tensor χ for monocrystals [6], it follows from relations (17) that tensor $\chi^{(e)}$ with components (17) has the form which is analogous to that for class $\overline{4}2m$ crystals. At normal incidence on a superlattice of a linearly polarized electromagnetic plane wave, which has wave vector perpendicular to the external fields $\mathbf{E}^0, \mathbf{H}^0$, the usual Faraday effect is absent, and the angle of polarization plane rotation is determined only by the induced effect [6]:

$$\phi = (\pi L/\lambda)(x\chi^{(1)} + (1-x)\chi^{(2)})E^0 H^0 \sin(\alpha + \beta) \qquad (18)$$

Here, L is the superlattice thickness, λ is the wavelength, $\chi^{(1)}$, $\chi^{(2)}$ are the components of the tensor χ in the first and the second layers [6], α (β) are the angles between the electric (magnetic) field vectors and the crystallographic axis X.

Let us consider the conditions when values of some effective magnetooptical parameters of the superlattice can exceed the analogous quantities of the superlattice monocrystal components. It follows from expressions (11) and (12), that satisfying the conditions $G_r^{(e)} > G_r^{(1)}, G_r^{(2)}$, where $r = 1, 2$, it is impossible at $0 \le x \le 1$ and for the layers of arbitrary crystallographic symmetry. Conditions $G_3^{(e)} > G_3^{(1)}, G_3^{(2)}$; $G_3^{(e)}(\varepsilon_{rr}^{(e)})^{-1/2} > G_3^{(1)}(\varepsilon_{11}^{(1)})^{-1/2}$, $G_3^{(2)}(\varepsilon_{11}^{(2)})^{-1/2}$ (the last two conditions are equivalent to the amplification of the Faraday rotation in the superlattice at wave normal parallel to the axis Z) are also not satisfied for diagonal tensors of the layers $\varepsilon^{(1)}, \varepsilon^{(2)}$. From relations (1), (4), (12) we get

$$G_3^{(e)} = <G_3> + x(1-x)\theta[x\varepsilon_{33}^{(2)} + (1-x)\varepsilon_{33}^{(1)}]^{-1} \qquad (19)$$

where $\theta = (\varepsilon_{31}^{(1)}) - \varepsilon_{31}^{(2)})(G_1^{(1)} - G_1^{(2)}) + (\varepsilon_{32}^{(1)} - \varepsilon_{32}^{(2)})(G_2^{(1)} - G_2^{(2)})$. From Eq. (19) it follows that the inequalities $G_3^{(e)} > G_3^{(1)} > G_3^{(2)}$ take place if the following conditions are simultaneously satisfied:

$$\begin{aligned} &\theta > \varepsilon_{33}^{(2)}(G_3^{(1)} - G_3^{(2)}), \\ &x > [1 - \frac{\varepsilon_{33}^{(2)}}{\varepsilon_{33}^{(1)}} + \frac{\theta}{\varepsilon_{33}^{(1)}(G_3^{(1)} - G_3^{(2)})}]^{-1}, \end{aligned} \qquad (20)$$

that can take place for non-diagonal tensors $\varepsilon^{(1)}$, $\varepsilon^{(2)}$. It follows from Eq. (19) that inequalities $a^{(e)} > a^{(1)} > a^{(2)}$ are fulfilled simultaneously under conditions

$$x > e\nu(1-d)(1-e)^{-1}(1-\nu)^{-1} \quad (1-e)(1-\nu) > 0 \qquad (21)$$

where $e = \varepsilon^{(1)}/\varepsilon^{(2)}$, $d = a^{(2)}/a^{(1)}$, $\nu = b^{(1)}/b^{(2)}$ and $a = \chi_{132}$, $b = \varepsilon^0$ or $a = \chi_{123}$, $b = \mu^0$.

4. Conclusion

In this paper, the effective magneto-optical tensors of superlattices have been calculated within the frame of the long wavelength approximation. These tensors characterize completely the linear Faraday effect (usual and induced) and magnetogyration in superlattices. It follows from expressions (9), (11), (15) that on the basis of superlattices it is possible to obtain new materials which have wide range of magnetooptical characteristics. Values of some effective magneto-optical parameters of superlattices can exceed the analogous quantities of superlattice components, which can be used for amplification of the induced gyrotropy effects. This effect is due to creation of additional translational symmetry when superlattices are formed.

References

1. Herman, M.A. (1989) *Semiconductor Superlattices*, Mir, Moscow.
2. Semchenko, I.V. (1990) Gyrotropic properties of superlattices in the long wavelength approximation, *Kristallografiya*, **35(5)**, 1047–1050.
3. Gaishun, V.E., Semchenko, I.V., and Serdyukov, A.N. (1993) Gyrotropic properties of superlattices with magnetic structure in the long wavelength approximation, *Kristallografiya*, **38(3)**, 24–27.
4. Plotnikov, A.E., Pudonin, F.A., and Stopachinskiy, V.B. (1987) Crystallization of superlattices $Si-SiO_2$ stimulated by uniaxial periodical potential, *Pis'ma v ZhETF*, **46(11)**, 443–446.
5. Aktsipetrov, O.A., Golovkina, V.N., Zayats, A.I. et al. (1995) Generation of anisotropic second optical harmonic in $Si - SiO_2$ superlattices, *Doklady RAN*, **340(2)**, 171–174.
6. Novikov, M.A. (1988) Effects of induced optical anisotropy in crystals. II. Faraday effect in constant electric field, *Kristallografiya*, **33(5)**, 1189–1193.
7. Djafari Rouhani, B. and Sapriel, J. (1986) Effective dielectric and photoelastic tensors of superlattices in the long wavelength regime, *Phys. Rev. B*, **34(10)**, 7114–7117.
8. Fedorov, F.I. (1976) *Theory of Gyrotropy*, Nauka i Tekhnika, Minsk.
9. Bely, V.N. and Serdyukov, A.N. (1974) About linear influence of magnetic field on optical activity, *Kristallografiya*, **19(6)**, 1279–1280.
10. Vlokh, O.G. (1984) *Phenomena of spatial dispersion in parametric crystal optics*, Vysshaya shkola, L'vov.
11. Okosi T. (Editor) (1990) *Fibre-optical transducers*, Energoatomizdat, Leningrad.
12. Bokut', B.V. and Serdyukov, A.N. (1978) *Principles of theoretical crystal optics, Part 3*, Gomel.

IV.

Scattering and Diffraction from Bi-Anisotropic Structures

ELECTROMAGNETIC SCATTERING BY THREE-DIMENSIONAL ARBITRARY SHAPED CHIRAL OBJECTS

A.G. DMITRENKO, A.I. MUKOMOLOV AND V.V. FISANOV
Tomsk State University,
Siberian Physical & Technical Institute,
1 Revolution Sq., Tomsk, 634050, Russia

Abstract. A numerical method for solving electromagnetic scattering problems for three-dimensional arbitrary-shaped chiral bodies is presented. The mathematical formulation of the method, a short description of the application package and some numerical results are given.

1. Introduction

During the last years the attention of several researchers was directed to studying the electromagnetic phenomena in regions fully or partially occupied by chiral media [1]–[3]. The difference of chiral media from isotropic magnetic-dielectric ones in view of electrodynamics appears in the form of the constitutive relations: the vectors of electric and magnetic inductions are connected with both the electric and magnetic fields. The wave processes in a chiral medium are characterized by two circularly polarized waves which propagate with different phase velocities. This property of the chiral medium gives a hope for the design of waveguide components and antennas with new properties. Progress in the development of artificial microwave chiral materials [4] reinforces the interest in studying all aspects of the electromagnetic interaction with chiral media including the scattering problems. In paper [5], the electromagnetic scattering by the axially symmetric chiral objects was considered. At present, it is possible to solve similar problems without severe limitations on the surface form. For this purpose the discrete sources method [6] can be used. In papers [7,8] a version of the discrete sources method was presented which allows to solve numerically the electromagnetic scattering problems for arbitrary shaped magneto-dielectric bodies. Here, this version is extended towards chiral bod-

A. Priou et al. (eds.), Advances in Complex Electromagnetic Materials, 179–188.

ies. The mathematical formulation, a short description of the application package and some numerical results are presented.

2. Formulation of the Problem and Method of its Solution

Let a homogeneous chiral body D_i with the permittivity ε_i, permeability μ_i and the chirality parameter β, bounded with the surface S, be located in an infinite homogeneous medium D_e with the parameters ε_e, μ_e. The system is excited by a time-harmonic field $\{\vec{E}_0, \vec{H}_0\}$. It is necessary to find the scattered field $\{\vec{E}_e, \vec{H}_e\}$ in D_e. Besides, the field $\{\vec{E}_e, \vec{H}_e\}$ is defined in D_e, and the field $\{\vec{E}_i, \vec{H}_i\}$ exists inside the body D_i.

These fields must satisfy the Maxwell equations

$$\nabla \times \vec{E}_e = i\omega\mu_e\vec{H}_e, \qquad \nabla \times \vec{H}_e = -i\omega\varepsilon_e\vec{E}_e \quad \text{in} \quad D_e \tag{1}$$

$$\nabla\times\vec{E}_i = i\omega\mu_i(\vec{H}_i+\beta\nabla\times\vec{H}_i), \quad \nabla\times\vec{H}_i = -i\omega\varepsilon_i(\vec{E}_i+\beta\nabla\times\vec{E}_i) \quad \text{in} \quad D_i \tag{2}$$

the boundary conditions

$$\vec{n} \times (\vec{E}_i - \vec{E}_e) = \vec{n} \times \vec{E}_0, \quad \vec{n} \times (\vec{H}_i - \vec{H}_e) = \vec{n} \times \vec{H}_0 \quad \text{on} \quad S \tag{3}$$

and the radiation conditions

$$\{\sqrt{\varepsilon_e}\vec{E}_e; \sqrt{\mu_e}\vec{H}_e\} \times \vec{R}/R + \{\sqrt{\mu_e}\vec{H}_e; -\sqrt{\varepsilon_e}\vec{E}_e\} = O(R^{-1}), \quad R \to \infty \tag{4}$$

where $\vec{n}$ is the normal unit vector to the surface S, R is the distance between the body D_i and the observation point.

The solution of this problem is constructed in the following way. Let us introduce two auxiliary surfaces $S_e = K_eS$ and $S_i = K_iS$ which are homothetic[1] to the surface S. The surface S_e with the homothetic coefficient $K_e < 1$ is placed inside D_i, the surface S_i with the homothetic coefficient $K_i > 1$ is placed outside D_i. Let us represent the unknown field $\{\vec{E}_i, \vec{H}_i\}$ in D_i as a sum of fields of fictitious elementary electric dipoles which are placed on the surface S_i in the points $\{M_{n,i}\}_{n=1}^{N_i}$ and are oriented tangentially to S_i. Using the dyadic Green function [2] for the infinite homogeneous chiral medium with the parameters ε_i, μ_i, β, we have the representations

$$\vec{E}_i(M) = \left(\frac{\omega^2\mu_i}{k_i}\right)\left(\frac{\gamma}{k_i}\right)^2 \sum_{n=1}^{N_i}[\gamma_1 \overset{=n,i}{G_1}(M, M_{n,i}) + \gamma_2 \overset{=n,i}{G_2}(M, M_{n,i})] \cdot \vec{p}_\tau^{\,n,i} \tag{5}$$

[1]In the Western literature, the term "homotopic" is used for curves to mean that there exists a continuous deformation between these curves. –*(Editors)*

$$\vec{H}_i(M) = (-i\omega)\left(\frac{\gamma}{k_i}\right)^2 \sum_{n=1}^{N_i}[\gamma_1 \overset{=n,i}{G_1}(M, M_{n,i}) - \gamma_2 \overset{=n,i}{G_2}(M, M_{n,i})] \cdot \vec{p}_\tau^{n,i}, \quad M \in D_i$$

where $k_i = \omega\sqrt{\varepsilon_i \mu_i}$, $\gamma^2 = k_i^2/(1 - k_i^2\beta^2)$; $\gamma_1 = k_i/(1 - k_i\beta)$, $\gamma_2 = k_i/(1 + k_i\beta)$ are the wave numbers of the left- and right polarized waves, $\vec{p}_\tau^{n,i}(n = 1, 2, \ldots, N_i)$ are unknown dipole moments consisting of two components $p_{\tau 1}^{n,i}$ and $p_{\tau 2}^{n,i}$ along the directions $\vec{e}_{\tau 1}^{n,i}$ and $\vec{e}_{\tau 2}^{n,i}$ selected in the tangential plane to S_i in the point $M_{n,i}$, $\overset{=n,i}{G_{1,2}}$ are the elements of the dyadic Green function defined by the expressions

$$\begin{aligned} \overset{=n,i}{G_1}(M, M_{n,i}) &= \frac{k_i}{2\gamma^2}[\gamma_1 \overline{\overline{I}} + \frac{1}{\gamma_1}\nabla\nabla + \nabla \times \overline{\overline{I}}]\Psi_1^i(M, M_{n,i}) \\ \overset{=n,i}{G_2}(M, M_{n,i}) &= \frac{k_i}{2\gamma^2}[\gamma_2 \overline{\overline{I}} + \frac{1}{\gamma_2}\nabla\nabla + \nabla \times \overline{\overline{I}}]\Psi_2^i(M, M_{n,i}) \end{aligned} \tag{6}$$

In Eqs. (6) $\overline{\overline{I}}$ is the unit dyadic, $\Psi_{1,2}^i(M, M_{n,i}) = \exp(i\gamma_{1,2}R_M^{n,i})/(4\pi R_M^{n,i})$ are the scalar Green functions, $R_M^{n,i}$ is the distance from the point $M_{n,i}$ to the point M in D_i. The scattered field $\{\vec{E}_e, \vec{H}_e\}$ in D_e is presented by an analogy with the inner field $\{\vec{E}_i, \vec{H}_i\}$ as the sum of fields of the tangentially oriented electric dipoles with unknown moments $\vec{p}_\tau^{n,e}(n = 1, 2, \ldots, N_e)$ placed on the auxiliary surface S_e in the points $\{M_{n,e}\}_{n=1}^{N_e}$ (N_e is the number of the dipoles):

$$\begin{aligned} \vec{E}_e(M) &= \frac{1}{\varepsilon_e}\sum_{n=1}^{N_e}(k_e^2\overline{\overline{I}} + \nabla\nabla)\Psi^e(M, M_{n,e}) \cdot \vec{p}_\tau^{n,e} \\ \vec{H}_e(M) &= (-i\omega)\sum_{n=1}^{N_e}(\nabla \times \overline{\overline{I}})\Psi^e(M, M_{n,e}) \cdot \vec{p}_\tau^{n,e}, \quad M \in D_e \end{aligned} \tag{7}$$

where $k_e = \omega\sqrt{\varepsilon_e \mu_e}$, $\Psi^e(M, M_{n,e}) = \exp(ik_e R_M^{n,e})/(4\pi R_M^{n,e})$, $R_M^{n,e}$ is the distance from the point $M_{n,e}$ to the point M in D_e, $\vec{p}_\tau^{n,e}$ consists of two components $p_{\tau 1}^{n,e}$ and $p_{\tau 2}^{n,e}$ along the directions $\vec{e}_{\tau 1}^{n,e}$ and $\vec{e}_{\tau 2}^{n,e}$ selected in the tangential plane to S_e in the point $M_{n,e}$. The fields (5) and (7) satisfy equations (2) and (1), respectively. Moreover, the field (7) satisfies the radiation conditions (4). In order to satisfy the boundary conditions (3), it is necessary to define in a suitable manner the unknown vector constants $\vec{p}_\tau^{n,i}(n = 1, \ldots, N_i)$, $\vec{p}_\tau^{n,e}(n = 1, \ldots, N_e)$. For this purpose let us use the collocation method. Let $M_j (j = 1, 2, \ldots, L)$ be collocation points on S. Then we receive the following system of linear algebraic equations with a complex matrix of $4L \times 2(N_e + N_i)$ dimension for unknown complex constants $p_{\tau 1}^{n,i}, p_{\tau 2}^{n,i}(n = 1, \ldots, N_i)$, $p_{\tau 1}^{n,e}, p_{\tau 2}^{n,e}(n = 1, \ldots, N_e)$:

$$\vec{n}^j \times (\vec{E}_i^j - \vec{E}_e^j) = \vec{n}^j \times \vec{E}_0^j, \quad \vec{n}^j \times (\vec{H}_i^j - \vec{H}_e^j) = \vec{n}^j \times \vec{H}_0^j, \quad j = 1, 2, \ldots, L \tag{8}$$

The solution of system (8) is found via minimization of the functional

$$\Phi = \sum_{j=1}^{L} \left\{ \left| \vec{n}^j \times (\vec{E}_i^j - \vec{E}_e^j) - \vec{n}^j \times \vec{E}_0^j \right|^2 + \right. \tag{9}$$

$$\left. +(\mu_e/\varepsilon_e) \left| \vec{n}^j \times (\vec{H}_i^j - \vec{H}_e^j) - \vec{n}^j \times \vec{H}_0^j \right|^2 \right\}$$

by the conjugate gradient method. The control of the solution accuracy is realized via the calculation of the value

$$Q = (\Phi^1/\Phi_0)^{1/2}, \qquad \Phi_0 = \sum_{m=1}^{L^1} \left\{ \left| \vec{n}^m \times \vec{E}_0^m \right|^2 + (\mu_e/\varepsilon_e) \left| \vec{n}^m \times \vec{H}_0^m \right|^2 \right\} \tag{10}$$

where Φ^1 is the value of the functional (9) on the set of intermediate points.

3. Numerical Results

The described method was realized as the application package in *FORTRAN*. The input data of the package are the exciting field, geometry of the scatterer, the constitutive parameters ε_i, μ_i, β, and also the method parameters: the homothetic coefficients K_e, K_i, the numbers of the dipoles N_e, N_i and the collocation points L. The interruption of the conjugate gradient iterative process is executed upon the condition that during ten consecutive iterations the relative variation of the functional (9) with each iteration does not exceed 0.001. By using this package the set of computer experiments was performed. These experiments were directed to the detection of the adequacy of the obtained results and some known results and also on the research of the chirality parameter influence on scattering properties of the different body shapes. For the scattering characteristics chosen for the analysis, the radar bistatic cross sections (BCS) for the co- and cross-polarized field components are given by

$$\sigma_{\theta\theta} = \lim_{R\to\infty} 4\pi R^2(|E_{e,\theta}|^2/|E_0|^2), \quad \sigma_{\theta\varphi} = \lim_{R\to\infty} 4\pi R^2(|E_{e,\varphi}|^2/|E_0|^2) \tag{11}$$

in which $E_{e,\theta}$ and $E_{e,\varphi}$ are the field components of (7) in the spherical coordinate system.

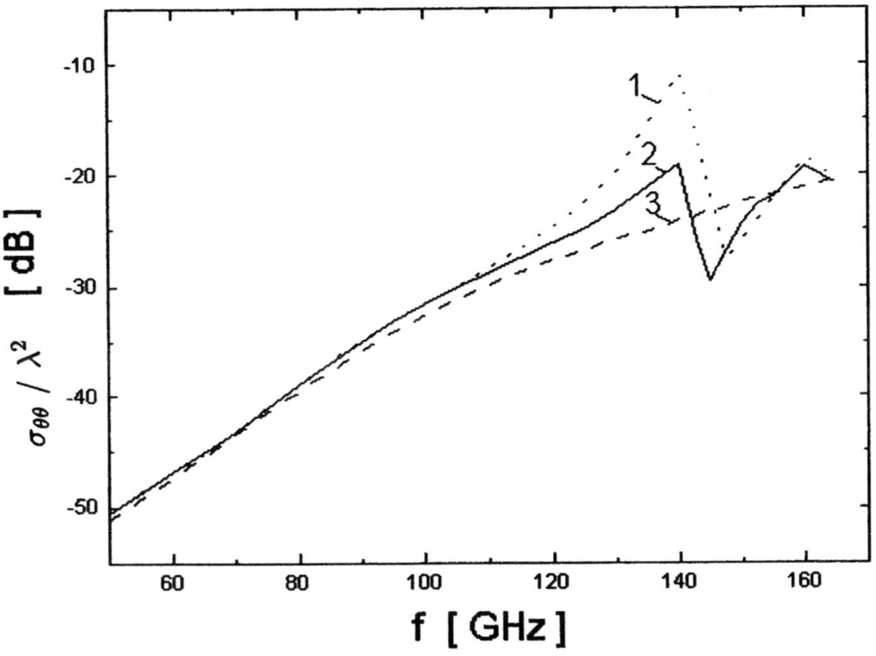

Figure 1. Comparisons: the results obtained by the method expounded above and the results for sphere from [5]

Figure 1 shows the results of comparison of the backscattering cross-section frequency dependence of the co-polarized component of scattered field from a sphere excited by a linearly polarized plane wave with that published in [5]. The following parameters are used: the radius of the sphere is 0.0002 m, the relative permittivity is $\varepsilon_i/\varepsilon_e = 5.0 + i0.1$, relative permeability is $\mu_i/\mu_e = 1.0$, and the chirality parameter is $\beta = 10^{-4}$ m. Curve 1 shows the results of paper [5] obtained by the T-matrix method, curve 2 corresponds to the results obtained by our program with $K_e = 0.3$, $K_i = 3.0$, $N_e = N_i = L = 32$ (uniform distribution of the dipoles and collocation points with angular distances between the neighbouring points $\Delta\theta = \Delta\varphi = 45°$). Here, the curve 3 shows the results for the achiral sphere ($\beta = 0$).

The comparison of curves 1 and 2 shows that there is good agreement between the results obtained by the two methods. The maximal discrepancy

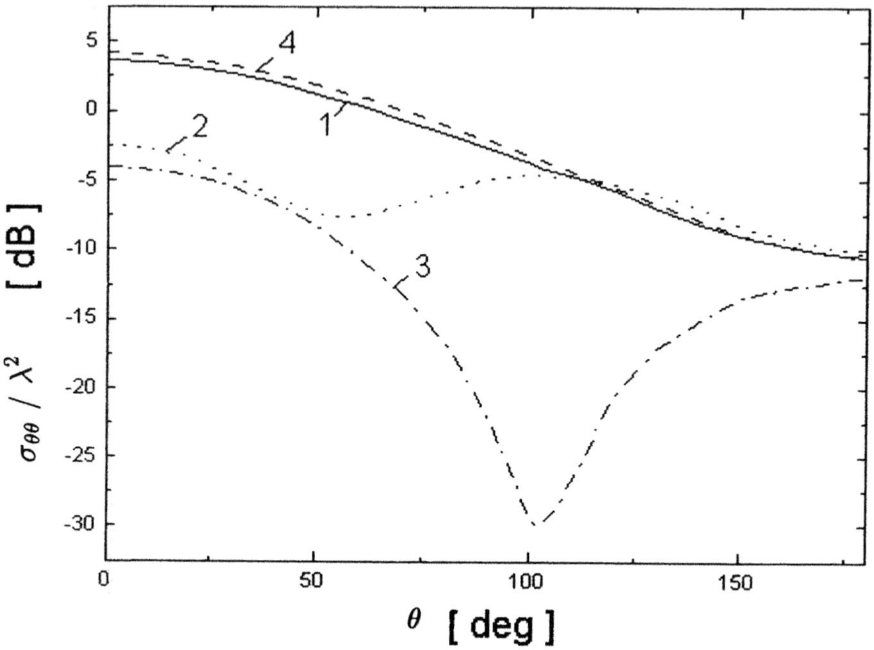

Figure 2. Bistatic cross section for the co-polarized scattering field component of sphere versus the observation angle θ for different values of the chirality parameter β

in numerical magnitudes takes place close to the frequency $f \approx 140$ GHz. This discrepancy may be explained by the increase of the computation error in both methods. The solution accuracy control based on the criterion (10) demonstrates that in the domain of $f < 120$ GHz the magnitude of Q does not exceed 3%, whence in the domain $f \approx 140$ GHz, the magnitude of Q increases to 12%. Analogous estimation of solution accuracy is not given in [5], but it may be supposed that similar effect takes place also in the T-matrix method because it is connected with the field structure complication inside the scatterer.

Figures 2 and 3 show the modification of BCS towards the "reinforcing" of the chirality parameter of the scatterer. The results refer to the sphere of radius $kr_0 = 1.5$, $k = 2\pi/\lambda$, where λ is the wavelength in free space. The following parameters of the sphere are used: $\varepsilon_i/\varepsilon_e = 4.0$, $\mu_i/\mu_e = 1.0$. The sphere is excited by linearly polarized plane wave propagating along the $z-$

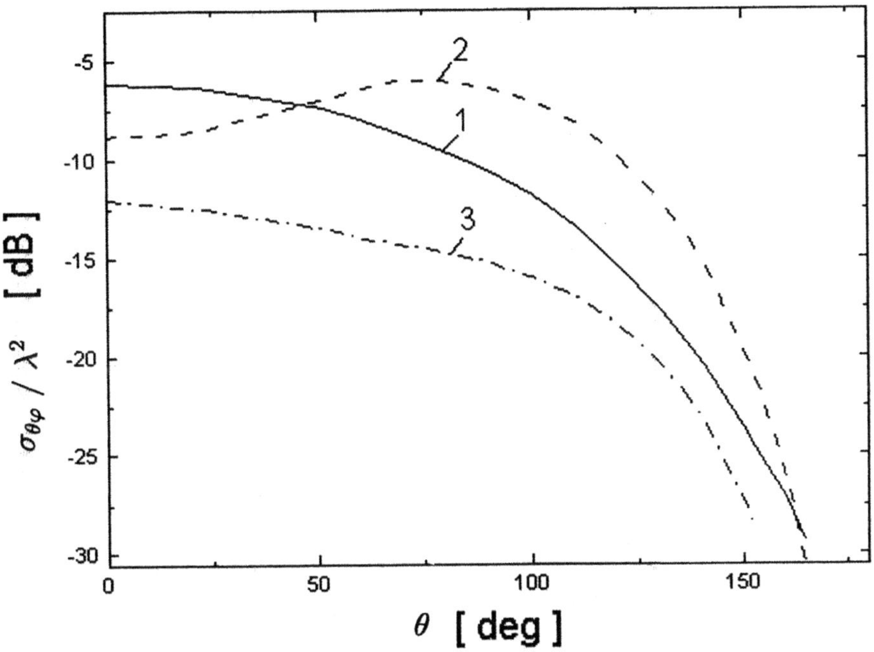

Figure 3. Bistatic cross section for the cross-polarized scattering field component of sphere versus the observation angle θ for different values of chirality parameter β

axis of Cartesian coordinate system with the origin in the centre of sphere, vector $\vec{E}_0$ of the plane wave is directed along the x–axis. Figure 2 concerns the co-polarized field component, Figure 3 concerns the cross-polarized field component. The results are given for x-z–plane, the angle θ is the angle between propagation vector and direction of observation in the x-z–plane. The curves 1, 2, 3, and 4 refer to the cases when $k_i\beta = 0.1; 0.3; 0.5$, and 0.0, respectively. The parameters of the method are: $K_e = 0.3$, $K_i = 3.0$, $N_e = N_i = L = 60$. The dipole and collocation points are allocated on the corresponding surfaces in the same manner: in each semi-section φ=const to be separated on the angular distance $\Delta\varphi = 36°$ are evenly distributed 6 dipoles (collocation points). Under these conditions the magnitude of the error functional (10) does not exceed 4%. The comparison of curves 1–4 reveals that chirality of the scatterer considerably influences the form of the scattering field diagrams. It is interesting to note that the growth of

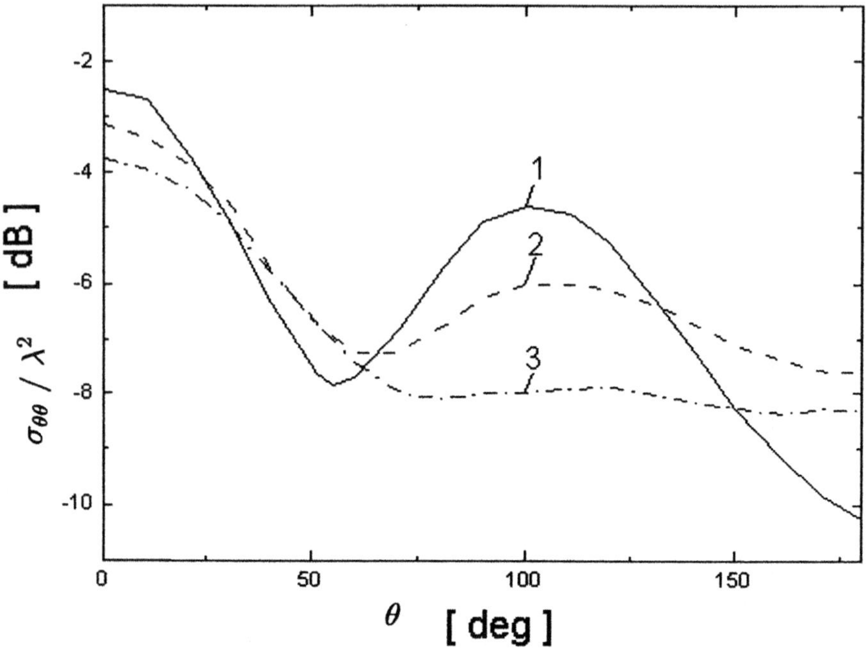

Figure 4. Bistatic cross sections for the co-polarized scattering field component of ellipsoid versus the observation angle θ

the magnitude of chirality parameter does not always lead to an increase of the level of the cross-polarized scattering field component (compare curves 1 and 3).

Finally, Figures 4 and 5 show the BCS in the x-z–plane of ellipsoids having the constitutive parameters $\varepsilon_i/\varepsilon_e = 4.0$, $\mu_i/\mu_e = 1.0$, $k_i\beta = 0.3$ but different ratio of semi–axes. Semi–axes ka, kb, and kc are directed along the x–, y–, and z–axes of Cartesian coordinate system with the origin in the centre of the ellipsoid and defined in the following manner: $ka = (1-\delta)kb$, $kb = 1.5$, $kc = (1+\delta)kb$. Such a way of definition of semi–axes allows to change the shape of the scatterer by changing δ only. If $\delta = 0$, the scatterer is a sphere. The ellipsoids are excited by a plane wave propagating along the z–axis so that the electric field $\vec{E}_0$ is oriented along the x–axis. The parameters of the method are chosen to be the same as for the results shown in Figures 2 and 3. The curves 1,2, and 3 refer to the cases when

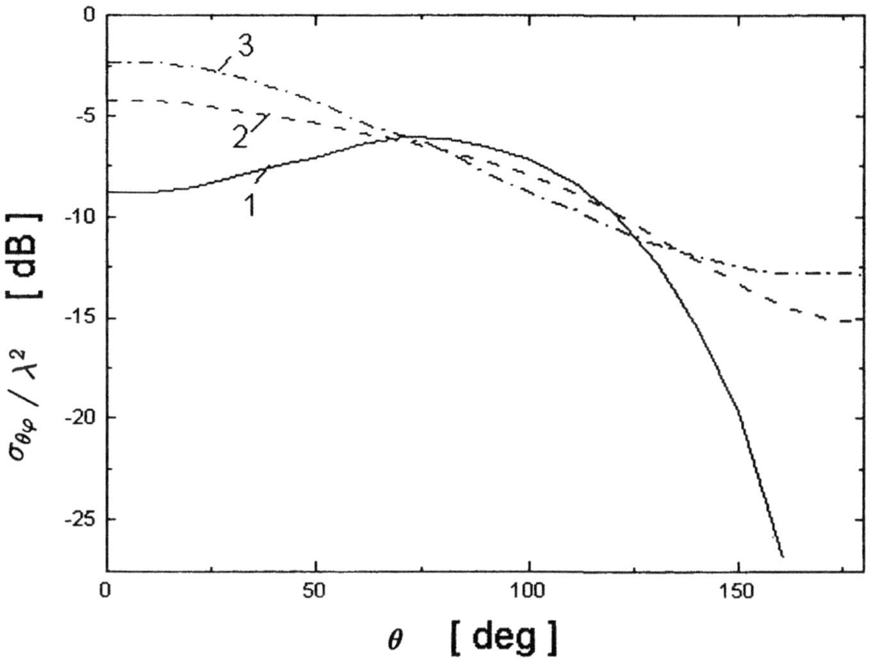

Figure 5. Bistatic cross sections for the cross-polarized scattering field component of ellipsoid versus the observation angle θ

$\delta = 0; 0.2$ and 0.3, respectively. Amplitude of the cross-polarized scattering field component in directions close to the forward scattering direction ($\theta = 0°$) is more sensitive to the form of the chiral scatterer than that of the co-polarized field component. This effect may be used for the detection of forms of chiral objects. However, the cross-polarized scattering field component decreases significantly in the directions close to the backscattering direction ($\theta = 180°$).

4. Conclusion

A numerical method for solving electromagnetic scattering problems for three-dimensional arbitrary shaped chiral bodies was presented. The mathematical formulation of the method, short description of the created application package and some numerical results were described. It was shown that the chirality of the scatterer considerably influences the form of the scattering field diagrams. It was also shown that amplitude of the cross-polarized scattering field component in directions close to the forward scattering direction is more sensitive to the shape of the chiral scatterer than that of the co-polarized field component.

References

1. Bassiri, S., Engheta, N., and Papas, C.H. (1986) Dyadic Green's function and dipole radiation in chiral media, *Alta Frequenza*, **55(2)**, 83–88.
2. Lakhtakia, A., Varadan, V.V., and Varadan, V.K. (1988) Field equations, Huygens principle, integral equations and theorems for radiation and scattering of electromagnetic waves in isotropic chiral media, *J. Opt. Soc. Am.*, **A5(2)**, 175–184.
3. Lindell, I.V., Tretyakov, S.A., and Oksanen, M.I. (1993) Vector transmission line and circuit theory for bi-isotropic layered structures, *J. Electromagn. Waves and Applications*, **7(7)**, 147–173.
4. Kuehl, S.A., Grove, S.S., Smith, A.G., and Theron, I.P. (1995) Manufacture of microwave chiral materials and their electromagnetic properties, in A. Sihvola et al. (eds.), *Proc. 4th Intern. Conference on Chiral, Bi-isotropic, and Bi-anisotropic Media*, Penn. State Univ., State College, USA, pp. 5–8.
5. Lakhtakia, A., Varadan, V.K., and Varadan, V.V. (1985) Scattering and absorption characteristics of lossy dielectric, chiral, nonspherical objects, *Applied Optics*, **24(23)**, 4146–4154.
6. Eremin, Yu.A., and Sveshnikov, A,G. (1992) *The Discrete Sources Method in Electromagnetic Diffraction Problems*, Izdatel'stvo MGU, Moscow (in Russian).
7. Dmitrenko, A.G., and Mukomolov, A.I. (1995) Numerical method for solution of electromagnetic scattering problems by a three dimensional arbitrary shaped magnetic dielectric body, *Radiotekhnika i Electronika*, **40(6)**, 875–880 (in Russian).
8. Dmitrenko, A.G. (1996) Numerical analysis of electromagnetic scattering by three dimensional arbitrary shaped bodies of different physical nature, in T. Wriedt et al. (eds.), *Proc. 1st Workshop on Electromagnetic and Light Scattering–Theory and Applications*, Universität Bremen, Bremen, pp. 71–74.

SCATTERING BY FREQUENCY SELECTIVE SURFACES SUPPORTED BY AN ISOTROPIC CHIRAL SLAB

T. EGE AND A.O. KOCA
University of Gaziantep
Electrical and Electronics Engineering Department
27310 Gaziantep, Turkey

Abstract. The plane wave scattering by a novel frequency selective surface (FSS), named Chiro-FSS, consisting of a two dimensional infinite periodic array of conducting elements printed on an isotropic chiral slab is analysed. Computed reflection and transmission characteristics of Chiro-FSS comprising of cross dipoles, circular rings, square loops and square patches are presented. The results indicate that Chiro-FSS exhibits novel spectral characteristics in the reflection and transmission bands.

1. Theory

Periodic arrays of conducting elements printed on a dielectric substrate have frequency selective properties and find many applications as dichroic surfaces for reflector antennas, radomes, artificial dielectrics and in optical and quasi-optical devices [1,2]. For a lossless structure, there exists a single frequency at which the periodic array of conducting elements allows the total reflection. The spectral response of a FSS depends on the geometry of the conducting elements, array configuration, surface parameters and angle and polarisation of the incident electromagnetic wave. In the Chiro-FSS, an isotropic chiral slab instead of dielectric slab is used to support the arrays of conducting elements. Therefore among other parameters, the chirality admittance of the slab is introduced as an extra degree of freedom and exploited to improve the spectral characteristics of the FSSs [3].

Recent years have witnessed a growth in the research that is focused on the chiral medium. For an isotropic chiral medium the constitutive relations with time convention $\exp(j\omega t)$ are given by [4]:

A. Priou et al. (eds.), Advances in Complex Electromagnetic Materials, 189–200.

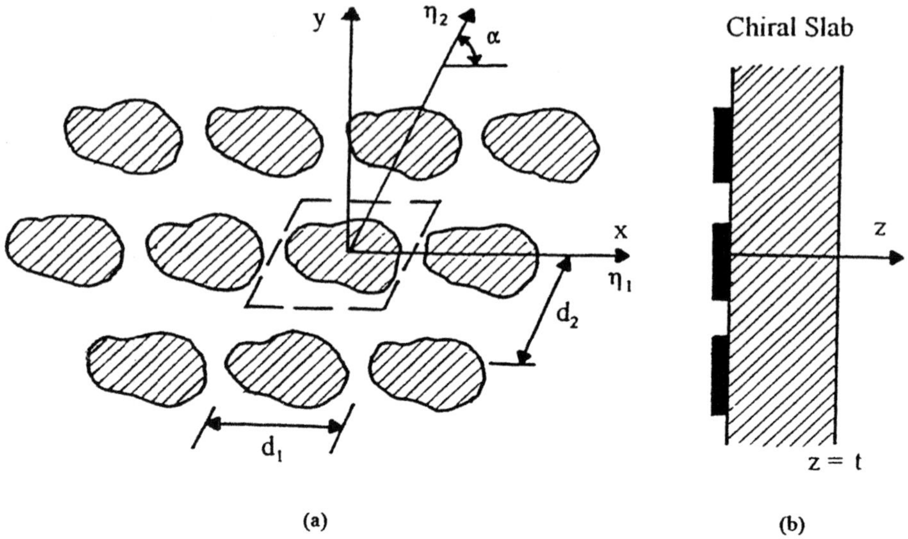

Figure 1. **Geometry of a Chiro-FSS. a) Top View, b) Side View**

$$\mathbf{D} = \epsilon\mathbf{E} - j\xi\mathbf{B} \tag{1}$$

$$\mathbf{H} = -j\xi\mathbf{E} + \frac{1}{\mu}\mathbf{B} \tag{2}$$

where ϵ, μ, and ξ are the permittivity, permeability and chirality admittance of the medium, respectively. In the analysis all constitutive parameters are assumed to be real quantities.

The problem of electromagnetic scattering by a Chiro-FSS is formulated as an integral equation in terms of the current distribution induced on a conducting element within a unit cell. In the analysis the incident plane wave is assumed to have an arbitrary incident angle and polarisation. For this, consider a two-dimensional infinite array of conducting elements printed on an isotropic chiral slab having a thickness of t as shown in Figure 1. The conducting elements have periodicities of d_1 and d_2 along the x axis and the skewed axis η_2, respectively.

The isotropic nature of the chiral slab allows us to apply the modal techniques which are frequently employed in the formulation of the FSSs having dielectric backing [5]. The first step in the formulation is to express the transverse components (x and y) of the scattered E and H fields as infinite sum of Right Circularly Polarised (RCP) and Left Circularly Polarised (LCP), TE and TM Floquet spatial harmonics propagating in the

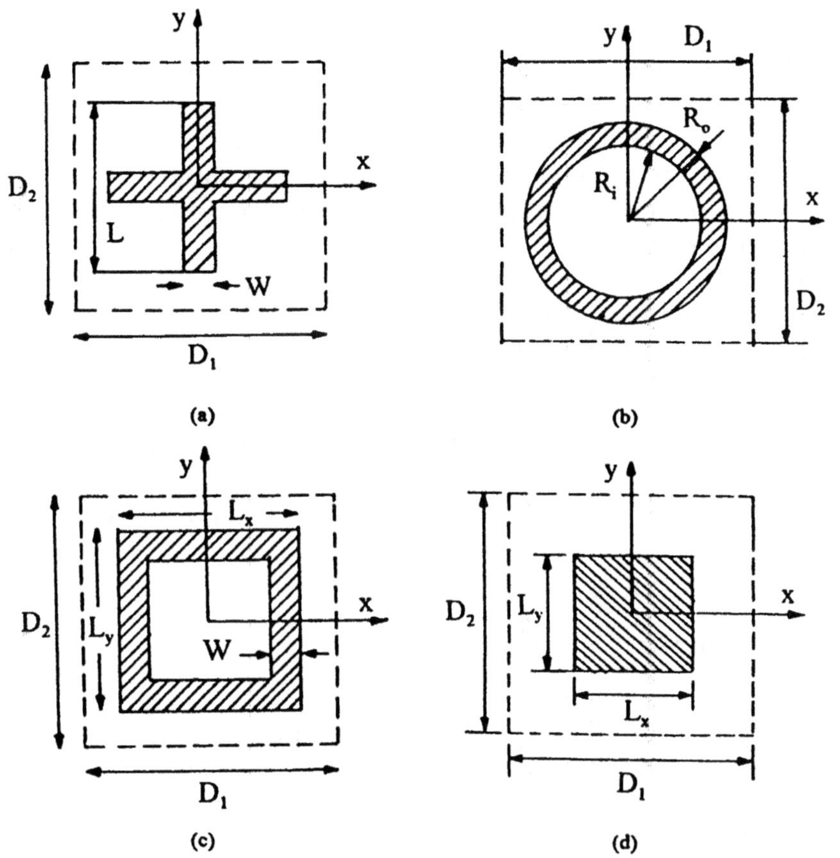

Figure 2. The Unit Cell Geometries.
a) Cross Dipole, b) Circular Ring, c) Square Loop, d) Square Patch.

three regions, *i.e.*, in the chiral slab and in the surrounding medium on either side.

Using the bounday conditions combined with the orthogonal property of the Floquet modes over a single periodic cell, the unknown modal coefficients of the scattered fields of the forward and backward travelling RCP and LCP, TM and TE waves are eliminated to obtain an expression for the transverse components of the scattered electric field in terms of the current density on a conducting element. Finally, the application of the boundary condition that the tangential component of the scattered electric field must cancel the tangential component of the incident electric field on the surface of a conducting element yields an integral equation for the unknown current distribution induced on the conducting element. To solve this resulting integral equation the current distribution is approximated by a finite sum of suitable functions having unknown coefficients. Substitution of the expression for the current density into the integral equation and then applying the Galerkin's testing procedure, the integral equation generates a system of linear algeabric equations with the current coefficients as unknowns. Once these current coefficients are obtained by matrix inversion, it is a relatively straightforward matter to obtain the reflection and transmission characteristics.

2. Numerical Results

In this Section, computed results of the scattering characteristics of four different types of Chiro-FSS configurations which is comprised of cross dipoles, circular rings, square loops, and square patches, respectively, are presented.

The unit cell geometry of each configuration considered is shown in Figure 2, and the corresponding current densities are taken as:

FOR CROSS DIPOLES

x-directed current,

$$\mathbf{J}_x = \sum_{m=1}^{M} \frac{a_m}{\sqrt{1-(2y/W)^2}} \begin{pmatrix} \cos \\ \sin \end{pmatrix} \left(\frac{m\pi}{L}x\right) \mathbf{u}_x, \qquad \begin{pmatrix} m = 1, 3, \cdots \\ m = 2, 4, \cdots \end{pmatrix} \tag{3}$$

y-directed current,

$$\mathbf{J}_y = \sum_{n=1}^{N} \frac{b_n}{\sqrt{1-(2x/W)^2}} \begin{pmatrix} \cos \\ \sin \end{pmatrix} \left(\frac{n\pi}{L}y\right) \mathbf{u}_y, \qquad \begin{pmatrix} n = 1, 3, \cdots \\ n = 2, 4, \cdots \end{pmatrix} \tag{4}$$

FOR CIRCULAR RINGS

$$\mathbf{J}_\phi = \left[\sum_{n=0}^{N} a_n \cos(n\phi) + \sum_{m=1}^{M} b_m \sin(m\phi)\right] \mathbf{u}_\phi \tag{5}$$

FOR SQUARE LOOPS

x-directed current,

$$\mathbf{J}_x = \frac{1}{\sqrt{1-(2y/W)^2}} \left[\sum_{m=1}^{M} a_m \cos\left(\frac{m\pi}{L_x}x\right) + \sum_{n=1}^{N} b_n \sin\left(\frac{n\pi}{L_x}x\right)\right] \mathbf{u}_x \tag{6}$$

y-directed current,

$$\mathbf{J}_y = \frac{1}{\sqrt{1-(2x/W)^2}}\left[\sum_{k=1}^{K} c_k \cos\left(\frac{k\pi}{L_y}y\right) + \sum_{l=1}^{L} d_l \sin\left(\frac{l\pi}{L_y}y\right)\right]\mathbf{u}_y \tag{7}$$

FOR SQUARE PATCHES

x-directed current,

$$\mathbf{J}_x = \sum_{r=0}^{R}\sum_{m=1}^{M} \frac{a_{m,r}}{\sqrt{1-(2y/L_y)^2}} T_r(2y/L_y) \begin{pmatrix} \cos \\ \sin \end{pmatrix} \left(\frac{m\pi}{L_x}x\right)\mathbf{u}_x \tag{8}$$

y-directed current,

$$\mathbf{J}_y = \sum_{s=0}^{S}\sum_{n=1}^{N} \frac{b_{n,s}}{\sqrt{1-(2x/L_x)^2}} T_s(2x/L_x) \begin{pmatrix} \cos \\ \sin \end{pmatrix} \left(\frac{n\pi}{L_y}y\right)\mathbf{u}_y \tag{9}$$

where in (8) and (9) the upper choice (cosines) are for odd values of m and n, and the lower ones (sines) have to be used for even values. The unit vectors are denoted by $\mathbf{u}$ with the direction as subindex.

Here, T_i is the Chebyshev function of the first kind, a_i, b_i, c_i, and d_i are the unknown coefficients to be determined. The radical terms in the denumerator are included to satisfy the edge conditions.

In order to validate the computed results, the conservation of power was checked and found to be excellent. Furthermore, the results corresponding to zero chirality were compared with the numerical and experimental results given in the literature [6]–[8], and it was found that the agreement is excellent. Note that in the limiting case, that is, when the chirality admittance approaches zero ($\xi \to 0$), the chiral slab becomes a dielectric one. For each structure, the convergence of the numerical results have also been studied by varying the number of the current functions and the Floquet harmonics.

In all the results presented here, the chiral slab is assumed to have a thickness of 5 mm with a dielectric constant $\epsilon_r = 1.08$. Three different values are taken for the chirality admittance; $\xi = 0.0$ (dielectric), $\xi = 0.0006$ mho, and $\xi = 0.0018$ mho. One should note that the chirality admittance is bounded by the intrinsic admittance of the medium, *i.e.*, $\xi \leq \sqrt{\epsilon/\mu}$ [4]. The Chiro-FSSs are illuminated by a normally incident TE wave. The variations of the reflection and transmission coefficients of the Chiro-FSSs comprised of the cross dipoles, circular rings, square loops, and the square patches arranged in a square lattice with frequency of the incident wave are shown in Figures 3–6, respectively. In order to show the effects of array geometry on the reflection and transmission characteristics of a Chiro-FSS, a ring array having the same dimensions as the rings used in Figure 4 but arranged in a triangular lattice is assumed and the corresponding results are presented in Figure 7.

As can be seen from the results, in all cases, for a small value of chirality admittance ($\xi = 0.0006$ mho) the spectral characteristics of a Chiro-FSS are similar to that of a conventional FSS (dielectric backed) which exhibits only one resonance in the band of single mode operation. But when the value of the chirality admittance is increased to $\xi = 0.0018$ mho, multiple resonances start to appear within the band of operation. This is an important feature of the Chiro-FSS. In all Chiro-FSS geometries presented here except the square patches, the first resonance frequency is lower in comparison to that of the conventional FSS. In the reflection characteristics the bandwidth of the higher resonances are always narrower than the bandwidth of the first resonance. Comparison of Figure 4 and Figure 7 reveals that the resonance bandwidth in the reflection and transmission characteristics of the triangular array configuration ($\alpha = 60°$) are wider than the resonance bandwidths of the Chiro-FSS arranged in a square lattice ($\alpha = 90°$). It can also be seen that the resonance frequencies are shifted in comparison with the square array configuration.

As illustrated in Figures 3–7, the reflected wave consists of only co-polarised (TE) component whereas the transmitted wave consists of both co- (TE) and cross-polarized (TM) components. Another interesting feature of a Chiro-FSS is that fully transmitted cross-polarized component may exists in the air on the other side as can be seen from Figures 3b–7b, thus, giving rise to mode conversion.

3. Conclusion

A new Frequency Selective Surface structure which consists of a two-dimensional infinite periodic array of perfectly conducting elements supported by a lossless isotropic chiral slab is proposed and analysed.

In contrast to dielectric-backed FSSs comprising of a single ring, cross dipole, square loop, and square patch within a unit cell, the computed results indicate that multiple resonances are possible even at normal incidence in Chiro-FSS structures. Results also indicate that the reflection and transmission characteristics of the Chiro-FSSs are very sensitive to the frequency of the incident plane wave. Another novel feature of these FSSs is that polarisation conversion is possible, that is, a TE wave incident on a Chiro-FSS can be converted into a TM wave (with total transmission) in air on the other side.

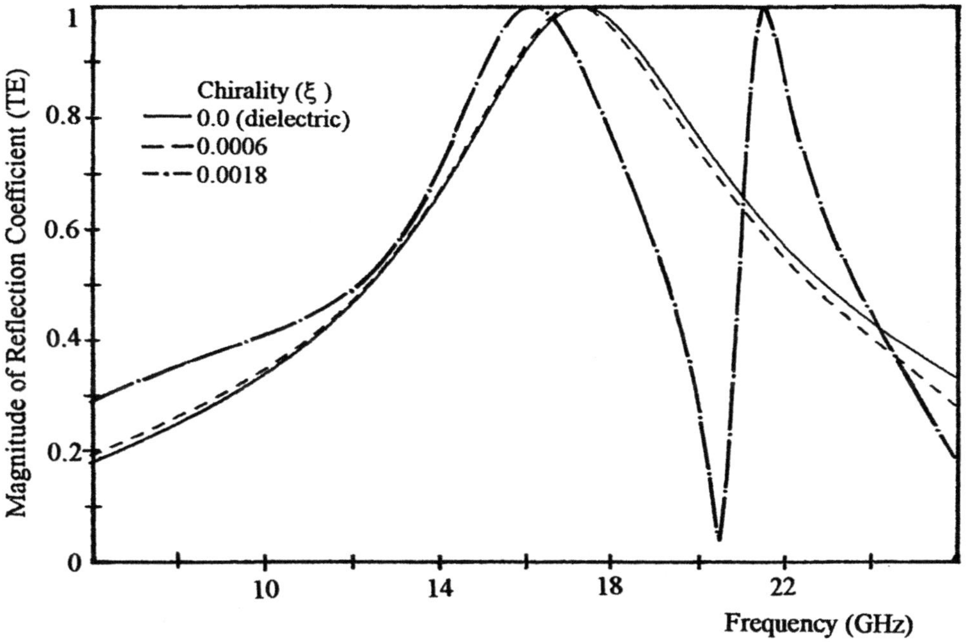

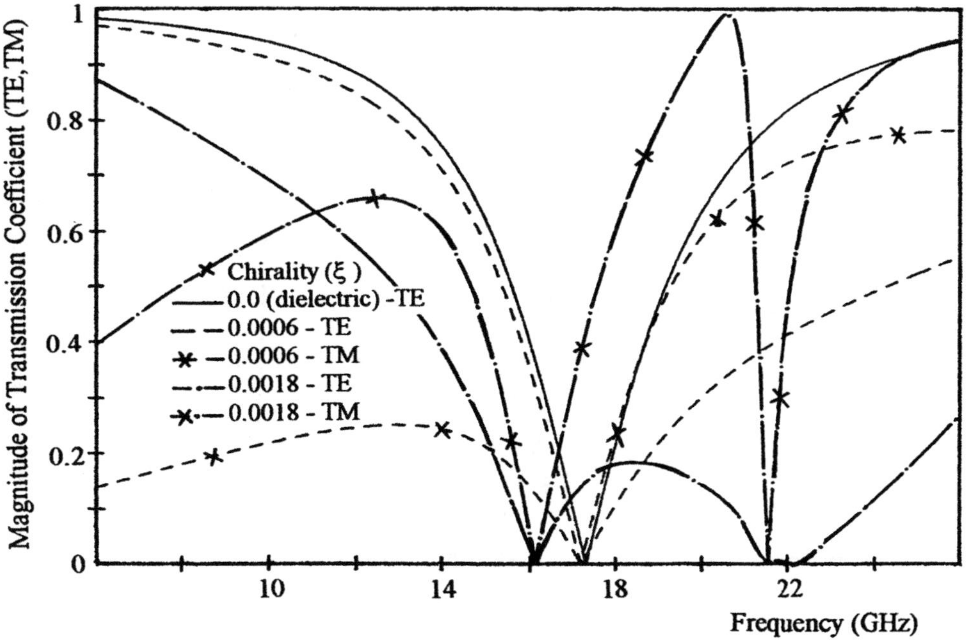

Figure 3. Computed results for array of cross dipoles; TE incidence, square lattice ($\alpha = 90°$), $d_1 = d_2 = 9.2\,\text{mm}$, $L = 8\,\text{mm}$, $W = 1\,\text{mm}$, permittivity $\epsilon_r = 1.08$, thickness $t = 5\,\text{mm}$. a) Reflection coefficients (TE), b) Transmission coefficients (TE, TM)

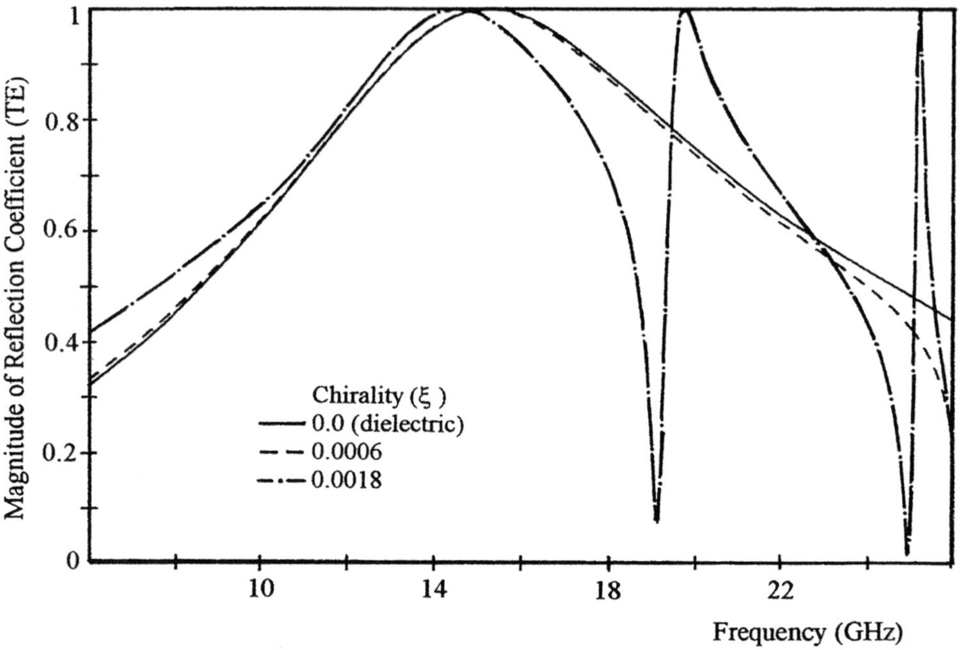

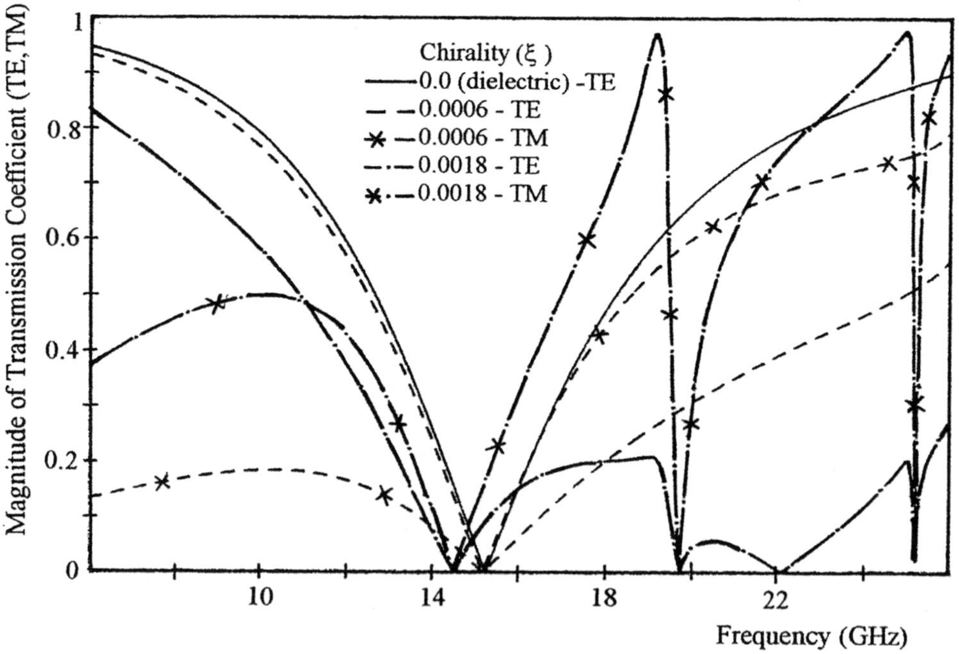

Figure 4. Computed results for an array of circular rings; TE incidence, square lattice ($\alpha = 90°$), $d_1 = d_2 = 10\,\text{mm}$, $R_i = 3.2\,\text{mm}$, $R_o = 4.2\,\text{mm}$, permittivity $\epsilon_r = 1.08$, and $t = 5\,\text{mm}$. a) Reflection coefficients (TE), b) Transmission coefficients (TE, TM)

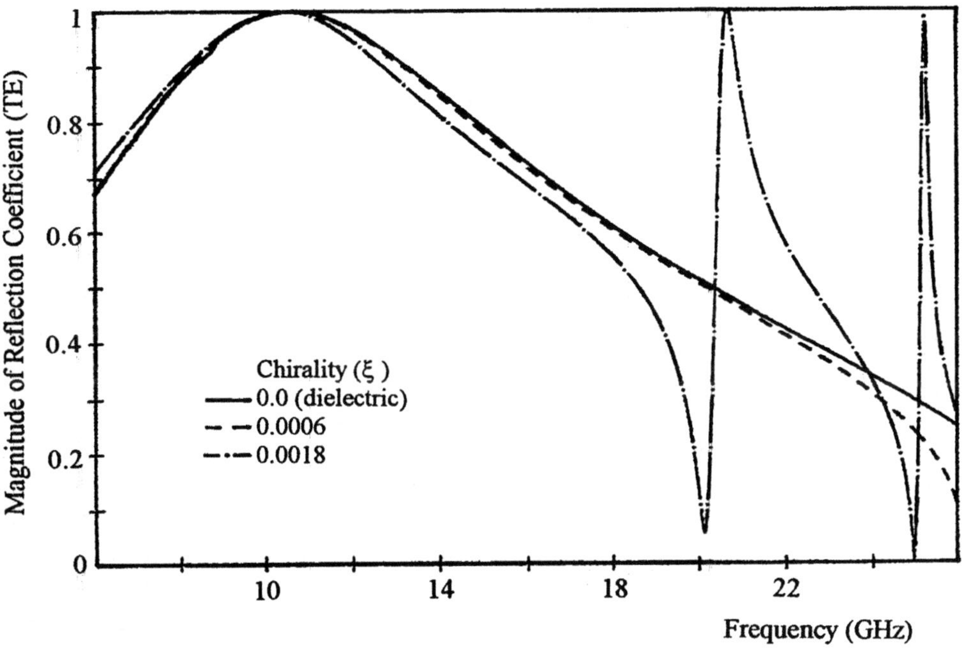

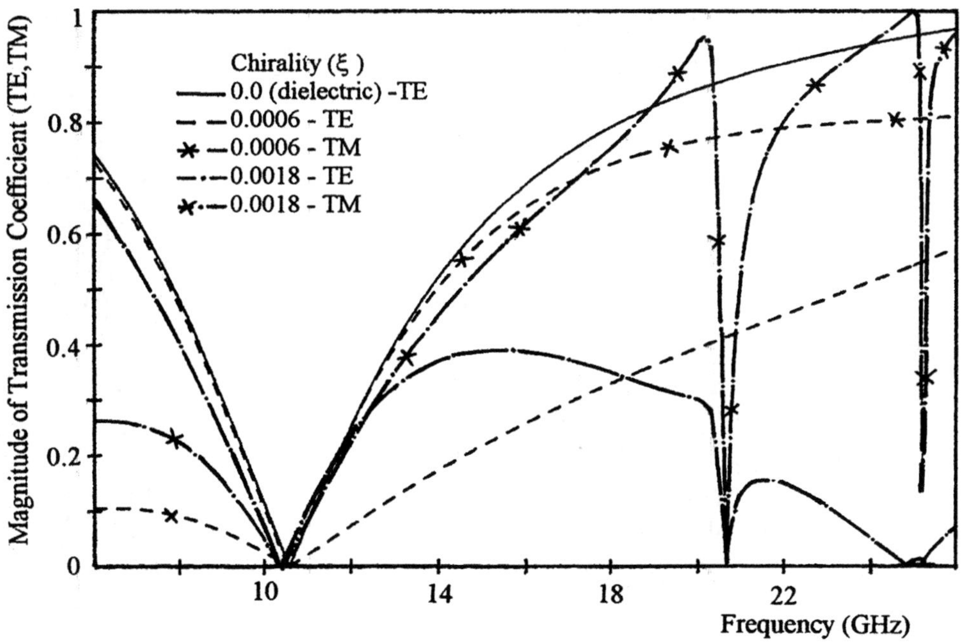

Figure 5. Computed results for array of square loops; TE incidence, square lattice ($\alpha = 90°$), $d_1 = d_2 = 10\,\text{mm}$, $L_x = L_y = 9.4\,\text{mm}$, $W = 1\,\text{mm}$, permittivity $\epsilon_r = 1.08$, and $t = 5\,\text{mm}$. a) Reflection coefficients (TE), b) Transmission coefficients (TE, TM)

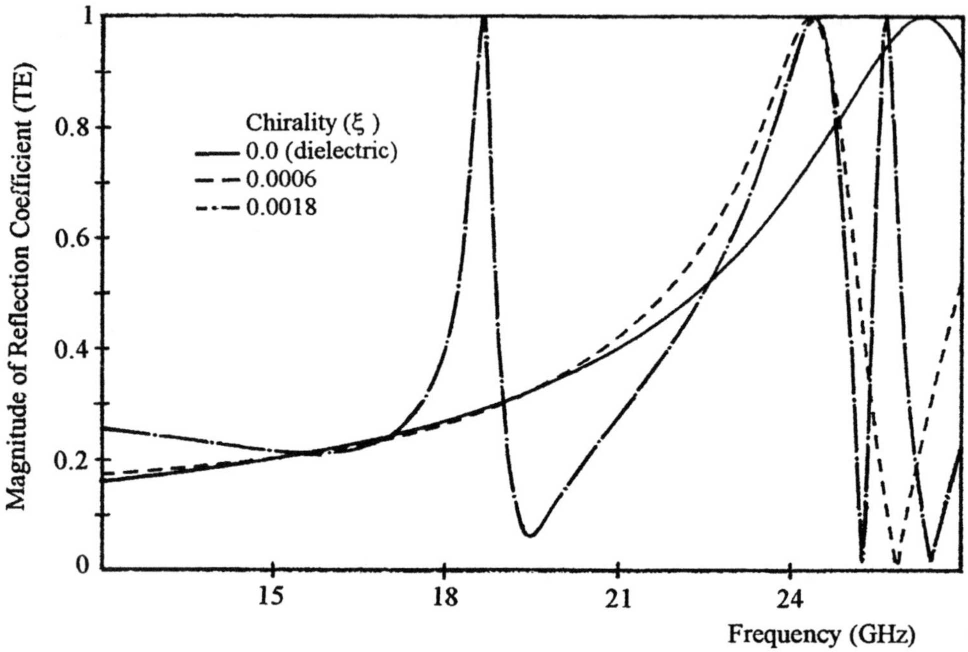

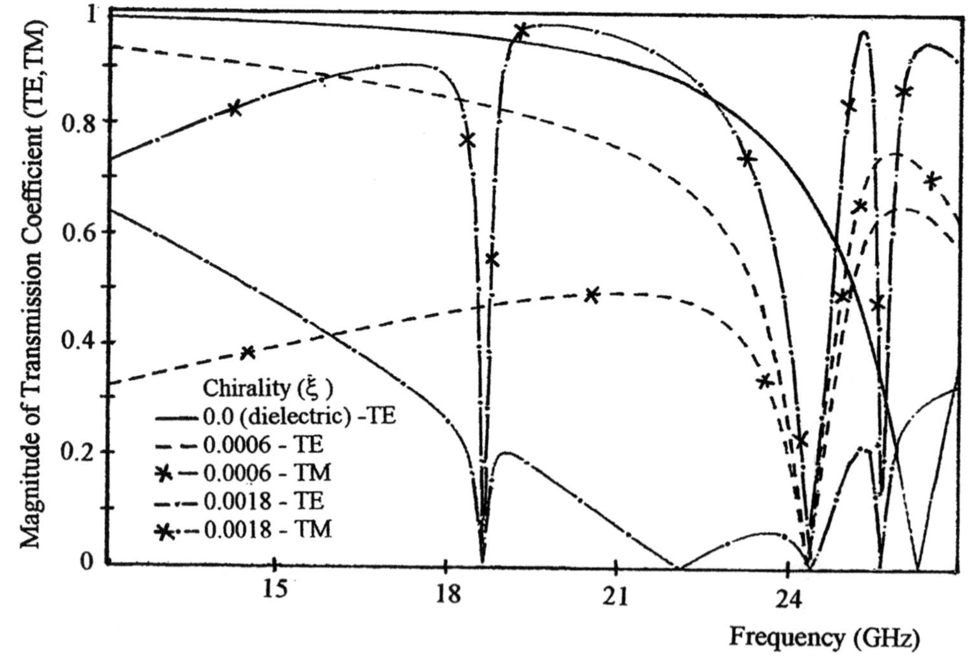

Figure 6. Computed results for an array of square patches; TE incidence, square lattice ($\alpha = 90°$), $d_1 = d_2 = 10\,\text{mm}$, $L_x = L_y = 4\,\text{mm}$, permittivity $\epsilon_r = 1.08$, thickness $t = 5\,\text{mm}$. a) Reflection coefficients (TE), b) Transmission coefficients (TE, TM)

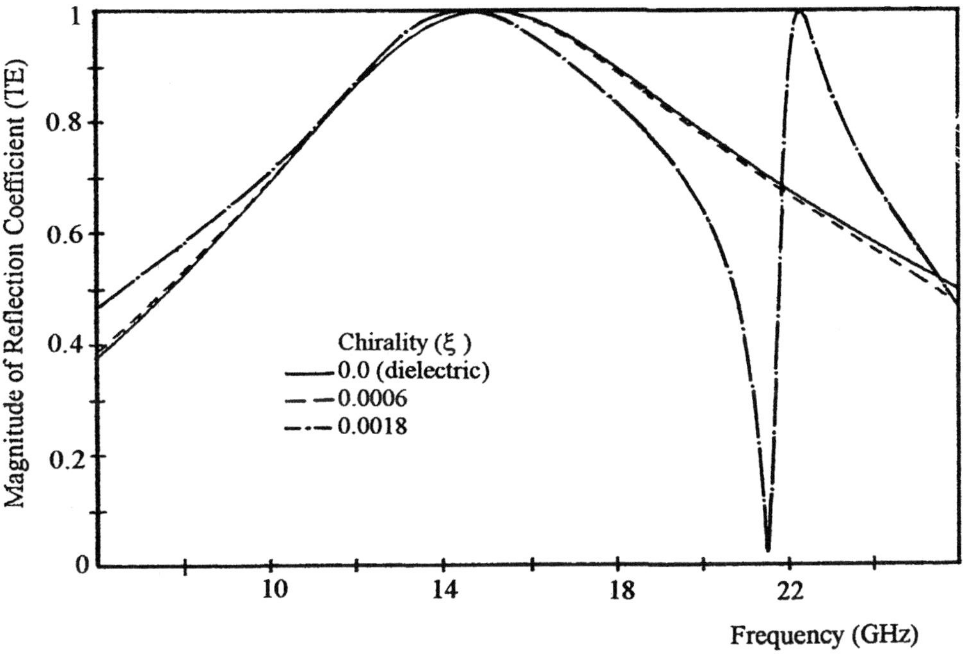

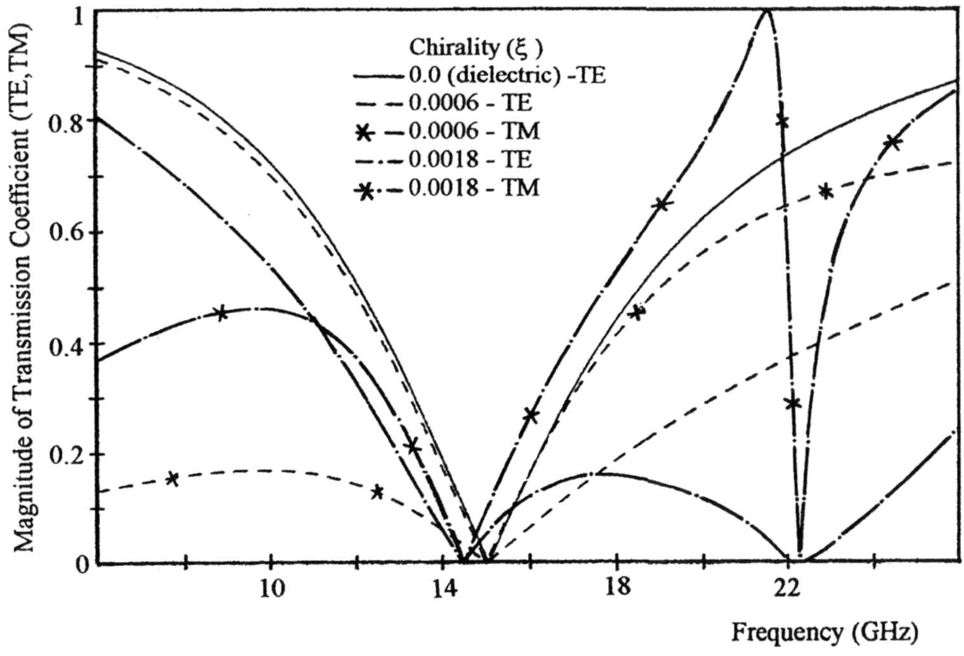

Figure 7. Computed results for an array of circular rings; TE incidence, triangular lattice ($\alpha = 60°$), $d_1 = d_2 = 10\,\text{mm}$, $R_i = 3.2\,\text{mm}$, $R_o = 4.2\,\text{mm}$, permittivity $\epsilon_r = 1.08$, and $t = 5\,\text{mm}$. a) Reflection coefficients (TE), b) Transmission coefficients (TE, TM)

References

1. Mittra, R., Chan, C.C., and Cwik, T. (1988) Techniques for analysing frequency selective surfaces–a review, *Proc. IEEE*, **76**, 1593-1615.
2. Agrawal, V.D. and Imbriali, W.A. (1979) Design of a dichroic Cassegrain subreflector, *IEEE Trans. Antennas Propagat.*, **27**, 466-473.
3. Koca, A.O. and Ege T. (1995) A novel frequency selective surface, *Microw. Opt. Tech. Lett.*, **10**, 157-160.
4. Engheta, N. and Jaggard, D.L. (1988) Electromagnetic chirality and its applications, *IEEE Antennas Propagat. Newsletter*, October 1988, 6-12.
5. Montgomery, J.P.O. (1975) Scattering by an infinite periodic array of thin conductors on a dielectric sheet, *IEEE Trans. Antennas Propagat.*, **23**, 70-75.
6. CSELT (1984) Study of subreflector for multi-frequency antennae, electrical design, *Final report*, ESA contract, No. 5355/83/NL/CM(SC).
7. Cwik, T. and Mittra, R. (1987) Scattering from a periodic array of freestanding arbitrarily shaped perfectly conductive or resistive patches, *IEEE Trans. Antennas Propagat.*, **35**, 1226-1234.
8. Wakabayashi, H., Kominami, M. Kusaka, H. and Nakashima, H. (1994), *IEE Proc. Microw. Antennas Propag.*, **141**, 477-482.

PLANE WAVE DIFFRACTION BY A WEDGE COATED WITH THIN BI-ISOTROPIC LAYERS

S.G. VASHTALOV
Department of Radiophysics, Tomsk State University,
36 Lenin Ave., Tomsk, 634050, Russia

AND

V.V. FISANOV
Siberian Physical & Technical Institute,
1 Revolution Sq., Tomsk, 634050, Russia

Abstract. In this paper the diffraction problem is considered for the plane electromagnetic wave incident on a perfectly conducting wedge, the faces of which are coated with thin layers of bi-isotropic materials. The presence of the layers is taken into account with the help of the approximate boundary conditions. These conditions contain the second derivative in the direction along the boundary. Therefore it is necessary to impose so-called contact conditions that prescribe certain relations between the wave field and its derivatives on the edge. The Malyuzhinets functional equations are obtained for the transform of the Sommerfeld–Malyuzhinets integral. These equations are considered with the help of the modified Malyuzhinets technique.

1. Introduction

Recent years have witnessed tremendous growth in the research activity focusing on the wave interaction with complex materials at microwave, millimeter wave, and infrared frequencies, namely, with chiral, bi-isotropic, and bi-anisotropic media [1]. Such materials can possibly be used for the creation of non-reflecting coverings. However, the calculation of the radar cross section of a scatterer with covering is very difficult.

Usually the covering is very thin, therefore it is possible to use generalized impedance boundary conditions for its simulation [2–4], which contain

A. Priou et al. (eds.), Advances in Complex Electromagnetic Materials, 201–206.

field derivatives above the first order. However, the standard boundary conditions are not sufficient on the edge to ensure uniqueness of solution in studying the scattering on the objects with edges. Therefore it is necessary to introduce so-called contact conditions that prescribe certain relations between the wave field and its derivatives on the edge [4–8]. A number of these conditions is determined only by the order of the highest derivative in the generalized impedance boundary conditions [4,7]. In this paper the diffraction problem is considered for the plane electromagnetic wave incident on a perfectly conducting wedge, the faces of which are coated with thin layers of bi-isotropic materials.

The presence of the layers is taken into account with the help of the approximate boundary conditions [9]. These conditions are obtained with exactness to the terms of the first-order layer thickness. For the E-polarized waves (vector **E** is parallel to the edge of the wedge) the standard Leontovich impedance boundary conditions can be applied. For the H-polarized waves (vector **H** is parallel to the edge) the generalized impedance boundary conditions of the second order are received. They contain the second derivative in the direction along the boundary. The diffraction of the H-polarized plane wave on such wedge is examined. The problem is considered with the help of the modified Malyuzhinets technique [4–6]. The Malyuzhinets functional equations are obtained for the transform of the Sommerfeld–Malyuzhinets integral. The general solution of these equations contains three unknown constants. The radiation condition and contact conditions on the edge of the wedge are used in order to find these constants [4–6], from which the system of linear algebraic equations is received. The asymptotic analysis of the solution is carried out. The coefficient of diffraction is obtained.

2. Formulation of the problem

Let us consider the diffraction of the H-polarized time-harmonic ($e^{-i\omega t}$ time dependence is assumed) electromagnetic plane wave

$$H_z^i(r,\varphi) = \exp[-ikr\cos(\varphi - \varphi_0)] \tag{1}$$

in an angular region $|\varphi| < \Phi \quad (0 < \Phi \leq \pi)$, where $k = \omega\sqrt{\varepsilon_0\mu_0}$, and ε_0, μ_0 are the free space permittivity and permeability, respectively. The faces of the wedge $\varphi = \pm\Phi$ are coated with thin layers of bi-isotropic materials, the thicknesses of which are $d_\pm$. Here and thereafter the symbol sequence of kind $\pm$ means that the upper(lower) sign corresponds to the face $\varphi = +\Phi$ ($\varphi = -\Phi$) of the wedge. In the polar coordinate system (r,φ) the generalized impedance boundary conditions [9] on the faces of the wedge can be expressed as

$$\left[\mp\frac{1}{ikr}\frac{\partial}{\partial\varphi}-p_\pm\left(1+\frac{1}{k^2N_\pm^2}\frac{\partial^2}{\partial r^2}\right)\right]H_z(r,\pm\Phi)=0 \tag{2}$$

where $p_\pm = ikd_\pm\mu_\pm$, $N_\pm^2 = n_\pm^2\cos^2\vartheta_\pm - \kappa_\pm^2$, $n_\pm^2 = \varepsilon_\pm\mu_\pm$, $\sin\vartheta_\pm = \chi_\pm/n_\pm$, $\varepsilon_\pm$ and $\mu_\pm$ are the complex permittivity and permeability of the layers, $\kappa_\pm \geq 0$ are the chirality parameters, $\chi_\pm \geq 0$ are the non-reciprocity parameters. For $\kappa_\pm = \chi_\pm = 0$ the boundary conditions (2) coincide with the boundary conditions for dielectric layers [2, 8]. The total field $H_z(r,\varphi)$ must satisfy the Helmholtz equation, the edge condition and the proper conditions at infinity. Nevertheless, these conditions are not sufficient to solve completely the present boundary value problem. It is necessary to add contact conditions. The general form of the contact conditions is given in papers [4, 5, 7]. The variants of the contact conditions of different orders are given in papers [5, 6, 8, 10, 11]. Only one contact condition is required for the boundary conditions (2), which is possible to write in the following form [8]:

$$\lim_{r\to 0}\left[\frac{1}{t_+}\frac{\partial}{\partial r}H_z(r,\Phi)-\frac{1}{t_-}\frac{\partial}{\partial r}H_z(r,-\Phi)\right]=0 \tag{3}$$

where $t_\pm = N_\pm^2/p_\pm$.

3. Solution method

According to the Malyuzhinets method, let us represent the total wave field in the form of the Sommerfeld–Malyuzhinets integral

$$H_z(r,\varphi)=\frac{1}{2\pi i}\int_\gamma e^{-ikr\cos\alpha}S(\alpha+\varphi)d\alpha \tag{4}$$

where the integration path γ looks like two loops in the complex α plane [12–14]. One of the loops goes from $\pi/2+i\infty$ above all singularities of $S(\alpha+\varphi)$ to $-3\pi/2+i\infty$ and the other is symmetrical to that with respect to the origin $\alpha=0$. To meet the conditions at the edge and at infinity, $S(\alpha)$ must be analytical inside the loops, must have a simple pole with a unit residue at $\alpha=\varphi_0$, and must be limited at infinity $\alpha=\infty$. From the boundary conditions (2), using Malyuzhinets theorem [13, 14], the system of functional equations is obtained:

$$a_\pm(\alpha)S(\alpha\pm\Phi)-a_\pm(-\alpha)S(-\alpha\pm\Phi)=2C_\pm\sin\alpha \tag{5}$$

where $C_\pm$ are arbitrary constants. Variable coefficients in the system (5)

$$a_\pm(\alpha)=\sin^2\alpha\mp t_\pm\sin\alpha+(N_\pm^2-1)=(\sin\alpha\pm\sin\Theta_1^\mp)(\sin\alpha\pm\sin\Theta_2^\mp) \tag{6}$$

are polynomials of $\sin\alpha$. The complex Brewster angles $\Theta^{\pm}_{1,2}$ can be expressed as

$$\begin{aligned}\sin\Theta^{+}_{1,2} &= -t_{+}/2 \pm \sqrt{(t_{+}/2)^2 - (N_{+}^2 - 1)} \\ \sin\Theta^{-}_{1,2} &= -t_{-}/2 \pm \sqrt{(t_{-}/2)^2 - (N_{-}^2 - 1)}\end{aligned} \tag{7}$$

The analysis of the expressions (7) shows that the Brewster angles satisfy the following inequalities in the wide region of variation of complex parameters:

$$0 < \operatorname{Re}\Theta^{\pm}_1 < \pi/2,\ \operatorname{Im}\Theta^{\pm}_1 < 0; \quad -\pi/2 < \operatorname{Re}\Theta^{\pm}_2 < 0,\ \operatorname{Im}\Theta^{\pm}_2 > 0 \tag{8}$$

The general solution of the system of equations (5) can be written in the following form

$$S(\alpha) = [\sigma(\alpha,\varphi_0) + C_{+}S_{+}(\alpha) + C_{-}S_{-}(\alpha) + C]\ \Psi(\alpha)/\Psi(\varphi_0) \tag{9}$$

where C is an arbitrary constant. The function

$$\sigma(\alpha,\varphi_0) = \frac{\mu\cos\mu\varphi_0}{\sin\mu\alpha - \sin\mu\varphi_0}, \quad \text{where} \quad \mu = \frac{\pi}{2\Phi} \tag{10}$$

is the meromorphic solution of the corresponding homogeneous system (5) with $a_{\pm}(\alpha) = 1$. The functions

$$S_{\pm}(\alpha) = \mp\frac{\mu}{2\pi i}\int_{-i\infty}^{+i\infty} \frac{\sin\mu\tau}{\cos\mu\tau \mp \sin\mu\alpha}\, \frac{\sin\tau d\tau}{a_{\pm}(\tau)\Psi(\tau\pm\Phi)} \tag{11}$$

are the partial solutions of the non-homogeneous system (5), regular in the strip $|\operatorname{Re}\alpha| < \Phi$. The function $\Psi(\alpha)$ is the main solution of the homogeneous system (5) without zeroes and poles in the strip $|\operatorname{Re}\alpha| < \Phi$. The form of this function depends on the signs of $\operatorname{Re}\Theta^{\pm}_{1,2}$ [4, 15]. Taking into consideration the inequality (8), the solution can be written in the form

$$\Psi(\alpha) = \frac{\Psi(\alpha+\Phi,\pi/2-\Theta^{+}_1)\ \Psi(\alpha-\Phi,\pi/2-\Theta^{-}_1)}{\Psi(\alpha+\Phi,\pi/2+\Theta^{+}_2)\ \Psi(\alpha-\Phi,\pi/2+\Theta^{-}_2)} \tag{12}$$

where $\Psi(\alpha,\Theta) = \Psi_{\Phi}(\alpha+\Theta)\,\Psi_{\Phi}(\alpha-\Theta)$, $\Psi_{\Phi}(\alpha)$ is a special meromorphic Malyuzhinets function [12, 14]. The function in (9) has poles in the strips $-\pi < \operatorname{Re}\alpha < -\Phi$ and $\Phi < \operatorname{Re}\alpha < \pi$. The residues in these poles correspond to plane waves and surface waves. However, not all the waves satisfy the radiation condition. Therefore it is necessary to choose the unknown constants $C_{\pm}$ and C so that such waves do not appear. Taking into consideration the inequality (8), it is possible to find two such poles of the function $S(\alpha)$: $\alpha_\pm = \pm(\Phi - \Theta_2^\pm)$. Assuming that the residues of the function $S(\alpha)$ in the poles $\alpha_\pm$ are equal to zero, we obtain two linear algebraic equations. The third equation is received from the contact condition (3) after substituting there the following expressions

$$\begin{aligned} \lim_{r\to 0} \frac{\partial}{\partial r} H_z(r, \pm\Phi) &= -\lim_{r\to 0} \frac{k}{2\pi} \int_\gamma e^{-ikr\cos\alpha} \cos\alpha S(\alpha \pm \Phi) d\alpha \\ &= \pm 2\left[C_\pm \mp t_\pm C/\Psi(\varphi_0)\right] \end{aligned} \tag{13}$$

Finally,the system of linear algebraic equations can be written in the matrix form

$$\begin{bmatrix} S_+(\alpha_+) & S_-(\alpha_+) & 1 \\ S_+(\alpha_-) & S_-(\alpha_-) & 1 \\ t_+ & t_- & 0 \end{bmatrix} \begin{bmatrix} C_+ \\ C_- \\ C \end{bmatrix} = \begin{bmatrix} -\sigma(\alpha_+, \varphi_0) \\ -\sigma(\alpha_-, \varphi_0) \\ 0 \end{bmatrix} \tag{14}$$

4. The diffracted field

When the path γ in (4) is closed using two steepest descent paths passing through $\alpha = \pm\pi$, the poles of $S(\alpha+\varphi)$ that lie within the strip $|\mathrm{Re}\,\alpha| < \pi$, are captured, giving rise to the incident and reflected waves of geometrical optics. In addition, one or more surface wave poles may be captured. The non-residue portion of (4) is the diffracted field, and this can be written as

$$H_z^{\mathcal{D}}(r,\varphi) = -\frac{1}{2\pi i} \int_{S(\varphi)} e^{ikr\cos(\alpha-\varphi)} W(\alpha) d\alpha \tag{15}$$

where $S(\varphi)$ is the steepest descent path passing through $\alpha = \varphi$, and

$$W(\alpha) = S(\alpha - \pi) - S(\alpha + \pi) \tag{16}$$

For $kr \gg 1$ stationary-phase evaluation of (15) gives

$$H_z^{\mathcal{D}}(r,\varphi) \simeq W(\varphi) \frac{e^{i(kr+\pi/4)}}{\sqrt{2\pi kr}}, \tag{17}$$

where $W(\varphi)$ is the diffraction coefficient. This equation (17) gives the (edge) diffracted field.

5. Conclusions

The diffraction of a time-harmonic H-polarized plane electromagnetic waves incident normaly to an edge of a perfectly conducting wedge covered with thin layers of the bi-isotropic materials has been considered. The boundary conditions on wedge faces are approximated using impedance conditions with the second-order field derivations. The general solution is derived in explicit form as the Sommerfeld–Malyuzhinets integral involving arbitrary constants. These constants are chosen from the contact condition that reflects the features of the structure near its edge. The unique closed-form solution which generalizes the Malyuzhinets solution for the impedance wedge to more complex boundary conditions has been presented.

References

1. Lindell, I.V., Sihvola, A.H., Tretyakov, S.A., and Viitanen, A.J. (1994) *Electromagnetic Waves in Chiral and Bi-Isotropic Media*. Artech House, Boston.
2. Weinstein, L.A. (1969) *The Theory of Diffraction and the Factorization Method*. Golem, Boulder.
3. Senior, T.B.A., and Volakis, J.L. (1989) Derivation and application of class of generalized boundary conditions, *IEEE Trans. on Antennas and Propagation*, **37**, No. 12, 1566–1572.
4. Osipov, A.V. (1994) General solution for a class of diffraction problems, *J.Phys.A: Math.Gen.*, **27**, L27–L32.
5. Malyuzhinets, G.D., and Tuzhilin, A.A. (1970) Diffraction of a plane sound wave by the thin semi-infinite elastic plate, *Zh.Vych.Mat.& Mat.Fiz.*, **10**, No. 5, 1210–1227 (in Russian).
6. Tuzhilin, A.A. (1973) Diffraction of a plane sound wave in the corner region of the surfaces which are ideally hard and slippery and coated with thin elastic plates, *Differentsialnye Uravnenija*, **9**, No. 10, 1875–1887 (in Russian).
7. Belinskii, B.P., Kouzov, D.P., and Cheltsova, V.D. (1973) About diffraction of the acoustic waves by plates, which are joined under right corner, *Appl.Math.Mech.*, **37**, No. 2, 291–299.
8. Rojas, R.G., Ly, H.C., and Pathak, P.H. (1991) Electromagnetic plane wave diffraction by a planar junction of two thin dielectric/ferrite half planes, *Radio Sci.*, **26**, No. 3, 641–660.
9. Tretyakov, S.A. (1994) Approximate boundary conditions for a thin bi-isotropic slab, *J. Communications Technology & Electronics*, **39**, No. 2, 184–192.
10. Osipov, A.V. (1995) Electromagnetic scattering by an arbitrary angled wedge with penetrable faces: analytical treatment using higher-order boundary conditions, in, *Proc.URSI Int.Symp.on Electromagnetic Theory*. St. Petersburg University Press, St. Petersburg, pp. 510–512.
11. Lyalinov, M.A., and Vardapetyan, L.G. (1995) Electromagnetic diffraction by a coated wedge for an arbitrary incident plane wave, in, *Proc. URSI Int.Symp.on Electromagnetic Theory*. St. Petersburg University Press, St. Petersburg, pp. 513–515.
12. Malyuzhinets, G.D. (1958) Excitation, reflection and emission of surface waves from a wedge with given face impedances, *Sov.Phys.Dokl.*, **3**, No. 4, 752–755.
13. Malyuzhinets, G.D. (1981) *Sommerfeld Integrals and Their Applications*. Rumb, Leningrad (in Russian).
14. Bobrovnikov, M.S., and Fisanov, V.V. (1988) *Wave Diffraction in Angular Regions*. Tomsk Univ. Publ., Tomsk (in Russian).
15. Tuzhilin, A.A. (1971) The theory of the Malyuzhinets functional equations. IV. Inhomogeneous functional equations with the periodic coefficients, *Differentsialnye Uravnenija*, **7**, No. 7, 1276–1287 (in Russian).

OPTIMIZATION APPROACH TO AN INVERSE PROBLEM FOR A STRATIFIED BI-ISOTROPIC SLAB

M. NORGREN
Department of Electromagnetic Theory
Royal Institute of Technology, S-100 44 Stockholm, Sweden

Abstract. An inverse problem for a stratified slab of a bi-isotropic medium is considered in the frequency domain. Considering obliquely impinging plane waves, an imbedding equation for the reflection coefficient matrix is derived through the concept of wave-splitting. An optimization approach is applied to reconstruct the constitutive parameters as functions of the position using band-limited reflection data from different angles of illumination. An exact and explicit expressions for the gradient is derived, which result in a one order faster computation compared with numerical perturbations of the objective functionals. The reconstruction algorithm (based upon a conjugate gradient method) is tested with both clean and noisy data.

1. Introduction

The inverse problem of parameter reconstruction in plane-stratified, or one-dimensional, media has many important applications. Such applications are for example characterization of materials which are essentially one-dimensional in their electromagnetic properties and the design of artificial stratified media for various applications e.g. polarization transformers and anti-reflection coatings.

The most general class of linear media in electromagnetics is the bianisotropic materials. A sub-class of bianisotropic materials is the bi-isotropic materials, for which several types of problems have been considered and solved recently; cf. e.g. [1, 2] for detailed reference lists. The *direct* scattering problem, for multilayered or stratified bi-isotropic media, has been studied in the frequency domain in e.g. [3]. The time-domain *inverse problem* for the bi-isotropic slab has been studied extensively in [4], where the

A. Priou et al. (eds.), Advances in Complex Electromagnetic Materials, 207–218.

scattering data due to a transient plane wave, are used for the reconstruction of the bi-isotropic susceptibility kernels.

However, in many situations such as parameter reconstruction from band-limited scattering data, design of frequency selective (anti-) reflection coatings, etc., it is of importance to study the inverse problem in the frequency domain. Optimization methods have shown to be useful when solving certain inverse problems, in both the frequency domain and in the time domain (cf. e.g. [5]-[8]); in [8] a frequency domain optimization approach is used for the reconstruction of the parameters of a nonuniform transmission line.

In this paper, the optimization approach in [8] is extended to solve an inverse problem for a stratified bi-isotropic slab. To apply an optimization approach to an inverse problem, one introduces a suitable objective functional first and then computes the gradient of this functional. Once the gradient of the objective functional has been computed, a conventional steepest descent or conjugate gradient method can be applied to minimize the objective functional in an iterative way. In the present paper, the imbedding equation for the reflection coefficient matrix is used as a starting point to derive an exact and explicit expression for the gradient through a suitable matrix formalism and by introducing suitable dual functions. The constitutive parameters are then reconstructed by an iterative algorithm, based on a conjugate gradient method [9].

2. Problem formulation and the direct solver

A stratified bi-isotropic material can be modeled by the following constitutive relations, in the frequency domain:

$$\vec{\mathbf{D}}(\vec{\mathbf{r}}) = \varepsilon_0 \varepsilon(z) \vec{\mathbf{E}}(\vec{\mathbf{r}}) + \sqrt{\varepsilon_0 \mu_0} \xi(z) \vec{\mathbf{H}}(\vec{\mathbf{r}}), \tag{1}$$

$$\vec{\mathbf{B}}(\vec{\mathbf{r}}) = \mu_0 \mu(z) \vec{\mathbf{H}}(\vec{\mathbf{r}}) + \sqrt{\varepsilon_0 \mu_0} \zeta(z) \vec{\mathbf{E}}(\vec{\mathbf{r}}), \tag{2}$$

where the constitutive parameters $\varepsilon(z), \mu(z), \xi(z)$ and $\zeta(z)$ are arbitrary functions of the position variable z. In (1) and (2), ε_0 and μ_0 (the permittivity and permeability in vacuum respectively) are introduced in order to make the parameters dimensionless.

2.1. THE INVERSE PROBLEM

We consider an electromagnetic plane wave with harmonic time dependence $\exp(\jmath\omega t)$ obliquely impinging from the left side $z < 0$ on a stratified bi-isotropic slab located in the region $z \in [0, l]$. The region $z < 0$ is filled with vacuum and the region $z > l$ is occupied by a homogeneous medium with known properties; cf. Fig. 1. The inverse problem is to determine one

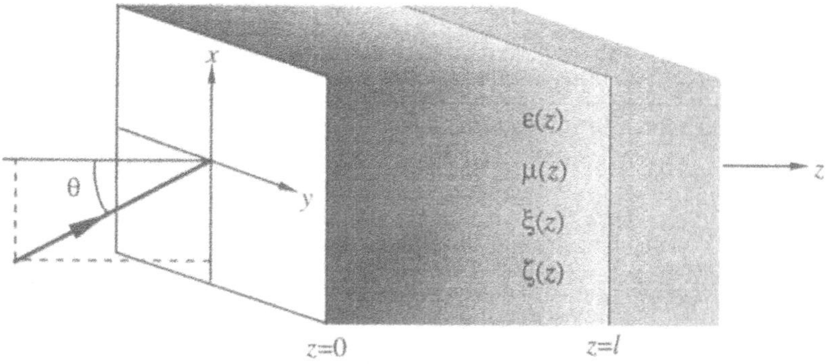

Figure 1. The scattering configuration.

or several of the constitutive parameters in the region $[0, l]$ as functions of the position z from the reflection and/or transmission data in a certain frequency band.

2.2. THE DIRECT SOLVER

In this paper we use an optimization approach to solve the inverse problem in an iterative way. To apply an optimization approach, we need a direct solver for the direct problem to calculate the gradient at each iteration. In this subsection the direct solver is presented.[1]

Maxwell's equations in the xyz coordinate system are

$$\nabla \times \vec{\mathbf{E}} = -\jmath\omega\vec{\mathbf{B}}, \qquad \nabla \times \vec{\mathbf{H}} = \jmath\omega\vec{\mathbf{D}}. \tag{3}$$

Without loss of generality, we can choose the xz-plane as the incidental plane, which from continuity of the tangential part of the impinging wave vector implies that

$$\nabla = -\jmath\omega\sqrt{\varepsilon_0\mu_0}\sin\theta\hat{\mathbf{x}} + \hat{\mathbf{z}}\frac{d}{dz}, \tag{4}$$

where θ is the angle of incidence; cf. Fig.1. Using (1)-(4), we arrive at the

[1]The direct solver for stratified *bianisotropic* media is described in [10], from a general point of view; here, the special bi-isotropic case is described in more detail.

following ODE (ordinary differential eqs.) system:

$$\frac{d}{dz}\begin{bmatrix} E_x \\ E_y \\ H_x \\ H_y \end{bmatrix} = \frac{\jmath\omega}{c_0}\begin{bmatrix} 0 & -\zeta+h\xi & 0 & -\eta_0\mu\,(1-h) \\ \zeta & 0 & \eta_0\mu & 0 \\ 0 & \eta_0^{-1}\varepsilon\,(1-h) & 0 & \xi-h\zeta \\ -\eta_0^{-1}\varepsilon & 0 & -\xi & 0 \end{bmatrix}\begin{bmatrix} E_x \\ E_y \\ H_x \\ H_y \end{bmatrix}, \tag{5}$$

where

$$h = \frac{\sin^2\theta}{\varepsilon\mu - \xi\zeta}, \tag{6}$$

and where $c_0 = (\varepsilon_0\mu_0)^{-1/2}$ and $\eta_0 = (\mu_0/\varepsilon_0)^{1/2}$ are the phase velocity and wave impedance in vacuum respectively. If we diagonalize the system (5) when $\varepsilon\,(z) = \mu\,(z) = 1$ and $\xi\,(z) = \zeta\,(z) = 0$, we obtain the eigenwaves in vacuum; thus, we can use e.g. the following wave-splitting [10]:

$$\begin{bmatrix} \mathbf{E}^+ \\ \mathbf{E}^- \end{bmatrix} \equiv \begin{bmatrix} \begin{bmatrix} E_1^+ \\ E_2^+ \end{bmatrix} \\ \begin{bmatrix} E_1^- \\ E_2^- \end{bmatrix} \end{bmatrix} = \frac{1}{2}\begin{bmatrix} 1 & 0 & 0 & \eta_0\cos\theta \\ 0 & 1 & -\eta_0/\cos\theta & 0 \\ 1 & 0 & 0 & -\eta_0\cos\theta \\ 0 & 1 & \eta_0/\cos\theta & 0 \end{bmatrix}\begin{bmatrix} E_x \\ E_y \\ H_x \\ H_y \end{bmatrix}, \tag{7}$$

With this wave-splitting it follows that in the region $z < 0$ the components E_1^+, E_1^-, of the split fields, become right- and left-moving TM modes respectively, and the components E_2^+, E_2^- become right- and left-moving TE modes respectively. Using the transformation (7) and its inverse together with (5), we arrive at the following system (which is Maxwell's equations in terms of the split-fields $\mathbf{E}^{\pm}$):

$$\frac{d}{dz}\begin{bmatrix} \mathbf{E}^+ \\ \mathbf{E}^- \end{bmatrix} = \frac{\jmath\omega}{2c_0}\begin{bmatrix} \boldsymbol{\alpha} & \boldsymbol{\beta} \\ \mathbf{b} & \mathbf{a} \end{bmatrix}\begin{bmatrix} \mathbf{E}^+ \\ \mathbf{E}^- \end{bmatrix}, \tag{8}$$

where

$$\left.\begin{aligned} \boldsymbol{\alpha} &= \begin{bmatrix} \alpha_{11} & \alpha_{12} \\ \alpha_{21} & \alpha_{22} \end{bmatrix}, \quad \mathbf{a} = \begin{bmatrix} -\alpha_{11} & \alpha_{12} \\ \alpha_{21} & -\alpha_{22} \end{bmatrix}, \\ \boldsymbol{\beta} &= \begin{bmatrix} \beta_{11} & \beta_{12} \\ \beta_{21} & \beta_{22} \end{bmatrix}, \quad \mathbf{b} = \begin{bmatrix} -\beta_{11} & \beta_{12} \\ \beta_{21} & -\beta_{22} \end{bmatrix}, \end{aligned}\right\} \tag{9}$$

and where

$$\left.\begin{aligned}
\alpha_{11} &= -\varepsilon\cos\theta - \frac{\mu(1-h)}{\cos\ theta}, & \alpha_{12} &= -\zeta + \xi\left(\cos^2\theta + h\right), \\
\alpha_{21} &= \zeta\left(1 + \frac{h}{\cos^2\theta}\right) - \frac{\xi}{\cos^2\theta}, & \alpha_{22} &= -\frac{\varepsilon(1-h)}{\cos\theta} - \mu\cos\theta, \\
\beta_{11} &= -\varepsilon\cos\theta + \frac{\mu(1-h)}{\cos\theta}, & \beta_{12} &= -\zeta - \xi\left(\cos^2\theta - h\right), \\
\beta_{21} &= \zeta\left(1 - \frac{h}{\cos^2\theta}\right) + \frac{\xi}{\cos^2\theta}, & \beta_{22} &= -\frac{\varepsilon(1-h)}{\cos\theta} + \mu\cos\theta,
\end{aligned}\right\}. \tag{10}$$

The reflection coefficient matrix (denoted $\mathbf{r}(z)$) for a sub-slab $[z,l]$, is defined through [10]

$$\mathbf{E}^-(z) = \mathbf{r}(z)\,\mathbf{E}^+(z) \equiv \begin{bmatrix} r_{11}(z) & r_{12}(z) \\ r_{21}(z) & r_{22}(z) \end{bmatrix} \mathbf{E}^+(z). \tag{11}$$

From (8) and (11), $\mathbf{r}(z)$ satisfies the following Riccati equation:

$$\frac{d}{dz}\mathbf{r} = \mathbf{b} + \mathbf{a}\mathbf{r} - \mathbf{r}\boldsymbol{\alpha} - \mathbf{r}\boldsymbol{\beta}\mathbf{r}. \tag{12}$$

As a consequence of the wave-splitting (7), the boundary value of the reflection coefficient matrix at $z = l$, i.e. $\mathbf{r}(l)$, is the reflection coefficient matrix for an incident wave directly on the background medium. Practically used backing materials are e.g. vacuum and perfect conductors which yield the following boundary values:

$$\begin{aligned} \mathbf{r}(l) &= \mathbf{0} &&: \text{ Vacuum background,} \\ \mathbf{r}(l) &= -\mathbf{I} &&: \text{ PEC background,} \end{aligned} \tag{13}$$

($\mathbf{I}$ is the 2×2 identity matrix). By integrating (12) in the $-z$ direction, starting with the boundary condition (13), we obtain the reflection coefficient matrix for the stratified bi-isotropic slab, i.e. $\mathbf{r}(0)$.
The true reflection coefficients (with respect to the total field strengths) are [10]

$$r_{MM} = r_{11}(0), \quad r_{ME} = \frac{r_{12}(0)}{\cos\theta}, \quad r_{EM} = r_{21}(0)\cos\theta, \quad r_{EE} = r_{22}(0), \tag{14}$$

and for a non-active medium it follows, from energy conservation, that

$$|r_{XY}| \le 1 \qquad X, Y \in \{E, M\}. \tag{15}$$

3. Optimization approach to the inverse problem

In the inverse problem, one can use an optimization approach to reconstruct various constitutive parameters. A crucial result of this paper is that the gradient of the objective functional can be calculated explicitly through the introduction of suitable dual functions; otherwise, one has to compute the gradient by numerical perturbations which results in a computation which is one order slower; see [8] for a comparison of the required CPU-times. In this section an optimization approach is presented.

Introduce p as a notation for all the constitutive parameters $\varepsilon, \mu, \xi, \zeta$. Define an objective functional as follows,

$$J(p) = \sum_{\omega=\omega_{\min}}^{\omega_{\max}} \sum_{\theta} \sum_{i,j\in\{1,2\}} W_{ij}(\omega,\theta) \left| r_{ij}(0;\omega,\theta) - r_{ij}^{m}(\omega,\theta) \right|^2, \tag{16}$$

where $[\omega_{\min}, \omega_{\max}]$ is the frequency band of interest and θ are different directions of the impinging wave vector. $r_{ij}^{m}(\omega,\theta)$ are the *measured* components of the reflection coefficient matrix, and $W_{ij}(\omega,\theta)$ are (positive) weighting functions describing the weight of the reflection data at different frequency points and at different directions of illumination. The minimum of $J(p)$ is zero if the inverse problem has a solution.

3.1. EXPLICIT EXPRESSION FOR THE GRADIENT

Let $\tilde{\mathbf{r}} = \mathbf{r}(z; p+\delta p)$ be the solution to (12) and (13) with the parameter vector $p+\delta p$. Then $\delta\mathbf{r} \equiv \tilde{\mathbf{r}} - \mathbf{r}$ satisfies the following equation and boundary condition:

$$\frac{d}{dz}\delta\mathbf{r} + \delta\mathbf{r}(\boldsymbol{\alpha} + \boldsymbol{\beta}\mathbf{r}) - (\mathbf{a} - \mathbf{r}\boldsymbol{\beta})\delta\mathbf{r} = \delta\mathbf{b} + \delta\mathbf{a}\,\mathbf{r} - \mathbf{r}\,\delta\boldsymbol{\alpha} - \mathbf{r}\,\delta\boldsymbol{\beta}\,\mathbf{r} + o(\delta p), \tag{17}$$

$$\delta\mathbf{r}(l) = \mathbf{0}, \tag{18}$$

where $\delta\boldsymbol{\alpha} = \boldsymbol{\alpha}(p+\delta p) - \boldsymbol{\alpha}(p)$ etc., and $o(\delta p)$ denotes higher order terms.

The corresponding increment of the functional $J(p)$ can then be written as

$$\begin{aligned} \delta J(p) &\equiv J(p+\delta p) - J(p) \\ &= \sum_{\omega=\omega_{\min}}^{\omega_{\max}} \sum_{\theta} \sum_{i,j\in\{1,2\}} 2\Re e\left\{ W_{ij}(\omega,\theta) \left[r_{ij}(0;\omega,\theta) - r_{ij}^{m}(\omega,\theta) \right]^{*} \delta r_{ij}(0;\omega,\theta) \right\} \\ &\quad + o(\delta p), \end{aligned} \tag{19}$$

where the superscript $*$ denotes the complex conjugate, and $\Re e$ denotes the real part.

Now, introduce rearranged versions of the quantities $\boldsymbol{\alpha}, \boldsymbol{\beta}, \delta\boldsymbol{\alpha}, \delta\boldsymbol{\beta}$ and $\delta\mathbf{r}$, from 2×2 matrices to 4×1-vectors:

$$\boldsymbol{\alpha} = \begin{bmatrix} \alpha_{11} & \alpha_{12} \\ \alpha_{21} & \alpha_{22} \end{bmatrix} \longrightarrow \overset{\leftrightarrow}{\boldsymbol{\alpha}} = \begin{bmatrix} \alpha_{11} \\ \alpha_{12} \\ \alpha_{21} \\ \alpha_{22} \end{bmatrix}, \quad \text{etc.,}$$

where 4×1-vectors are denoted by leftright-arrows on top. Hence, equation (17) can be rewritten into the following conventional form:

$$\frac{d}{dz} \overleftrightarrow{\delta \mathbf{r}} + \mathbf{C} \overleftrightarrow{\delta \mathbf{r}} = \mathbf{A} \overleftrightarrow{\delta \boldsymbol{\alpha}} + \mathbf{B} \overleftrightarrow{\delta \boldsymbol{\beta}} + o(\delta p), \tag{20}$$

where the 4×4 matrices $\mathbf{A}, \mathbf{B}$ and $\mathbf{C}$ are

$$\mathbf{A} = -\frac{\jmath\omega}{2c_0} \begin{bmatrix} 2r_{11} & -r_{21} & r_{12} & 0 \\ r_{12} & r_{11} - r_{22} & 0 & r_{12} \\ r_{21} & 0 & r_{22} - r_{11} & r_{21} \\ 0 & r_{21} & -r_{12} & 2r_{22} \end{bmatrix}, \tag{21}$$

$$\mathbf{B} = -\frac{\jmath\omega}{2c_0} \begin{bmatrix} r_{11}^2 + 1 & r_{11}r_{21} & r_{11}r_{12} & r_{12}r_{21} \\ r_{11}r_{12} & r_{11}r_{22} - 1 & r_{12}^2 & r_{12}r_{22} \\ r_{11}r_{21} & r_{21}^2 & r_{11}r_{22} - 1 & r_{21}r_{22} \\ r_{12}r_{21} & r_{21}r_{22} & r_{12}r_{22} & r_{22}^2 + 1 \end{bmatrix}, \tag{22}$$

$$\mathbf{C} = \frac{\jmath\omega}{2c_0} \begin{bmatrix} (\boldsymbol{\alpha} + \boldsymbol{\beta}\mathbf{r})^{\mathsf{T}} - (\boldsymbol{\vartheta} - \mathbf{r}\boldsymbol{\beta})_{11}\mathbf{I} & -(\boldsymbol{\vartheta} - \mathbf{r}\boldsymbol{\beta})_{12}\mathbf{I} \\ -(\boldsymbol{\vartheta} - \mathbf{r}\boldsymbol{\beta})_{21}\mathbf{I} & (\boldsymbol{\alpha} + \boldsymbol{\beta}\mathbf{r})^{\mathsf{T}} - (\boldsymbol{\vartheta} - \mathbf{r}\boldsymbol{\beta})_{22}\mathbf{I} \end{bmatrix}. \tag{23}$$

Introduce a *vector* dual function

$$\overleftrightarrow{\mathbf{U}} (z; \omega, \theta) \equiv \begin{bmatrix} U_{11} \\ U_{12} \\ U_{21} \\ U_{22} \end{bmatrix} (z; \omega, \theta), \tag{24}$$

determined from the following ODE and boundary condition:

$$\frac{d}{dz} \overleftrightarrow{\mathbf{U}} - \mathbf{C}^{\mathsf{T}} \overleftrightarrow{\mathbf{U}} = \mathbf{0}, \tag{25}$$

$$U_{ij}(0; \omega, \theta) = 2W_{ij}(\omega, \theta) \left[r_{ij}(0; \omega, \theta) - r_{ij}^m(\omega, \theta) \right]^*, \quad i, j \in \{1, 2\}. \tag{26}$$

The increment in the objective functional can thus be written as

$$\delta J(p) = \sum_{\omega = \omega_{\min}}^{\omega_{\max}} \sum_{\theta} \Re e \left\{ \overleftrightarrow{\mathbf{U}}^{\mathsf{T}} (0; \omega, \theta) \overleftrightarrow{\delta \mathbf{r}} (0; \omega, \theta) \right\} + o(\delta p). \tag{27}$$

From (20) and (25), we have

$$\frac{d}{dz} \left(\overleftrightarrow{\mathbf{U}}^{\mathsf{T}} \overleftrightarrow{\delta \mathbf{r}} \right) = \overleftrightarrow{\mathbf{U}}^{\mathsf{T}} \left(\mathbf{A} \overleftrightarrow{\delta \boldsymbol{\alpha}} + \mathbf{B} \overleftrightarrow{\delta \boldsymbol{\beta}} \right) + o(\delta p), \tag{28}$$

which yields (by integrating (28) over $[0, l]$ and using the boundary condition (18))

$$\overleftrightarrow{\mathbf{U}}^{\mathsf{T}} (0; \omega, \theta) \overleftrightarrow{\delta \mathbf{r}} (0; \omega, \theta) = -\int_0^l \overleftrightarrow{\mathbf{U}}^{\mathsf{T}} \left(\mathbf{A} \overleftrightarrow{\delta \boldsymbol{\alpha}} + \mathbf{B} \overleftrightarrow{\delta \boldsymbol{\beta}} \right) dz + o(\delta p). \tag{29}$$

Therefore, we can write the increment of the objective functional as

$$\delta J(p) = -\int_0^l dz\, \Re e \sum_{\omega=\omega_{\min}}^{\omega_{\max}} \sum_{\theta} \left\{ \overleftrightarrow{\mathbf{U}}^{\mathsf{T}} \left(\mathbf{A}\, \delta\overleftrightarrow{\boldsymbol{\alpha}} + \mathbf{B}\, \delta\overleftrightarrow{\boldsymbol{\beta}} \right) \right\} + o\,(\delta p)\,. \tag{30}$$

The incremental vectors $\delta\overleftrightarrow{\boldsymbol{\alpha}}$ and $\delta\overleftrightarrow{\boldsymbol{\beta}}$ in (30) are

$$\delta\overleftrightarrow{\boldsymbol{\alpha}} = \frac{\partial\, \overleftrightarrow{\boldsymbol{\alpha}}}{\partial \varepsilon}\delta\varepsilon + \frac{\partial\, \overleftrightarrow{\boldsymbol{\alpha}}}{\partial \mu}\delta\mu + \frac{\partial\, \overleftrightarrow{\boldsymbol{\alpha}}}{\partial \xi}\delta\xi + \frac{\partial\, \overleftrightarrow{\boldsymbol{\alpha}}}{\partial \zeta}\delta\zeta + o\,(\delta p)\,, \tag{31}$$

$$\delta\overleftrightarrow{\boldsymbol{\beta}} = \frac{\partial\, \overleftrightarrow{\boldsymbol{\beta}}}{\partial \varepsilon}\delta\varepsilon + \frac{\partial\, \overleftrightarrow{\boldsymbol{\beta}}}{\partial \mu}\delta\mu + \frac{\partial\, \overleftrightarrow{\boldsymbol{\beta}}}{\partial \xi}\delta\xi + \frac{\partial\, \overleftrightarrow{\boldsymbol{\beta}}}{\partial \zeta}\delta\zeta + o\,(\delta p)\,, \tag{32}$$

where the partial derivatives follow from (10) and (6). Inserting (31) and (32) into (30), the functional increment can be written as

$$\delta J(p) = -\int_0^l dz\, \Re e \sum_{\omega=\omega_{\min}}^{\omega_{\max}} \sum_{\theta} \{F_\varepsilon \delta\varepsilon + F_\mu \delta\mu + F_\xi \delta\xi + F_\zeta \delta\zeta\} + o\,(\delta p)\,, \tag{33}$$

where the scalar functions $F_\varepsilon, F_\mu, F_\xi$ and F_ζ are

$$F_\varepsilon = \overleftrightarrow{\mathbf{U}}^{\mathsf{T}} \left(\mathbf{A} \frac{\partial\, \overleftrightarrow{\boldsymbol{\alpha}}}{\partial \varepsilon} + \mathbf{B} \frac{\partial\, \overleftrightarrow{\boldsymbol{\beta}}}{\partial \varepsilon} \right), \quad \text{etc.} \tag{34}$$

The first-order term of (33) is in an inner-product form; to identify the gradients we must take into consideration that the constitutive parameters are in general complex-valued functions of z. If we restrict the analysis to a lossless bi-isotropic medium, the parameters $\varepsilon\,(z)$ and $\mu\,(z)$ become real-valued while the parameters $\xi\,(z)$ and $\zeta\,(z)$ are complex-conjugates of each other viz.

$$\xi\,(z) = \chi\,(z) - \jmath\kappa\,(z)\,, \qquad \zeta\,(z) = \chi\,(z) + \jmath\kappa\,(z)\,, \tag{35}$$

where $\kappa\,(z)$ (the chirality) and $\chi\,(z)$ (the non-reciprocity) are real-valued parameters [11]. Thus, we wish to identify the following inner-product form:

$$\begin{aligned} \delta J(p) &= \int_0^l \frac{\partial J}{\partial p}\,(z) \cdot \delta p\,(z)\; dz + o\,(\delta p) \\ &\equiv \left\langle \frac{\partial J}{\partial \varepsilon}, \delta\varepsilon \right\rangle + \left\langle \frac{\partial J}{\partial \mu}, \delta\mu \right\rangle + \left\langle \frac{\partial J}{\partial \kappa}, \delta\kappa \right\rangle + \left\langle \frac{\partial J}{\partial \chi}, \delta\chi \right\rangle + o\,(\delta p)\,, \end{aligned} \tag{36}$$

where $\dfrac{\partial J}{\partial p}\,(z)$ is the gradient of the objective functional. Hence, substituting $\delta\xi = \delta\chi - \jmath\delta\kappa$ and $\delta\zeta = \delta\chi + \jmath\delta\kappa$ into (33) we can identify the gradients of the objective functional in (16), with respect to the *real-valued* parameters ε, μ, κ and χ, as

$$\frac{\partial J}{\partial \varepsilon}(z) = -\Re e \sum_{\omega=\omega_{\min}}^{\omega_{\max}} \sum_{\theta} F_{\varepsilon}(z;\omega,\theta)\,, \tag{37}$$

$$\frac{\partial J}{\partial \mu}(z) = -\Re e \sum_{\omega=\omega_{\min}}^{\omega_{\max}} \sum_{\theta} F_{\mu}(z;\omega,\theta)\,, \tag{38}$$

$$\frac{\partial J}{\partial \kappa}(z) = -\Re e \sum_{\omega=\omega_{\min}}^{\omega_{\max}} \sum_{\theta} \jmath\left(F_{\zeta}(z;\omega,\theta) - F_{\xi}(z;\omega,\theta)\right), \tag{39}$$

$$\frac{\partial J}{\partial \chi}(z) = -\Re e \sum_{\omega=\omega_{\min}}^{\omega_{\max}} \sum_{\theta} \left(F_{\xi}(z;\omega,\theta) + F_{\zeta}(z;\omega,\theta)\right). \tag{40}$$

The functions $F_{\varepsilon}, F_{\mu}, F_{\xi}$ and F_{ζ} may be considered as primitive gradients;[2] if the constitutive parameters ε, μ, ξ and ζ are functions of more fundamental parameters (which certainly is the case for dispersive bi-isotropic media; cf. e.g. [12]) the new gradients are easily identified by insertion into (33) of the expanded expressions for $\delta\varepsilon, \delta\mu, \delta\xi$ and $\delta\zeta$.

Note that the above gradients for the left-sided excitation are functions of only the z variable. Also note that if the scattering data for right-sided excitation, as well as the the scattering data for left-sided excitation, are used in an inverse problem, an analogous term should be added in the definition of the objective functional (16). Consequently in that case the above expressions for the gradients will be modified with additional terms (which are similar to the right-hand-side expressions of (37)-(40)).

4. Numerical reconstruction

We consider a lossless and non-dispersive bi-isotropic slab with constitutive parameters ε, μ, κ and χ, given by the solid lines in Fig.2. The slab is backed by a perfect conductor at $z = l$. We use double-sided reflection data (hence the slab must be turned around between the measurements) from the directions $\theta \in \{0^\circ, 50^\circ\}$.

To enhance the convergence rate, it appears better to start iterations using scattering data at low frequencies and with a relatively sparse discretization in $z \in [0, l]$, to obtain a rough reconstruction of the parameter profiles. Then one gradually refines the grid in z and uses higher frequencies to obtain the fine structures of the parameter profiles. In this example, the final frequency range is 73 equidistant points in the band $f \in [0.2c_0/l,\ 2.0c_0/l]$

[2]Instead of F_{ξ} and F_{ζ}, we can consider $F_{\kappa} \equiv \jmath(F_{\zeta} - F_{\xi})$ and $F_{\chi} \equiv F_{\xi} + F_{\zeta}$ as primitive gradients; cf. (39) and (40).

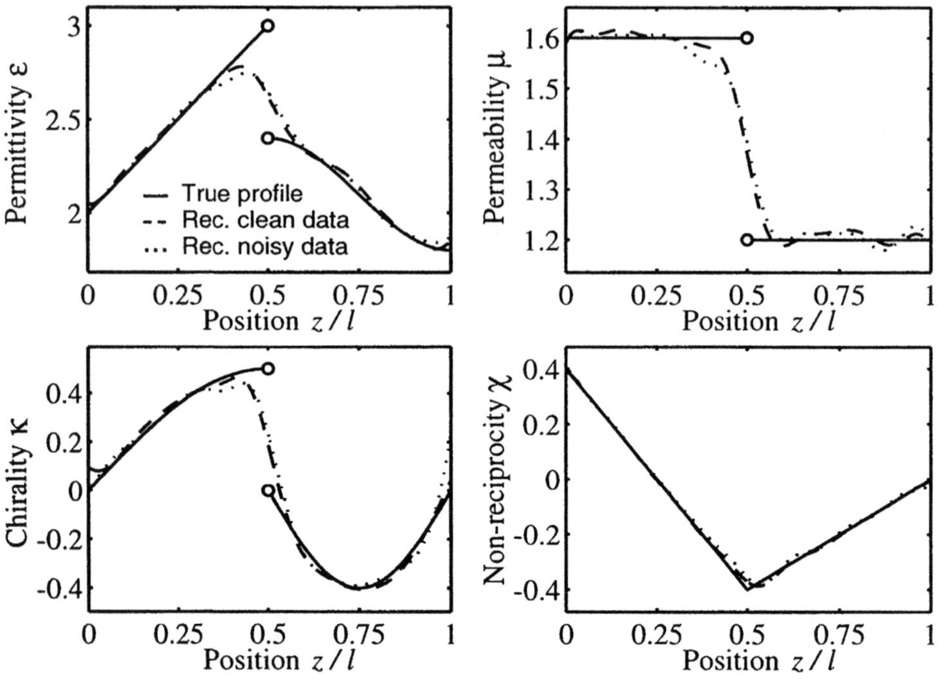

Figure 2. **Reconstruction of all constitutive parameters for a stratified bi-isotropic medium, using double sided reflection data. The solid lines are for the true profiles, the dashed line are the reconstructions using clean reflection data and the dotted lines are the reconstructions using noise-contaminated reflection data.**

and the final grid size is 201 points, in z. The weighting functions for the reflection data are

$$W_{11} = W_{22} = 100^{-(\omega/\omega_{\max})^2},$$
$$W_{12} = [\cos\theta]^{-2}\, 100^{-(\omega/\omega_{\max})^2}, \quad W_{21} = [\cos\theta]^{2}\, 100^{-(\omega/\omega_{\max})^2},$$

where $\omega_{\max} = 2c_0/l$; the $\cos\theta$-terms in W_{12} and W_{21} yields an equal weight for the true reflection coefficients; cf (14).

The reconstructed values of the parameters, after 200 iterations, are shown in Fig.2 by the dashed lines; the initial guesses of the parameters are $\varepsilon(z) = 2.5\varepsilon_0$, $\mu(z) = \mu_0$ and $\kappa(z) = \chi(z) = 0$. To improve the reconstructions of the discontinuities in ε, μ and κ, at $z = 0.5l$, (a profile with discontinuities is more difficult to reconstruct than a smooth profile using the present algorithm), one may refine the grid in z and use higher frequencies. To test the stability of the algorithm, Gaussian noise with a standard deviation $\sigma = 0.05$ is added on both the real and the imaginary parts of all components in the reflection coefficient matrix $\mathbf{r}^m(\omega, \theta)$. The dotted lines in

Fig.2 show the corresponding reconstructions using the noise-contaminated reflection data. We see that, the reconstructed profiles become essentially the same, using clean or noise-contaminated data.

5. Discussion and conclusion

An optimization approach to the electromagnetic inverse problem of parameter reconstruction in a stratified bi-isotropic slab has been described. In the lossless and non-dispersive case, the reconstruction algorithm exhibits encouraging performance and shows stability against Gaussian noise. Although only reflection data have been used in the present paper, it is straight forward to include transmission data in the objective functional. However, it appears that, as in [8], transmission data introduce local minima in the objective functional (hence, the analysis for the transmission coefficient matrix has not been described).

Using relevant dispersion models for the constitutive parameters, the reconstruction algorithm has potential usefulness when solving a wide class of optimal-design problems; see e.g. [13] for design using nonuniform transmission lines. Finally, it should be noted that the methodology for the derivation of an explicit gradient through the concept of dual functions, outlined in this paper, can be easily extended to the general bianisotropic case.

Acknowledgment

The partial support of the Swedish Research Council for Engineering Sciences is gratefully acknowledged.

References

1. Lindell, I.V., Sihvola, A.H., Tretyakov, S.A., and Viitanen, A.J. (1994) *Electromagnetic Waves in Chiral and Bi-isotropic Media*, Artech House.
2. Cory, H. (1995) Chiral devices – an overview of canonical problems, *J. Electromagn. Waves Applic.* **9**, 805-829.
3. He, S. (1992) A time-harmonic Green function technique and wave propagation in a stratified nonreciprocal chiral slab with multiple discontinuities, *J. Math. Phys.* **33**, 4103-4110.
4. Rikte, S. (1995) *Propagation of transient electromagnetic waves in stratified bi-isotropic media and related inverse scattering problems.*, Ph.D. thesis, Department of Electromagnetic Theory, Lund, Sweden.
5. Bandler, J.W. and Chen, S.H. (1988) Circuit optimization: the state of the art, *IEEE Trans. Microwave Theory Tech.* **36**, 424-443.
6. Pesque, J.J., Bouche, D.P. and Mittra, R. (1992) Optimization of multilayer antireflection coatings using an optimal control method, *IEEE Trans. Microwave Theory Tech.* **40**, 1789-1796.
7. He, S. and Kabanikhin, S.I. (1995) An optimization approach to a three-dimensional acoustic inverse problem in the time domain, *J. Math. Phys.* **36**, 4028-4043.

8. Norgren, M. and He, S. (1996) An optimization approach to the frequency domain inverse problem for a nonuniform LCRG transmission line, *IEEE Trans. Microwave Theory Tech.* **44** (in press).
9. Polak, E. (1971) *Computational Methods in Optimization*, Academic Press, New York and London, p. 306.
10. Norgren, M. and He, S. (1995) A general scheme for the electromagnetic reflection and transmission for composite structures of complex media, *Proc. IEE, Part H: Microwaves, Antennas and Propagation* **142**, 52-56.
11. Sihvola, A.H. and Lindell, I.V. (1991) Bi-isotropic constitutive relations, *Microw. Opt. Technol. Lett.* **4**, 295.
12. Bahr, A.J. and Clausing, K.R. (1994) An approximate model for artificial chiral material, *IEEE Trans. Antennas and Propagation* **42**, 1592-1599.
13. Norgren, M. and He, S. (1996) Optimal designs for nonuniform LCRG transmission lines, *J. Electromagn. Waves Applic.* **10**, 1113-1127.

THE INFLUENCE OF INDUCED CHIRAL PROPERTIES ON THE TRANSFORMATION OF ACOUSTIC WAVES POLARIZATION IN PIEZOELECTRIC SEMICONDUCTORS

I.V. SEMCHENKO AND S.A. KHAKHOMOV
Department of Physics, Gomel State University
Sovyetskaya Str. 104, 246699, Gomel, Belarus

1. Introduction

The possibility to control acoustic waves polarization by electric field inducing artificial spiral anisotropy was shown by Bely and Sevruk [1]. Later on, the same authors [2] showed a possibility of rotating acoustic anisotropy forming in dielectric crystals under external electric field influence. Detailed study of acousto-electron interaction in semiconductors was given in papers [3, 4] at the *Bianisotropics'93* and *Chiral'94* meetings. The case of resonance interaction when acoustic wave frequency ω coincides with the rotating electric field frequency Ω was considered in article [5] and in paper [6]. In this work, various regimes of non-resonance interaction of acoustic waves with rotating electric field are considered.

2. Calculation and Discussion

The availability of charge carriers in semiconductors allows to influence the character of interaction of acoustic waves with external electric fields [7, 8]. Charge carriers interact with the electric field of waves in crystal. The character of this interaction can be changed by external constant electric field with strength E'. Owing to the external electric field influence, the drift of electrons with velocity $v = -\mu E'$ (μ is the mobility of electrons) becomes possible [9, 10].

We have considered the case when a ferroelectric ceramic (for example, of $BaTiO_3$) is placed into an electric field that rotates around axis X with frequency Ω. The properties of the ceramic can be described by the material equations taking into account the piezoelectric effect:

A. Priou et al. (eds.), Advances in Complex Electromagnetic Materials, 219–226.

$$\sigma = c\gamma + \varepsilon_0 \varepsilon g \mathbf{E}_0 \mathbf{E} \tag{1}$$

$$\mathbf{D} = \varepsilon_0 \varepsilon \mathbf{E} - \varepsilon_0 \varepsilon g \mathbf{E}_0 \gamma \tag{2}$$

Here σ, γ, c are tensors of tension, deformation, and elastic constant, g is a tensor of rank four, gE_0 is a tensor of rank three which takes into account the piezoelectric effect induced by the rotating electric field, ε_0 is the permittivity of free space, ε is the relative permittivity of the medium,

$$\mathbf{E} = \mathbf{E}_1 + \mathbf{E}'$$

$\mathbf{E}_1$ is the longitudinal electric field created by the piezoelectric effect, $\mathbf{E}'$ is the external constant longitudinal electric field, and

$$E_{01} = 0, \quad E_{02} = -E_0 \sin \Omega t, \quad E_{03} = E_0 \cos \Omega t$$

are the components of rotating electric field.

When the incident acoustic wave frequency ω corresponds to the following relation

$$|\delta^{*2}|((\omega - \Omega)^2 + \Omega^2)^2 \ll 4\Omega^2(\omega - \Omega)^2 \tag{3}$$

the eigenwaves of the crystal have a form of circularly polarized waves, and a rotation of the plane of polarization of ultrasound takes place. The rotation angle can be calculated as

$$\Phi = \Re \frac{k_1(\omega + \Omega) - k_2(\omega - \Omega)}{2} = -\Re \frac{\omega^4 \delta^{*2} L}{8V\Omega(\omega^2 - \Omega^2)} \tag{4}$$

After the substitutions $\delta^* \rightarrow \delta/\overline{\Lambda}, V \rightarrow \sqrt{\overline{\Lambda}/\rho}, \Omega \rightarrow q\sqrt{\overline{\Lambda}/\rho}$ this expression transforms to the expression for rotating ability of ceramic cylinder with spiral anisotropy of the period $P = 2\pi/q$ [1], which is analogical to the de Vries formula for rotating ability of cholesteric liquid crystals [11]. This fact is an evidence of analogy between chiral properties of crystal with time periodic structure and crystal with space periodic structure. The specific rotation of polarization plane $\gamma = \Phi/L$ can reach a few radn/cm for ceramics of $BaTiO_3$ (see Fig. 1). Calculations were made with the values of parameters [12, 13]: $m* = 0.0145$ m$_e$, m$_e$ is the electron mass, $\Omega = 10^9$ Hz, $\nu = 10^{13}$ s^{-1}, $T = 300$ K, $c = 10^{11}$ N/m^2, $\rho = 5.7 \cdot 10^3$ kg/m^3, $\delta = 10^{-2}$.

The influence of rotating electric field can be a cause of circular dichroism, *i.e.* predominating absorption of right- or left-polarized circular acoustic wave. The analytic expression for circular dichroism is

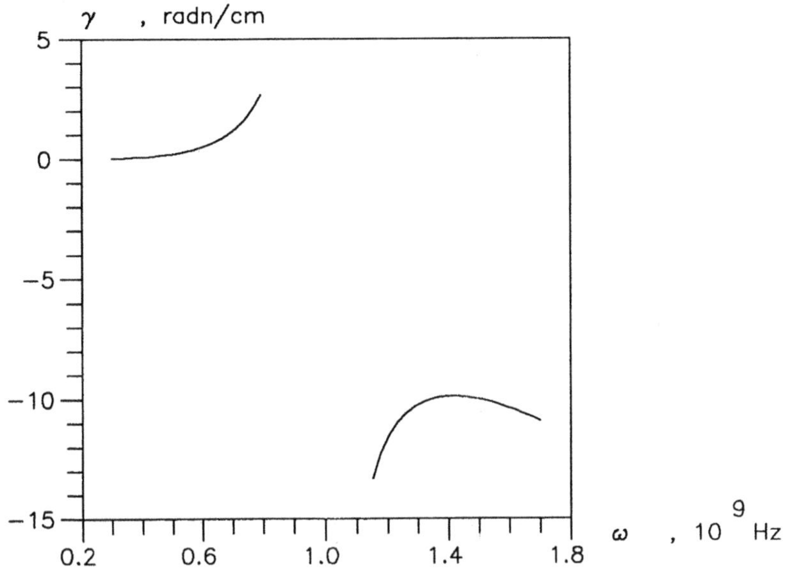

Figure 1. Dependence of the specific rotation of polarization plane on the incident wave frequency ω: $\Omega = 10^9$ Hz, $\nu = 10^{13}$ s^{-1}, $T = 300$ K, $c = 10^{11}$ N/m^2, $\rho = 5.7 \cdot 10^3$ kg/m^3, $\delta = 10^{-2}$

$$D = \Im \frac{k_1(\omega+\Omega) - k_2(\omega-\Omega)}{2} = \frac{-\omega^4 \delta^2 \omega_p^2 \Im(Q(\omega_0)) L}{4V\Omega(\omega^2-\Omega^2)[\Re(Q(\omega_0))^2 + \Im(Q(\omega_0))^2]} \tag{5}$$

The analysis of the expression led us to the conclusion about the difference of the dichroism for dielectrics and semiconductors. For dielectric crystal the dichroism is an even function of frequency disturbation $\omega - \Omega$ and it has the maximum on the resonance frequency $\omega = \Omega$ (Fig. 2). For semiconductors the dichroism is an odd function of frequency disturbation $\omega - \Omega$ and has zero value on the resonance frequency $\omega = \Omega$ (Fig. 3). Besides, dichroism changes sign when the direction of the outside longitudinal electric field changes. This is an evidence of predominating absorption of right- or left-polarized component of the acoustic wave at the opposite directions of the external electric field.

In the above expressions the following notations are used: $k_{1,2}$ are the wave numbers of proper modes of acoustic field into crystal, $\omega_0 = \omega - \Omega$, $\delta^* = \delta(1 - \omega_p^2/Q(\omega_0))$, $Q(\omega_0) = -\omega_0^2 - i|\omega_0|(\nu + ikv) + \omega_p^2 + Dk^2(\omega_0) +$

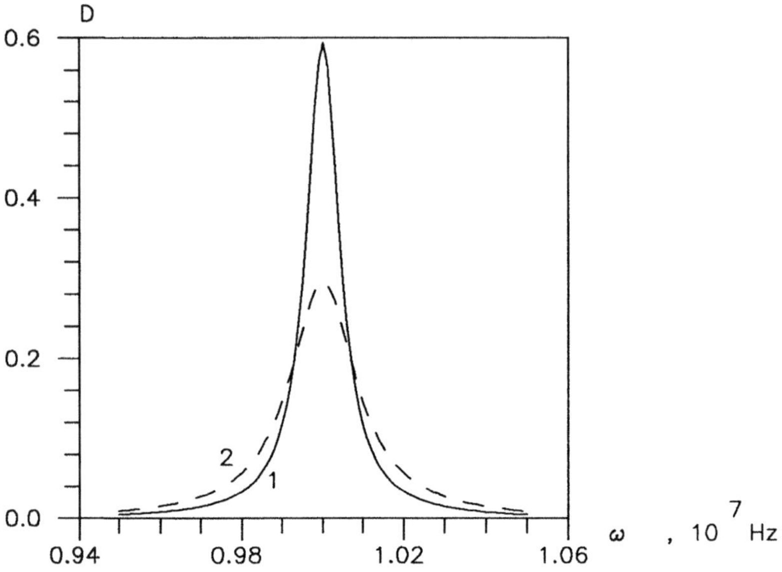

Figure 2. Circular dichroism dependence on the incident wave frequency ω in a dielectric: 1: medium viscosity is 1000 units of international system of units, 2: medium viscosity is 2000 units of international system of units; $L = 0.02$ m, $\Omega = 10^6$ Hz, $T = 300$ K, $c = 10^{11}$ N/m^2, $\rho = 5.7 \cdot 10^3$ kg/m^3, $\delta = 10^{-2}$.

$ikv\nu$, $\delta = \varepsilon_0 \varepsilon g^2 E_0^2 / 2\rho V^2$ is the relative change of electron velocity under external field influence, $V = \sqrt{c/\rho}$ is the velocity of acoustic wave without electric field influence, ρ is the density of crystal, $D_1 = D/\nu = k_B T/\nu m^*$ is the diffusion coefficient, k_B is Boltzmann constant, T is the absolute temperature, m^* is the effective mass of electron, $\omega_p = \sqrt{e^2 N_0 / \varepsilon_0 \varepsilon m^*}$ is the plasma frequency, ν is the electron collisions frequency, L is the crystal thickness.

If the frequency of the incident acoustic waves ω obeys the relation

$$4\Omega^2 \ll |\delta^{*2}|\omega^2 \tag{6}$$

the regime of interaction is possible when acoustic field in crystal has a form of two interconnected waves with opposite circular polarizations and different frequencies:

$$\begin{aligned} \mathbf{u} = & \frac{u_0^{\pm}}{2} \mathbf{n}_{\pm} [\exp[ik_1(\omega \pm \Omega)z] + \exp[ik_2(\omega \pm \Omega)z]] \exp[-i\omega t] \\ & - \frac{u_0^{\mp}}{2} \mathbf{n}_{\mp} [\exp[ik_1(\omega \pm \Omega)z] + \exp[ik_2(\omega \pm \Omega)z]] \exp[-i(\omega \pm \Omega)t] \end{aligned} \tag{7}$$

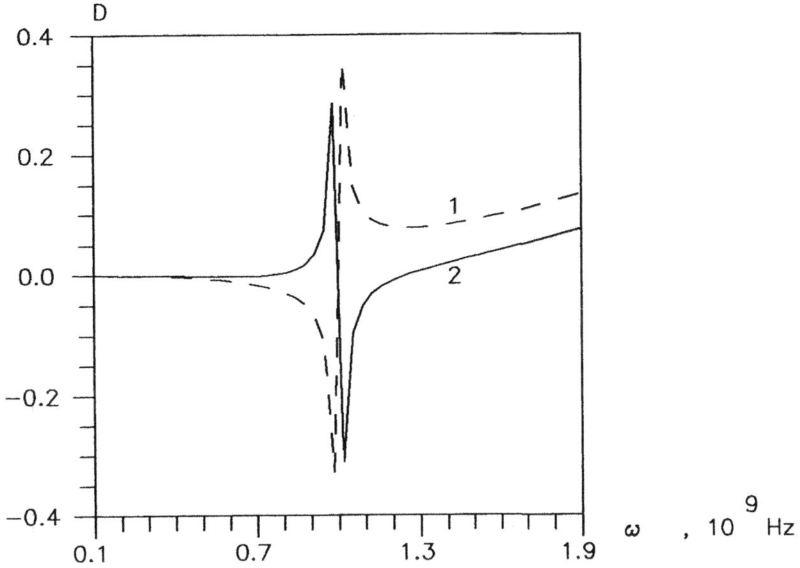

Figure 3. Circular dichroism dependence on the incident wave frequency ω in semiconductor: $1 : v = -V$, $2 : v = V$; $L = 10^{-3}$ m, $\Omega = 10^9$ Hz, $\nu = 10^{13}$ s^{-1}, $T = 300$ K, $c = 10^{11}$ N/m^2, $\rho = 5.7 \cdot 10^3$ kg/m^3, $\delta = 10^{-2}$

The propagation of ultrasound is accompanied by a mutual exchange of energy between these waves. When the crystal thickness Z is equal to

$$Z = \frac{\pi V}{\omega_0 \delta \left(1 - [\omega_p^2 \Re(Q(\omega_0))] \cdot [\Re(Q(\omega_0))^2 + \Im(Q(\omega_0))^2]^{-1}\right)} \tag{8}$$

the incident wave

$$\mathbf{u}_0^{\pm} = u_0^{\pm} \mathbf{n}_{\pm} [\exp[ik(\omega)z - i\omega t] \tag{9}$$

can be completely transformed to the wave with the opposite circular polarization and frequency $\omega \pm \Omega$. The relative change of velocity of acoustic wave in ferroelectric ceramics under the external electric field of 10 KV/cm can reach 10 per cent [14, 15]. Then the complete transformation in $BaTiO_3$ ceramics takes place at the thickness $Z \approx 0.2$ cm. The dependence of the thickness Z on the longitudinal electric field influence is presented in Fig. 4. The amplitude of transformed wave

$$A = \frac{u_T^{\mp}}{u_0^{\pm}} = \exp[-\Im(k_1(\omega \pm \Omega)Z] - \exp[-\Im(k_2(\omega \pm \Omega)Z] \tag{10}$$

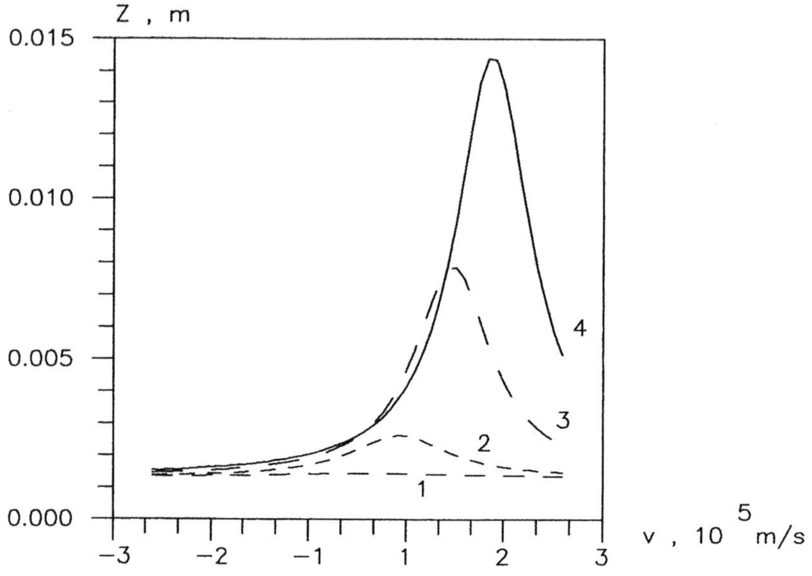

Figure 4. Dependence of the effective thickness Z on the direction and velocity of electron drift: $1 : \omega_d/\omega_c = 0.1$, $2 : \omega_d/\omega_c = 1$, $3 : \omega_d/\omega_c = 5$, $4 : \omega_d/\omega_c = 10$; $\Omega = 10^9$ Hz, $\nu = 10^{13}$ s^{-1}, $T = 300$ K, $c = 10^{11}$ N/m^2, $\rho = 5.7 \cdot 10^3$ kg/m^3, $\delta = 10^{-2}$

increases under the following two conditions. First, the drift direction coincides with velocity direction of the incident wave, and second, conductivity effects predominate over diffusion effects. The amplitude of the transformed wave can reach the values 10–15 times larger than the amplitude of the incident wave. The effect is illustrated in Fig. 5, where the parameter values are the same as in the previous figures. The condition of dominating conductivity effects means ($\omega_d \geq \omega_c$), where $\omega_d = V^2\nu/D$ is the diffusion frequency, $\omega_c = \omega_p^2/\nu$ is the frequency of conductivity relaxation. In the other case, *i.e.*, when the direction of electron drift and direction of propagation of acoustic waves are opposite each other, the amplification of transformed wave is impossible.

3. Conclusion

In this paper the analogy between the properties of two structures was shown: the acoustic chiral properties of crystals with rotating structures, and the corresponding properties of spiral structures induced by the electric field. Moreover, there were demonstrated the distinguishing characteristics between the acoustic dichroism of dielectrics and that of semiconductors,

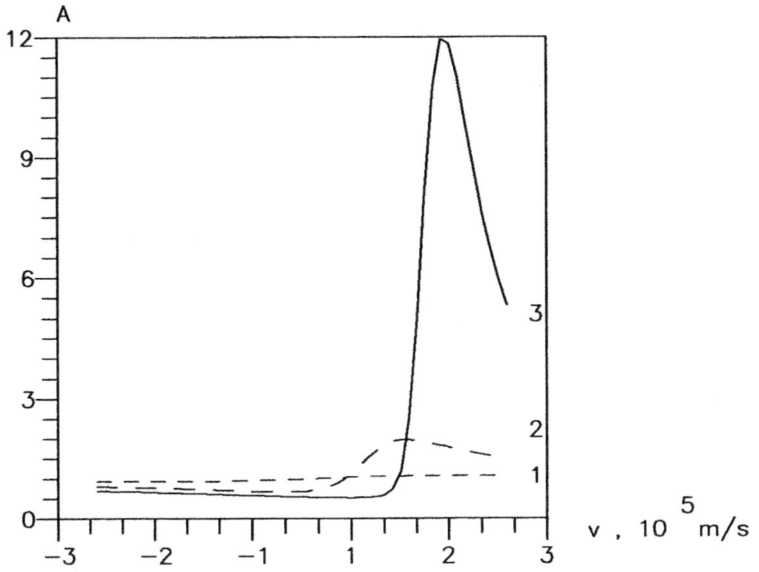

Figure 5. Dependence of the amplitude of the transformed wave on the direction and velocity of electron drift: $1 : \omega_d/\omega_c = 0.1$, $2 : \omega_d/\omega_c = 1$, $3 : \omega_d/\omega_c = 5$; $L = 0.03$ m, $\Omega = 10^9$ Hz, $\nu = 10^{13}$ s^{-1}, $T = 290$ K, $c = 10^{11}$ N/m^2, $\rho = 5.7 \cdot 10^3$ kg/m^3, $\delta = 10^{-2}$

when these materials are placed into a rotating electric field.

The obtained results can be used for the design of the devices that can rotate the polarization plane of ultrasound. Another property of these structures that could be exploited, is their ability to transform the polarization and frequency, and, simultaneously, to amplify the magnitude of acoustic waves.

References

1. Bely, V.N., and Sevruk, B.B. (1983) Control of polarization of acoustic waves by the electric field inducing spiral anisotropy, *Akust. Zhurnal* **57(2)**, 157–161.
2. Bely, V.N., and Sevruk, B.B. (1987) Parametric electro-acoustic effects in crystal with rotating acoustic anisotropy induced by the external electric field, *Zhurnal Tekhn. Fiziki* **29(2)**, 336–340.
3. Semchenko, I.V., Sevruk, B.B., and Khakhomov, S.A. (1993) Acousto-electron interaction in conductor crystal of ferroelectric ceramics in the condition of inducing of piezoelectric, anisotropic and gyrotropic properties by the rotating electric field, in A. Sihvola et al. (eds.), *"Bianisotropics'93"* Proceedings, Gomel, Belarus, 93–96.
4. Khakhomov, S.A., Semchenko, I.V. and Sevruk, B.B. (1994) Acousto-electron interaction in piezoelectric semiconductors: chiral properties induced by the rotating electric field, in F. Mariotte and J. Parneix (eds.), *"Chiral'94"* Proceedings, Perigueux, France, 445–450.

5. Semchenko, I.V., Sevruk, B.B., and Khakhomov, S.A. (1994) Electro-acoustic interaction in ferroelectric ceramics in presence of the rotating electric field, *Kristallografiya* **39(6)**, 1088–1091.
6. Khakhomov, S.A., and Semchenko, I.V. (1995) The amplification of ultrasound in semiconductors by the electron drift in condition of forming of chiral properties by the rotating electric field, in A. Sihvola et al. (eds.), *"Chiral'95"* Proceedings, Pennsylvania, USA, 139–142.
7. Pustovoyt, V.I. (1969) Interaction of electron stream with elastic waves of lattice, *Uspekhi Fizich. Nauk* **97(2)**, 257–276.
8. Gurevich, V.L. (1968) The theory of acoustic properties of piezoelectric semiconductors, *Fizika i Tekhnika Poluprovodnikov* **2(11)**, 1557–1592.
9. Pekar, S.I. (1965) Electron-phonon interaction proportional to external field and sound amplification in semiconductors, *Zhurnal Eksperim. i Teor. Fiziki* **49(2)**, 621–669.
10. Gulyaev, Y.V. (1967) To the question about electron-phonon interaction proportional to external field, *Fizika Tverdogo Tela* **9(6)**, 1816–1818.
11. Belyakov, V.A., and Sonin, A.S. (1982) *Cholesteric liquid crystals optics.* Nauka, Moscow.
12. Mazon, W. (Editor) (1969) *Physical acoustics*, Mir, Moscow.
13. Taker, G., and Rampton, V. (1975) *Hypersound in the solid-state physics.* Mir, Moscow.
14. Pekar, S.I., Demidenko, A.A., Zdebsky, A.P. et al. (1976) Investigations of electrostrictive constants of rank one and two in the substances with strong permittivity. *Doklady Akademii Nauk SSSR. Ser. Fizika* **230(5)**, 1089–1091.
15. Zhabitenko, N.K., and Kucherov, I.V. (1978) Investigations of electric field influence on the velocity of elastic waves propagation in isotropic solid states. *Ukrainskii Fizich. Zhurnal* **23(2)**, 263–266.

ELECTRODYNAMICS OF CHIROWAVEGUIDES: MODE ORTHOGONALITY, MODE EXCITATION, MODE INTERACTIONS

E. O. KAMENETSKII
Tel Aviv University
Faculty of Engineering
Department of Electrical Engineering -Physical Electronics
Tel Aviv 69978, Israel

Abstract. A number of problems related to chirowaveguides have been investigated and reported. Most of them are devoted to calculation of dispersion characteristics of chirowaveguides. In a variety of material and geometric effects in such waveguides, the necessity of conceptual analysis is obvious. The aim of this paper is to formulate some principles in electrodynamics of chirowaveguides.

1. Introduction

An analysis of waveguide problems based on a solution of a full vectorial form of Maxwell's equations is very important for many applications on microwaves and optics. Notation using six-vectors appears very applicable for such an analysis. Six-vector notation has been used for describing electromagnetic problems in anisotropic [1-3] and bianisotropic [3,4] media. Felsen and Marcuvitz applied six-vectors for waveguides based on anisotropic media [1].

Among a number of methods to solve waveguide problems in a full vectorial form, one can apply the so-called spectral method based on Galerkin's technique using a functional basis of normal modes described by six-vectors [5,6]. The spectral method enables us to take into account singularities caused by discontinuities of excitation current and is a general method to solve the problems of wave excitation and wave interaction in waveguides based on complex (anisotropic, chiral, bianisotropic) and composite (discretely inhomogeneous) media.

A. Priou et al. (eds.), Advances in Complex Electromagnetic Materials, 227–238.

In this paper we shall use six-vector notation to formulate some general principles in electrodynamics of chirowaveguides. These principles are the following.

A. Mode Orthogonality [7,8]. Relations of mode orthogonality are useful to analyze a structure of mode fields in chirowaveguides. These relations are necessary to solve mode excitation problems. On the basis of orthogonality relations, we shall show that in chirowaveguides the polarization of complex modes differs from the polarization of propagating and evanescent modes.

B. Mode Excitation [6]. In a general case of an excitation problem in chirowaveguides, special attention has to be paid to the role of the longitudinal components of excitation (electric and magnetic) currents. To solve an excitation problem in chirowaveguides, one may use a complete set basis of transverse components of mode fields. In nonchiral waveguides, decomposition is realized for $\vec{E}, \vec{H}$ fields. In contrast, in chirowaveguides we have to decompose $\vec{D}, \vec{B}$ fields.

C. Mode Interactions [9-13]. Chirality may provide an interaction between two eigenpolarization modes (TE and TM) of a conventional nonchiral waveguide. In this paper, mode interaction in chirowaveguides is considered, taking into account the so-called induced polarization effects caused by discontinuities of longitudinal polarization currents.

2. Mode orthogonality and field structure

Vector formulated orthogonality relations in chirowaveguides are directly derived from Maxwell's equations. Fig. 1 shows a cylindrical waveguide with an arbitrary cross-sectional shape which is filled with chiral material characterized by the chirality admittance ν (the cross-section $S^{(1)}$) and nonchiral material (the cross-section $S^{(2)}$). The contour ℓ separates chiral and nonchiral media. The external contour L surrounding the waveguide cross section S ($S = S^{(1)} + S^{(2)}$) is an electric or magnetic wall. In the case of an open waveguide, the boundary L is at infinity.

For every region q ($q = 1, 2$) of the waveguide, we write homogeneous Maxwell's equations for the time-harmonic fields ($e^{i\omega t}$) [14]:

$$M^{(q)}\vec{U}^{(q)} = 0 \tag{1}$$

where

$$M^{(q)} = \begin{pmatrix} i\omega\left[\epsilon^{(q)} + \mu(\nu^{(q)})^2\right] & -\left[(\nabla\times) - \omega\mu\nu^{(q)}\right] \\ (\nabla\times) - \omega\mu\nu^{(q)} & i\omega\mu \end{pmatrix} \tag{2}$$

are Maxwell's operators (here we have $\nu^{(1)} = \nu,\ \nu^{(2)} = 0$),

$$\vec{U}^{(q)} = \begin{pmatrix} \vec{E}^{(q)} \\ \vec{H}^{(q)} \end{pmatrix} \tag{3}$$

is a six-vector field.

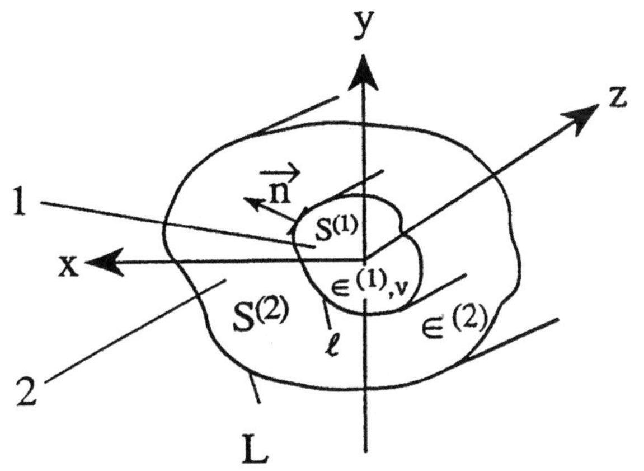

Figure 1. A cylindrical waveguide. 1-Chiral medium, 2-Nonchiral medium

For two modes; m of the main boundary problem and n of the conjugate boundary problem, we have a vector formulated orthogonality relation [7,8]

$$(\gamma_m + \tilde{\gamma}_n^*) \sum_{q=1}^{2} \int_{S^{(q)}} \left(Q \hat{\vec{U}}_m^{(q)} \right) \cdot \left(\tilde{\vec{U}}_n^{(q)} \right)^* ds = 0 \tag{4}$$

where

$$Q = \begin{pmatrix} 0 & -\vec{e}_z \times \\ \vec{e}_z \times & 0 \end{pmatrix} \tag{5}$$

$\vec{e}_z$ is the unit vector along the longitudinal z-axis.

Two modes are orthogonal for propagation constants $\gamma_m + \gamma_{\tilde{n}}^* \neq 0$. If, however, $\gamma_m + \gamma_{\tilde{n}}^* = 0$, we will denote $n = \tilde{m}$ and consider this mode as a conjugate mode to mode m. For conjugate modes, the norm is defined as

$$N_m \equiv N_{m\tilde{m}} = \sum_{q=1}^{2} \int_{S^{(q)}} \left(Q \hat{\vec{U}}_m^{(q)} \right) \cdot \left(\hat{\vec{U}}_{\tilde{m}}^{(q)} \right)^* ds$$

$$= \sum_{q=1}^{2} \int_{S^{(q)}} \left(\hat{\vec{E}}_m^{(q)} \times \hat{\vec{H}}_{\tilde{m}}^{(q)*} + \hat{\vec{E}}_{\tilde{m}}^{(q)*} \times \hat{\vec{H}}_m^{(q)} \right) \cdot \vec{e}_z ds \quad (6)$$

In expressions (4) and (6), we use the fields of mode membrane functions (functions dependent on transverse coordinates).

The vector formulated orthogonality relation (4) is correct for propagating modes in closed or open waveguides, and for evanescent and complex modes in closed waveguides. On the complex plane, pairs of conjugate modes are symmetrical with respect to the β axis ($\gamma = \alpha + i\beta$). Fig. 2 shows the positions of complex modes of lossless reciprocal chirowaveguide. Modes with γ_m and $-\gamma_m^*$ (or $-\gamma_m$ and γ_m^*) are conjugate modes.

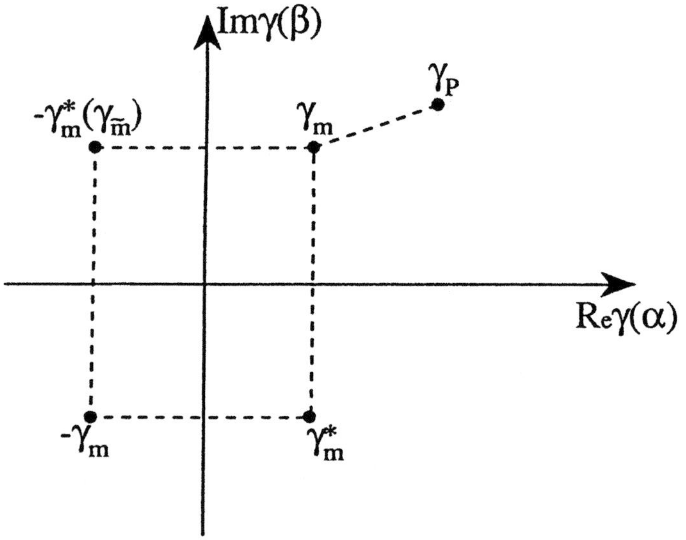

Figure 2. Positions of complex wave number on a complex plane for shielded lossless isotropic waveguide.

We apply now the vector formulated orthogonality relation (4) for a case of a two-parallel-plate waveguide partially loaded with an isotropic chiral slab (Fig. 3). One obtains as a result [8]:

$$\sum_{q=1}^{2} \int \left[\left(\tilde{\gamma}_n^* - \gamma_m \right) E_{mx}^{(q)} \tilde{E}_{nx}^{(q)*} \right.$$

$$-k_d^2\left(\frac{1}{\tilde{\gamma}_n^*}-\frac{1}{\gamma_m}\right)E_{m_y}^{(q)}\tilde{E}_{n_y}^{(q)*}\Bigg]\,dy \quad = \quad \delta_{mn}N_m \tag{7}$$

where $\delta_{mn}=1$ if $\gamma_m+\tilde{\gamma}_n^*=0$ and $\delta_{mn}=0$ if $\gamma_m+\tilde{\gamma}_n^*\neq 0$. The norm N_m is defined as

$$N_m=-\frac{i2}{\omega\mu}\sum_{q=1}^{2}\int\left[\gamma_m E_{m_x}^{(q)}E_{\tilde{m}_x}^{(q)*}-\frac{k_d^2}{\gamma_m}E_{m_y}^{(q)}E_{\tilde{m}_y}^{(q)*}\right]dy \tag{8}$$

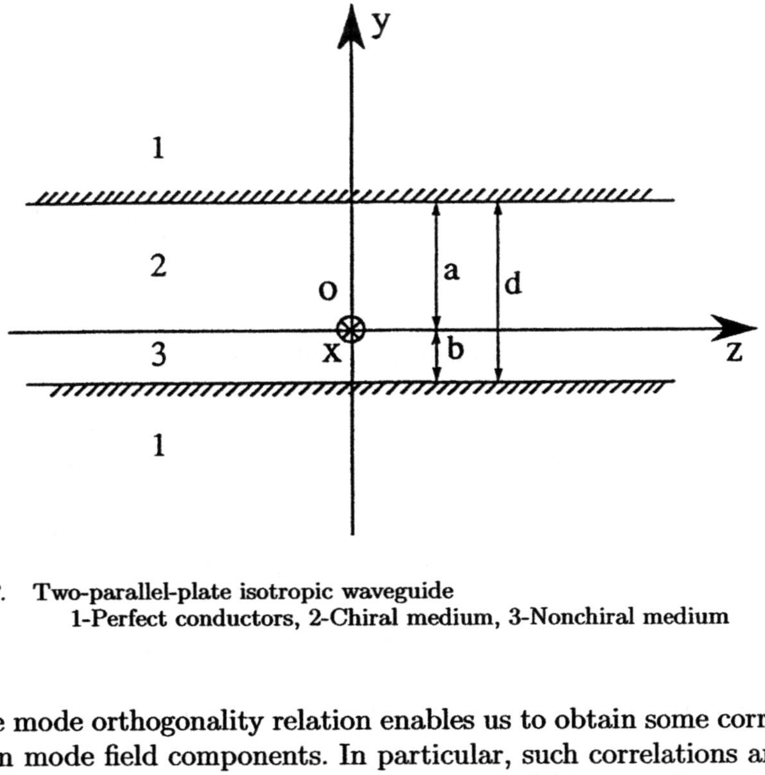

Figure 3. **Two-parallel-plate isotropic waveguide**
1-Perfect conductors, 2-Chiral medium, 3-Nonchiral medium

The mode orthogonality relation enables us to obtain some correlations between mode field components. In particular, such correlations are interesting for complex modes. To obtain the norm (8) as a real quantity, and thus to provide an active power flow by two complex modes, one has to require

$$E_{\tilde{m}_x}=\frac{i\gamma_m}{\mid\gamma_m\mid}E_{m_x},\quad E_{\tilde{m}_y}=\frac{i\mid\gamma_m\mid}{\gamma_m}E_{m_y} \tag{9}$$

Analogous correlations we have for the magnetic fields of the modes.

We can consider relations (9) (and analogous relations for the magnetic fields) as a special type of reflection symmetry for complex modes which provide an active power flow through a chirowaveguide. Such a type of reflection symmetry has to be provided by a special discontinuity at the end of a complex-mode waveguide and is different from reflection symmetry in a plane perpendicular to the z-axis in an infinite waveguide. The last type of reflection symmetry was described in [15].

On the basis of (9), it was shown in [8] that in parallel-plate chirowaveguides, complex modes have an elliptical polarization of the type

$$\frac{E_{m_x}^{(q)}}{E_{m_y}^{(q)}} = a + ib \tag{10}$$

where a and b are real coefficients. At the same time for propagating modes in chiral slab-waveguides, we have the following type of elliptical polarization [14]:

$$\frac{E_{m_x}}{E_{m_y}} = iA \tag{11}$$

where A is a real quantity.

For an isotropic chirowaveguide, one has the following vector equation [14]:

$$\nabla \times \nabla \times \vec{E} - 2\omega\mu\nu \nabla \times \vec{E} - k_d^2 \vec{E} = 0 \tag{12}$$

For vector $\vec{H}$ we have an analogous equation. For two arbitrary modes m and p (Fig. 2) which satisfy the vector-formulated eigenvalue problem, equation (12) (together with the analogous equation for vector $\vec{H}$) enables us to obtain the scalar-formulated orthogonality relation [8]. Common consideration of the vector-formulated and scalar-formulated orthogonality relations gives correlations between parameters of the polarization ellipse and propagation constants ($\gamma = \alpha + i\beta$). For complex modes m described by (10), we have

$$\frac{a}{b} = -\frac{\gamma_m}{\beta_m} \tag{13}$$

For two propagating modes m and n described by (11), we have

$$\left.\begin{aligned} \beta_m \beta_n A_m A_n + k_d^2 = 0 \\ \beta_m^2 - \beta_n^2 + 2\omega\mu\nu(\beta_n A_n - \beta_m A_m) = 0 \end{aligned}\right\} \tag{14}$$

3. Mode Excitation

The excitation problem in chirowaveguides is described by the inhomogeneous Maxwell's equations

$$M^{(q)} \vec{U}^{(q)} = \vec{f} \tag{15}$$

where $M^{(q)}$ and $\vec{U}^{(q)}$ are defined by (2) and (3), and

$$\vec{f} = -\begin{pmatrix} \vec{j}^{el} \\ \vec{j}^{mag} \end{pmatrix} \tag{16}$$

is a vector-function of electric and magnetic currents.

To solve the inhomogeneous Maxwell's equations (15), one can use Galerkin's method. The following conditions have to be satisfied for the traditional Galerkin's method [16]: a) weight functions belong to the same set as trial functions, b) trial and weight functions have to be linearly independent c) trial and weight functions have to belong to a complete system of functions, d) trial and weight functions have to satisfy boundary conditions. Conditions a) and b) are the main conditions which determine Galerkin's method. Conditions c) and d) are correlated with an effectiveness of the method.

$\{\vec{E}, \vec{H}\}$ six-vectors composed by transverse fields of the modes form a complete orthonormal functional basis on a waveguide cross section. Such a functional basis was used as a system of trial and weight functions to solve the excitation problem in nonchiral waveguides [5]. In the presence of longitudinal currents, only transverse fields may be decomposed by a system of mode transverse fields (a system of trial functions) [1,5]:

$$\vec{U}_{\perp} = \sum_{m} a_m(z) \hat{\vec{U}}_{m_\perp} \tag{17}$$

where $a_m(z)$ is a scalar coefficient dependent on waveguide longitudinal coordinate z, $\hat{\vec{U}}_{m_\perp}$ is a membrane function of mode transverse fields. To use the system of mode transverse fields as weight functions as well, necessary transformations in inhomogeneous Maxwell's equations have to be made.

For a nonchiral waveguide, one has the following excitation equation:

$$\frac{da_m(z)}{dz} + \gamma_m a_m(z) = \frac{1}{N_m} \int_S \vec{F} \cdot \hat{\vec{U}}^*_{m_\perp} ds \tag{18}$$

where $\vec{F}$ is a vector-function of electric and magnetic currents which are composed only with transverse components [5].

As it was shown in [6], $\{\vec{E}, \vec{H}\}$ six-vectors composed by the transverse fields of the modes cannot be used as a system of trial and weight functions to solve the excitation problem in chirowaveguides. In inhomogeneous Maxwell's equations written in respect to $\vec{D}$ and $\vec{B}$ six-vector

$$\vec{V}^{(q)} = \begin{pmatrix} \vec{D}^{(q)} \\ \vec{B}^{(q)} \end{pmatrix} \tag{19}$$

in every region q ($q = 1, 2$), may be represented as [6]

$$\vec{V}^{(q)} = \vec{V}_{\perp}^{(q)} + \vec{V}_{\parallel}^{(q)} = \sum_m a_m(z) \hat{\vec{V}}_{m_\perp}^{(q)} + \sum_m a_m(z) \hat{\vec{V}}_{m_\parallel} + \vec{V}_{ex} \quad (20)$$

where

$$\hat{\vec{V}}_{m_{\perp,\parallel}}^{(q)} = \begin{pmatrix} \hat{\vec{D}}_{m_{\perp,\parallel}}^{(q)} \\ \hat{\vec{B}}_{m_{\perp,\parallel}}^{(q)} \end{pmatrix} \quad (21)$$

are membrane vector-functions (six-vectors) of transverse and longitudinal mode fields,

$$\vec{V}_{ex} = \frac{i}{\omega} \begin{pmatrix} \vec{j}_{\parallel}^{el} \\ \vec{j}_{\parallel}^{mag} \end{pmatrix} \quad (22)$$

is the additional longitudinal field in the excitation region caused by longitudinal electric and magnetic currents.

One can see that only transverse part of the six-vector $\vec{V}$ may be decomposed by a system of six-vectors $\vec{V}_{m_\perp}$. The fields $\vec{D}_{m_\perp}$ and $\vec{B}_{m_\perp}$ do not give an orthonormal basis. Nevertheless, they are linearly independent, satisfy boundary conditions and may be used as a system of trial functions. It was shown in [6] that $\{\vec{D}_{m_\perp}, \vec{B}_{m_\perp}\}$ six-vectors may also be used as a system of weight functions. To use these vectors as weight functions, one has to transform Maxwell's equations written in respect to $\vec{D}$ and $\vec{B}$ vectors. As a result, we have the following modes excitation equation for a chiral waveguide (see Fig. 1) [6]:

$$\sum_{q=1}^{2} \sum_m \left\{ \left[\frac{da_m(z)}{dz} + \gamma_m a_m(z) \right] \int_{S_{(q)}} \left(Q \hat{\vec{V}}_{m_\perp}^{(q)} \right) \cdot \left(\hat{\vec{V}}_{n_\perp}^{(q)} \right)^* ds \right\}$$

$$= \sum_{q=1}^{2} \int_{S_{(q)}} \vec{G}^{(q)} \cdot \left(\hat{\vec{V}}_{n_\perp}^{(q)} \right)^* ds \quad (23)$$

where

$$\vec{G} = - \begin{pmatrix} \mu \vec{j}_{\perp}^{el} - i\mu\nu^{(q)} \vec{j}_{\perp}^{mag} - \frac{i}{\omega} \nabla \times \vec{j}_{\parallel}^{mag} \\ i\mu\nu^{(q)} \vec{j}_{\perp}^{el} + \left[\epsilon^{(q)} + \mu(\nu^{(q)})^2 \right] \vec{j}_{\perp}^{mag} + \frac{i}{\omega} \nabla \times \vec{j}_{\parallel}^{el} \end{pmatrix} \quad (24)$$

is a source vector function composed of transverse components.

In our analysis of an excitation problem, we used the same systems of trial and weight functions and applied projective relations on the cross-section of a chirowaveguide. In the presence of longitudinal currents in chirowaveguides, excitation of mode n (the excitation integral in the right-hand side of eqn. (23)) causes an excitation of all modes in the spectrum. The efficiency of mode excitation is determined by the mode overlap integrals (the integrals on the left-hand side of eqn. (23)).

4. Mode Interactions

The theory of excitation of chirowaveguides was used in [6] to develop a coupled-mode formalism for mode coupling caused by the differences $\Delta\epsilon$ and $\Delta\nu$ from the nominal distributions respectively $\epsilon^{(0)}$ and $\nu^{(0)}$. In our analysis, we restrict ourselves to a case when $\Delta\epsilon = 0$ and $\nu^{(0)} = 0$, that is we shall consider an interaction of modes of a conventional nonchiral waveguide caused by chirality. The fact that chirality may provide an interaction between two eigenpolarization modes (TE and TM) of a conventional nonchiral waveguide was analyzed theoretically and confirmed experimentally [9-11].

The coupled-mode formalism used in [10, 11] is based on the reciprocity relation disregarding the longitudinal parts of electrical and magnetic polarization currents. In the paper [13], mode interactions in chirowaveguides were analyzed, taking into account longitudinal components of polarization currents. Special consideration was given to the so-called induced polarization effects caused by discontinuities of longitudinal polarization currents on the boundary of chiral–nonchiral media.

Let nonchiral dielectric materials with permittivities $\epsilon^{(1)}$ and $\epsilon^{(2)}$ compose a conventional guide structure with the geometry corresponding to Fig. 1. In such a case one can consider wave processes in a chirowaveguide as an excitation of a nonchiral waveguide caused by polarization currents:

$$\vec{j}^{el} = \omega\mu\nu\vec{H} \tag{25}$$

$$\vec{j}^{mag} = -\omega\mu\nu\vec{E} \tag{26}$$

These expressions were obtained on the basis of expression (1) using the conditions of small chirality ($\epsilon^{(1)} \gg \mu\nu^2$).

An integration by parts of the integral in the right-hand side of eqn. (18) gives [12]

$$\int_s \vec{F}\cdot\hat{\vec{U}}^*_{m_\perp} ds = \int_s \left(\vec{f}_\perp\cdot\hat{\vec{U}}^*_{m_\perp} + \vec{f}_\parallel\cdot\hat{\vec{U}}^*_{m_\parallel}\right) ds + \oint_\ell U_{ex}\cdot\left(R\hat{\vec{U}}^*_{m_\perp}\right) dl \tag{27}$$

where

$$\vec{U}_{ex} = \frac{i}{\omega}\begin{pmatrix} \frac{1}{\epsilon}\vec{j}^{el}_\parallel \\ \frac{1}{\mu}\vec{j}^{mag}_\parallel \end{pmatrix} \tag{28}$$

$$\vec{f}_{\perp,\parallel} = -\begin{pmatrix} \vec{j}^{el}_{\perp,\parallel} \\ \vec{j}^{mag}_{\perp,\parallel} \end{pmatrix}, \qquad \vec{f} = \vec{f}_{\perp} + \vec{f}_{\parallel} \tag{29}$$

$$R = -\begin{pmatrix} 0 & \vec{n}\times \\ -\vec{n}\times & 0 \end{pmatrix} \tag{30}$$

$\vec{n}$ is the unit vector along the external normal to the contour ℓ.

Taking into account (25) and (26) for polarization currents, one obtains, on the basis of (18) and (27) the following coupled-mode equations [13]:

$$\frac{da_m(z)}{dz} + \gamma_m a_m(z) = \sum_P (C_{mp} + K_{mp}) a_p(z) \tag{31}$$

where

$$\begin{aligned} C_{mp} &= -\frac{\omega\mu\nu}{N_m}\int_{S^{(1)}} \Big[\hat{\vec{H}}_{p_\perp}\cdot\hat{\vec{E}}^*_{m_\perp} - \hat{\vec{E}}_{p_\perp}\cdot\hat{\vec{H}}^*_{m_\perp} + \hat{\vec{H}}_{p_\parallel}\cdot\hat{\vec{E}}^*_{m_\parallel} \\ &- \hat{\vec{E}}_{p_\parallel}\cdot\hat{\vec{H}}^*_{m_\parallel} - i\nu\left(\hat{\vec{E}}_{p_\parallel}\cdot\hat{\vec{E}}^*_{m_\parallel} + \frac{\mu}{\epsilon^{(1)}}\hat{\vec{H}}_{p_\parallel}\cdot\hat{\vec{H}}^*_{m_\parallel}\right)\Big] ds \end{aligned} \tag{32}$$

$$\begin{aligned} K_{mp} &= -i\frac{\nu}{N_m}\oint_{\ell}\Big[\hat{\vec{E}}_{p_\parallel}\times\hat{\vec{E}}^*_{m_\perp} + \frac{\mu}{\epsilon^{(1)}}\hat{\vec{H}}_{p_\parallel}\times\hat{\vec{H}}^*_{m_\perp} \\ &- \frac{i\mu\nu}{\epsilon^{(1)}}\left(\hat{\vec{E}}_{p_\parallel}\times\hat{\vec{H}}^*_{m_\perp} - \hat{\vec{H}}_{p_\parallel}\times\hat{\vec{E}}^*_{m_\perp}\right)\Big]\cdot\vec{n}dl \end{aligned} \tag{33}$$

In comparison with the results obtained by Pelet and Engheta [10] and by Cory and Tamir [11], we have additional terms by taking into account the field longitudinal components. The first group of these terms is described by the integral over the cross-section $S^{(1)}$ and is derived from the reciprocity relation in a general use form. The second group is represented by the contour integral. This group of terms (the coefficients K_{mn}) is caused by the so-called induced polarization effects.

The induced polarization effect was analyzed in dielectric (nonchiral) waveguides and it was shown that such an effect may give an asymmetry in the coupling of propagating modes [12]. Similar effect takes place in chirowaveguides [13].

For two propagating modes, the following energy relation takes place [13]:

$$\begin{aligned} N_p\frac{d\,|\,a_p\,|^2}{dz} + N_q\frac{d\,|\,a_q\,|^2}{dz} &= N_p\,|\,a_p\,|^2\,(K_{pp}+K^*_{pp}) + N_q\,|\,a_q\,|^2\,(K_{qq}+K^*_{qq}) \\ &+ a_p a_q^*(N_q K_{qp} + N_p K^*_{pq}) + a_q a_p^*(N_p K_{pq} + N_q K^*_{qp}) \end{aligned} \tag{34}$$

Energy relation (34) shows that full exchange of power between two normal modes propagating in a lossless conventional waveguide does not take place, even in the case of phase synchronism. Moreover, one can also draw another important conclusion. The coefficients K_{pp} and K_{qq} may be complex quantities (see (33)). Therefore, for one propagating mode we may have exponential decrease of the amplitude in a lossless waveguide..

To understand the reason for such phenomena, let us extend the analysis of the energy relation to the infinite functional space of propagating modes. In such a case we have as a result [13]:

$$\sum_{m=1}^{\infty} N_m \frac{d \mid a_m \mid^2}{dz} = 0 \tag{35}$$

we can see that full exchange of active power between propagating modes in lossless waveguides takes place if propagating modes form a complete set basis.

When we consider two propagating modes in a lossless waveguide, we have a coupling not only between these two modes, but also with the other modes of the spectrum. Because of the contour K (see (33)), we have an interaction between propagating and evanescent modes in a closed waveguide.

Let us have only one propagating mode in a nonchiral waveguide. One obtains the complex propagating constant for this mode because of the chirality parameter. In closed lossless waveguides, it takes place because of an interaction between this mode and the continuum of evanescent modes caused by induced polarization effects.

5. Conclusion

In this paper we made an attempt to consider some general aspects in electrodynamics of chirowaveguides. Our analysis is based on a solution of a full vectorial form of Maxwell's equations using a functional basis of normal modes described by six-vectors.

In the analysis of mode orthogonality relations in chirowaveguides, special attention was given to the problem of complex modes. It was shown that the polarization of complex modes differs from the polarization of propagating and evanescent modes. A special type of reflection symmetry was obtained for complex modes. This symmetry shows the correlation between field components for two complex modes which transfer an active power flow in chirowaveguides.

The theory of mode excitation in chirowaveguides, based on the so-called spectral method, shows that in the presence of longitudinal currents, we have an excitation of all modes in the spectrum. This theory was realized by using $\{\vec{D}, \vec{B}\}$ six-vectors of mode transverse fields as a system of trial and weight functions.

Mode interactions in a nonchiral waveguide caused by the chirality parameter were considered, taking into account the field longitudinal components. The main attention was paid to so-called induced polarization effects caused by discontinuities of longitudinal polarization currents. Such an effect may give an asymmetry in the coupling of propagating modes.

Acknowledgment

This work was partly supported by the Ministry of Science in Israel.

References

1. Felsen, L.B. and Marcuvitz, N.(1973) *Radiation and Scattering of Waves*, Englewood Cliffs, Prentice-Hall.
2. Krowne, C.M. (1984) Electromagnetic theorems for complex anisotropic media, *IEEE Trans. Antennas Propagat.* **32**, 1224-1230.
3. Altman, C. and Suchy, K. (1991) *Reciprocity, Spatial Mapping and Time Reversal in Electromagnetics*, Dordrecht, Kluwer Academic Publishers.
4. Lindell, I.V., Sihvola, A.H. and Suchy, K. (1995) Six-vector formalism in electromagnetics of bi-anisotropic media, *J. Electromagn. Waves Appl.* **9**, 887-903
5. Kamenetskii, E.O. (1994) Spectral method for modes excitation problem in anisotropic waveguides, *IEEE Trans. Microw. Theory Techn.* **42**, 1685-1689.
6. Kamenetskii, E.O. (1995) Theory of excitation of chirowaveguides, *IEE Proc. H* **142**, 240-244, 368.
7. Engheta, N. and Pelet, P. (1990) Mode orthogonality in chirowaveguides, *IEEE Trans. Microw. Theory Techn.* **38**, 1631-1634.
8. Kamenetskii, E.O.(1996) Mode orthogonality relations and field structure in chirowaveguides, *IEEE Trans. Microw. Theory Techn.* **44**, 465-469.
9. Chien, M., Kim, Y. and Grebel, H. (1989) Mode conversion in optically active and isotropic waveguides, *Optics Lett.* **14**, 826-828.
10. Pelet, P. and Engheta, N. (1990) Coupled-mode theory for chirowaveguides, *J. Appl. Phys.* **67**, 2742-2745.
11. Cory, H. and Tamir, T. (1992) Coupling processes in circular open chirowaveguides, *IEE Proc. H* **139**, 165-170.
12. Kamenetskii, E.O. (1996). Induced polarization effects in coupling processes of waveguide modes, *IEEE Trans. Microw. Theory Techn.* **44**, 572-584.
13. Kamenetskii, E.O. (1966) Mode interaction in chirowaveguides, *J. Appl. Phys.* **79**, 8899-8905.
14. Pelet, P. and Engheta, N. (1990) The theory of chirowaveguides, *IEEE Trans. Antennas Propagat.* **38**, 90-98.
15. Mariotte, F. and Engheta, N. (1993) Reflection and transmission of guided electromagnetic waves at an air-chiral interface and at a chiral slab in a parallel-plate waveguide, *IEEE Trans. Microw. Theory Techn.* **41**, 1895-1906.
16. Fletcher, C.S.J. (1984) *Computational Galerkin Methods*, Springer-Verlag, New York.

ELECTROMAGNETIC SURFACE WAVES AT A PLANE BOUNDARY OF SEMI-INFINITE FARADAY CHIRAL MEDIA

V.V. FISANOV AND D.A. MARAKASOV
Tomsk State University,
Siberian Physical & Technical Institute,
1 Revolution Sq., Tomsk, 634050, Russia

Abstract. The characteristics of the bulk and surface plane waves in the Faraday chiral media are investigated for the case in which the biasing magnetostatic field is impressed perpendicularly to the propagation plane and along the boundary of the medium. The full-wave analysis of the dispersion relations for propagation coefficients is fulfilled for different values of the chirality parameter.

1. Introduction

The Faraday chiral medium belongs to the class of bi-anisotropic media. This concept has been introduced Engheta et al. [1] as a generalization of the Post-Jaggard constitutive relations for the isotropic chiral medium. Faraday chiral medium is known in two modifications: as *chiroplasma* and as *chiroferrite*. Like for any other material medium, various surface effects are important in order to gain insight into wave phenomena in the Faraday chiral media. Propagation of surface waves along plane regular boundary are among them. For these complicated chiral media one may expect appearance of interface and boundary guided waves analogously to that in gyrotropic media. Such waves (termed also as *surface polaritons*) are known for the interface of a magnetized ferrite and an isotropic medium [2,3]. They have been studied thoroughly in magnetoplasmas both for the gaseous and solid-state models. The comprehensive analysis has shown that surface waves are possible in the magnetoactive plasma medium which is bounded with either an interface with an isotropic half-space [4-7], or an unpenetrable boundary (an electric wall, an impedance wall) [8,9], under different directions of the imposed static magnetic field. Due to an additional de-

A. Priou et al. (eds.), Advances in Complex Electromagnetic Materials, 239–248.

gree of freedom in comparison with the simple isotropic chiral medium, the Faraday chiral media may be used for the chirality control [1] and may have other potential applications. Therefore the blending effects of chirality and anisotropy, which are inherent in these media, deserve study.

In this paper we look for electromagnetic surface waves propagating along plane interface between a chiroplasma or a chiroferrite and an isotropic achiral medium for the case when the static magnetic field is parallel to the boundary and perpendicular to the propagation direction. The eigenwaves of unbounded Faraday chiral media are determined at first. After that, the basic features of surface waves on the electric and magnetic walls in chiroplasma and chiroferrite half-spaces are determined.

2. Chiroplasma

The constitutive relations [1]

$$\vec{D} = \varepsilon_\infty \hat{\varepsilon} \cdot \vec{E} + i\xi \vec{B}, \qquad \vec{H} = i\xi \vec{E} + \mu^{-1} \vec{B} \tag{1}$$

imply that the relative dielectric tensor $\hat{\varepsilon}$ corresponds to the static magnetic field impressed in the z–direction of a rectangular coordinate system

$$\hat{\varepsilon} = \varepsilon(\hat{x}\hat{x} + \hat{y}\hat{y}) - ig(\hat{x}\hat{y} - \hat{y}\hat{x}) + \varepsilon_z \hat{z}\hat{z} \tag{2}$$

and its elements

$$\varepsilon = (\Omega^2 - 1 - R^2)/(\Omega^2 - R^2), \quad g = R/[\Omega(\Omega^2 - R^2)], \quad \varepsilon_z = (\Omega^2 - 1)/\Omega^2 \tag{3}$$

only depend on the normalized frequencies Ω (the ratio of the wave and electron plasma frequencies) and R (the ratio of the cyclotron and plasma frequencies). The medium handedness is shown obviously by the chirality admittance ξ.

2.1. BULK WAVES

The dispersion equation for the two-dimensional ($\partial/\partial z = 0$) free plane waves of the form

$$\exp\{i[k_\infty(\alpha x + \beta y) - \omega t]\} \tag{4}$$

where α and β are the components of the normalized transverse wave vector (so that $\eta = \sqrt{\alpha^2 + \beta^2}$) and $k_\infty = \omega\sqrt{\mu\varepsilon_\infty}$, follows from the relations (1) and the Maxwell equations as the biquadratic equation

$$\eta^4 - (\kappa + \varepsilon_z + a^2)\eta^2 + \kappa\varepsilon_z = 0 \tag{5}$$

It is noteworthy that the equation coefficients contain the second powers of the normalized propagation coefficients of ordinary waves ε_z and extraordinary waves $\kappa = \varepsilon - g^2/\varepsilon$ of the achiral magnetoplasma. Furthemore, the coefficient by η^2 includes the dimensionless chirality parameter $a^2 = 4\mu\xi^2/\varepsilon_\infty$ as an addend. There are pass and stop bands on the plane of the frequency parameters $(\Omega,\ R)$ for two characterisic hybrid waves η_+ and η_- as solutions of (5). For example, the regions of total transparency for both the eigenwaves in the achiral magnetoplasma $1 < \Omega < \Omega_U$ and $\Omega > \Omega_2$ are added here by the triangular region which is bounded by the line $\Omega = \Omega_1$ from above, by the curve $D = 0$ from below, where

$$D = (\kappa + a^2 - \varepsilon_z)^2 + 4a^2\varepsilon_z \tag{6}$$

is the discriminant of (5). That region occupies the interval $1/\sqrt{1 + a^2/4} < \Omega < 1$ for $R = 0$. The aforementioned critical frequencies are

$$\Omega_{1,2} = 0{,}5(\sqrt{R^2 + 4} \mp R), \quad \Omega_U = \sqrt{R^2 + 1} \tag{7}$$

where Ω_U is the upper hybrid resonant frequency for transverse propagation in a magnetoplasma.

It is worth noting that in the isotropic limit, when R approaches zero, these modes become the right-circularly and the left-circularly polarized Beltrami waves.

2.2. SURFACE WAVES

Let $y = 0$ be a boundary of a half-space of the chiroplasma medium. The procedure leading to the dispersion equation for surface waves guided along a plane interface which separates the chiroplasma medium from another isotropic achiral medium is quite straightforward. The expression for the field in the chiral domain should be rewritten so as to take into consideration the both eigenmodes. It is necessary to take a linear superposition of both bulk waves with the common longitudinal wave number α and with the transversal wave numbers $\beta_\pm = \sqrt{\eta_\pm^2 - \alpha^2}$. In the half-space $y < 0$ a suitable superposition of eigenwaves is taken likewise. The boundary conditions state that the tangential components of the total electric and magnetic fields are continuous across the interface, which leads to a set of linear homogeneous algebraic equations. To ensure a nontrivial solution its determinant must vanish, which gives the dispersion equation specifying α. In the general case the dispersion equation is involved, and we confine ourselves with two extreme cases of the perfectly conducting boundary $y = 0$: the electric wall (the tangential electric field vanishes, the permittivity of the lower medium tends to infinity) and the magnetic wall (the tangential magnetic field vanishes, the permeability of the lower medium tends to infinity).

2.2.1. *Electric Wall*

The dispersion equation for the surface wave on the electric wall has the following form:

$$(\beta_+ - ig\alpha/\varepsilon)(\kappa - \eta_-^2) = (\beta_- - ig\alpha/\varepsilon)(\kappa - \eta_+^2) \tag{8}$$

We see from (8) that the surface wave is unidirectional, and its direction of propagation is reversed along with the reversion of the magnetostatic field. The analytical solution of (8) reads:

$$\alpha = -\sqrt{\varepsilon}|\sqrt{\varepsilon - \kappa} + \sqrt{\varepsilon - \varepsilon_z}|/\sqrt{(\sqrt{\varepsilon - \kappa} + \sqrt{\varepsilon - \varepsilon_z})^2 - a^2} \mathrm{sgn}(R) \tag{9}$$

and if the chiral parameter vanishes, the propagation constant coincides with that of the TEM surface wave [10]. This wave exists only for the frequencies smaller than the gyrofrequency and under the condition

$$a^2 < R^2(1 + \sqrt{R^2 + 1 - \Omega^2})^2/[\Omega^2(R^2 - \Omega^2)(R^2 + 1 - \Omega^2)] \tag{10}$$

see Figure 1.

The dispersion diagram for the normalized propagation coefficient of this combined surface wave (Figure 2) shows that the slope of the curves is negative if a^2 is sufficiently large.

Both parts of the hybrid surface wave decay exponentially at different rates $\mathrm{Im}(\beta_+)$ and $\mathrm{Im}(\beta_-)$ along the direction normal to the plane $y = 0$. The rates are equal within the bounds when the discriminant (6) $D < 0$. Such a wave is called "the generalized surface wave", but in achiral gyroelectric media such solutions exist under different conditions, e.g., if the static magnetic field is perpendicular to the boundary [3].

2.2.2. *Magnetic Wall*

The dispersion relation for surface waves on the magnetic wall may be presented as

$$A\beta_+ + B\beta_- + C\alpha = 0 \tag{11}$$

where

$$A = -(2\varepsilon_z + a^2/2) + \eta_-^2(2\kappa + a^2/2)/\kappa \tag{12}$$

$$B = 2\varepsilon_z + a^2/2 - \eta_+^2(2\kappa + a^2/2)/\kappa \tag{13}$$

$$C = iga^2(\eta_+^2 - \eta_-^2)/(2\varepsilon\kappa) \tag{14}$$

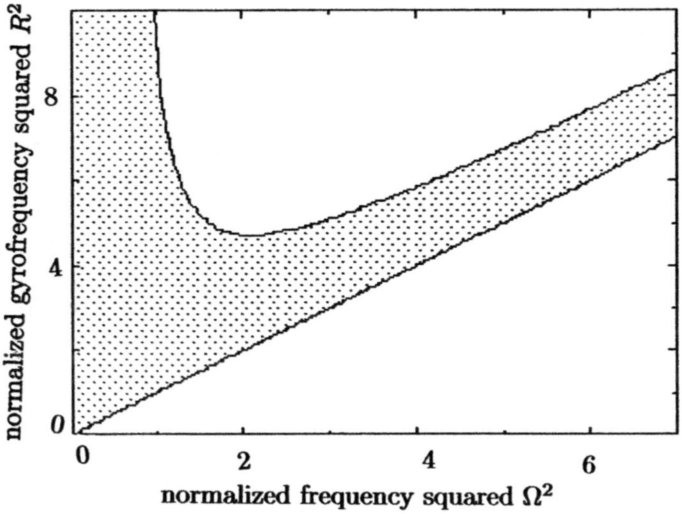

Figure 1. Surface wave on the electric wall in the chiroplasma, $a^2 = 4$. The region of existence is painted over.

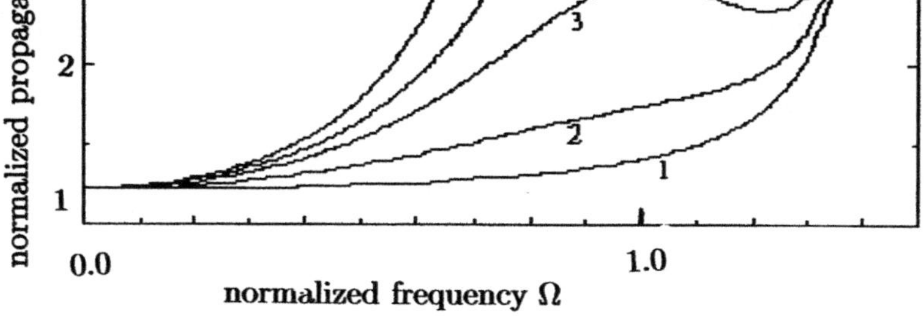

Figure 2. The dispersion diagram of the surface wave on the electric wall in the chiroplasma, $R^2 = 2$; 1—$a^2 = 0$, 2—$a^2 = 2$, 3—$a^2 = 4$, 4—$a^2 = 5$, 5—$a^2 = 6$

Analytical solution of (11) and its direct analysis is not easy because many cumbersome manipulations are necessary. Therefore we adopt a combined method of investigation. At first, the frequency regions are singled out, in which the sign of the equation coefficients and the sign of the discriminant of the related biquadratic equation are the same, and the resonances of the propagation coefficient ($|\alpha| \to \infty$) are found. The quadrant of the (Ω^2, R^2) plane is divided into several domains. There are domains, where the related algebraical equation has positive solution (e.g., we have 15 such domains if $a^2 < 2$). Then we reveal by means of numerical calculations the regions that contain the sought solution of (11). The unidirectional surface wave may exist in two areas of the frequency plane. One such wave propagates at the frequencies smaller than the plasma frequency in the domain that is confined by the curves

$$R^2 = [\Omega^2 - 4/(a^2 + 4)]^2/\Omega^2 \tag{15}$$

and

$$R^2 = (\Omega^2 - 1)^2/\Omega^2 \tag{16}$$

where (15) is the line of resonance for the surface wave. The indicated domain is divided by the curve

$$R^2 = R_0^2 = (1 - \Omega^{-2})[a^4\Omega^4 - 16(\Omega^2 - 1)^2]/[a^4\Omega^2 - 16(\Omega^2 - 1)] \tag{17}$$

into two subdomains in which the directions of the surface wave propagation are the opposite. Namely, the signs of α and R coincide provided $R^2 > R_0^2$. This wave becomes the generalized one in that part of the existence domain, where the discriminant (6) is negative. The surface wave also exists at the frequencies larger than the plasma frequency in the domain located between the line $R^2 = \Omega^2 - 1$ and the curve (15) where the signs of α and R coincide.

3. Chiroferrite

We use the constitutive relations [1]:

$$\vec{D} = \varepsilon\vec{E} + i\xi\vec{B}, \qquad \vec{H} = i\xi\vec{E} + (\mu_\infty\hat{\mu})^{-1} \cdot \vec{B} \tag{18}$$

and assume that the effects of chirality are obviously modelled by the chirality admittance ξ. Some symbols of the chiroferrite section differ from that above in their meanings, as that will be understandable from the context. The diagonal element μ and the off-diagonal component κ of the relative permeability tensor $\hat{\mu}$ are the functions of the medium frequency parameters:

$$\mu = 1 + R/(R^2 - \Omega^2), \quad \kappa = \Omega/(R^2 - \Omega^2) \tag{19}$$

where Ω and R are the ratios of the wave frequency and the frequency of the gyromagnetic resonance to the frequency of the saturation magnetization, respectively. All derivations of the chiroferrite section are parallel to the chiroplasma case and quite straightforward.

3.1. BULK WAVES

The dispersion equation for the relative propagation coefficient is given by

$$\eta^4 - (\chi + \chi a^2 \mu_z + \mu_z^2)\eta^2 + \chi \mu_z = 0 \tag{20}$$

where μ_z is another diagonal element of $\hat{\mu}$ $\chi = (\mu^2 - \kappa^2)/\mu$ is the square of the propagation coefficient for the extraordinary wave in the magnetized ferrite, and $a^2 = 4\mu_\infty \xi^2/\varepsilon$ is the normalized chirality parameter. The propagation coefficient is normalized to $k_\infty = \omega\sqrt{\mu_\infty \varepsilon}$. Unlike (5) for the chiroplasma case, where there are four distinct cases (both $\eta_\pm$ are real, both $\eta_\pm$ are imaginary, one of $\eta_\pm$ is real, and both $\eta_\pm$ are complex), the discriminant of (20) is always positive. Consequently, the chiroferrite plane wave spectrum is simpler than that of the chiroplasma version. Two free plane waves propagating in an unbounded chiroferrite transversely to the static magnetic field are hybrid waves of the elliptical polarization. The left-handed wave exists at all frequencies and may be called quasi-ordinary one. The right-handed (quasi-extraordinary) wave has the stop band $\Omega_1 < \Omega < \Omega_2$, where

$$\Omega_1 = \sqrt{R(R+1)}, \qquad \Omega_2 = R + 1 \tag{21}$$

see the dispersion curves on Figure 3. Note that the time-average Poynting vector of the plane eigenwaves has a component proportional to the chirality admittance ξ, directed along the static magnetic field for both types of the Faraday chiral medium.

3.2. SURFACE WAVES

Consider the conditions of the propagation of surface waves along a perfectly conducting plane boundary $y = 0$ of a chiroferrite half-space.

3.2.1. *Electric Wall*

The dispersion relation for the propagation coefficient can be found as

$$\beta_+(\mu_z - \eta_-^2) - \beta_-(\mu_z - \eta_+^2) = 0 \tag{22}$$

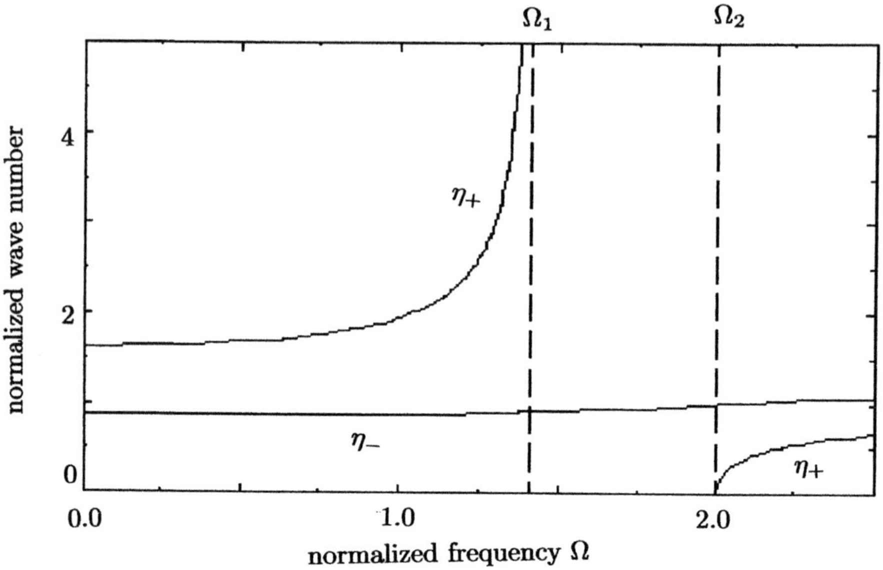

Figure 3. Dispersion curves of the bulk eigenwaves in the chiroferrite, $\mu_z = 1$; $a^2 = 0.2$; $R = 1$

where $\beta_\pm = i\sqrt{\alpha^2 - \eta_\pm^2}$, and for a surface wave $\alpha^2 \geq \eta_\pm^2$. The equation (22) does not contain linear terms with respect to α. Consequently, the surface wave is bidirectional. The analytical solution of (22) reads

$$\alpha^2 = \mu_z(\chi - \mu_z)/[\chi - \mu_z + \chi a^2 \mu_z] \tag{23}$$

The combined analysis of (22) and (23) shows that the surface wave exists only in the frequency band $\Omega_1 < \Omega < \Omega_2$. The propagation coefficients of the surface wave and the quasi-ordinary bulk wave are equal on the ends of the band (21).

3.2.2. *Magnetic Wall*

Here, again, the dispersion relation of the surface wave is given by (11), where now

$$A = -(2 + a^2\mu_z/2) + (2/\chi + a^2/2)\eta_-^2 \tag{24}$$

$$B = 2 + a^2\mu_z/2 - (2/\chi + a^2/2)\eta_+^2 \tag{25}$$

$$C = 2i\kappa(\eta_+^2 - \eta_-^2)/(\mu\chi) \tag{26}$$

The study of the critical lines location on the plane of the frequency parameters and the analysis of the features of the dispersion equation show that the surface wave existence domain is enclosed within the range

$$R(R+1) < \Omega^2 < R^2 + [(2-\mu_z)a^4 - 16]R/[(1-\mu_z)a^4] + 1/(1-\mu_z) \quad (27)$$

which is intersected by the curve

$$\Omega = R + a^2/(a^2+4) \quad (28)$$

This curve is the resonance line for the surface wave that propagates in the lower part of (27), in which connection $\mathrm{sgn}(\alpha\Omega) = -1$. Another surface wave propagates in the opposite direction and weakly slows down with the growth of the frequency in the band (27). The presence of two different unidirectional surface waves was established earlier for an impedance plane in the magnetoactive plasma [9].

4. Discussion

The features of the wave phenomena in case of the transverse propagation with respect to the static magnetic field are rather similar in both types of the Faraday chiral media. These complex media show the increased ability to support surface waves than their achiral counterparts. Waves in chiroplasmas have quite different features, e.g., there is a possibility for existance of the generalized surface wave. Unidirectionality is a notable characteristic of the surface waves in the chiral Faraday media if their boundary plane is parallel and the propagation direction is perpendicular to the impressed static magnetic field. The surface wave reverses together with reversion of the magnetostatic field, but the direction of propagation does not depend on the sign of the chirality parameter ξ. The symmetric surface wave along the electric wall in the chiroferrite is the only exclusion, which is connected with the absence of the longitudinal component of the electric field in the bulk eigenwaves.

Ishimaru [11] has formulated the so called "thermodynamic paradox" for the surface wave in the magnetoplasma. The marked property of unidirectionality also enables us to offer two versions of this paradox in the lossless chiroplasma medium with the boundary formed by a connection of electric and magnetic walls. There is a possibility to choose the medium parameters so that the only surface wave of a certain wall can propagate (in the direction of the connection edge) without attenuation and both bulk waves and surface wave of another wall can not exist. The presence of the paradox points out insufficiency of the lossless chiroplasma model for certain wave diffraction problems in wedge-shaped regions.

5. Conclusion

We have investigated the spectrum of the transversely propagating characteristic bulk and surface waves of the Faraday chiral media (the chiroplasma and the chiroferrite) bounded by perfectly conducting plane surfaces both of the electric and magnetic kinds. The propagation direction of surface waves is perpendicular to the direction of the external static magnetic field which is parallel to the boundary. The results form the basis for consideration of the radiation and diffraction problems in the semi-bounded chirogyrotropic media. Some notable features such as pass and stop bands, unidirectionality of surface waves, etc. may be taken into account in the design of novel microwave devices and components.

References

1. Engheta, N., Jaggard, D.L., and Kowarz, M.W. (1992) Electromagnetic waves in Faraday chiral media, *IEEE Trans. on Antennas and Propagation*, **40**, 367-374.
2. Gincburg, M.A. (1958) Surface waves on the boundary of the gyrotropic medium, *Nauchnye Doklady Vysshej Shkoly. Radiotekhnika i Elektronika*, **3**, 38-47 (in Russian).
3. Hartstein, A., Burstein, E., Maradudin, A.A., Brewer, R., and Wallis, R.F. (1973) Surface polaritons on semi-infinite gyromagnetic media, *J. Phys. C: Solid State Phys.*, **6**, 1266-1276.
4. Gincburg, M.A. (1960) Surface waves on the boundary of a plasma in a magnetic field, *Trudy IZMIR AN*, **17**, 208–215 (in Russian).
5. Chiu, R.W., and Quinn, J.J. (1972) Magneto-plasma surface waves in solids, *Nuovo Cimento B*, **10(1)**, 1-20.
6. Wallis, R.W., Brion, J.J., Burstein, E., and Hartstein, A. (1974) Theory of surface polaritons in semiconductors, *Phys. Rev. B*, **9**, 3424-3427.
7. Kondratenko, A.N. (1985) *Surface and Bulk Waves in the Bounded Plasma*. Energoatomizdat, Moscow (in Russian).
8. Azarenkov, N.A., Kondratenko, A.N., and Ostrikov, K.N. (1993) Surface waves in plasma-metal structures (a review), *Izvestiya VUZ. Radiofizika*, **36**, 335-389 (in Russian).
9. Fisanov, V.V. (1970) Excitation of surface waves on the impedance plane in anisotropic plasma, *Izvestiya VUZ. Radioelektronika*, **13**, 53-58 (in Russian).
10. Seshadri, S.R. (1962) Excitation of surface waves on a perfectly conducting screen covered with anisotropic plasma, *IRE Trans. on Microwave Theory and Techniques*,**10**, 573-578.
11. Ishimaru, A. (1963) Unidirectional waves in anisotropic media, in E.C. Jordan (ed.), *Electromagnetic Theory and Antennas*. Pergamon Press, New York, pp. 591–601.

CHIRO-PLASMA SURFACE WAVE

H. TORRES-SILVA
Facultad de Ingeniería, Dpto. de Electrónica
Casilla 6-D, Av. 18 de Septiembre 2222, Arica-Chile

Abstract. In this paper we analyse the behavior of electromagnetic waves propagating along the interface between a linear chiro-dielectric medium and vacuum. The gyrodielectric tensor of a chiro-plasma is used to investigate the dispersion relation for bulk waves and surface magnetoplasmon polaritons in metals or doped semiconductors modified by the chirality factor. For bulk waves, which propagate parallel to the magnetic field, the medium is a cold magnetized chiral-plasma, which presents a new mode conversion. For waves which propagate perpendicular to the magnetic field, there is no mode conversion but a lower band of forbidden frequencies disappears when the chirality parameter increases. For surface waves, one notes that the dispersion relation obtained is non-reciprocal, i.e. positive and negative values of the wavenumber k are not equivalent. A new feature is immediately apparent, namely the dispersion curve consist of two parts with a gap between them. The low portion starts from the origin, rises just to the right of the light line, $\omega = kc$, bends over, and terminates when the curve intersects the dispersion curve for bulk magnetoplasmons (bulk polaritons). The upper branch starts on the line defined by $\epsilon^*_{xx} = 0$, rises, and then approaches the asymptotic frequency for unretarded surface magnetoplasmons defined by the equation $1 + \epsilon^*_{xx} - i\epsilon_{xz} = 0$. Here ϵ^*_{xx} is modified by the chiral parameter. In the large wave vector, unretarded limit, the electric vector of the upper branch executes a circular motion in the sagittal plane. At the small wavevector extremity of this branch, the electric vector is plane polarized perpendicular to the surface and this branch stops before it reaches the light line. Also we note that there are frequencies at which both a bulk wave and a surface wave can propagate, a situation which does not exist in the absence of a magnetic field and the chiral parameter. Calculations of the dispersion relation have been carried out for n-type InSb.

A. Priou et al. (eds.), Advances in Complex Electromagnetic Materials, 249–258.

1. Theoretical Foundations

We propose the following constitutive relations for chiral-plasma media [1,2]

$$\vec{D} = \epsilon_0 \overleftrightarrow{\epsilon} \cdot \vec{E} + t_1 \vec{H} \quad (1)$$

$$\vec{B} = \mu \vec{H} + t_2 \vec{E}. \quad (2)$$

where the magnetoelectric scalars t_1, t_2 account for the chiral effects.[1] The E-field vector Helmholtz equation is derived by inserting constitutive relation (Equations (1) and (2)) into Maxwell's equations so that

$$\nabla \times \vec{E} = -i\omega\mu_0 \vec{H} - i\omega t_2 \vec{E} \quad (3)$$

$$\nabla \times \vec{H} = i\omega\epsilon_0 \overleftrightarrow{\epsilon} \cdot \vec{E} + i\omega t_1 \vec{H}. \quad (4)$$

Here, the plasma is included in the permittivity tensor $\overleftrightarrow{\epsilon}$.

Defining $\vec{E}$ as

$$\vec{E} = \vec{E}_0 e^{-i\vec{k}\cdot\vec{r}}, \quad (5)$$

we obtain

$$-\vec{k} \times \vec{k} \times \vec{E}_0 + \omega(t_2 - t_1)\vec{k} \times \vec{E}_0 - \omega^2 \mu_0 \epsilon_0 \left(\overleftrightarrow{\epsilon} - \frac{t_1 t_2}{\mu_0 \epsilon_0} \right) \vec{E}_0 = 0. \quad (6)$$

Putting $\vec{E}_0$ into rectangular coordinates we obtain three components system of equations which determine the eigenvectors, and the determinant of the coefficient component matrix M_k will determine the eigenvalues, thereby yielding the ω against $\vec{k}$ dispersion diagram in phase-space. Writing $\det(M_k) = 0$, with $k_y = 0$ and with symmetry about the y-axis we obtain

$$f(\theta) = F(n^2, \omega, \epsilon_{xx}, \epsilon_{yy}, t, \epsilon_{xz}, k). \quad (7)$$

For a semiconductor chiroplasma, $t_1 = it\sqrt{\mu_0\epsilon_0}$ and $t_2 = -it\sqrt{\mu_0\epsilon_0}$. This equation is then the general dispersion relation for waves propagating in a cold collisionless homogeneous chiro-semiconductor-plasma in a uniform magnetic field [3]. For a given plasma frequency ω_p, a given cyclotron frequency ω_c, a given wave frequency ω and a given direction of propagation θ, Equation (7) can be solved for the index of refraction n, having the chirality as parameter t.

[1]For chiral media, the relation $t_1 = -t_2$ holds. *–(Editors)*

Setting $\theta = \pi/2$, we obtain the following dispersion relations:

$$k_X = \pm \frac{\sqrt{A - \sqrt{B}}}{\sqrt{2(\epsilon_{xx} - t^2)}} \tag{8}$$

and

$$k_O = \pm \frac{\sqrt{A + \sqrt{B}}}{\sqrt{2(\epsilon_{xx} - t^2)}}, \tag{9}$$

where

$$A = \frac{\omega^2}{c^2} \left[\epsilon_{xx}^2 - \epsilon_{xz}^2 + \epsilon_{xx}\epsilon_{yy} + t^2(\epsilon_{xx} - \epsilon_{yy}) - 2t^4 \right] \tag{10}$$

and

$$\begin{aligned} B \;=\; & \frac{\omega^4}{c^4} \left[\left((\epsilon_{xx}^2 - \epsilon_{xz}^2) - \epsilon_{xx}\epsilon_{yy} \right)^2 + \right. \\ & t^2(6\epsilon_{xx}^3 - 6\epsilon_{xx}\epsilon_{xz}^2 - 2\epsilon_{xx}\epsilon_{yy}^2 + 12\epsilon_{xx}^2\epsilon_{yy} - 2\epsilon_{xz}^2\epsilon_{yy}3) + \\ & t^4(-15\epsilon_{xx}^2 + 8\epsilon_{xz}^2 - 18\epsilon_{xx}\epsilon_{yy} + \epsilon_{yy}^2) + \\ & \left. 8t^6(\epsilon_{xx} + \epsilon_{yy}) \right] \end{aligned} \tag{11}$$

where

$$\epsilon_{xx} = 1 - \frac{w_p^2}{w^2 - w_c^2} \tag{12}$$

$$\epsilon_{xz} = \frac{w_{ce}}{w} \frac{w_p^2}{w^2 - w_c^2} \tag{13}$$

$$\epsilon_{yy} = 1 - \frac{w_p^2}{w^2}, \tag{14}$$

$$\omega_p^2 = \frac{n_e e^2}{\epsilon_0 m_e} \tag{15}$$

$$\omega_{c0} = \frac{eB_0}{m_e}. \tag{16}$$

It should be pointed out that the electric field of the extraordinary wave, k_X is perpendicular to the magnetic field and the electric field of the ordinary wave, k_O, is parallel to the magnetic field.

In Figure 1 we present the effect of the parameter t on the dispersion relations for the case $\theta = \pi/2$. In this figure the ordinary and extraordinary waves are indicated by circles and stars, respectively. When $t = 0.05$, for $\theta = \pi/2$, the effect of the parameter is very small. We can observe that the dispersion relations are a little modified, but the parameter is not able to break up the forbidden regions that exist when $t = 0$. When $t = 0.5$, the dispersion relations present very different curves with respect to the curves for $t = 0$. The difference in the way the t parameter acts in the parallel and perpendicular directions is due to the kind of equations we have. For parallel propagation the t parameter appears as a linear term and in the Equations (8) and (9) the t parameter appears just inside a square root [6]. Also, we see that for $\theta = \pi/2$ the parameter t does not lead to the conversion of modes, as it happens for $\theta = 0$. However, at $t = 0.5$ the dispersion relation for the ordinary wave mode is flat and the reflection point is shifted. Now the ordinary and extraordinary wave mode have not the same value of ω when k is very large. The more important effect is on the lower branch of the extraordinary wave mode because there are no more bands of forbidden frequencies, and the reflection point vanishes [2,5].

2. Surface Magnetoplasmon Polaritons

The gyrodielectric tensor of chiro-plasma can be used to investigate the dispersion relation for surface magnetoplasmon polaritons in metals or doped semiconductors. The external magnetic field is again taken to be in the $-y$ direction [4].

We first consider TM modes and focus our attention on the electric vector $\mathbf{E}$ which, after eliminating the magnetic field from Maxwell's curl equations, satisfies the equation (6) with $t_1 = -t_2$.

For gyrodielectric media with elements for the dielectric tensor $\overleftrightarrow{\epsilon}$ and the displacement $\vec{D}$ given by $\vec{D} = \epsilon_0 \cdot \overleftrightarrow{\epsilon} \cdot \vec{E} - it\sqrt{\mu_0\epsilon_0}\vec{H}$, the divergence equation

$$\nabla \cdot \vec{D} = 0 \tag{17}$$

does not lead to the conclusion $\nabla \cdot \vec{E} = 0$.

For a hybrid wave propagating in the x-direction with wavenumber k and frequency ω, we can write the solution to Eq. (6) in the form

$$\vec{E} = (E_{2x}, 0, E_{2z})\, e^{ikx} e^{\alpha_2 z} e^{-i\omega t} \tag{18}$$

where α_2 is the decay constant. Substituting Eq. (18) into Eq. (6), we obtain

$$(\alpha_2\epsilon_{xx} + ik\epsilon_{xz})\, E_{2z} - (\alpha_2\epsilon_{xz} - ik\epsilon_{xx})\, E_{2x} = t^2(ikE_{2x} + \alpha_2 E_{2z}) \tag{19}$$

$$(1 + it)E_{1z} = \epsilon_{zx}E_{2x} + (\epsilon_{zz} - t^2)E_{2z} \tag{20}$$

The condition for a nontrivial solution to Eqs. (19) and (20) is obtained by setting the determinant of the coefficients of E_{2x} and E_{2z} to zero to given the result [5],

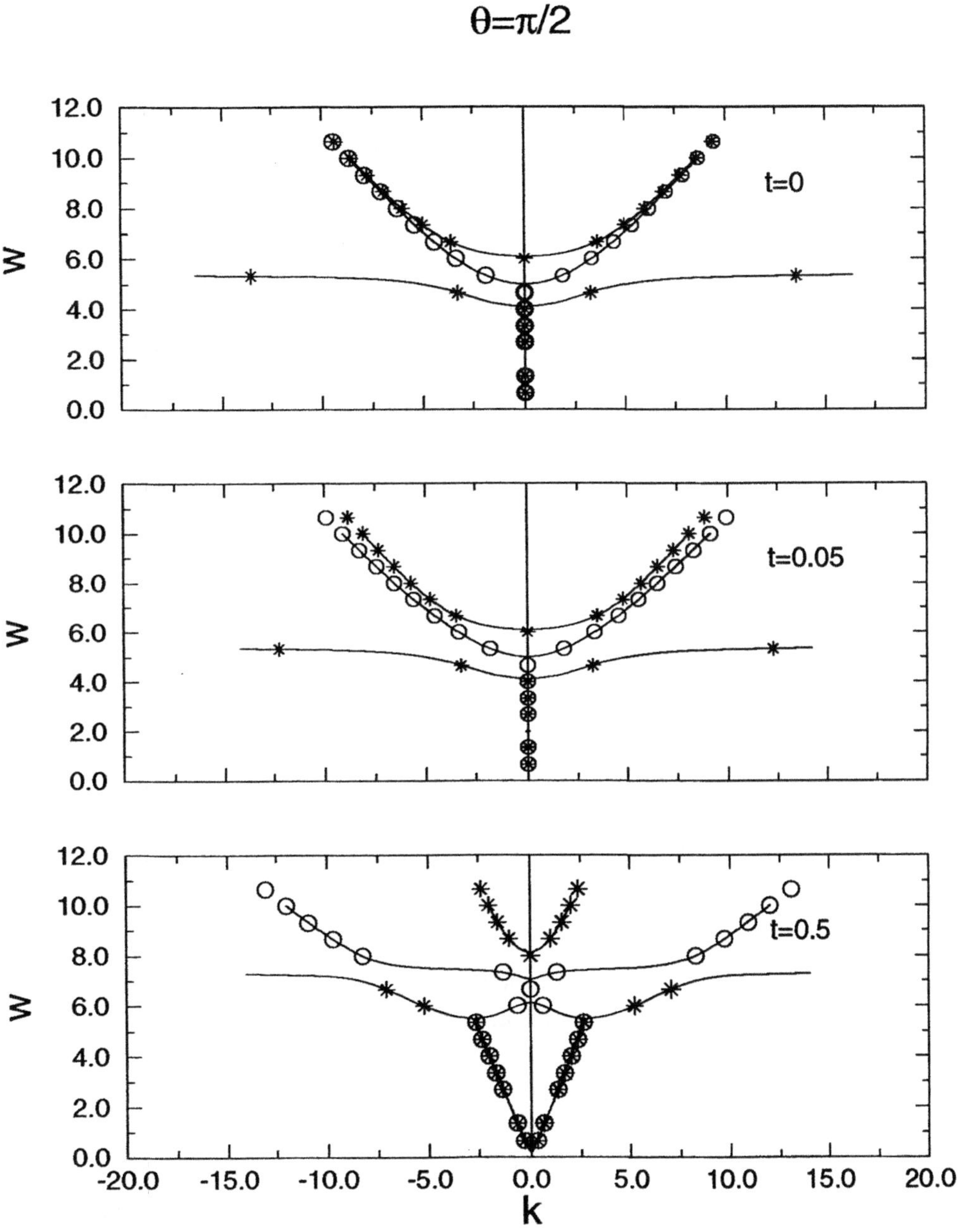

Figure 1. Dispersion relations for various values of the parameter t when the direction of propagation is perpendicular to the magnetic field ($\theta = \pi/2$). The curves indicated by circles and stars correspond to the ordinary and extraordinary waves, respectively.

$$\alpha_2^2 = k^2 - k_0^2 \epsilon_v^* \tag{21}$$

where

$$\epsilon_v^* = \epsilon_{xx} - t^2 + \frac{\epsilon_{xz}^2}{\epsilon_{xx} - t^2} \tag{22}$$

is the Voigt dielectric constant.

In the vacuum region, $z \geq 0$, E can be written in the form

$$\mathbf{E} = (E_{1x}, 0, E_{1z})\, e^{ikx} e^{\alpha_1 z} e^{-i\omega t} \tag{23}$$

The next task is to satisfy the electromagnetic boundary conditions at the surface, which we take to be

$$E_{1x} = E_{2x} \tag{24}$$

$$D_{1z} = D_{2z} \tag{25}$$

We use the divergence equation, Eq. (17), to establish the relationship between E_{1x} and E_{1z} and between E_{2x} and E_{2z}. The result is

$$E_{2z} = \frac{\alpha_2 \epsilon_{xz} - ik\epsilon_{xx} + ikt^2}{\alpha_2 \epsilon_{xx} + ik\epsilon_{xz} - \alpha_2 t^2} E_{2x} \tag{26}$$

$$E_{1z} = \frac{ik}{\alpha_1} E_{1x} \tag{27}$$

Eliminating D_{1z} and D_{2z} from Eq. (25) in favor of E_{1z} and E_{2z} using the appropriate forms of the relation $\mathbf{D} = \epsilon_0 \cdot \overleftrightarrow{\epsilon} \cdot \mathbf{E} - it\sqrt{\mu_0 \epsilon_0}\mathbf{H}$, we find that

$$\alpha_2 + ik\frac{\epsilon_{xz}}{\epsilon_{xx}^*} + \frac{\alpha_1 \epsilon_v^*}{1 + it} = 0 \tag{28}$$

which is the dispersion relation for surface chiromagnetoplasmon polaritons.

One notes that this dispersion relation is non-reciprocal, i.e. positive and negative values of the wavenumber k are not equivalent. In the unretarded limit, $t = 0, \alpha_1 \to |k|$, $\alpha_2 \to |k|$, and the dispersion relation reduces to

$$1 + \epsilon_{xx} - i\epsilon_{xz}\,\mathrm{sgn}\,(k) = 0 \tag{29}$$

This is a cubic equation in ω. For small ω_c/ω_p, one obtains the approximate value for the unretarded surface magnetoplasmon frequency

$$\omega_{sm} \simeq \omega_{sp} + \frac{1}{2}\omega_c\,\mathrm{sgn}\,(k) \tag{30}$$

where ω_{sp} is the zero field non-retarded surface plasmon frequency given by

$$\omega_{sp} = \omega_p \left[1 + \frac{1}{\epsilon_\infty}\right]^{-\frac{1}{2}} \tag{31}$$

Calculations of the dispersion relation for this configuration have been carried out for nInSb. The case $k > 0$ with $t = 0.1$ and $\omega_c/\omega_p = 0.5$ is plotted in Figure 2. A very interesting feature is immediately apparent, namely the dispersion curve consist of two parts with a gap between them. The low portion starts from the origin, rises just to the right of the light line, $\omega = kc$, bends over, and terminates when the curve intersects the dispersion curve for bulk chiromagnetoplasmons (bulk polaritons) defined by $\alpha_2^2 = k^2 - \omega^2\epsilon_v^*/c^2 = 0$. The upper branch starts on the line defined by $\epsilon_{xx}^* = 0$, rises, and then approaches the asymptotic frequency for unretarded surface chiromagnetoplasmons defined by the equation $1 + \epsilon_{xx}^* - i\epsilon_{xz} = 0$. In the large wave vector, unretarded limit, the electric vector of the upper branch executes a circular motion in the sagittal plane.[2] At the small wavevector extremity of this branch, the electric vector is plane polarized perpendicular to the surface. Note that this branch stops before it reaches the light line.

The reduced wave vector at which the upper branch starts is specified by the equation

$$\varsigma_s^2 = \frac{1+\eta^2}{1-(1+\eta^2)\,/\,(\epsilon_\infty^2\eta^2)} \tag{32}$$

where $\varsigma_s = |k|\,c/\omega_p$ and $\eta = \omega_c/\omega_p$. If ς_s^2 is to be finite and positive, then ϵ_∞ and η must satisfy the inequality

$$\epsilon_\infty > \frac{(1+\eta^2)}{\eta} = \frac{\left[\omega_c^2 + \omega_p^2\right]}{\omega_c} \tag{33}$$

That a gap can exist in the dispersion curve is evident from a consideration of Figure 3, where ϵ_v is plotted against ω. There is a region to the left of the line $\epsilon_{xx}^* = 0$ where ϵ_v^* is large and positive , so that α^2 is negative for finite wave vectors and no surface wave exists. If the unretarded surface magnetoplasmon frequency lies above the line $\epsilon_{xx}^* = 0$, as it does for InSb with $\eta = 0.5$, then a surface wave exists both above and well below this line, and a gap must exit just below this line. To have a surface-wave branch above the line $\epsilon_{xx}^* = 0$, the ratio ω_c/ω_p must exceed a critical value specified by Eq. (33). For InSb, this critical value is 0.064.

For a value of ϵ_∞ just below the right-hand side of Eq. (33), the upper branch lies below the line $\epsilon_{xx}^* = 0$ and to the right of the lower bulk dispersion curve. As ϵ_∞ decreases, the gap rapidly closes , and only a single branch remains.

[2]Here "sagittal plane" means the plane perpendicular to the wave vector. –*(Editors)*

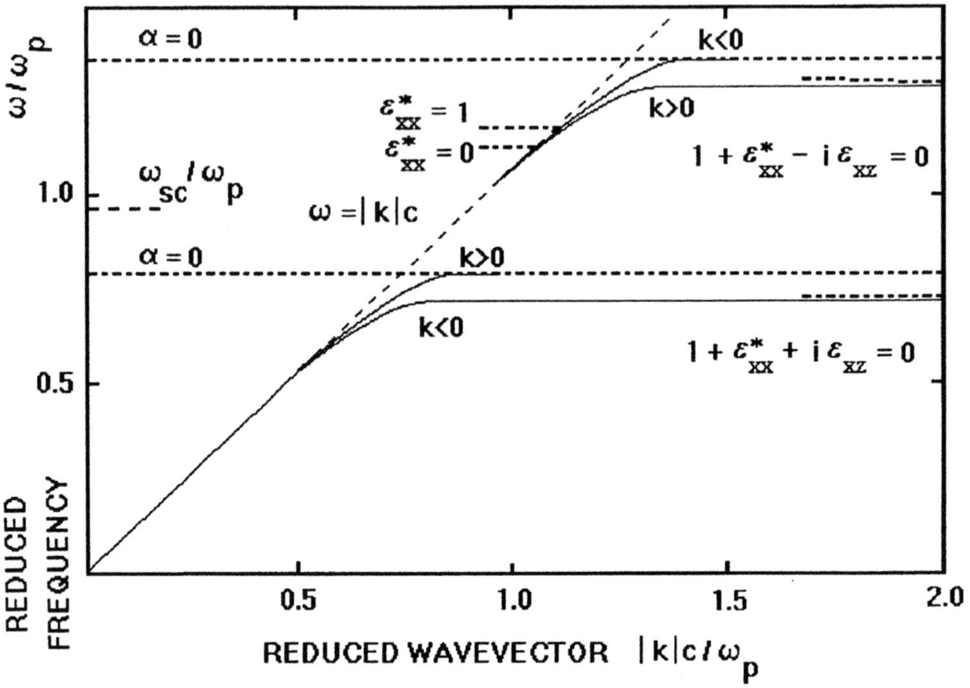

Figure 2. Dispersion curves for surface chiromagnetoplasmon polaritons in n-type InSb. The dashed lines labelled $\alpha = 0$ are bulk polariton dispersion curves with $t = 0.1$.

Turning now to the case of $k < 0$, we plot in Figure 2 the surface-polariton dispersion curve for n-type InSb with $\eta = 0.5$ and $k < 0$. It is clear that the situation is qualitatively different from that where $k > 0$. We now have a complete lower branch running from the origin out to the asymptotic value specified by the equation $1 + \epsilon_{xx}^* - i\epsilon_{xz} = 0$. In addition, however, we have an upper branch which stars at the light line, where $\epsilon_{xx}^* = 1$, rises to right of the light line, and ends when it meets the upper bulk-polariton dispersion curve, $\alpha = 0$. The upper branch for $k < 0$ exists whenever $\epsilon_\infty > 1$ and $\eta > 0$. It may be remarked that, for both $k > 0$ and $k < 0$, there are frequencies at which both a bulk wave and a surface wave can propagate, a situation which does not exist in the absence of a magnetic field.

Acknowledgements

The author is grateful for financial supports to Comisión Nacional de Ciencia y Tecnología (Fondecyt, grant 1950963, Chile).

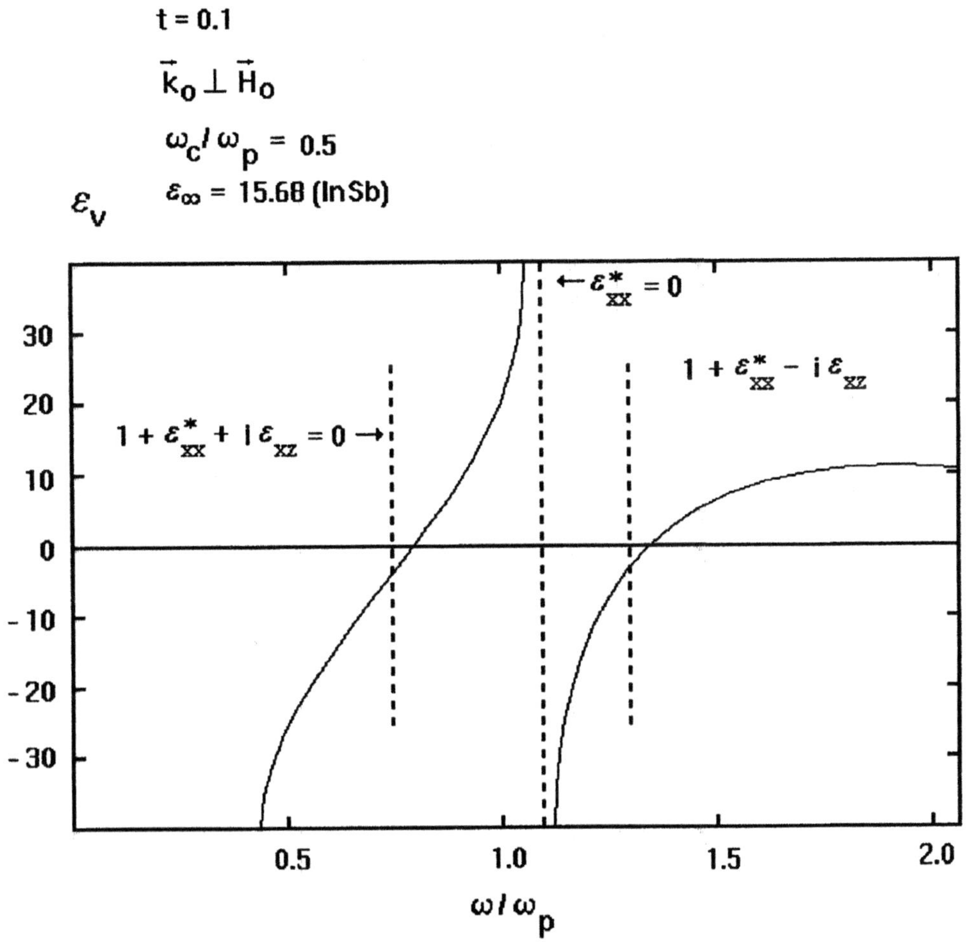

Figure 3. Voigt chirodielectric constant ϵ_v^* versus frequency.

References

1. Torres-Silva H., Sakanaka P.H and Reggiani N. (1995) Electromagnetic properties of a chiral-plasma-media, CB mag'95 – *Congresso Brasileiro de Eletromagnetismo*, Vol. I, p. 201, Brasil.
2. Torres-Silva H., Reggiani N. and Sakanaka P. H. (1996) The effect of chirality on a plasma medium, Revista Mexicana de Física, **42**, No. 6, p. 1480.
3. Engheta N., Jaggard D.L and Kowarz M.W. (1992) Electromagnetic waves in Faraday chiral media, *IEEE Transactions on Antennas and Propagation* **40**, 367-374.
4. Li J. C. Liu and Jaggard D. L. (1992) Chiral layers on planar surface, *J. Electromagnetic Waves Applicat.*, special issue, **6**, No. 5/6, 651–667.

5. Wallis R. F et al, (1990) Nonlinear electromagnetic surface Waves on gyrotropic media, *Nonlinear Waves in Solid State Physics*, vol. 1, p. 51, Plenum Press, New York.
6. Saadoun M.M.I. and Engheta N. (1994) Theoretical study of variation of propagation constant in a cylindrical waveguide due to chirality: chiro-phase shifting, *IEEE Transactions on Microwave Theory and Techniques*, **42**, No. 9, 1690–1694.

V.

Electromagnetic Modelling of Complex Particles

ON ELECTROMAGNETIC THEORY OF ARTIFICIAL NONCHIRAL AND CHIRAL MEDIA WITH RESONANT PARTICLES (A REVIEW)

M.V. KOSTIN AND V.V. SHEVCHENKO
Institute of Radioengineering and Electronics
Russian Academy of Sciences
11 Mokhovaya str., Moscow, 103907, Russia

Abstract. An analytical review of papers by the authors devoted to design and development of composite media based on circular conducting elements is presented. The review includes:

1. The theory of artificial diamagnetics on the base of non-magnetic circular conducting elements (ring metallic particles).
2. Extension of this theory on composite media with broken circular elements, that is the transition to the theory of artificial paramagnetics.
3. Theory of media containing elements in the form of single broken rings and in the form of double broken rings, including concentric rings.
4. Relation between the effective permeability of a composite medium and susceptibility of a single inclusion.
5. Conditions for existence of a magnetic permeability resonant maximum and analysis of the frequency of the maximum dependence on the electric parameters of a single ring.
6. Theory of artificial chiral electromagnetic medium on the base of spherical helically conducting particles.

1. Introduction

All the papers to be reviewed in the following can be classified in three groups. The first group includes papers on investigations of the permeability dispersion properties of the artificial diamagnetic media based on solid circular conducting elements. Calculations and examinations of the permeability dispersion properties of the artificial paramagnetic media on the base of broken circular elements, single and double coupled inclusions are

A. Priou et al. (eds.), Advances in Complex Electromagnetic Materials, 261–270.

put into the second group. In the third group of papers there are investigations of artificial chiral media on the base of spherical helically conducting particles (SHCPs).

In the works of the first and the second groups, composite media that are formed by a combination of four cubic lattices at the intersections of which annular inclusions are located, were considered. The axes of inclusions are assumed to be parallel. Such a combination makes it possible to increase by factor of 4 the concentration of elements in a unit volume compared with a lattice of spheres.

In order to evaluate the effective permeability it was assumed that the behavior of the composite medium in quasi-static electromagnetic field is similar to the behavior of non-polar dielectrics, the molecular dipole moment of which appears because of the action of the external field. Then, the magnetic dipole moment of the composite element can be obtained and the effective permeability of the structural medium is determined from the Onsager formula [1–4] or from the Lorenz–Lorentz formula [5–8]. Because in the structure being considered all of the rings are oriented so that their axes are parallel, the medium is anisotropic. Its permeability tensor is diagonal and the relative components are equal to $\mu_x = \mu_y = 1,\ \mu_z = \mu$.

In the third group of investigations we present the results for the permittivity, permeability, and chirality parameter of an isotropic artificial medium, formed by SHCPs whose axes are randomly oriented. These results have been corrected as compared with [10–12]. The effective constitutive parameters for this medium are calculated using a generalization of the Lewin self-consistent method [13].

2. Results for the Ring Element Media

The following results were published in the first group of papers [1–5].

The quasistatic theory for calculation of the complex permeability and permittivity of the artificial composite medium with conducting (metallic) rings as elements of the spatial structures is constructed. It is shown that such a medium has diamagnetic properties and large magnetic losses ($\tan\delta_\mu = 0.3 - 0.4$) in a wide frequency band in the microwave range while containing only a small volume amount of metal (less than 0.1%) [1–5]. Let us point out that these are not the resonant losses. The dielectric losses of this composite medium are insignificant compared with the magnetic losses [5].

The theoretically predicted large magnetic losses in the considered composite media have been confirmed experimentally [3,5]. The experiment was performed for a composite medium with thin-film rings. The annular elements are produced by a photolithography method from a metallic film, deposited onto a thin-film dielectric substrate. A stack of films with an annular structure creates a volume dielectric medium with metallic inclusions in the form of the elements of the properties of a magnetodielectric as a whole.

Next, we turn to the second group of papers [6–8].

The quasistatic theory is generalized for composite media with broken conducting annular elements and with double coupled elements [6–8]. It is shown that in the resonant frequency range the composite medium with broken metallic rings has paramagnetic properties and large magnetic losses ($\tan\delta_\mu \approx 1$ [6]). The media with concentric and double coupled broken annular elements have $\tan\delta_\mu \approx 1$ while the metal volume fraction is smaller than 0.1% [7].

For the composite media being considered the frequency bands with large magnetic losses are obtained in dependence on medium parameters. The frequency band positions in the microwave range can be changed and controlled by matching of the medium parameters.

3. Theory of Ring-Element Media

According to the Lorenz–Lorentz formula

$$\mu = \frac{\chi}{1-\chi/3} \tag{1}$$

where $\chi = N\chi_0$, N is the concentration (the number of elements in a unit volume), χ_0 characterises properties of a single element.

For an annular conducting element χ_0 is the proportionality coefficient between the magnetic moment $\mathbf{m}$ of an element and the local magnetic field $\mathbf{H}_l$ in the following manner: $\mathbf{m} = \chi_0\mathbf{H}_l$.

For closed and broken annular elements

$$m = IS \tag{2}$$

where I is the electric current, induced by the magnetic field, $S = \pi b^2$, and b refers to the disk spanned by ring.

The value of χ can be represented as

$$\chi = -\gamma(i\omega L)\begin{cases} \dfrac{1}{R+i\omega L} & \text{for closed rings} \\ \dfrac{1}{R+i\left(\omega L - \dfrac{1}{\omega C}\right)} & \text{for broken rings} \end{cases} \tag{3}$$

where ω is the angular frequency; R is the ring resistance; $L = L_0 + M$, $L_0 = \mu_0 b(\log 8b/d - 1/2)$, d is the diameter of metal strand, and $M =$

$\mu_0 bf(2b/l)/4\pi$. The quantities L_0 and M are the internal and mutual inductances of the elements, the function $f(2b/l)$ is tabulated in [9], l is the lattice period; C is the capacitance (C is determined through the Q–factor, see (5)), $\gamma = \pi^2 b^3 N/q$; $q = L/\mu_0 b$; (μ_0 is the vacuum permeability, $0 \leq \gamma \leq 1$, $\gamma \approx 1$ corresponds to the concentration limit. The function $\mu''(\omega)$ has a maximum at $\omega = \omega_\mu$, where

$$\omega_\mu = \alpha \begin{cases} \alpha \dfrac{R}{L} & \text{for closed rings} \\ \omega_0 \sqrt{\sqrt{\beta^2 + 3} - \beta} & \text{for broken rings} \end{cases} \tag{4}$$

$$\alpha = \frac{1}{\sqrt{1 + \gamma/3}} \qquad \beta = 1 - \frac{1}{2}\sqrt{\frac{\alpha}{Q}}$$

$$\omega_0 = \frac{1}{\sqrt{LC}} \qquad Q = \frac{1}{R}\sqrt{\frac{L}{C}} \tag{5}$$

The typical dispersion dependencies for media on the base of closed and broken rings are represented in Figure 1, where $\gamma = 0.65$, $Q = 0$ for solid rings (lines 1), $Q = 0.67$ (lines 2), and $Q = 2$ (lines 3) for broken rings. The broken rings are presented in Figure 2.

For concentric and double coupled broken ring elements the magnetic moment is

$$m = m_1 + m_2 \tag{6}$$

where

$$m_1 = I_1 S_1 \qquad m_2 = I_2 S_2 \tag{7}$$

$I_{1,2}$ are the currents, induced by the magnetic field, $S_{1,2} = \pi b_{1,2}^2$ are the areas of the rings.

For this case

$$\chi = -\gamma_1 (i\omega L_1) \frac{Z_2 + n^4 Z_1 - 2n^2 Z_c}{Z_1 Z_2 - Z_c^2} \tag{8}$$

where $Z_{1,2}$ are the impedances of the elements, Z_c is the coupling impedance:

$$Z_{1,2} = R_{1,2}\left(1 + i\left(\omega \frac{L_{1,2}}{R_{1,2}} - \frac{1}{\omega}\frac{R_{1,2}}{L_{1,2}} Q_{1,2}^2\right)\right) \qquad Z_c = i\omega M$$

$$L_{1,2} = \mu_0 b_{1,2}\left[\log(8b_{1,2}/d) - 1/2\right] \qquad M = \frac{\mu_0\sqrt{b_1 b_2}}{4\pi}\begin{Bmatrix} \phi_c(n) \\ -\phi_a(n,p) \end{Bmatrix}$$

$R_{1,2}$ are the ring resistances, $Q_{1,2}$ are the quality factors, $n = b_2/b_1$, $p = l_0/b_1$, l_0 is the distance between the ring centers, $\gamma_1 = \pi^2 b_1^3 N/q_1$, $q_1 = L_1/\mu_0 b_1$.

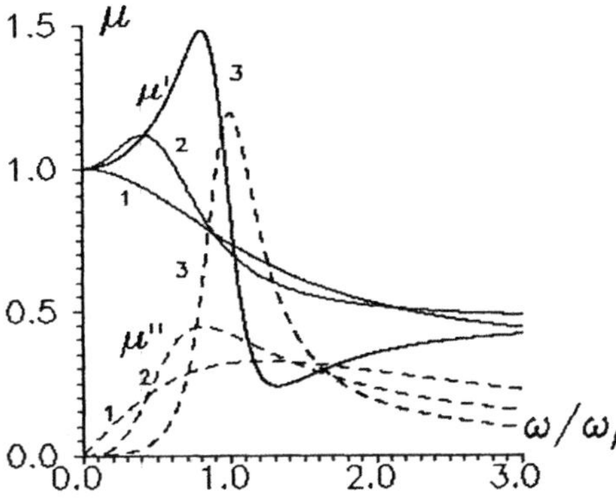

Figure 1.

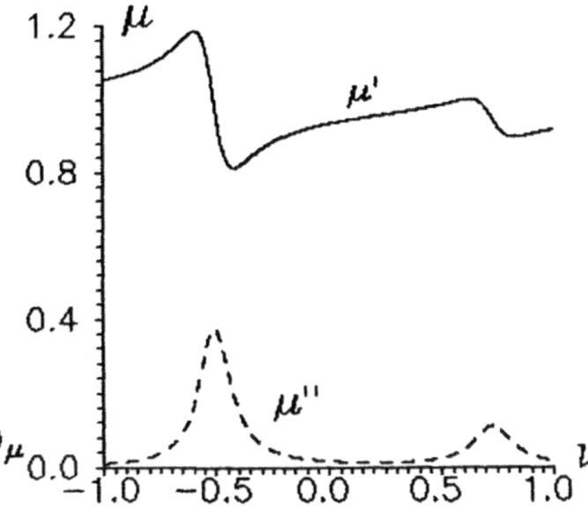

Figure 3 (a).

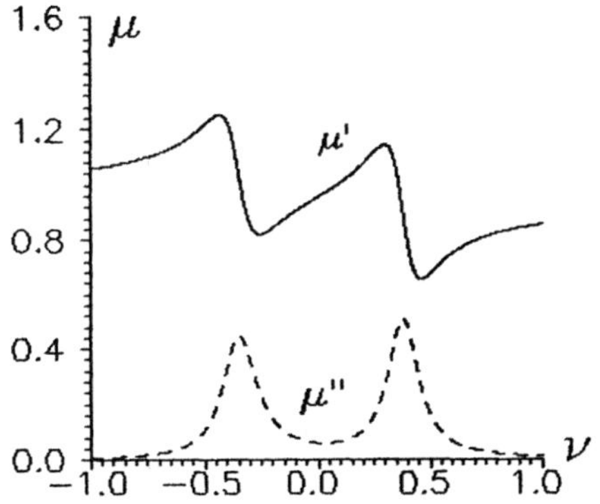

Figure 3 (b).

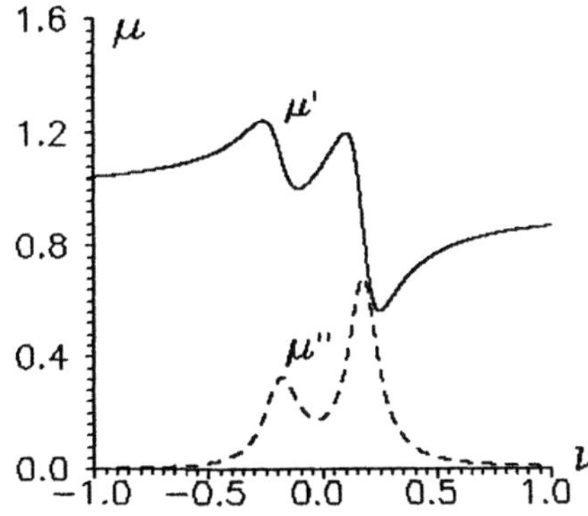

Figure 3 (c).

Figure 1. — Figure 3. **Permeability dispersion of the ring materials**

Figure 2. Geometry of broken rings

The functions $\phi_c(n)$, $\phi_a(n,p) > 0$, and they are applicable to concentric and non-concentric double ring elements, respectively [9].

1. If Q_1, $Q_2 \leq 1$, the function $\mu''(\omega)$ has a single maximum for

$$\omega_{\max} \approx \sqrt{\omega_{\mu 1}\omega_{\mu 2}} \tag{9}$$

2. If Q_2, $Q_1 \gg 1$ and $\omega_{\mu 1}1 = \omega_{\mu 2} = \omega_\mu$, the behaviour of the function $\mu''(\omega)$ can have two maxima at $\omega_{\max 1}$ and $\omega_{\max 2}$, which satisfy the relations

$$\frac{\omega_{\max 1,2}}{\omega_\mu} \approx 1 \pm \frac{\kappa}{2} \tag{10}$$

where

$$\kappa = \sqrt{k_c^2 - \frac{1}{Q_1 Q_2}} \tag{11}$$

$$k_c = \frac{1}{2\pi}\frac{\mu_0 b_1}{\sqrt{L_1 L_2}}\phi_c \tag{12}$$

3. If Q_2, $Q_1 \gg 1$, $Q_1 \geq Q_2$, and $\omega_{\mu 1} > \omega_{\mu 2}$, the function $\mu''(\omega)$ can have two maxima

$$\frac{\omega_{\max 1,2}}{\omega_{\mu 1,2}} = 1 \pm \frac{\kappa}{2} \quad \text{for } |\nu_{\mu 2}| \ll \kappa \ll 1 \tag{13}$$

$$\frac{\omega_{\max 1,2}}{\omega_{\mu 1,2}} = 1 \pm \frac{\kappa}{2|\nu_{\mu 2}|} \quad \text{for } \kappa^2 \ll |\nu_{\mu 2}| \ll 1 \tag{14}$$

where

$$|\nu_{\mu 2}| = 2\frac{|\omega_{\mu 2} - \omega_{\mu 1}|}{\omega_{\mu 1}} \tag{15}$$

The last result is demonstrated in Figure 3, where $\nu = \omega/\sqrt{\omega_{\mu1}\omega_{\mu2}} - \sqrt{\omega_{\mu1}\omega_{\mu2}}/\omega$. The curves are calculated for $Nb_1^3 = 0.03$, $Q_1 = 15$, $Q_2 = 10$; $n = 0.8$, $k_c = 0.45$ (a); $n = 0.6$, $k_c = 0.24$ (b); $n = 0.4$, $k_c = 0.24$ (c). For composite media with double non-concentric elements the function $\mu''(\omega)$ has one maximum like for media with single annular elements, if $\omega_{\mu1} = \omega_{\mu2}$, because $\phi_a \ll 1$ [9], and $\mu''(\omega)$ has two maxima if $\omega_{\mu1} \neq \omega_{\mu2}$.

4. Theory of the SHCP Medium

Next, a cubic lattice structure with spherical helically conducting particles (SHCPs) [10] is considered as an artificial chiral medium. The effective constitutive parameters of this medium are calculated analytically by the generalized Lewin self-consistent method [13].

Let the constitutive equations are

$$\mathbf{D} = \epsilon(\mathbf{E} + \rho\nabla\times\mathbf{E}) \qquad \mathbf{B} = \mu(\mathbf{H} + \rho\nabla\times\mathbf{H}) \tag{16}$$

where ϵ, μ, and ρ are the permittivity, permeability and the chirality parameter of the medium. The propagation constants of the circularly polarized (right and left) plane waves in this medium are [14,15]

$$k_\pm = \frac{k}{1 \pm k\rho} \tag{17}$$

where $k = \omega\sqrt{\epsilon\mu}$, ω is the angular frequency. Like for the usual medium the reflection coefficient from a plane interface with the chiral medium is

$$\Gamma = \frac{\hat{\zeta} - \hat{\zeta}_1}{\hat{\zeta} + \hat{\zeta}_1} \tag{18}$$

where $\hat{\zeta} = \sqrt{\mu/\epsilon}$. Here $\hat{\zeta}_1 = \sqrt{\mu_1/\epsilon_1}$ is the wave impedance of the outside medium.

Using the results of the diffraction problem of circularly polarized waves on a small spherical helically conducting particle (SHCP) [11, 12] and averaging the Hertz vectors on the random orientations of the particle axis in space we obtain

$$\langle\overline{\Pi^e}\rangle = a^3\frac{\exp(-ik_1r)}{r}\left[(\alpha^e\langle E_x\rangle - i\beta\hat{\zeta}_1\langle H_x\rangle)\overline{x} + (\alpha^e\langle E_y\rangle - i\beta\hat{\zeta}_1\langle H_y\rangle)\overline{y}\right]$$

$$\langle\overline{\Pi^m}\rangle = a^3\frac{\exp(-ik_1r)}{r}\left[(\alpha^m\langle H_x\rangle + i\beta\hat{\zeta}_1^{-1}\langle E_x\rangle)\overline{x} + (\alpha^m\langle H_y\rangle + i\beta\hat{\zeta}_1^{-1}\langle E_y\rangle)\overline{y}\right] \tag{19}$$

where

$$\begin{pmatrix} \alpha^e & -i\beta \\ i\beta & \alpha^m \end{pmatrix} = \begin{pmatrix} 1 + \delta_1^2\gamma^2/(2R) & -i\delta_1\gamma/(2R) \\ i\delta_1\gamma/(2R) & -(1 - 1/R)/2 \end{pmatrix} \tag{20}$$

and $k_1 = \omega\sqrt{\epsilon_1\mu_1}$, $r = \sqrt{(x-\xi)^2 + (y-\eta)^2 + (z-\zeta)^2}$. Here, the matrix (20) is corrected compared with [10—12]. Also, $\langle E_{x,y}\rangle$, $\langle H_{x,y}\rangle$ are the values of the incident fields at the point $(\xi,\ \eta,\ \zeta)$, which is the particle location. x, y are the transverse coordinates, and z is the longitudinal coordinate with respect to the wave propagation direction, $\gamma = \tan\alpha$, α is the angle between the conductivity direction (loxodromy) and meridians on the particle surface, $\delta_1 = k_1 a$, a is the particle radius, R is the resonance factor [11,12].

Using (17)–(19) and the generalized Lewin theory we have

$$\frac{k}{k_1} = \frac{1}{N} \qquad \frac{\hat{\zeta}}{\hat{\zeta}_1} = \frac{M+N}{L+N} \qquad k_1\rho = \frac{3f\beta}{D}$$

$$\frac{\epsilon}{\epsilon_1} = \frac{1}{N}\frac{L+N}{M+N} \qquad \frac{\mu}{\mu_1} = \frac{1}{N}\frac{M+N}{L+N} \tag{21}$$

where

$$L = \frac{A}{D} \qquad M = \frac{B}{D} \qquad N = \sqrt{\frac{C}{D} + k_1^2\rho^2}$$

$$A = (1+2f\alpha^e)(1-f\alpha^m)+2f^2\beta^2 \quad B = (1-f\alpha^e)(1+2f\alpha^m)+2f^2\beta^2 \tag{22}$$

$$C = (1-f\alpha^e)(1-f\alpha^m) - f^2\beta^2 \quad D = (1+2f\alpha^e)(1+2f\alpha^m) - 4f^2\beta^2$$

$$f = \frac{4\pi a^3}{3s^3}$$

and s is the side length of the cubical cell.

The dependencies of $\epsilon(\delta_1\gamma)/\epsilon_1$, $\mu(\delta_1\gamma)/\mu_1$, $k_1\rho(\delta_1\gamma)$ are represented in Figure 4 for $f = 0.1$, $\gamma = 3$, $\gamma = 5$, and $\epsilon_1 = \epsilon_2 = \epsilon_0$, $\mu_1 = \mu_2 = \mu_0$, where ϵ_2, μ_2 and ϵ_0, μ_0 are the parameters of the medium inside the particle and vacuum, respectively. In that case [12]

$$R = 3\left(1 - \frac{\delta_1^2\gamma^2}{2}\right) + i\delta_1^3(1 + \delta_1^2\gamma^2)$$

In the limiting cases we have: for $\delta_1\gamma \to 0$

$$\frac{\epsilon}{\epsilon_1} = 1 + \frac{3f}{1-f} \qquad \frac{\mu}{\mu_1} = 1 - \frac{3f}{3+f} \qquad (k_1\rho = 0) \tag{23}$$

and for $\delta_1\gamma \to \infty$

$$\frac{\epsilon}{\epsilon_1} = 1 + \frac{6f}{3-2f} \qquad \frac{\mu}{\mu_1} = 1 - \frac{3f}{2+f} \qquad (k_1\rho = 0) \tag{24}$$

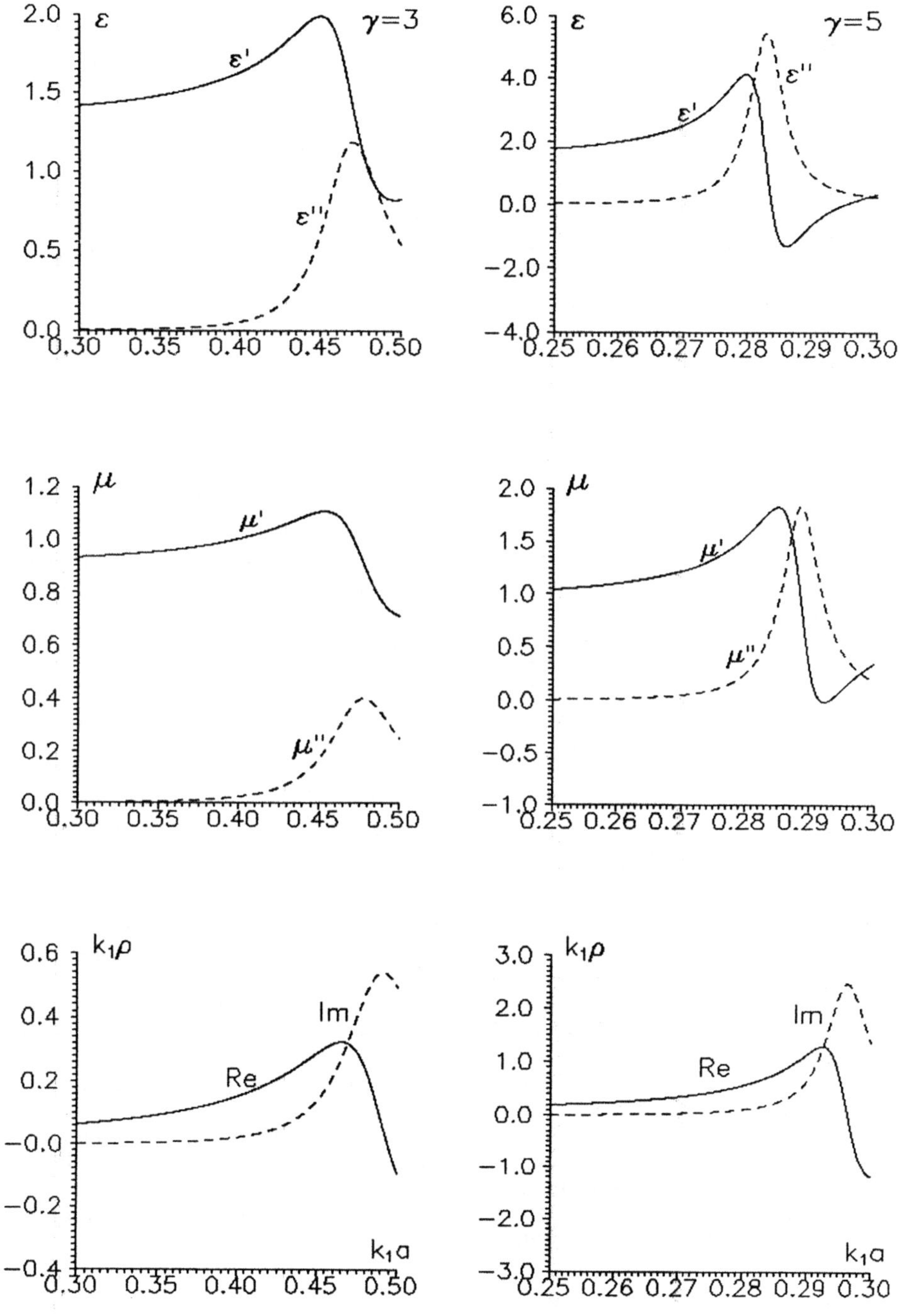

Figure 4. **Parameter dispersion of the SHCP medium**

Acknowledgment

This work was supported by the Russian Fund of Fundamental Research, Grant No. 95-02-06070a.

References

1. Kostin, M.V., Shevchenko, V.V. (1988) Artificial magnetic material based on ring currents, *Sov. J. of Communications Technology and Electronics* **33**(12), 38-42.
2. Kostin, M.V. (1990) Artificial magnetic on the base of ring–film elements, *Sov. J. of Communications Technology and Electronics* **35**(7), 131-134.
3. Kostin, M.V., Kazantsev, Yu.N., Kraftmakher, G.A., Shevchenko V.V. (1990) Artificial diamagnetic with magnetic losses, *Pisma v Zhurnal Tekhnicheskoi Fiziki* **16**(13), 22-25.
4. Kostin, M.V., Shevchenko, V.V. (1992) Artificial magnetic based on ring currents, Proc. XI *International Conference on Microwave Ferrites,* Alushta, the Ukraine, Oct. 1992 **1**, 10-13.
5. Kostin, M.V., Shevchenko, V.V. (1992) Theory of artificial magnetics substances based on ring currents, *Sov. J. of Communications Technology and Electronics* **38**(5), 78-83.
6. Kostin, M.V., Shevchenko, V.V. (1993) Artificial magnetics based on circular film elements, in A. Sihvola, S. Tretyakov, I. Semchenko (eds.), Proc. of *Bianisotropics'93*, International Seminar on electrodynamics of chiral and bianisotropic media, Gomel, Belarus, Oct. 1993. Report 159, Electromagnetics Laboratory, Helsinki University of Technology, Dec. 1993, 32-35.
7. Kostin, M.V., Shevchenko, V.V. (1994) Artificial magnetics based on double circular elements, in F. Mariotte and J. Parneix (eds.) Proc. of *Chiral'94*, 3rd International Workshop on Chiral, Bi-isotropic and Bi-anisotropic Media, Périgueux, France, May 1994, 49-56.
8. Kazantsev, Yu.N., Kostin, M.V., Kraftmakher, G.A., Ponomarenko, V.I., Shevchenko, V.V. (1994) Artificial paramagnetic, *J. of Communications Technology and Electronics* **39**(15), 78-81.
9. Kalantarov, P.L., Zeitlin, L.A. (1983) Inductance Calculation, Energoatomizdat, Leningrad.
10. Kostin, M.V., Shevchenko, V.V. (1995) A novel model of artificial chiral electromagnetic medium, in A. Sihvola et al. (eds.), Proc. of *Chiral'95*, The Pennsylvania State University, USA, Oct. 1995, 64-66.
11. Shevchenko, V.V. (1995) A novel model of a chiral electromagnetic particle, in A. Sihvola et al. (eds.), Proc. of *Chiral'95*, The Pennsylvania State University, USA, Oct. 1995, 42-45.
12. Shevchenko, V.V. (1995) Diffraction on a small chiral particle, *J. of Communications Technology and Electronics* **40**(15), 75-85.
13. Lewin, L. (1975), Theory of Waveguides, Newnes-Butterworth, London.
14. Holinder, R.D., Varadan, V.V., Varadan, V.K. (1991) Eigenmodes in a chiral waveguide containing an isotropic chiral material, *Radio Science* **26**(5), 1335-1344.
15. Shevchenko, V.V. (1993) Modes in chiral fibre lightguide, *Telecommunications and Radio Engineering* **48**(6), 120-123.

THE RELATION BETWEEN CO- AND CROSS-POLARIZABILITIES OF SMALL CONDUCTIVE BI-ANISOTROPIC PARTICLES

S.A. TRETYAKOV, C.R. SIMOVSKI, A.A. SOCHAVA
Radiophysics Department
St. Petersburg State Technical University
195251 St. Petersburg, Polytekhnicheskaya 29, Russia

Abstract. Polarization of small complex-shaped conductive particles by external time-harmonic electric and magnetic fields is considered. An original approach to excitation of particles in quasistatic magnetic fields is suggested. The thin-wire approximation for the induced current is assumed, i.e., the theory applies to arbitrary shapes formed by thin conductive wires. It is shown that under a certain restriction on the current distribution functions the co- and cross-polarizabilities are not independent, but related to each other. The physical meaning of the relation as well as its importance for the practical design of complex composites are discussed.

1. Introduction

In the design of novel composite materials, especially for applications in microwave engineering, small bi-anisotropic inclusions are often used. Especially much attention is devoted to artificial chiral materials with inclusions in form of small spirals or wire-and-loop particles, see, e.g., [1]. Other complex chiral and non-chiral shapes are also of practical interest. For example, omega composites with Ω–shaped inclusions are promising for some microwave applications. Polarizabilities of individual inclusions, after appropriate averaging, define the effective properties of bi-anisotropic composites.

Consider a small conductive particle located in a uniform isotropic dielectric (free space, to simplify the notations) excited by external time-harmonic electric and magnetic fields. To facilitate the analysis, we assume that the particle is made of a thin wire (whose diameter is much smaller

A. Priou et al. (eds.), Advances in Complex Electromagnetic Materials, 271–280.

than the wire length) which forms an arbitrary three-dimensional figure. This assumption allows us to consider the induced current as a vector function of a single scalar position variable (arc length, measured by s). In the equations below, the wire is assumed to have open ends, although the theory easily extends to closed figures as well.

The characteristic particle size a (the volume occupied by the particle is of the order of a^3) is assumed to be small as compared to the wavelength λ associated with the frequency ω of the external fields time variation ($\lambda = 2\pi c/\omega$, where c is the speed of light). However, the total wire length L may be comparable to λ. This is practically the most interesting case, because the interaction with the field becomes strong in the resonance area.

Small particles of complex shape are usually characterized by four dyadic polarizabilities, which define the bi-anisotropic relations between induced electric and magnetic dipole moments $\mathbf{p}$, $\mathbf{m}$ and external fields $\mathbf{E}$, $\mathbf{H}$ in the particle center:

$$\begin{aligned} \mathbf{p} &= \overline{\overline{a}}_{ee} \cdot \mathbf{E} + \overline{\overline{a}}_{em} \cdot \mathbf{H} \\ \mathbf{m} &= \overline{\overline{a}}_{me} \cdot \mathbf{E} + \overline{\overline{a}}_{mm} \cdot \mathbf{H} \end{aligned} \tag{1}$$

Due to the reciprocity principle, the dyadic polarizabilities satisfy

$$\overline{\overline{a}}_{em} = -\overline{\overline{a}}_{me}^{T} \qquad \overline{\overline{a}}_{ee} = \overline{\overline{a}}_{ee}^{T} \qquad \overline{\overline{a}}_{mm} = \overline{\overline{a}}_{mm}^{T} \tag{2}$$

where T denotes the transpose operation.

Using a simple quasistatic model of symmetrical wire-and-loop particles [1], it was shown that the polarizabilities of such particles are related as $a_{ee}a_{mm} = a_{em}a_{me}$. In that study, the particles were assumed to be uniaxial objects, so that the polarizabilities along all the directions but the particle axis could be neglected. That was an example of a particle whose polarizabilities were not independent quantities. Here we study the polarization processes for arbitrary-shaped wire scatterers and establish a similar relation between the polarizability dyadics in the general case.

2. Theory

2.1. EXCITATION IN QUASISTATIC FIELDS

In the studies of the polarization of small scatterers in electromagnetic fields it is usually assumed that the exciting fields are uniform in space, although varying in time.[1] This approach can lead to simple and physically

[1]This is incompatible with the Maxwell equations, and it is implicitly assumed that the fields are uniform only inside the volume occupied by the particle.

sound results when the particle has a loop portion, so that the magnetic field excitation can be easily treated by the calculation of the magnetic flux through the loop area. For arbitrary-shaped particles we should adopt a more realistic model of the fields, taking into account that time-varying electromagnetic fields must exhibit variations in space as well.

2.1.1. *External fields*

In fact, when calculating the currents excited by (nearly-)uniform magnetic fields, one usually adopts the quasistatic approximation (see, e.g., [2]). Assuming the magnetic induction to be uniform in space, one makes use of one of the Maxwell equations in order to evaluate the electric field (which drives the currents in conductors):

$$\nabla \times \mathbf{E} = -j\omega\mathbf{B} \tag{3}$$

Since the magnetic induction is uniform, the curl is a constant function. If there is a closed loop formed by conductive wire, it is easy to evaluate the induced electromotive force using the Stokes theorem. For arbitrary-shaped inclusions, that is not a trivial task.

Here we suggest a simple and physically clear approach, which allows to treat excitation of arbitrary (sufficiently small) particles in electromagnetic fields.

To model the processes of both the electric-field and magnetic-field excitations, we assume that the particle is positioned in the fields of a standing wave formed by two plane electromagnetic waves of equal amplitudes travelling in the opposite directions (of a certain axis z). To see how the particle reacts on electric fields we position the particle in the maximum of the total electric field of the waves. The electric field of the standing wave depends on the space position according to the following relation:

$$\mathbf{E}(z) = \mathbf{e}_0 E \cos(kz) \tag{4}$$

where $\mathbf{e}_0$ is the unit vector in the direction of the electric field, E is its complex amplitude measured in the electric field maximum, whose position is at $z = 0$. $k = \omega\sqrt{\epsilon_0\mu_0}$ is the wavenumber. If the particle size is small compared to the wavelength, and we keep only the first-order terms $O(ka)$ in our model (remember that a is the characteristic size of the scatterer), we can neglect the space variation, and assume $\mathbf{E}(z) = \mathbf{e}_0 E = \text{const}(z)$. This is possible simply because the Taylor expansion of the $\cos(kz)$ has no linear term.

Now, position the particle in a maximum of the magnetic field of the standing wave in order to see how the particle is excited in magnetic fields. The currents in conducting particles are always excited by electric fields, thus, we should evaluate the electric field of the wave. The electric field in the vicinity of a maximum of the total magnetic field of the standing wave is given by

$$\mathbf{E}(z) = j\eta_0 H \mathbf{z}_0 \times \mathbf{h}_0 \sin(kz) \tag{5}$$

where $\eta_0 = \sqrt{\mu_0/\epsilon_0}$, and $\mathbf{h}_0$ is the unit vector in the direction of the magnetic field. The currents induced in the wire in the field (5) can be calculated using the conventional methods. Within the frame of the first-order approximation, we replace $\sin(kz)$ by its first term of the Taylor expansion. Obviously, the effect of the magnetic-field excitation is a first-order effect in ka.

To persuade oneself that the excitation by "quasiuniform" magnetic fields is indeed correctly described as the excitation by the associated electric field, one can consider a simple example of a rectangular loop. It is an easy matter to see that the electromotive force $\mathcal{E} = -j\omega\mu_0 HS$ (S is the loop area) equals to the integral $\int \mathbf{E} \cdot d\mathbf{s}$ of the electric field (5) around the loop. Moreover, in this example the electromotive force does not depend on the z-axis co-ordinate provided that the axis belongs to the loop plane.

The suggested model allows to study the magnetic-field excitation of arbitrary-shaped small particles.

2.1.2. *Discussion*

Coming back to the discussion of the quasistatic approximation (Eq. (3)), we stress again that in our approach we assume that the magnetic field is not uniform, but it is a function of one of the space co-ordinates (z). In accordance, the electric field must depend on that co-ordinate (and all the spatial derivatives along the other directions must vanish). In this case, it follows from (3) that the electric field strength is a linear function of one of the spatial co-ordinates. This agrees with (5) (the argument of the sin-function is small under our assumptions).

The problem of the excitation of small conductive particles in quasistatic electric and magnetic fields has been attacked here by positioning the particle in the fields of a standing wave. As was already noted, in the case if the particle contains a loop portion, the magnetic-field excitation can be easily treated in the conventional way, by calculating the electric field circulation around the loop. In that case, the present approach gives the same result. Moreover, the result does not depend on the direction, in which the standing wave "propagates".

However, in the general case of arbitrary-shaped particles, the currents induced by magnetic fields depend on the direction along which the field varies in space.[2] That means that the very description of the interaction in terms of the magnetic polarizability loses its physical sense. One should rather speak about particle reaction on the spatial derivatives of the electric field. It is to be noted here that introducing in the inclusion model (1) dependencies on both the magnetic field and the spatial derivatives of the electric field can possibly lead to mistakes.

2.2. POLARIZATION IN EXTERNAL ELECTRIC FIELDS

Let a wire particle be excited by an external time-varying electric field $\mathbf{e}_0 E$ ($\mathbf{e}_0$ is the unit vector along the field direction, and E is the complex amplitude of the field). The current density induced on the particle is a linear function of the external electric field, which can be written as

$$\mathbf{j}(s) = \frac{d\mathbf{s}}{ds} \int_L \mathbf{A}(s, s') \cdot \mathbf{e}_0 E ds' \tag{6}$$

Here we assume that the current flows along a certain line (whose direction at every wire point is given by the unit vector $d\mathbf{s}/ds$). At the ends of the current line (conductive wire) the current vanishes. The general linear relation (6) can be rewritten in terms of the antenna theory [3] as

$$\mathbf{j}(s) = g_e(s) \frac{d\mathbf{s}}{ds} \frac{1}{Z} \int_L f_e(s) d\mathbf{s} \cdot \mathbf{e}_0 E \tag{7}$$

where $f_e(s)$ is the current distribution function for the wire excited by a unit point voltage source at s. It is normalized so that its value at a certain point on the wire (arbitrary chosen as the terminal position) is unity. The function $g(s)$ is the current distribution function for excitation by external electric field, again normalized in the same sense. The coefficient Z has the meaning of an equivalent impedance. It depends on the particle shape, size, and the frequency. The induced charge density distribution is then

$$\rho(s) = -\frac{1}{j\omega} \frac{dg_e(s)}{ds} \frac{1}{Z} \int_L f_e(s) d\mathbf{s} \cdot \mathbf{e}_0 E \tag{8}$$

The electric dipole moment of the particle is

$$\mathbf{p} = \int_L \rho(s) \mathbf{r}(s)\, ds = \frac{1}{j\omega Z} \int_L g_e(s) d\mathbf{r}(s) \int_L f_e(s) d\mathbf{s} \cdot \mathbf{e}_0 E \tag{9}$$

The last relation defines the electric polarizability dyadic:

[2]The standard approach cannot be directly applied to arbitrary particles, since the solution of (3) in general contains constants which cannot be found without additional physical assumptions.

$$\bar{\bar{a}}_{ee} = \frac{1}{j\omega Z} \int_L g_e(s) d\mathbf{r}(s) \int_L f_e(s) ds \tag{10}$$

The same current generates the magnetic dipole moment $\mathbf{m}$:

$$\mathbf{m} = \frac{1}{2} \int_L \mathbf{r}(s) \times \mathbf{j}(s)\, ds \tag{11}$$

and the cross-polarizability dyadic is

$$\bar{\bar{a}}_{me} = \frac{1}{2Z} \int_L g_e(s)\mathbf{r}(s) \times ds \int_L f_e(s) ds \tag{12}$$

2.3. POLARIZATION IN EXTERNAL MAGNETIC FIELDS

The relations for the magnetic-field excitation can be written in a similar fashion, since the particle is excited by the associated electric field (5). The difference is that the electric field depends on the position z, and, even more important, that the induced current distribution can be different from that above. Denote the current distribution function in the present case as $f_h(s)$.

The polarizability dyadics take the forms

$$\bar{\bar{a}}_{em} = -\frac{\eta_0}{\omega Z} \int_L g_h(s) d\mathbf{r}(s) \int_L f_h(s) \sin(kz)\mathbf{z}_0 \times ds \tag{13}$$

$$\bar{\bar{a}}_{mm} = -\frac{j\eta_0}{2Z} \int_L g_h(s)\mathbf{r}(s) \times ds \int_L f_h(s) \sin(kz)\mathbf{z}_0 \times ds \tag{14}$$

Note that from the reciprocity principle it follows that

$$\int_L g_h(s)\mathbf{r}(s) \times ds = \int_L f_h(s) \sin(kz)\mathbf{z}_0 \times ds \tag{15}$$

$$\frac{1}{2} \int_L g_e(s) ds \int_L f_e(s)\mathbf{r}(s) \times ds = \frac{\eta_0}{\omega} \int_L g_h(s) d\mathbf{r}(s) \int_L f_h(s) \sin(kz)\mathbf{z}_0 \times ds \tag{16}$$

3. The relation between the polarizabilities

Consider the dot products

$$\bar{\bar{a}}_{ee}^T \cdot \bar{\bar{a}}_{mm} = -\frac{\eta_0}{2\omega Z^2} \int_L f_e(s) ds \int_L g_e(s) d\mathbf{r}(s)$$
$$\cdot \int_L g_h(s)\mathbf{r}(s) \times ds \int_L f_h(s) \sin(kz)\mathbf{z}_0 \times ds \tag{17}$$

and

$$\overline{\overline{a}}_{me}^T \cdot \overline{\overline{a}}_{em} = -\frac{\eta_0}{2\omega Z^2} \int_L f_e(s) ds \int_L g_e(s) \mathbf{r}(s) \times ds$$

$$\cdot \int_L g_h(s) d\mathbf{r}(s) \int_L f_h(s) \sin(kz) \mathbf{z}_0 \times ds \tag{18}$$

One can note that the only difference between the above quantities (17) and (18) is that the integrals containing the functions $g_e(s)$ and $g_h(s)$ are interchanged.

If the current distribution functions along the wire satisfy the relation

$$\int_L g_e(s) d\mathbf{r}(s) \cdot \int_L g_h(s) \mathbf{r}(s) \times ds = \int_L g_h(s) d\mathbf{r}(s) \cdot \int_L g_e(s) \mathbf{r}(s) \times ds \tag{19}$$

the polarizability dyadics are related to each other as

$$\overline{\overline{a}}_{ee}^T \cdot \overline{\overline{a}}_{mm} = \overline{\overline{a}}_{me}^T \cdot \overline{\overline{a}}_{em} = -\overline{\overline{a}}_{em}^2 \tag{20}$$

The simplest example when the relation (20) holds is the case when the current distribution functions f_e and f_h are just the same.

3.1. DISCUSSION

First, we note that the magnetoelectric coupling is indeed always zero if one of the co-polarizabilities $\overline{\overline{a}}_{ee}$ or $\overline{\overline{a}}_{mm}$ is zero. This can be seen from simple geometrical considerations: if a particle can produce a magnetic dipole under external electric fields, so there exists a non-zero cross-polarizability, it must include a portion which can be polarized by external magnetic fields, or, an element with a close path for the current. Similar considerations can be applied to the electric field excitation case, and they show that the particle must also be electrically polarizable.

It is well known that the field coupling phenomena, such that both the electric and magnetic fields point in the same direction, can occur in chiral configurations only. In all non-chiral structures the trace of the cross-polarizability dyadics must vanish. This conforms to the relation (20). For example, consider planar structures: induced electric and magnetic dipole moments will be orthogonal vectors, whence $\overline{\overline{a}}_{ee}^T \cdot \overline{\overline{a}}_{mm} = \overline{\overline{a}}_{me}^T \cdot \overline{\overline{a}}_{em} = \overline{\overline{0}}$, as should be.

4. Example

Consider an asymmetric configuration of a circular loop connected to a straight wire dipole, see Figure 1. Here the two parts are connected at a certain point x $(-l < x < l)$, and for non-zero x the relation (20) is not true, for the uniaxial approximation of the polarizability dyadics. As was stressed above, the validity of the relation (20) depends on the current distribution functions for excitations in electric an magnetic fields. In this example, these two distribution functions are indeed different: the current excited by the electric field has a maximum in the particle center, but the current, excited by the magnetic field, has its maximum at x. The analytical expression for the non-zero value of the left-hand side of (20) is given in [4]. Here, in Figures 2 and 3 we illustrate the current distribution functions for this particle.

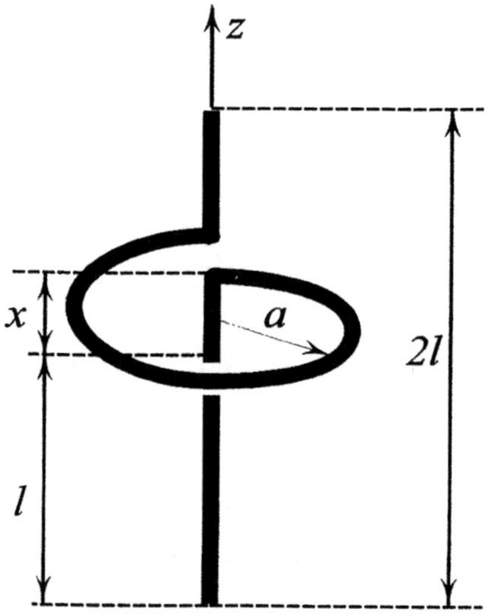

Figure 1. Geometrical configuration of an asymmetric chiral particle ($x \neq 0$)

5. Conclusion

The relation (20), which holds for small particles formed by conductive wires under certain symmetry restrictions, appears to be important in practical considerations. First, it clearly shows that the effective medium parameters of chiral mixtures are not independent. The permittivity (determined basically by the electric polarizability of inclusions $\overline{\overline{a}}_{ee}$), permeability (dependent on $\overline{\overline{a}}_{mm}$), and the effective chirality parameter (in low-density mixtures that is proportional to the average of $\overline{\overline{a}}_{em}$) are all connected, so that all three parameters have resonances at common resonance frequencies. Moreover, one can not, say, increase the chirality parameter without affecting the permittivity and permeability. Physically that is a consequence of the fact that it is the same current excited in an inclusion what determines all the polarization phenomena.

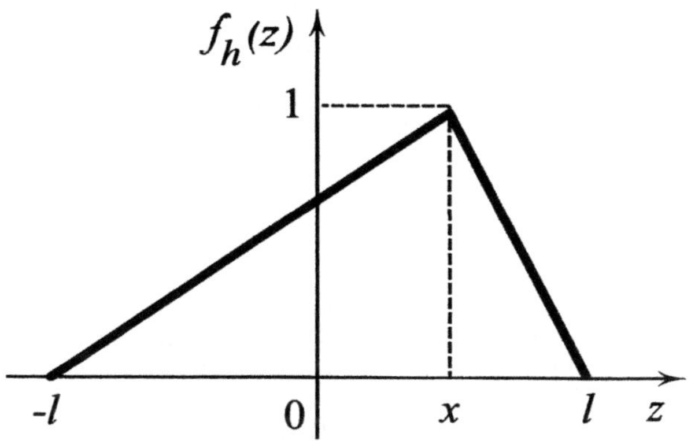

Figure 2. Current distribution along a short radiating asymmetric dipole

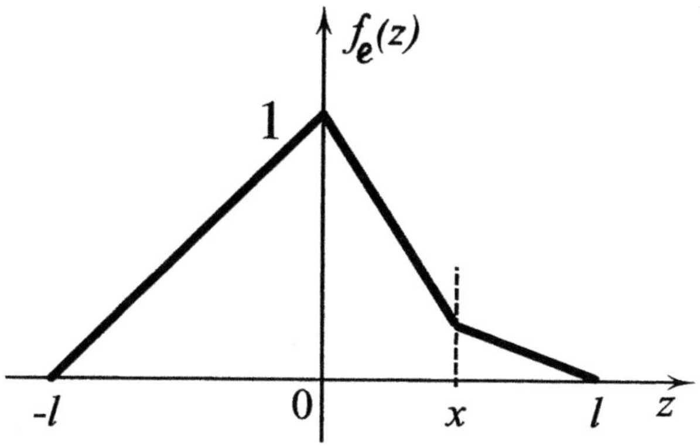

Figure 3. Current distribution in a receiving asymmetric short dipole antenna

Second, the relation (20) is important in analysing mixtures of complex bi-anisotropic inclusions, since exactly the difference between the right- and left-hand sides of (20) comes into the Maxwell Garnett averaging relations for chiral mixtures [1]. (20) leads to the conclusion that the second-order terms in the Maxwell Garnett relation vanish, so that the effective permittivity and permeability in that approximation do not depend of the position-averaged chirality parameter of a single inclusion.

On the other hand, we stress again that the relation between the polarizabilies is not universal, but it actually reflects a certain symmetry of particles. There exist such scatterers that the relation (20) does not hold, as our example shows.

References

1. Lindell, I.V., Sihvola, A.H., Tretyakov, S.A., Viitanen, A.J. (1994) *Electromagnetic Waves in Chiral and Bi-isotropic Media,* Boston and London: Artech House.
2. Landau, L.D., Lifshits, E.M. (1984) *Electrodynamics of Continuous Media,* Second edition, Oxford, Pergamon Press.
3. Schelkunoff, S.A., and Friis, H.T. (1952) *Antennas: Theory and Practice,* New York, John Wiley & Sons.
4. Tretyakov, S.A., Sochava, A.A., Simovski, C.R. (1996) Influence of chiral shapes of individual inclusions on the absorption in chiral composite coatings, *Electromagnetics,* **16**, 113-127.

MODELLING COMPOSITE MEDIA INCLUDING CHIRAL OR PSEUDO-CHIRAL SCATTERERS

L. BIDEO(†-‡), S. BOLIOLI(†), P.F. COMBES(‡)
(†) CERT/ONERA, 2 avenue E. BELIN,
31055 TOULOUSE, FRANCE
(‡) Université P. SABATIER, 118 route de NARBONNE,
31062 TOULOUSE, FRANCE

1. Introduction

Chiral media present two basic properties : the optical activity and the circular dichroism. In the microwave range, such media may be manufactured by including metallic helices in a dielectric host. Other similar effects on the reflection coefficient can be obtained by using omegas (pseudo-chiral) instead of helices. The problem consists in the optimisation of the helices' size (radius and step) and their distribution (density and orientation versus the incident wave vector), according to the application.
The models based on diffusion (RAYLEIGH-BOHREN) and the generalised MAXWELL-GARNETT [1],[2] approach are the most commonly used to describe the electromagnetic properties of such chiral or pseudo-chiral composites. Generally, they give better results for scatterers with dimensions small compared to the wavelength, and for sparse concentration. In such formalisms, it is not easy to describe accurately the interactions between elements. The fully numerical technique based on the method of moments has been developed and applied to these cases. To develop this model, new scalar GREEN's functions have been written, taking into account the dielectric host and the air-composite interfaces.
Whatever the global modelling, one needs first to understand the behaviour of the single particle. Concerning the helix it has been showed [3],[4],[5] that a one turn element, with a pitch equal to the diameter, provides an almost isotropic distribution of the scattered energy, as well as an important chiral performance. This efficiency has been related to the excitation and resonance conditions.
It has been shown that random distributions are not optimised for crosspolarization generation. Hence, our interest concerns only regular arrays. It is possible to describe the properties of large arrays using the results computed for smaller ones and asymptotic relations. In that conditions, it has been possible to optimise the helices distribution to get higher crosspolarization.

A. Priou et al. (eds.), Advances in Complex Electromagnetic Materials, 281–291.

2. Regular arrays of helices

2.1. INTRODUCTION

We have considered metallic elements imbedded in an homogeneous medium (with permittivity and permeability ε_d and μ_d, respectively.)

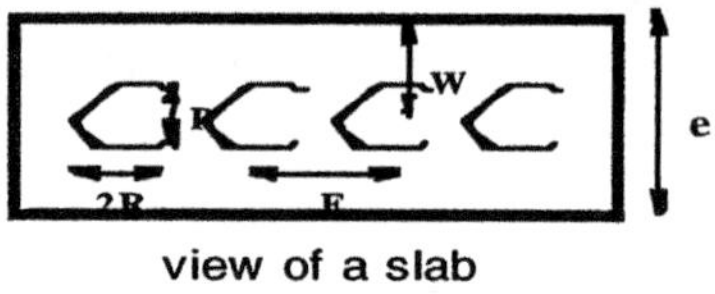

Figure 1 : Chiral slab

In the figure, « e » is the thickness of the slab, « W » the array depth inside the dielectric, « E » the distance between helices, « P » the pitch of the helix and « R » the radius of the helix.

2.2. THEORETICAL BASIS : [6],[7],[8],[9],[10],[11]

The incident electric field $\vec{E}_{inc}$ induces electric currents $\vec{J}$ on the elements. These currents will then radiate a secondary field called the scattered field $\vec{E}_{scat}$. The total field is the sum of these two components :

$$\vec{E}_{tot}(\vec{r}) = \vec{E}_{inc}(\vec{r}) + \vec{E}_{scat}(\vec{r}) \tag{1}$$

where $\vec{r}$ is the space vector from the origin to the observation point. The incident electric field is assumed to be known while the scattered field is unknown. In the frequency domain, this field can be written in terms of magnetic vector potential $\vec{A}$, due to the currents $\vec{J}$ on the structures, and of the electric scalar potential, due to the charges ρ :

$$\vec{E}_{scat}(\vec{r}) = -j\omega\vec{A}(\vec{J},\vec{r}) - \vec{\nabla}\phi(\rho,\vec{r}) \tag{2}$$

Enforcing the boundary condition that the tangential component of the electric field must vanish on the surface of the conductor, equation (1) becomes :

$$\vec{E}^{tan}_{inc}(\vec{r}) = -\vec{E}^{tan}_{scat}(\vec{r}) \tag{3}$$

The Electrical Field Integral Equation (EFIE) for the unknown current $\vec{J}$ is of POCKLINGTON's type :

$$\vec{E}^{tan}_{inc}\left(\vec{r}\right) = \left[j\omega\mu \int_v \vec{J}\left(\vec{r}'\right) G_0\left(\vec{r},\vec{r}'\right) dv - \frac{1}{j\omega\varepsilon} \vec{\nabla} \int_v \vec{\nabla} \cdot \vec{J}\left(\vec{r}'\right) G_0\left(\vec{r},\vec{r}'\right) dv \right]^{tan}_{inc} \qquad (4)$$

where $G_0\left(\vec{r},\vec{r}'\right)$ is the scalar GREEN's function in free space, $\vec{r}$ is the space vector from origin and observation point and $\vec{r}'$ the space vector from origin to source point.

The method of moments (MOM) is employed to convert the EFIE into a matrix equation, which can be algebraically solved to determine the unknown current on the elements. However, the GREEN's function in free space can be replaced by the function specific of the medium. To calculate this function, the method of images is used :

$$G\left(\vec{r},\vec{r}'\right) = G_0\left(\vec{r},\vec{r}'\right) + F\left(\vec{r},\vec{r}'\right) \qquad (5)$$

The analytical expression of F depends on the medium containing the observation point:
Inside the dielectric slab :

$$\vec{F}(\vec{r},\vec{r}') = K_2.\frac{e^{-jkR_3}}{R_3} + \sum_{i=1}^{\infty} K_1^i.K_2^{i-1}\frac{e^{-jkR_1}}{R_1} + \sum_{i=1}^{\infty} K_1^i.K_2^i\left(\frac{e^{-jkR_2}}{R_2} + \frac{e^{-jkR_4}}{R_4}\right) + \sum_{i=1}^{\infty} K_1^i.K_2^{i+1}\frac{e^{-jkR_5}}{R_5} \qquad (6)$$

Outside the dielectric slab :

$$\vec{F}(\vec{r},\vec{r}') = -K_2.\frac{e^{-jkR_0}}{R_0} + \sum_{i=1}^{\infty} K_1^i.K_2^{i-1}.(1-K_2).\frac{e^{-jkR_1}}{R_1} + \sum_{i=1}^{\infty} K_1^i.K_2^i.(1-K_2).\frac{e^{-jkR_4}}{R_4} \qquad (7)$$

Where

$$K_1 = \frac{\sqrt{\varepsilon_d\mu_1} - \sqrt{\varepsilon_1\mu_d}}{\sqrt{\varepsilon_d\mu_1} + \sqrt{\varepsilon_1\mu_d}} \qquad (8)$$

$$K_2 = \frac{\sqrt{\varepsilon_d\mu_2} - \sqrt{\varepsilon_2\mu_d}}{\sqrt{\varepsilon_d\mu_2} + \sqrt{\varepsilon_2\mu_d}}$$

and :
R_0 : Amplitude of the space vector from the true source located at z=a inside the dielectric to the observation point.
R_1 : Amplitude of the space vector from the image sources located at z=-a+2ie to the observation point.
R_2 : Amplitude of the space vector from the image sources located at z=a-2ie to the observation point.
R_3 : Amplitude of the space vector from the image source located at z=-a to the

observation point.
R_4 : Amplitude of the space vector from the image sources located at z=a+2ie to the observation point.
R_5 : Amplitude of the space vector from the image sources located at z=-a-2ie to the observation point.
« i » being the image source index (see Figure 2)

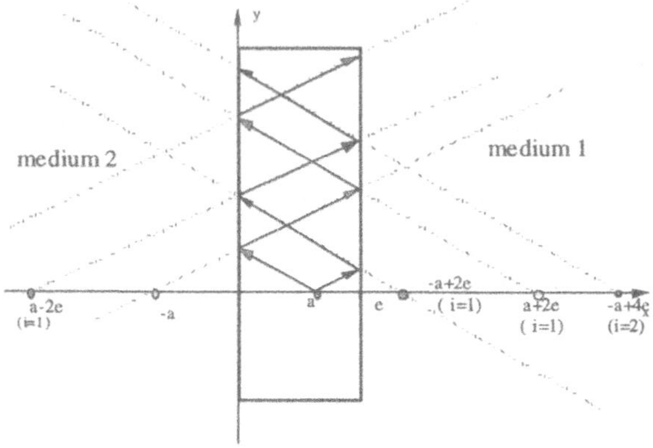

Figure 2 : Method of images : true source and image sources.

The dyadic GREEN's function would give a more rigorous, but much heavier, representation of the fields distribution, describing the transverse surface waves. We consider here only the diffracted waves, and as it will de shown, the results obtained with the scalar one are quite satisfactory.

3. Optical activity criterion

To choose the appropriate helix in term of the crosspolarization component of the electric scattered field, we introduce a criterion named the Optical Activity Criterion, derived from the one defined previously by Ougier [4] :

$$\text{O.A.C.} = \frac{\sigma_{cross}}{\lambda^2}\sin\left(\arctan\left(\sqrt{\frac{\sigma_{cross}}{\sigma_{co}}}\right)\right) \tag{9}$$

where σ_{co} and σ_{cross} are proportional to the square of the components in crosspolarization or copolarization of the electric field, for the backscattered wave or for the forwardscattered wave. As far as a single layer of helices is considered, the rotation angle is smaller than 90°, and there is no ambiguity on the argument of the Arctg function. On the figure 4, the AOC is plotted as a function of the ratio Pitch/Radius, for the six basic excitations presented in figure 3. The symbol « E » indicates the direction of the electric field and « k » the direction of the wave vector

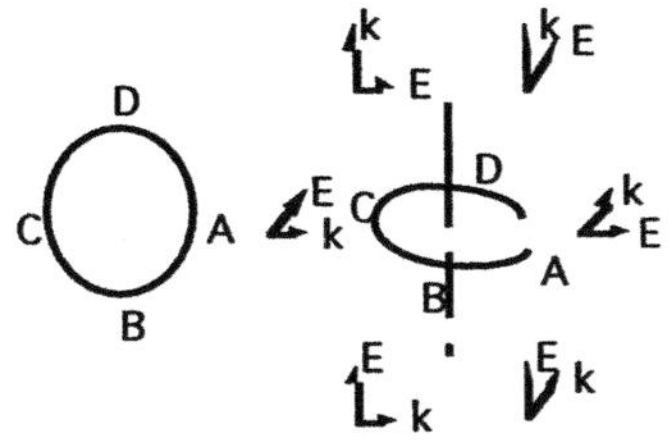

Figure 3 : The six different basic excitations.

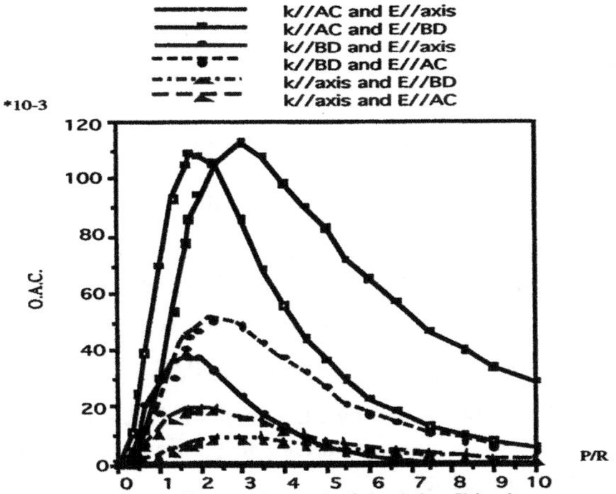

Figure 4 : Optical Activity Criterion

The maximum of the AOC appears when the pitch is approximately twice the radius. The worst value occurs when the direction of the wave vector is parallel to the axis. However, an important crosspolarization is produced when $\vec{k}$ is perpendicular to the axis.

4. Influence of the distance E

Here we consider an array of 8*8 helices. Its characteristics (λ, R, P, E) are given on figure 5.

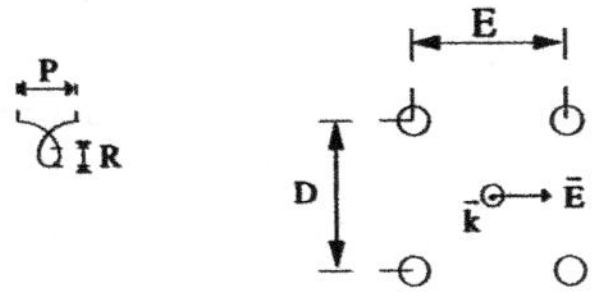

$$\begin{cases} \lambda = 40\text{mm} \\ R = \lambda/13 \\ P = 2\lambda/13 \\ E = 3\lambda/4 \end{cases}$$

single helix array (top view)

Figure 5 : Distribution of the array

The normalised Radar Cross Section (RCS) presented on figure 6 is computed from the backscattered field, as a function of the ratio E/λ

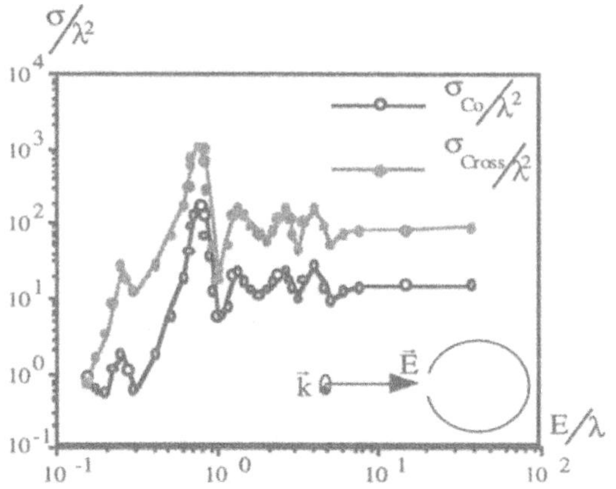

Figure 6 : Normalised R.C.S

According to this plot, we can point out the following results :

- The asymptotic value of the RCS is equal to its value for a single helix multiplied by N^2, N being the total number of scatterers.
- The interactions become insignificant when E is above 7λ.
- The higher value is given for $E/\lambda \cong 0{,}75$
- A small resonance can be observed for $E/\lambda \cong 0{,}25$

The plots corresponding to the forwardscattered field are similar, except for this last resonance which does not occur.

One must notice that when E/λ increases, the physical size of the array increases also. This plot is useful to understand how an array of a given number of helices works. In the real case, the size of the element (coating, target,...) is generally fixed, and it is more convenient to optimise the density.

Therefor we introduce the Distance Optimisation Criterion (D.O.C.) :

$$\text{D.O.C.} = \text{O.A.C.}\left(\frac{\lambda}{E}\right)^4 \qquad (10)$$

It represents the A.O.C. normalised by the square of the helices density (expressed in number per λ^2). One should remind that the A.O.C. is proportional to R.C.S. in crosspolarization, which is itself proportional to the square of the helices number. In order to compare with the previous case, the D.O.C.'s fluctuations versus E/λ are plotted on the figure 7. If the direction of the wave vector is collinear to the axis, the best efficiency corresponds to E equals to 3λ/4. A much better one can be reached with $E \cong 0{,}225\lambda$ if $\vec{k}$ is parallel to the segment AC.

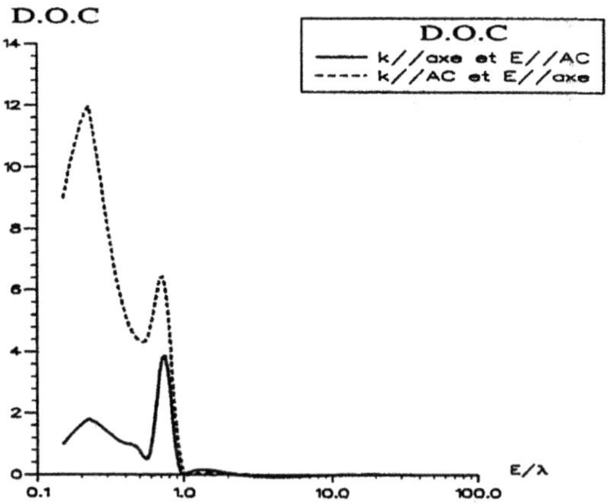

Figure 7 : Criterion of Optimisation of the Distance

5. Experimental validations

The transmission coefficient of an array of 2∗2 helices in free space (see figure 8 the draft of one helix with a queue) was measured to get an experimental validation of the theoretical results. The equipment used is a free space set-up made of 2 confocal ellipsoids which produce a local plane wave in the confocal zone. A sample composed of a dielectric (rigid polyurethane foam, P10, from MARPLEX) supporting the array is measured in transmission in copolarisation as well as in crosspolarisation. The theoretical coefficient is calculated using the formula 11 in order to take into account the dielectric slab :

$$S_{21} = \frac{\left|\vec{E}_{P10} + \vec{E}_{scat}\right|}{\left|\vec{E}_{inc}\right|} \tag{11}$$

$\vec{E}_{scat}$ derives from simulations, $\vec{E}_{P10}$ derives from the measurement of the only dielectric plate.

A close comparison between theoretical and experimental results is demonstrated. For the crosspolarisation wave (figure 8 b), the expected peak at 7,1GHz (the total wire length of an element equals 21,13mm ≅ λ/2) is noticed.

6. Influence of depth of the array inside the dielectric slab on the cross polarisation component of the backscattered field

On figure 9, we have plotted the fluctuations of the reflected power as a function of the array depth inside the dielectric. In order to avoid dependence on the host medium, the depth is considered as the electrical depth « n'.W » where n' is the real part of the medium index, and the attenuation in the medium is compensated by multiplying the reflected power by $e^{2(k_0.n''.W)}$ which corresponds to the inverse of the losses for the forward and backward paths.

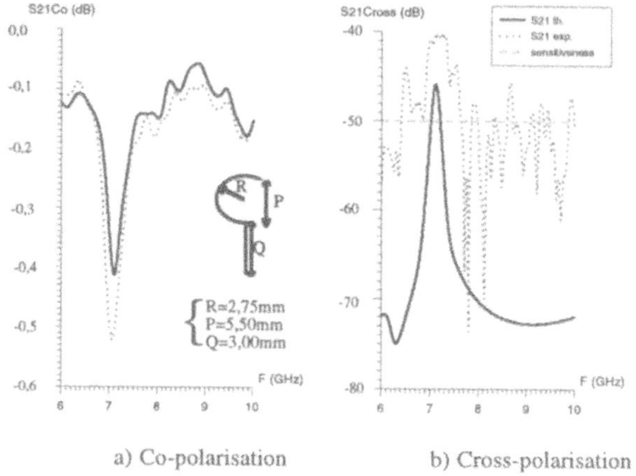

Figure 8 : The array's attenuation in transmission.

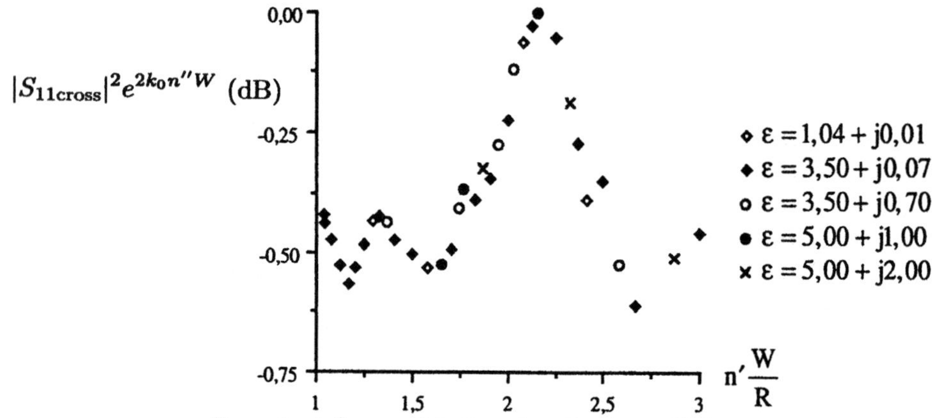

Figure 9 : Influence of depth on the reflection coefficient

The optimum electrical depth « n'.W » is slightly higher than twice the helix radius. It means that for high permittivity media (case of some absorptive coatings), the helices have to be placed close to the interface.

7. Example of experimental validation

The medium is composed by a polyurethane foam with a permittivity ε equal to 1,04+j0,0012. The thickness is fixed to 10,4mm (thickness resonance equal to helix resonance). The radius and pitch of the helices are respectively R=1,6mm and P=1,2mm. They are made with industrial springs and so the pitch is not exactly equal to the diameter. The regular array is composed by 8x8 helices. The mesh is ExE = 4,6x4,6mm^2. The depth inside the dielectric is W = 3,5mm (Figure 1).
On figures 11 and 13, we present the comparison between theoretical and experimental reflection coefficients.
The electric field is collinear to the axis and the direction of the wave vector is parallel to the segment AC (figure 10).

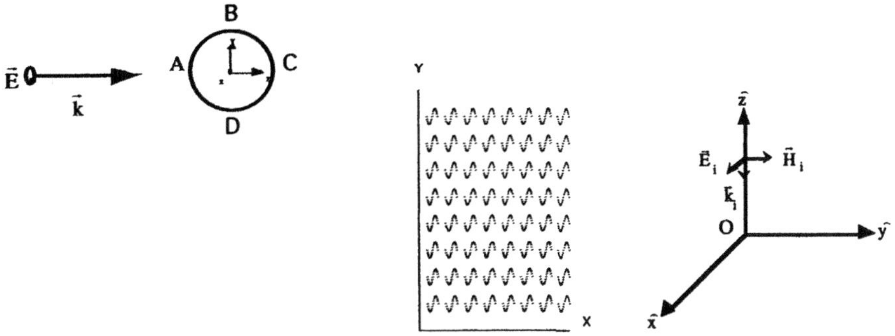

Figure 10 : **Incident wave**

The reflection coefficient in copolarization (figure 11 a) is smaller than the reflection coefficient in crosspolarization (figure 11 b) at the resonant frequency. The copolarization component of the backscattered field due to the array is combined with the same due to the air/dielectric interface with opposite phase.

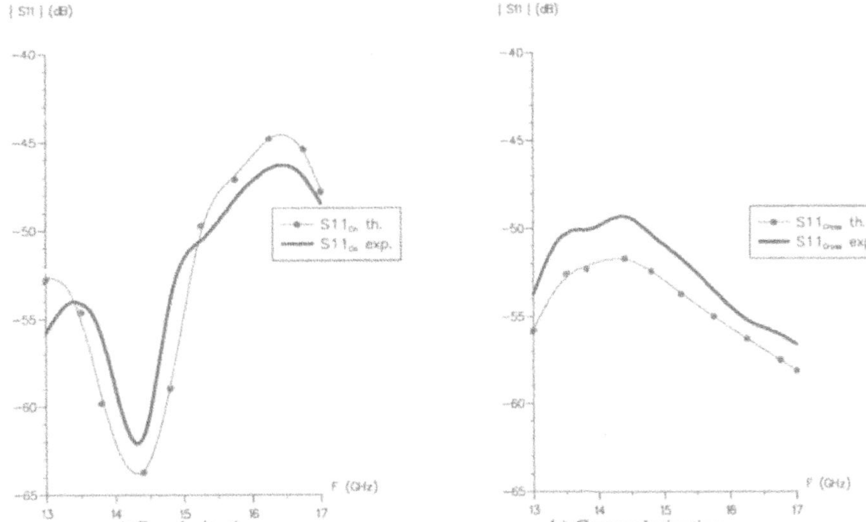

a) Copolarization b) Crosspolarization

Figure 11 : **Reflection coefficient of the composite slab (k//AC and E//axis).**

Now, the incident electric field rotates 90° (figure 12).

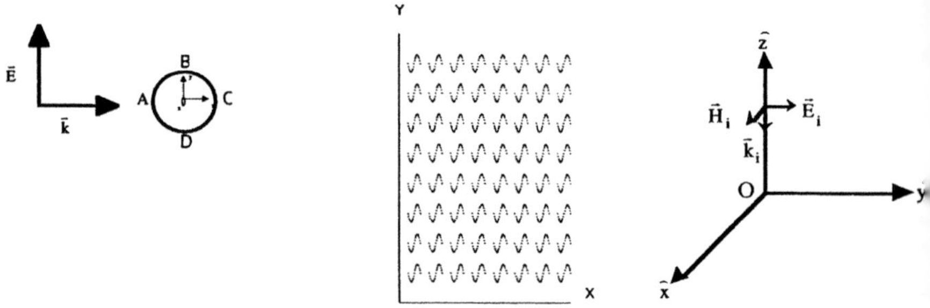

Figure 12 . Incident wave

The composite gives almost the same reflection coefficient in crosspolarisation (figure 13 b). The reflection coefficient in copolarization increases (figure 13 a). In both cases, we show a good agreement between theory and experiments.

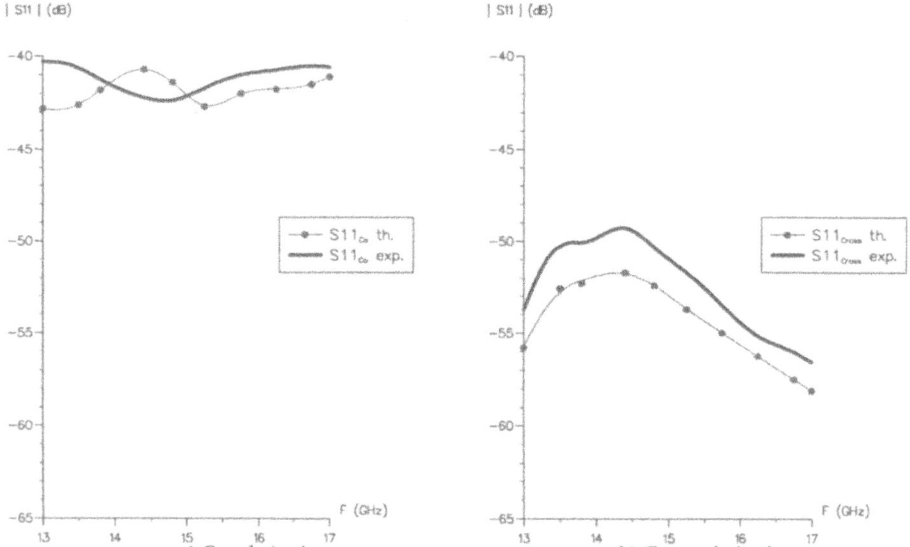

a) Copolarization b) Crosspolarization

Figure 13 . Reflection coefficient of the composite slab (k//AC and E//BD)

8. Conclusion

The MOM, using the scalar GREEN's function of the parallel plate, gives good results to describe the electromagnetic fields scattered by an array of helices imbedded in a dielectric. The measurements performed on different structures showed a good agreement between theoretical and experimental results. As a next step, different limit conditions , like a metal backed composite slab or a Salisbury's screen using a lossy dielectric as a spacer, could be taken into account.

REFERENCES

1. F. GUERIN (1995) Contribution à l'étude théorique et expérimentale des matériaux composites chiraux et bianisotropes dans le domaine mico-onde, *thèse de l'Université de LIMOGES.*

2. S.A. TRETYAKOV, F. MARIOTTE, K.R. SIMOVSKI, T.G. KHARINA, S. BOLIOLI (1994) Antenna model for individual chiral and omega scatterers, *Chiral'94, p 41-48, Périgueux, FRANCE, May 18-20*

3. S. OUGIER, I. CHENERIE, A. PRIOU (1993) Measurement of chiral media method and practical examples, *PIERS, Pasadena Californie, USA, July 1993.*

4. S. OUGIER, I. CHENERIE, S. BOLIOLI (1993) How to tailor and orient metallic helices to get microwave chirality , *Proceeding of "Bianisotropics'93", p 54-59, Gomel, BELARUS, October 12-14 1993.*

5. T.ROBIN, B. SOUILLARD, R. ACCOLAS, P. THERET, S. BOLIOLI (1994) Microwave properties of helix-loaded layers : comparison between theory and experiments, *Chiral'94, p 229, Périgueux, FRANCE, May 18-20*

6. R. F. HARRINGTON (1968) Field Computation by Moment Methods, *Macmillan Compagny, New York*

7. J. C. VEIHL (1993) Antenna analysis in complex environments, *Thesis for MS degree, University of Illinoi*

8. E. K. N. YUNG, R. S. K. WONG (1995) Analysis of a wire antenna of arbitrary shape, *Journal of Electromagnetic Waves and applications, Vol. 9, n° 7/8, p 855-869.*

9. K. W. WHITES (1995) Full-wave computation of constitutive parameters for lossless composite chiral materials, *IEEE transaction on antennas and propagation, Vol. 43, n° 4, April*

10. M. E. PETERS (1995) Method of Moments analysis of anisotropic artificial media composed of dielectric wire objects, *IEEE transactions on microwave theory and techniques, Vol. 43, n°9, September*

11. P. K. KOIVISTO, J. C. E. STEN (1995) Quasi-static image method applied to bi-isotropic microstrip geometry, *IEEE transactions on microwave theory and techniques, Vol. 43, n°1, January*

MUTUAL COUPLING IN ARRAYS OF PLANAR CHIRAL STRUCTURES

L.R. ARNAUT
UMIST – Department of Electrical Engineering and Electronics
P.O. Box 88, Manchester M60 1QD, United Kingdom

Abstract. Interaction is analysed in arrays consisting of two special classes of planar chiral particles. The interaction six-dyadic for such particles shows formal equivalence with the constitutive dyadics of the individual particles and with the constitutive dyadics of the effective medium. Modified dipolarisability dyadics are obtained from a hybrid analytical–numerical approach which show the effect of particle interaction on the constitutive dyadics of the medium.

1. Plano - chiral and Plano - semi-chiral Media

Plano - chiral and plano - semi-chiral media are classes of bianisotropic media whose microstructure shows a reduced degree of chirality (*ie* an increased degree of symmetry) as compared to conventional (3-dimensional) chiral structures. Both types emerge when the concept of chirality is reconsidered in multi-dimensional space [1, 4]. Plano - chiral structures are defined as being planar, axi-dissymmetric and centro-asymmetric. In contrast, plano - semi-chiral structures are planar and axi-dissymmetric but centro-symmetric.

From the invariance and non-invariance of these media under affine spatial reflection transformations [10], it was shown that the constitutive dyadics of the effective media constructed from collections of such microstructures take the following form [4]. In a Cartesian reference frame (u, v, w), where the unit vector $\underline{1}_w$ is along the normal to the plane of the structures, plano - chiral media satisfy:-

$$\underline{\underline{\epsilon}} = \begin{bmatrix} \epsilon_{uu} & \epsilon_{uv} & 0 \\ \epsilon_{vu} & \epsilon_{vv} & 0 \\ 0 & 0 & \epsilon_{ww} \end{bmatrix}, \quad \underline{\underline{\xi}} = \begin{bmatrix} 0 & 0 & \xi_{uw} \\ 0 & 0 & \xi_{vw} \\ \xi_{wu} & \xi_{wv} & 0 \end{bmatrix} \tag{1}$$

A. Priou et al. (eds.), Advances in Complex Electromagnetic Materials, 293–309.

$$\underline{\underline{\zeta}} = \begin{bmatrix} 0 & 0 & \zeta_{uw} \\ 0 & 0 & \zeta_{vw} \\ \zeta_{wu} & \zeta_{wv} & 0 \end{bmatrix}, \quad \underline{\underline{\mu}} = \begin{bmatrix} \mu_{uu} & \mu_{uv} & 0 \\ \mu_{vu} & \mu_{vv} & 0 \\ 0 & 0 & \mu_{ww} \end{bmatrix} \tag{2}$$

with $\underline{\underline{\xi}}, \underline{\underline{\zeta}} \neq \underline{\underline{0}}$ and $\underline{D} \stackrel{\Delta}{=} \underline{\underline{\epsilon}}\cdot\underline{E}+\underline{\underline{\xi}}\cdot\underline{H},\ \underline{B} \stackrel{\Delta}{=} \underline{\underline{\zeta}}\cdot\underline{E}+\underline{\underline{\mu}}\cdot\underline{H}$. Examples of plano - chiral structures include single-arm spirals, irregular multi-arm spirals, irregular multi-leg swastikas, ... (Fig 1a). The term irregular here refers to unequal size of adjacent arms or legs of a single structure, or unequal angles between them.

Plano - semi-chiral media are defined as exhibiting an additional point centre of symmetry. The latter is a sufficient condition for the medium to be simply anisotropic [4]. Consequently, plano - semi-chiral media are characterised by fewer constitutive parameters:-

$$\underline{\underline{\epsilon}} = \begin{bmatrix} \epsilon_{uu} & \epsilon_{uv} & 0 \\ \epsilon_{vu} & \epsilon_{vv} & 0 \\ 0 & 0 & \epsilon_{ww} \end{bmatrix}, \quad \underline{\underline{\xi}} = \underline{\underline{\zeta}} = \underline{\underline{0}}, \quad \underline{\underline{\mu}} = \begin{bmatrix} \mu_{uu} & \mu_{uv} & 0 \\ \mu_{vu} & \mu_{vv} & 0 \\ 0 & 0 & \mu_{ww} \end{bmatrix} \tag{3}$$

Examples include regular multi-arm spirals, regular multi-leg swastikas, ... (Fig 1b). For both plano - chiral and plano - semi-chiral structures discussed in this paper, $\underline{\underline{\mu}}_t = \mu_o(\underline{1}_x\underline{1}_x + \underline{1}_y\underline{1}_y)$; for the plano - chiral structures, $\xi_{wu} = \xi_{wv} = \zeta_{uw} = \zeta_{vw} = 0$.

Plano - chiral and plano - semi-chiral media may find application as polarisation-agile frequency-selective surfaces (FSSs), ultra-thin twist polarisers, radomes, reflectors, ... We note in passing that the periodic spacing of patches in classical FSSs causes the appearance of grating-lobes at higher frequencies [8]. These unwanted lobes limit the upper operational frequency. A FSS or any other synthetic screen that consists of randomly positioned patches does not exhibit such lobes, and their properties can be analysed using multipole and bianisotropic effective medium models.

2. Interaction Six-dyadic

As shown in [5], the interaction W_{12} of the electric and magnetic dipole moments $\underline{p}_e$ and $\underline{p}_m$ for a particle 2 onto those for a particle 1, induced by the respective source fields $(\underline{E}_2, \underline{H}_2)$ and $(\underline{E}_1, \underline{H}_1)$, can be expressed in terms of an interaction six-dyadic $\left[\underline{\underline{M}}\right]$ that contains the dipolarisability dyadics $\underline{\underline{p}}_{kl}$ $(k, l = e, m)$ of both particles, as defined by:-

$$\begin{bmatrix} \underline{p}_e \\ \underline{p}_m \end{bmatrix} = \begin{bmatrix} \epsilon_o \underline{\underline{p}}_{ee} & \sqrt{\mu_o\epsilon_o}\,\underline{\underline{p}}_{em} \\ \sqrt{\mu_o\epsilon_o}\,\underline{\underline{p}}_{me} & \mu_o \underline{\underline{p}}_{mm} \end{bmatrix} \cdot \begin{bmatrix} \underline{E} \\ \underline{H} \end{bmatrix} \tag{4}$$

and their separation distance vector $\underline{d} = d\underline{1}_d$, measured from 1 to 2:-

$$W_{12} = -\frac{\exp(-j\underline{k}\cdot\underline{d})}{4\pi d^3}\begin{bmatrix} \underline{E}_1^T & \underline{H}_1^T \end{bmatrix}\cdot\left[\underline{\underline{M}}\right]\cdot\begin{bmatrix} \underline{E}_2 \\ \underline{H}_2 \end{bmatrix} \tag{5}$$

with

$$\left[\underline{\underline{M}}\right] \triangleq \begin{bmatrix} \underline{\underline{M}}_{11} & \underline{\underline{M}}_{12} \\ \underline{\underline{M}}_{21} & \underline{\underline{M}}_{22} \end{bmatrix} \tag{6}$$

and $\left[\underline{\underline{M}}\right] = \left[\underline{\underline{M}}^{near}\right] + (jkd)\left[\underline{\underline{M}}^{boundary}\right] + (jkd)^2\left[\underline{\underline{M}}^{far}\right]$. For arbitrary and independent directions and magnitudes of the source fields, the six-dyadic $\left[\underline{\underline{M}}\right]$ can be used to compute the interaction energy. Here we focus on the near-field interaction ($kd \ll 1$), for which:-

$$\underline{\underline{M}}_{11}^{near} = \epsilon_o\left[\underline{\underline{p}}_{ee_1}^T\cdot(3\underline{1}_d\underline{1}_d - \underline{\underline{I}})\cdot\underline{\underline{p}}_{ee2} + \underline{\underline{p}}_{me_1}^T\cdot(3\underline{1}_d\underline{1}_d - \underline{\underline{I}})\cdot\underline{\underline{p}}_{me2}\right] \tag{7}$$

$$\underline{\underline{M}}_{12}^{near} = \sqrt{\mu_o\epsilon_o}\left[\underline{\underline{p}}_{ee_1}^T\cdot(3\underline{1}_d\underline{1}_d - \underline{\underline{I}})\cdot\underline{\underline{p}}_{em2} + \underline{\underline{p}}_{me_1}^T\cdot(3\underline{1}_d\underline{1}_d - \underline{\underline{I}})\cdot\underline{\underline{p}}_{mm2}\right] \tag{8}$$

$$\underline{\underline{M}}_{21}^{near} = \sqrt{\mu_o\epsilon_o}\left[\underline{\underline{p}}_{em_1}^T\cdot(3\underline{1}_d\underline{1}_d - \underline{\underline{I}})\cdot\underline{\underline{p}}_{ee2} + \underline{\underline{p}}_{mm_1}^T\cdot(3\underline{1}_d\underline{1}_d - \underline{\underline{I}})\cdot\underline{\underline{p}}_{me2}\right] \tag{9}$$

$$\underline{\underline{M}}_{22}^{near} = \mu_o\left[\underline{\underline{p}}_{em_1}^T\cdot(3\underline{1}_d\underline{1}_d - \underline{\underline{I}})\cdot\underline{\underline{p}}_{em2} + \underline{\underline{p}}_{mm_1}^T\cdot(3\underline{1}_d\underline{1}_d - \underline{\underline{I}})\cdot\underline{\underline{p}}_{mm2}\right] \tag{10}$$

where the superscript T denotes the transpose of the vector or dyadic. The more elaborate dyadics for far-zone ($kd \gg 1$) and boundary-zone ($kd \sim 1$) interaction are listed in [5].

If $\underline{d}$ is parallel to one of the co-ordinate axes, then $(3\underline{1}_d\underline{1}_d - \underline{\underline{I}})$ is diagonal with no vanishing diagonal elements. Then the dyadic products in Eqns (7)–(10) take the following form:-

$$\underline{\underline{p}}_{ee_1}^T\cdot\left(3\underline{1}_d\underline{1}_d - \underline{\underline{I}}\right)\cdot\underline{\underline{p}}_{ee2} \sim \underline{\underline{p}}_{me_1}^T\cdot\left(3\underline{1}_d\underline{1}_d - \underline{\underline{I}}\right)\cdot\underline{\underline{p}}_{me2} \sim \begin{bmatrix} \bullet & \bullet & 0 \\ \bullet & \bullet & 0 \\ 0 & 0 & 0 \end{bmatrix} \tag{11}$$

$$\underline{\underline{p}}_{ee_1}^T\cdot\left(3\underline{1}_d\underline{1}_d - \underline{\underline{I}}\right)\cdot\underline{\underline{p}}_{em2} \sim \underline{\underline{p}}_{me_1}^T\cdot\left(3\underline{1}_d\underline{1}_d - \underline{\underline{I}}\right)\cdot\underline{\underline{p}}_{mm2} \sim \begin{bmatrix} 0 & 0 & \bullet \\ 0 & 0 & \bullet \\ 0 & 0 & 0 \end{bmatrix} \tag{12}$$

$$\underline{\underline{p}}_{em_1}^T\cdot\left(3\underline{1}_d\underline{1}_d - \underline{\underline{I}}\right)\cdot\underline{\underline{p}}_{ee2} \sim \underline{\underline{p}}_{mm_1}^T\cdot\left(3\underline{1}_d\underline{1}_d - \underline{\underline{I}}\right)\cdot\underline{\underline{p}}_{me2} \sim \begin{bmatrix} 0 & 0 & 0 \\ 0 & 0 & 0 \\ \bullet & \bullet & 0 \end{bmatrix} \tag{13}$$

$$\underline{\underline{p}}_{em_1}^T\cdot\left(3\underline{1}_d\underline{1}_d - \underline{\underline{I}}\right)\cdot\underline{\underline{p}}_{em2} \sim \underline{\underline{p}}_{mm_1}^T\cdot\left(3\underline{1}_d\underline{1}_d - \underline{\underline{I}}\right)\cdot\underline{\underline{p}}_{mm2} \sim \begin{bmatrix} 0 & 0 & 0 \\ 0 & 0 & 0 \\ 0 & 0 & \bullet \end{bmatrix} \tag{14}$$

where the tilde signifies formal correspondence. Substituting these forms into Eqn (6), it follows that $\left[\underline{\underline{M}}^{near}\right]$ has the same form as the effective medium six-dyadic Eqns (1)–(2) with $\xi_{wu} = \xi_{wv} = \zeta_{uw} = \zeta_{vw} = 0$, or Eqn (3). The latter, in turn, has the same appearance as the dipolarisability six-dyadic, $\left[\underline{\underline{p}}\right]$. Thus, the field components that contribute to the near-field interaction can be immediately identified from the form of the constitutive six-dyadic itself. For plano - semi-chiral structures, the expressions for $\underline{\underline{M}}_{11}$ and $\underline{\underline{M}}_{22}$ contain a single term only and $\underline{\underline{M}}_{12} = \underline{\underline{M}}_{21} = \underline{\underline{0}}$. Again it is found that the constitutive six-dyadic has the same form as the interaction six-dyadic.

Figs 2 and 3 show $\left[\underline{\underline{p}}\right]$ and the total $\left[\underline{\underline{M}}\right]$ for two co-planar one-turn single-arm spirals defined by $r = r_o(\alpha^\theta - 1)$ with $r_o = \alpha = 10\,\text{mm}$, $0 \leq \theta \leq 2\pi$ with aligned anisotropies, and made of cylindrical copper wire of radius 0.1 mm. The dominant elements of the $\underline{\underline{M}}_{ij}$ $(i, j = 1, 2)$ are, although small, within the numerical accuracy.

3. Mixing

Mixing, which may be described by a variety of mixing rules (see *eg* [12]), gives rise to less interdependence between effective medium parameters in the case of plano - chiral and plano - semi-chiral media as compared to 3D - chiral media. For example, the Maxwell–Garnett formulas for bianisotropic composites contain dot products of the form $\underline{\underline{p}}_{ij} \cdot \underline{\underline{p}}_{kl}$ $(i, j, k, l = e, m)$ to second and higher order in the particle concentration [11]. In conventional (3D -) helix-based chiral media, neither of these dyadic products vanishes in general because all $\underline{\underline{p}}_{kl}$ contain a dominant term of the form $p\underline{1}_z\underline{1}_z$ [2, 3]. For plano - chiral media, $\underline{\underline{p}}_{ee} \cdot \underline{\underline{p}}_{mm} = \underline{\underline{p}}_{ee} \cdot \underline{\underline{p}}_{em} = \underline{\underline{p}}_{me} \cdot \underline{\underline{p}}_{mm} = \underline{\underline{0}}$, hence microscopic electric, magnetic and magneto-electric quantities do no longer couple in mixing formulas. For plano - semi-chiral media, $\underline{\underline{p}}_{em} \cdot \underline{\underline{p}}_{me} \equiv \underline{\underline{0}}$ in addition. Thus, the effective medium parameters $\underline{\underline{\epsilon}}$, $\underline{\underline{\xi}}$, $\underline{\underline{\zeta}}$, $\underline{\underline{\mu}}$ show a lower degree of dependence on microscopic parameters of different nature in plano - chiral and plano - semi-chiral media as compared to conventional chirals. This smaller interdependence is clearly advantageous in the tailoring and design of high-density chiral composites with specified medium parameters.

4. Modified Dipolarisability Dyadics for Arrays of Interacting Particles

The perturbations of the internal field due to particle interaction can be taken into account by incorporating these changes into the dipolarisabilities of the particles, either numerically [3] or analytically [5]. This is possible because the field perturbations themselves depend on the constitutive parameters of neighbouring particles. Therefore, the depolarisation field can be computed explicitly in terms of the particle and configurational properties. This enables particle interaction to be incorporated explicitly into mixing theories.

An approximation for the total interaction can be made by considering all pair-wise interaction in the medium, which accounts for the total first-order coupling. For a pair of particles i and $\bar{i}$ [5]:-

$$\left[\Delta \underline{\underline{p}}^{(\infty)}_{i,\bar{i}}\right] = \left[\underline{\underline{p}}_i\right] \cdot \left[\underline{\underline{I}} - \left[\underline{\underline{p}}_i\right]^{-1} \cdot \left[\Delta \underline{\underline{p}}^{(1)}_i\right] \cdot \left[\underline{\underline{p}}_{\bar{i}}\right]^{-1} \cdot \left[\Delta \underline{\underline{p}}^{(1)}_{\bar{i}}\right]\right]^{-1} \cdot \left[\underline{\underline{p}}_i\right]^{-1} \cdot \left[\Delta \underline{\underline{p}}^{(1)}_i\right] \quad (15)$$

The explicit dyadic expressions of $\left[\Delta \underline{\underline{p}}^{(1)}_i\right]$ in terms of the $\underline{\underline{p}}_{kl}$ and arbitrary $\underline{d}$ are listed in [5]. The perturbation six-dyadic $\left[\Delta \underline{\underline{p}}^{(1)}_i\right]$ represents the first-order coupling between two particles. For the central particle at $(0,0,0)$ in an infinite 3D array, this amounts to:-

$$\left[\Delta \underline{\underline{p}}^{(\infty)}_{000}\right] = \sum_{i,j,l=-\infty}^{+\infty} \left[\Delta \underline{\underline{p}}^{(\infty)}_{000,ijl}\right], \qquad (i,j,l) \neq (0,0,0) \qquad (16)$$

where $\left[\Delta \underline{\underline{p}}^{(\infty)}_{000,ijl}\right]$ represents the total first–order pair-wise perturbation for the particle at $(0,0,0)$ due to radiation by the particle at (i,j,l). This perturbation is then generalised for all particles of the array, normalised by the number of particles and used to replace the original dipolarisability dyadics:-

$$\left[\underline{\underline{p}}_{ijl}\right] \rightarrow \left[\underline{\underline{p}}_{ijl}\right] + \left[\Delta \underline{\underline{p}}^{(\infty)}_{ijl}\right] \qquad (17)$$

These perturbed or modified dipolarisability dyadics may then be used with existing mixing formulae to account for particle interaction.

Since the summation is performed explicitly the result is not restricted to periodic arrays only, nor is it limited to the assumption of the local field being uniform. The particle positioning may be perturbed by replacing the cell size a for a periodic lattice by $\langle a\rangle + \Delta a$ for a randomised lattice, with Δa being a random deviation (displacement) vector with uniform distribution.

5. Effect of Particle Interaction on Effective Medium Parameters

In a first example, a planar 99×99×1 array is analysed consisting of aligned single-turn one-arm spirals (plano - chiral) as described above. All multipole moments associated with each spiral are assumed to act at its origin. In such a model, the moments are permitted to be spaced arbitrarily closely. Fig 2 shows the dipolarisability dyadics, obtained with the method of counterpropagating waves with averaged direction of propagation [2, 6]. This is an algorithm based on a moment method computation of the distribution of currents that are induced in the individual spirals by a set of appropriately chosen counterpropagating plane waves.

Figs 4, 5 and 6 show the nine dominant (nonzero) elements of the four dipolarisability dyadics, with incorporation of interaction $(\underline{\underline{p}}_{kl} + \Delta\underline{\underline{p}}_{kl})$ for a configuration of two spirals spaced a distance $\underline{d} = (8.15\,\mathrm{mm})\underline{1}_x$, $\underline{d} = (8.15\,\mathrm{mm})\underline{1}_x + (8.15\,\mathrm{mm})\underline{1}_y$ and $\underline{d} = (8.15\,\mathrm{mm})\underline{1}_y$, respectively, along with those for a single isolated spiral ($\Delta\underline{\underline{p}}_{kl} = \underline{\underline{0}}$, *ie* $|\underline{d}| \to \infty$). For the two-particle system, the interaction is strongly dependent on both the magnitude and direction of $\underline{d}$. However, for cubic periodic ($\Delta a = 0$) or totally randomised ($\Delta a/\langle a\rangle = 1$) large arrays, the pair-wise particle interaction has been verified to cancel after summation, *ie* $\langle\Delta\underline{\underline{p}}_{kl}\rangle = \underline{\underline{0}}$, hence $\underline{E}_{near} = \underline{0}$ [9]. Fig 7 shows the particle interaction in a partially randomised 99×99×1-array with $\Delta a/\langle a\rangle = 0.2$, for which the interaction is seen to largely vanish as well.

Fig 8 shows the nine dominant (non-zero) elements of the four constitutive dyadics for the effective medium of this array using a Maxwell–Garnett model with explicit inclusion of particle interaction. The depolarisation dyadic of the spirals, $\underline{\underline{L}}$, has been approximated by $\underline{1}_z\underline{1}_z$. The host medium was chosen to be Kapton film ($\epsilon_{host}/\epsilon_o = 3.4 - j0.008$, $\mu_{host}/\mu_o = 1$). All parameters are with reference to the host medium, *ie* $\underline{\underline{\epsilon}}_r = \underline{\underline{\epsilon}}/\epsilon_{host}$, $\underline{\underline{\xi}}_r = \underline{\underline{\xi}}/\sqrt{\mu_{host}\epsilon_{host}}$, $\underline{\underline{\zeta}}_r = \underline{\underline{\zeta}}/\sqrt{\mu_{host}\epsilon_{host}}$ and $\underline{\underline{\mu}}_r = \underline{\underline{\mu}}/\mu_{host}$. The contribution of dyadic quadrupole moments (see *eg* [7, 13]), as defined by $\underline{\underline{q}}_e = \underline{\underline{\underline{q}}}_{ee} \cdot \underline{E} + \underline{\underline{\underline{q}}}_{em} \cdot \underline{H}$ and $\underline{\underline{q}}_m = \underline{\underline{\underline{q}}}_{me} \cdot \underline{E} + \underline{\underline{\underline{q}}}_{mm} \cdot \underline{H}$ in a dipole–quadrupole model of the medium [14], was taken into account in the computation of the effective medium dyadics. Here $\underline{\underline{\underline{q}}}_{kl}$ are the quadrupolarisability tryadics[1] (tensors of rank three) of the particle. For general bianisotropic particles (not necessarily chiral or non-magnetic), the contributions of $\underline{\underline{\underline{q}}}_{em}$, $\underline{\underline{\underline{q}}}_{me}$ and $\underline{\underline{\underline{q}}}_{mm}$ are to be assessed in addition to the contribution of $\underline{\underline{\underline{q}}}_{ee}$. Below the first resonance, the dominant elements of $\underline{\underline{\underline{q}}}_{ee}$ were found to contribute to the

[1]The term *triadic* appears also in the literature. –*(Editors)*

electric polarisation to the same order as the dipolar magneto-electric coupling: $\left\|\left|\frac{k}{2}\underline{\underline{q}}_{ee}\right\|\right| / \left\|\underline{p}_{em}\right\| \leq 0.45$. The contributions of $\left\|\left|\frac{k}{2}\underline{\underline{q}}_{em}\right\|\right| / \left\|\underline{p}_{em}\right\|$, $\left\|\left|\frac{k}{2}\underline{\underline{q}}_{me}\right\|\right| / \left\|\underline{p}_{me}\right\|$ and $\left\|\left|\frac{k}{2}\underline{\underline{q}}_{mm}\right\|\right| / \left\|\underline{p}_{me}\right\|$, whose respective ranges were found not to exceed 0.25, 0.20 and 0.08 below the first resonance, are negligible in this particular example. In the near-field zone, the quadrupole and higher-order moments may contribute significantly to the particle interaction, and this requires separate investigation.

Next, a partially randomised 99×99×3 array was considered, consisting of aligned regular half-turn four-arm spirals (plano - semi-chiral), defined by $r = r_o(\alpha^\theta - 1)$ with $r_o = \alpha = 10\,\text{mm}$, $0 \leq \theta \leq \pi$, and made of cylindrical copper wire of radius 0.1 mm. The transverse spacing is uniformly distributed with $\Delta a/\langle a\rangle = 0.2$; the spacing in the z-direction, normal to the three sheets of spirals, is held constant at 1.5 mm. The dominant elements of the effective medium dyadics are shown in Fig 9. For the modified dipolarisabilities, the same interaction phenomena have been observed as for the one-arm spirals. The array is seen to emulate a non-reciprocal anisotropic medium, for the computed values of the magneto-electric coupling are within the numerical noise and $\left[\underline{\underline{\epsilon}}_r\right]_{xx} = \left[\underline{\underline{\epsilon}}_r\right]_{yy}$, $\left[\underline{\underline{\epsilon}}_r\right]_{xy} = -\left[\underline{\underline{\epsilon}}_r\right]_{yx}$. Changing the handedness of the spiral was verified to invert the signs of $\left[\underline{\underline{\epsilon}}_r\right]_{xy}$ and $\left[\underline{\underline{\epsilon}}_r\right]_{yx}$.

6. Conclusions

The analysis of plano - chiral and plano - semi-chiral structures shows that these geometries enable the design of flat and ultra-thin bianisotropic media. Compared to conventional chiral structures, plano - chiral and plano - semi-chiral structures are found to exhibit less interdependence between dielectric, magnetic and magneto-electric parameters in mixing formulas.

The numerical study of the mutual coupling between particles shows that the interaction of nearby particles largely vanishes in cubic and in partially or totally randomised arrays, although the pair-wise interaction itself may be significant and strongly dependent on the relative positions of the particles. Hence in any practical situation of partial randomisation of the particles' positions, the internal field ($\underline{E}^i$, $\underline{H}^i$) [9] may be considered as being equal to the homogenised depolarisation field. Therefore, any discrepancies between measured and predicted constitutive parameters of high-density composites are to be attributed to differences in the incident field ($\underline{E}$, $\underline{H}$) (scattering losses) for the particles rather than to the spatial non-uniformity of ($\underline{E}^i$, $\underline{H}^i$) for each particle.

If the particle concentration increases, then multi-particle interaction (higher-order multiple scattering) becomes significant. An extension of the de-embedding technique to more than two particles is then required. Higher-order multipole moments must be considered if the distance between particles further decreases, in order to obtain more accurate predictions of the near-field interaction between realistic particles. As the frequency increases, interparticle and intraparticle scattering effects become more significant.

References

1. Arnaut, L R (1994) *Analysis and design of lossy chirals and biisotropics,* PhD dissertation, UMIST, Ch 8.
2. Arnaut, L R (1995) Computation of Dipolarisability Tensors of Helix-Based Bianisotropic Media, *Proc 2nd "Journées Maxwell" – International Conference on Electromagnetics (6–9 Jun 1995, Bordeaux, France),* 319– 327.
3. Arnaut, L R and Davis, L E (1995) Effect of Mutual Coupling Between Aligned Helices on the Constitutive Dyadics of High-Density Bianisotropic Composites, in A Sihvola *et al* (eds), *Proc "Chiral'95" International Conference (11–14 Oct 1995, State College, PA),* 80–86.
4. Arnaut, L R (1996) Chirality in multi-dimensional space with application to electromagnetic characterisation of plano - chiral, plano - semi-chiral and axi - chiral media, UMIST Report, Part V/3.
5. Arnaut, L R and Davis, L E (to appear) Mutual coupling between bianisotropic particles: a theoretical study, in Kong, J A (ed), *Progress In Electromagnetics Research (PIER)* **15**, Monograph Series, Elsevier. See also Arnaut, L R (to appear) Interaction between bianisotropic particles, *J Electr Waves Applic.*
6. Brewitt-Taylor, C R (1994) Modelling of helix-loaded chiral radar-absorbing layers, Research Report (1994), DRA Malvern, UK, 6–7.
7. Graham, E B and Raab, R E (1990) Light propagation in cubic and other anisotropic crystals, *Proc Roy Soc London A* **430**, 593–614.
8. Huang, J, Wu, T K and Lee, S W (1994) Tri-band FSS with circular ring elements, *IEEE Trans Antennas Propag* **AP–42(2)**, 166–175.
9. Jackson, J D (1975) *Classical Electrodynamics,* Wiley, New York, Ch 4.
10. Lindell, I V (1992) *Methods for Electromagnetic Field Analysis,* Clarendon, Oxford, Ch 4.
11. Sihvola, A H (1994) On polarizability properties of bianisotropic spheres with non-complete magnetoelectric dyadics, *Microw Opt Tech Lett* **7(14)**, 658–661.
12. Sihvola, A H and Kong, J A (1988) Effective permittivity of dielectric mixtures, *IEEE Trans Geoscience and Remote Sensing* **26(4)**, 420–429; Sihvola, A H and Kong, J A (1989) Corrections, *ibid* **27(1)**, 101–102.
13. Theron, I P and Cloete, J H (1995) Microwave optical activity of an artificial non-magnetic uniaxial chiral crystal, in A Sihvola *et al* (eds), *Proc "Chiral'95" International Conference (11–14 Oct 1995, State College, PA),* 76–79.
14. Van Bladel, J (1991) *Singular Electromagnetic Fields and Sources,* Clarendon, Oxford, Ch 2.

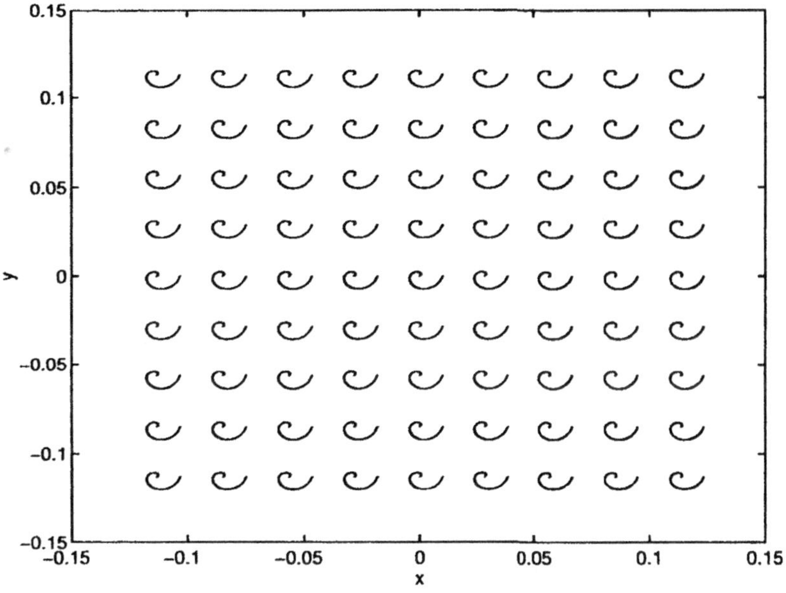

(a)

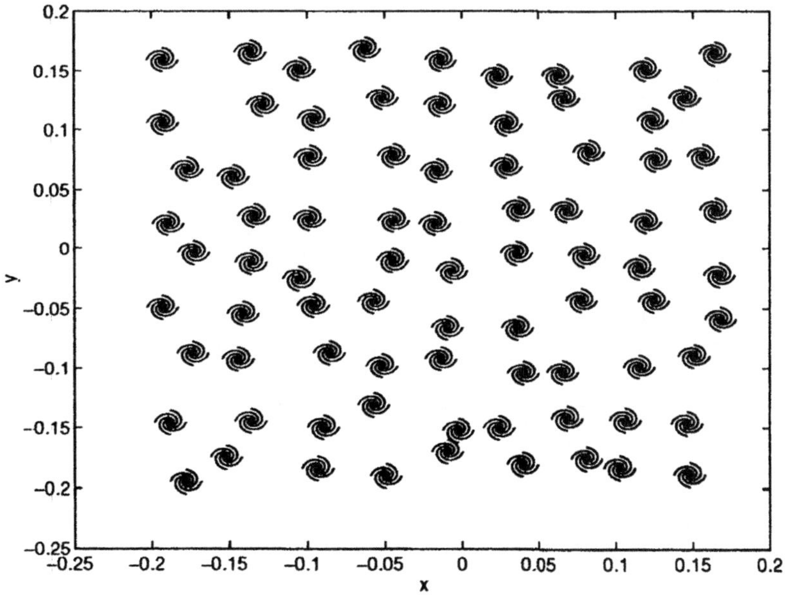

(b)

Figure 1. (a) Periodic planar 9×9-array of one-turn one-arm spirals ($a = 0.028$m) and (b) randomised planar 9×9-array of half-turn four-arm spirals ($\Delta a/\langle a \rangle = 0.013\text{m}/0.042\text{m} = 0.3$).

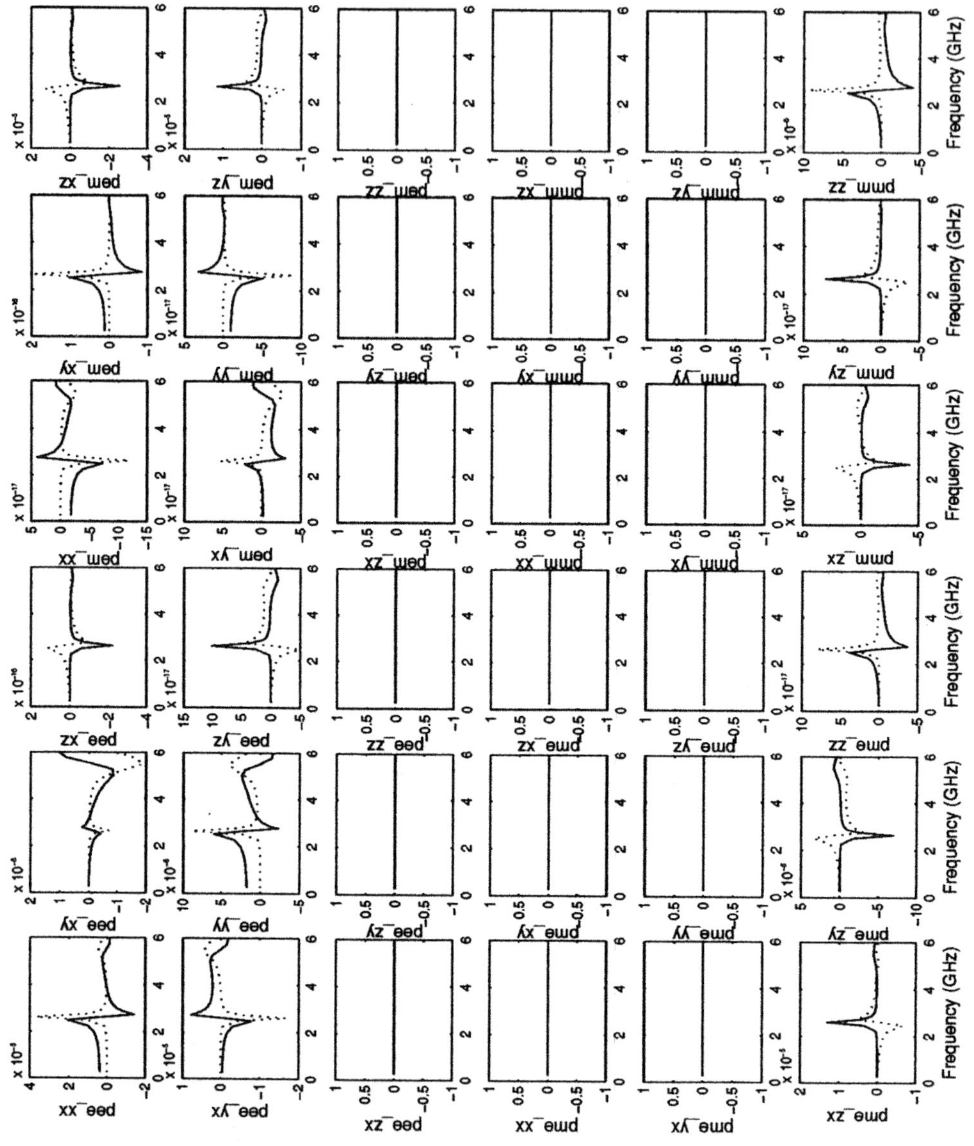

Figure 2. Complex dipolarisabilities for single isolated single-turn one-arm spiral. — = real part, $\cdots$ = imaginary part.

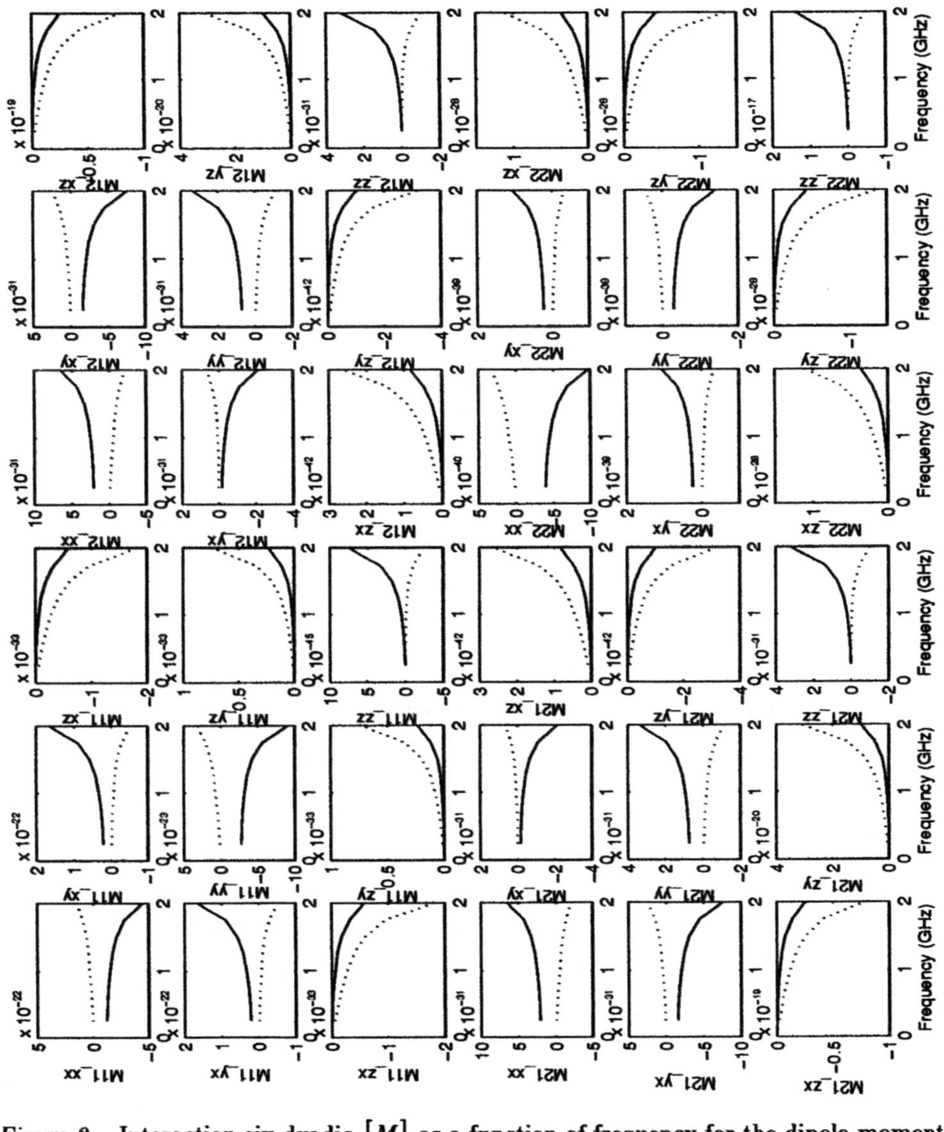

Figure 3. Interaction six-dyadic $\lfloor \underline{\underline{M}} \rfloor$ as a function of frequency for the dipole moments of two identical aligned co-planar single-turn one-arm spirals ($d = 10\text{mm}\ \underline{1}_x$). — = real part, ··· = imaginary part.

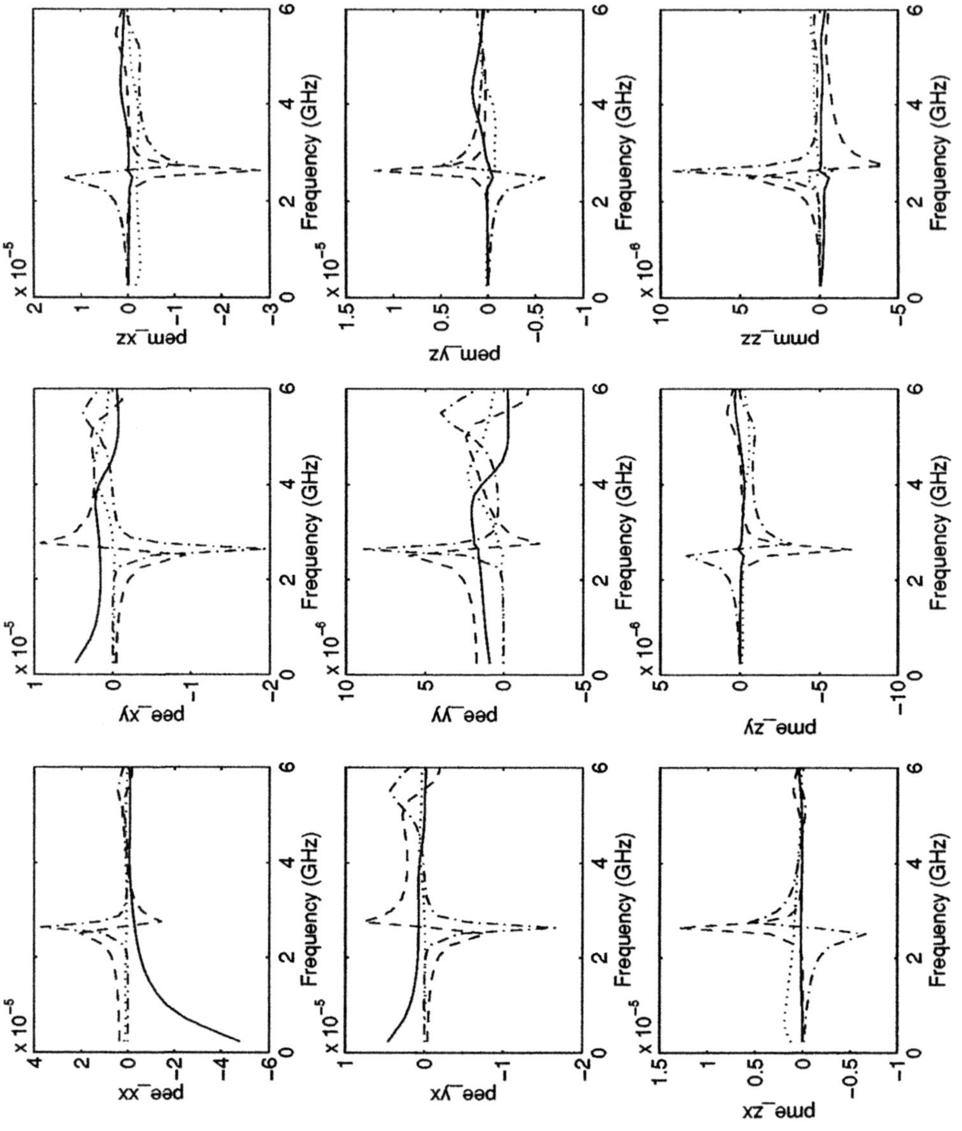

Figure 4. Complex dipolarisabilities for a pair of single-turn one-arm spirals with $\underline{d} = (8.15\text{mm})\underline{1}_x$ (dominant dipolarisability elements only). Inclusive of dipole moment interaction (modified dipolarisabilities): — = real part, $\cdots$ = imaginary part; exclusive of dipole moment interaction (isolated spirals): – – = real part, – · – = imaginary part.

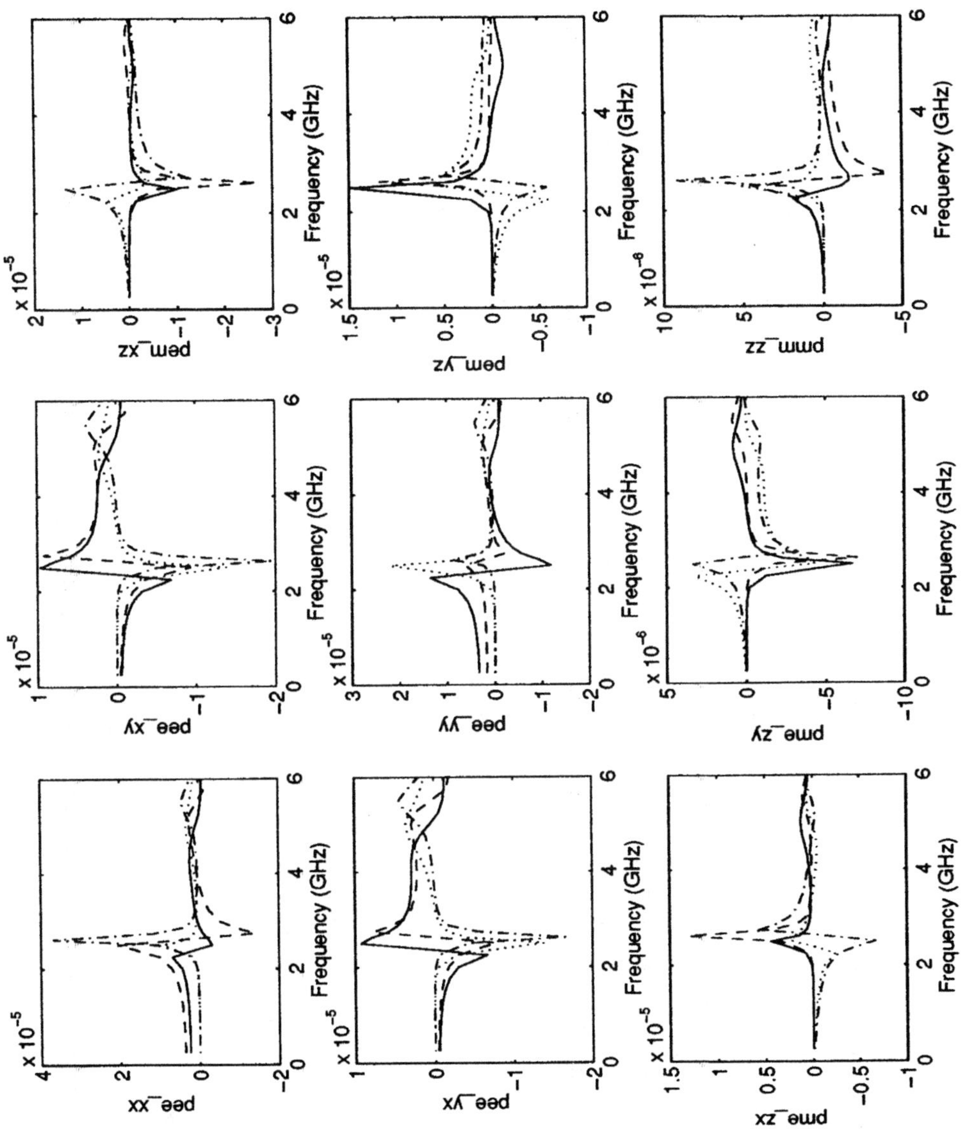

Figure 5. Complex dipolarisabilities for a pair of single-turn one-arm spirals with $\underline{d} = (8.15\text{mm})\underline{1}_x + (8.15\text{mm})\underline{1}_y$ (dominant dipolarisability elements only). Inclusive of dipole moment interaction (modified dipolarisabilities): — = real part, $\cdots$ = imaginary part; exclusive of dipole moment interaction (isolated spirals): – – = real part, – · – = imaginary part.

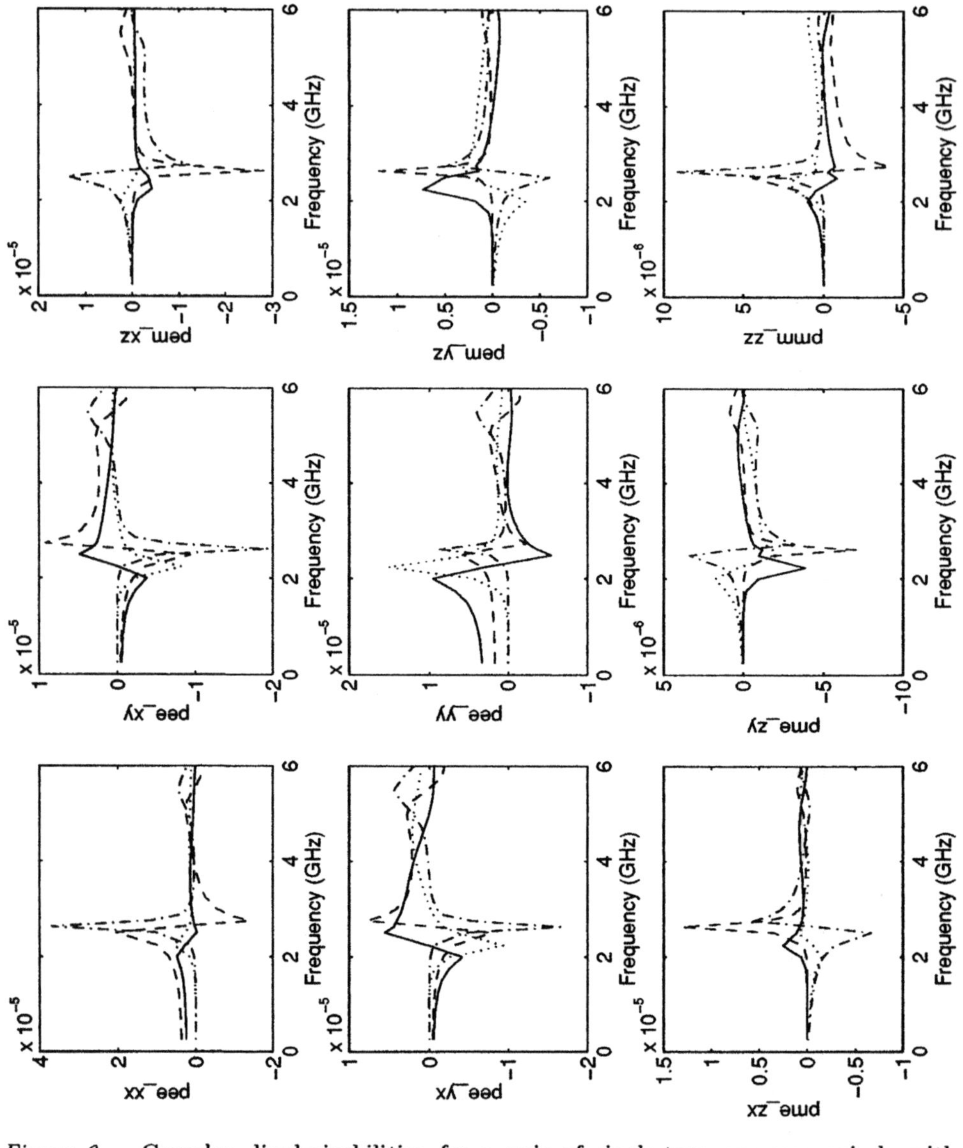

Figure 6. Complex dipolarisabilities for a pair of single-turn one-arm spirals with $\underline{d} = (8.15\text{mm})\underline{1}_y$ (dominant dipolarisability elements only). Inclusive of dipole moment interaction (modified dipolarisabilities): — = real part, $\cdots$ = imaginary part; exclusive of dipole moment interaction (isolated spiral): – – = real part, – · – = imaginary part.

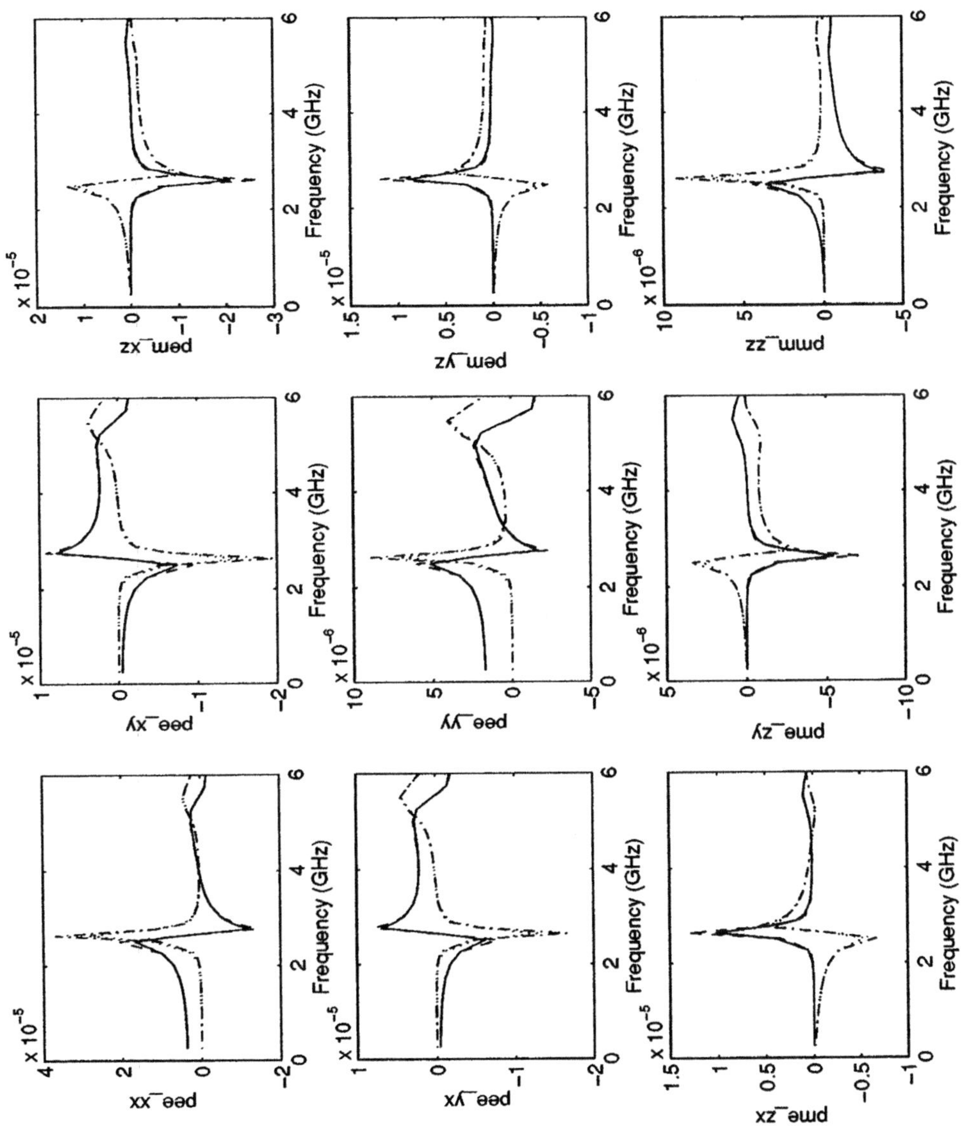

Figure 7. Complex dipolarisabilities for a partially randomised 99×99×1-array of single-turn one-arm spirals with $\Delta a/\langle a\rangle = 0.2$, $\langle a\rangle = 8.15$mm (dominant dipolarisability elements only). Inclusive of dipole moment interaction (modified dipolarisabilities): — = real part, ··· = imaginary part; exclusive of dipole moment interaction (isolated spirals): – – = real part, – · – = imaginary part.

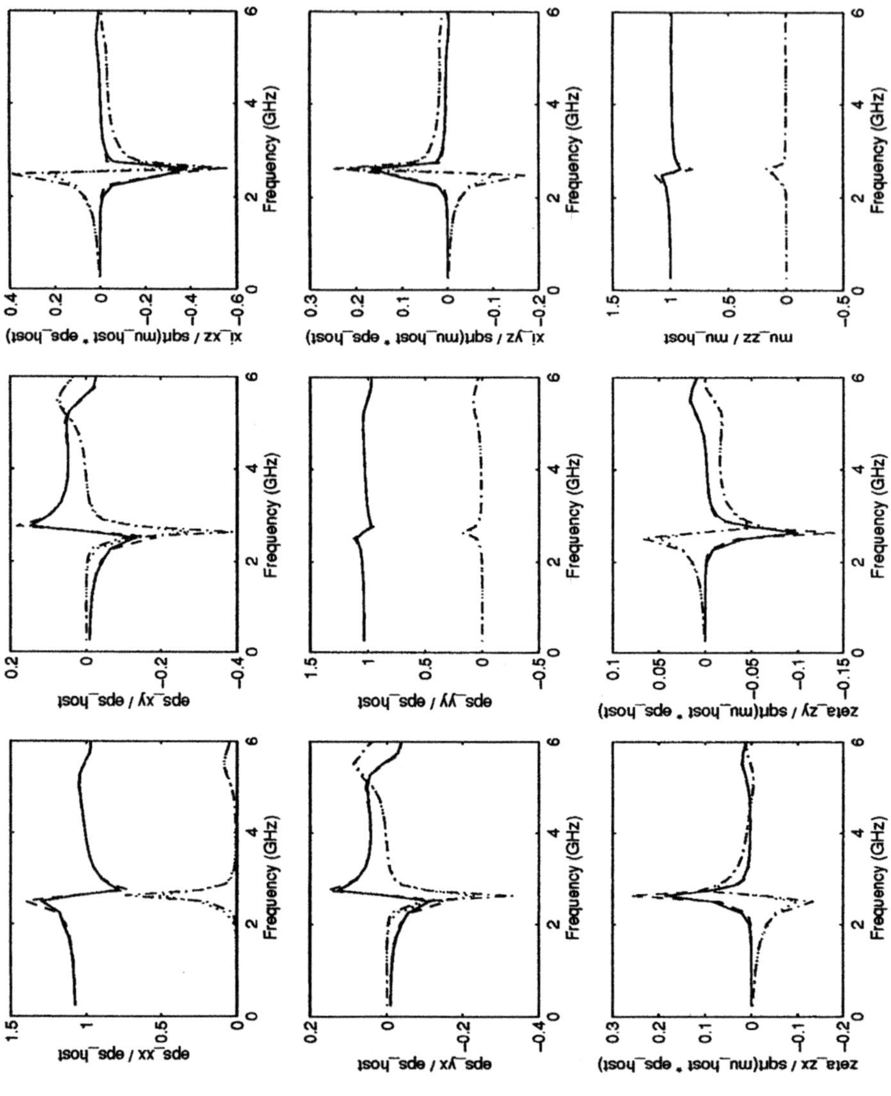

Figure 8. **Maxwell–Garnett estimates of complex normalised effective medium parameters $\underline{\underline{\epsilon}}_r$, $\underline{\underline{\xi}}_r$, $\underline{\underline{\zeta}}_r$ and $\underline{\underline{\mu}}_r$ for partially randomised randomised 99×99×1-array of one-turn one-arm spirals with $\Delta a/\langle a\rangle = 0.2$, $\langle a\rangle = 8.15$mm (dominant elements only). Inclusive of dipole moment interaction: — = real part, ··· = imaginary part; exclusive of dipole moment interaction: – – = real part, – · – = imaginary part.**

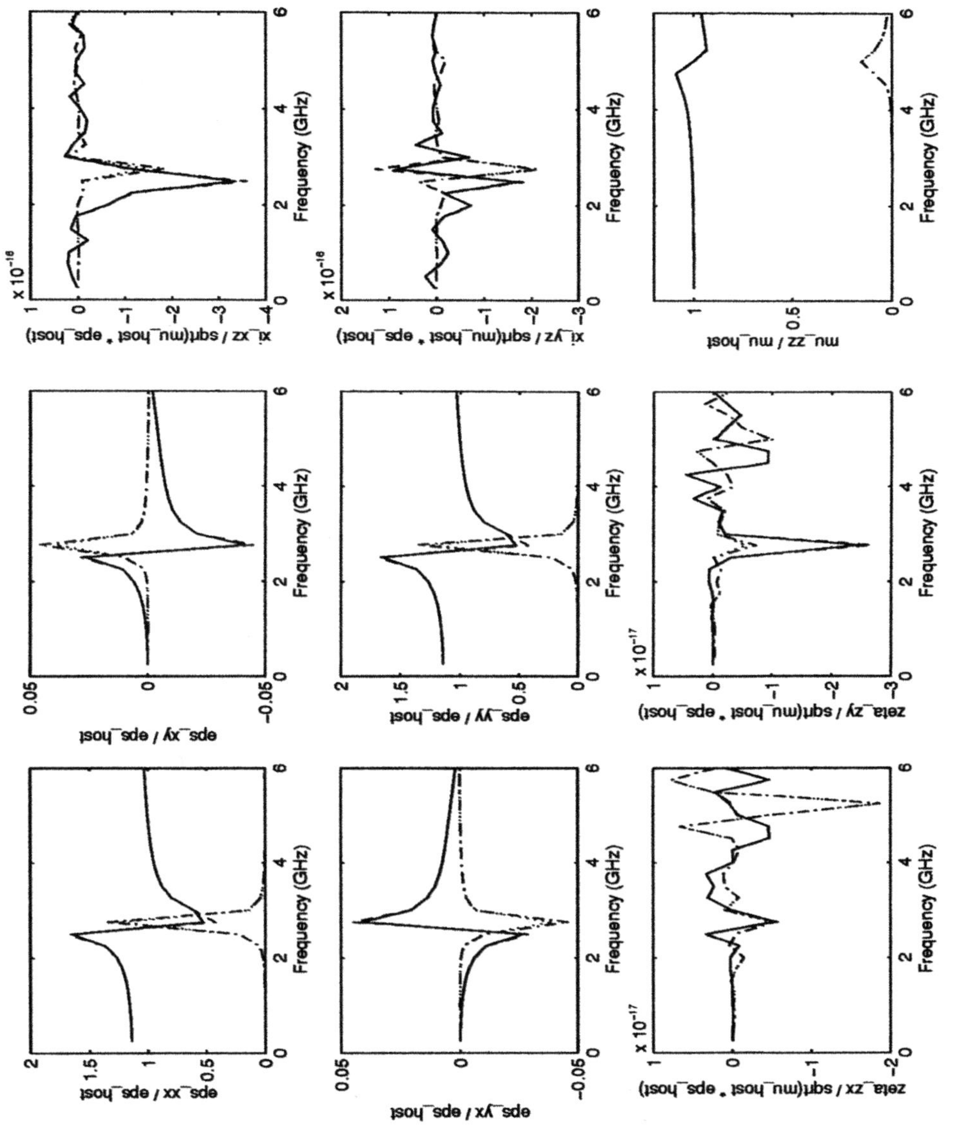

Figure 9. Maxwell–Garnett estimates of complex normalised effective medium parameters $\underline{\underline{\epsilon}}_r$, $\underline{\underline{\xi}}_r$, $\underline{\underline{\zeta}}_r$ and $\underline{\underline{\mu}}_r$ for randomised 99×99×3-array of regular half-turn four-arm spirals with $\Delta a/\langle a\rangle = 0.2$, $\langle a\rangle = 8.15$mm (dominant elements only). Inclusive of dipole moment interaction: — = real part, ··· = imaginary part; exclusive of dipole moment interaction: - - = real part, - · - = imaginary part.

VI.

Experiments, Measurements, Applications

STEREO-GLASSES WITH COMPOSITE AND CHIRAL LIQUID CRYSTALS

V.V. BELYAEV
Central R&D "Cometa", 5, Velozavodskaya st., Moscow 109280 Russia

Y.V. DEVYATKIN
Microcosm Ltd, Moscow

V.G. NAZARENKO
Institute for Physics of NASU, 46, pr. Nauki, Kiev 252650 Ukraine

G.B. NOSOV AND A.S. SONIN
Organic Intermediates and Dyes Institute, 1/4, B. Sadovaya st., Moscow, 103787, Russia

1. Introduction

Chiral liquid crystalline (LC) phases (cholesteric and smectic C*) are now widely used as dopants for stabilization of LC orientation in thermotropic films, fast response electro-optical shutters, and in twist and supertwist display devices. The chirality is a consequence of the asymmetry of molecular rigid skeleton or separate flexible substituent which builds up the skeleton. Such a structure provides an increase of effective elasticity in nemato-cholesteric mixtures and, hence, a significant reduction of switch time of LC electro-optical response. In smectic C* phase the asymmetry causing the chirality is one of the reasons of occurrence of spontaneous polarization in LC material. It offers a possibility to reduce the operating voltage and switch time.

Polymer dispersed liquid crystals (PDLC) present another type of promising materials. The PDLC is an inhomogeneous medium consisting of a porous polymer matrix filled by small LC droplets. The surface elastic interaction of LC with polymer walls results in significant reduction of switch time and occurrence of light scattering. It is well-known that the PDLC devices do not require polarizers, that allows to increase the light efficiency of various optical systems.

The goal of the paper is to study the properties of different electro-optical shutters on the base of chiral LC and PDLC. The shutters are designed for stereovision. This application requires fast response, high optical transmission and good modulation ability. These are the properties which are considered here.

2. Requirements for Electrooptical Shutters for Stereoglasses

Stereoglasses for monitors of a personal computer (PC) provide volume vision and reconstruction of virtual reality for educational, military, training and other purposes.

A. Priou et al. (eds.), Advances in Complex Electromagnetic Materials, 313–316.

The minimal frame frequency in PC is 30 Hz. It means that the light modulation frequency of each shutter both for the right and left eyes should be also 30 Hz. The phase shift of the open state of the shutter equals to a half of the time period, i.e., 16.7 ms. The transmission in the open state T_{max} should be as large as possible. When it is smaller than 30 %, the brightness of monitor's screen can not be effectively used. The reduction of shutter's transmission in the closed state $K=T_{max}/T_{min}$ should be more than 30. At lower K values the right eye sees the traces of the left picture. The same situation is for the left eye. The shape of each transmission optical pulse has to be close to the rectangular one. It demands the short switch time (t_{on} or t_{off}). Otherwise, each eye will see the traces of undesired picture or top lines of a frame will seem dark. Usual glasses with liquid crystal electro-optic shutters require polarizers (see, e.g., [1]). For this reason they have low transmission and alternative light flux on each eye. It induces fatigue of an operator [2]. From this point of view the light scattering effects seem to be promising.

3. Design of LC Cells and Their Parameters

The main ergonomic parameters of various types of LC electro-optical shutters on the base of chiral and composite systems were measured, namely, the transmission in the open and closed states, and switch-on and switch-off times.

Four types of the LC electro-optical shutters were tested in this paper. The effects investigated are the following:

a) the effect of "surface stabilized mode" (Clark-Lagerwall effect) in ferroelectric smectic C* LC [3],

b) the twist-effect in LC materials with low frequency inversion of the sign of dielectric anisotropy,

c) the effect of fast response in 270° twisted nemato-cholesteric layers,

d) the effect of electrically induced clearing of opaque PDLC films.

All LC cells are sandwich structures consisting of glassy substrates with transparent ITO electrodes on the internal surfaces of the substrates. The electrodes are coated by polymeric films to provide LC planar orientation.

The cells with ferroelectric LC were manufactured in Physical Lebedev Institute (Moscow). The cells have relatively large thickness (15 micrometers) and, as a consequence, weak sensitivity to gap inhomogeneity and colorless transmission. The smectic layers in ferroelectric LC are orthogonal to the planes of the substrates.

The comparative long electro-optical response in non-polar nematic LC is connected with rebuilding of the whole domain and is of the order of hydrodynamic time. In contrast, in ferroelectric LC another type of molecular reorientation is observed. Under the action of the applied electric field the molecules in each layer are rotated on a cone base so that their dipole moments are always parallel to the direction of the electric field. The rotation time is inversely proportional to the eigenfrequency of rotational oscillations about the long molecular axis. Hence, the ferroelectric LC substances have the shortest switch times from 0.1 to 1 ms at voltage 10-50 V.

The twist cells with LC the material ZhK-999 (Organic Intermediates and Dyes Institute, Dolgoprudnyi, Moscow Region) have the frequency of inversion of the sign of dielectric anisotropy equal to 7 kHz at room temperature. The value of the dielectric anisotropy for ZhK-999 varies from +2.5 at 100 Hz to -2.2 at 20 kHz. To reduce the switch-off time a special driving regime is necessary. The driving pulse consists of two main parts. The first one is a DC positive (or negative) pulse. It is followed by the second AC (sinusoidal with frequency 50 kHz) pulse. Such pulse form results in a significant reduction of relaxation time for this materials [4]. At the amplitude of AC pulse being equal to 50 V, the switch-off time reduces from 30 ms to 0.7 ms. In comparison with the usual cells driven without high-frequency pulses, the current flowing through the cell is lager (up to 10 mA instead of 1 mA). As a consequence we must allow significant power consumption.

The 270° twisted cells (R&D Volga, Saratov) have 3.5 micrometer in thickness. The special choice of the chiral dopant concentration provides the thickness-to-cholesteric pitch ratio equal to approximately 0.52. As a consequence, the effective elasticity of a LC material sufficiently increases. The other physical parameters of the LC material are chosen to result in significant (up to 10 times) reduction of the switch-off time in comparison with the commonly used 90° cells. In this case the driving voltage is minimal (12 V). Hence, the cells are compatible with PC serial ports, and one does not need any additional voltage source for operation of such glasses.

For the PDLC both the switch time and driving voltage depend on the dimension and shape of the LC droplets. However, these dependencies have different tendencies. To reduce the switch-off time one should create small pores. This in turn results in an increase of the operating voltage. The reduction of the operating voltage could be achieved by a proper choice of the polymer material and control of chemical surface potential. For example, the use of polyacrilate components yields the voltage reduction in 3 times.

In the absence of the supplied voltage the PDLC cell is opaque. After applying the voltage pulse, the cell becomes transparent. Thus, the PDLC shutter has two states. In the first one, the shutter is transparent, and the operator can see the picture. In the second state the shutter is opaque and the operator can see only diffuse white light without any picture. The intensity of the transmitted light measured directly behind the cell is nearly the same in both cases. However, the operator does not see any traces of an undesired picture since the spatial spectrum of the image is changed.

The results of the tests of various shutters are shown in Table 1. The transmission and reduction coefficients values for the PDLC were measured at the 30 cm distance between the cell and the photoreceptor.

Electro-optical effect	Open state trans-mission, %	Reduction of light flux	Driving voltage, V	Switch-off time, ms
Surface stabilized mode in ferroelectric smectic C*	10-17	8-10	10-20	0.2
Twist-effect with inversion of DA sign (f_1~100 Hz, f_2~50 kHz)	20-34	20-30	50-80	1-2
270°-twisted layers	30-38	20-30	15	0.8-3
Electrical field induced clearing of PDLC films	60-80	20-30	50-80 100-250	5-7 3-8

Table 1. Test results for various shutters

4. Comparison of Various Types of Stereo-Glasses

The samples of stereo-glasses with various LC cells were manufactured. The cells were positioned into a rim and connected by wires with the driving electronic scheme. The scheme includes also a synchronization block.

Parameters of different glasses were compared. Stereo-glasses with ferroelectric LC have the best switch times, but worse transmission and homogeneity. They have rather expensive technology, especially for the case of higher transmission (to 34 %). Twist-cells with inversion of the sign of the dielectric constant have the most complicated driving with high power consumption. 270°-twisted nemato-cholesteric layers have sufficiently high speed of response and the most simple driving without an additional electric supply. The PDLC cells have the best transmission. It allows to use the brightness of monitor screen with the maximal efficiency. The main advantage of the PDLC based glasses is the absence of the fatigue due to small variation of light intensity on operator's eyes.

5. References

1. Belyaev, S. et al. (1995) in Proc. of *"Asia Display'95"*, pp. 155-158.
2. Yamazaki, T. et al. (1989) in Proc. of *"Japan Display'89"*, pp. 606-609.
3. Clark, N.A., Lagerwall, S.T. (1980) *Appl.Phys.Lett.* **36**, 899-901.
4. Belyaev, V., Chigrinov, V. (1978) *Sov. Phys.Crystallography*, **23**, 454-457.

MANUFACTURE OF MICROWAVE CHIRAL MATERIALS AND THEIR ELECTROMAGNETIC PROPERTIES

S.A. KUEHL, S.S.GROVÉ AND E. KUEHL
Institute for Polymer Science,
University of Stellenbosch, Stellenbosch, South Africa

AND

M. BINGLE AND J.H. CLOETE
Department of Electrical and Electronic Engineering,
University of Stellenbosch, Stellenbosch, South Africa

Abstract. Since *Chiral'95*, where certain features of our new method of producing chiral materials were introduced [1], our manufacturing capabilities have increased dramatically resulting in a large selection of novel materials which have been partially characterised. We now wish to reveal all details of this method. We call our new method of producing chiral materials the Supported Microfilament Helix (SMFH) Technique [2]. The method involves winding fine conductive or resistive filaments around a supporting mandrel line. This wound continuous line serves as a chiral precursor from which a wide variety of configurations can be made. The different electromagnetic, size and strength characteristics of the mandrel line coupled with the large variety of conductive filaments which can be wound around the line to produce chiral precursors, enlarges the scope of novel chiral materials. This should provide a better understanding of the phenomenon as well as offering a more efficient method of producing these materials for science and engineering applications.

1. The Supported Microfilament Helix (SMFH) Technique [2]

A chiral material precursor is constructed by winding fine conductive or resistive filament around a supporting mandrel line in a continuous process and passing this line through a coating device which deposits an adhesive onto the wound line. The adhesive bonds the conductive filament to the

A. Priou et al. (eds.), Advances in Complex Electromagnetic Materials, 317–332.

supporting mandrel. The type of conductive filament, its pitch and internal diameter can be varied at will while both right and left handed wound line can be produced. The conductive filament need only have Ohmic conductive properties and is thus not limited to solid metal wire. Neither is there any stringent limit as to the diameter of this conductive filament. The mandrel line, which remains within the helical winding as a core, can be chosen to exhibit desirable mechanical or electromagnetic properties aside from being supportive in nature. The continuous length of wound line can then be processed for the production of chiral media. Typically, this continuous line can be either chopped to produce segments to be dispersed randomly in some host medium, or cast as a uniaxially aligned medium from which panels can be split off to produce an oriented lattice of helices of a specific length. Figure 1 below shows an overall schematic drawing of the SMFH device as well as a chopping device used to process the line. Section 1 is the supply spool of mandrel line fitted with a friction tensioner on the supporting shaft. Section 2 is the rotating spindle (winder) driven by a stepper motor. The supply spool of microfilament is fixed within this housing. Section 3 is the coating system which delivers the adhesive solution. Section 4 is the drying area. Section 5 is the drive spool which pulls the line through the entire device using a two roll pinch tensioner all driven by a stepper motor. Alternatively, the chopping device (cutter), employing an overlocker blade driven by a solenoid is shown next to the drive Section (5) and can be used in a separate step to process the line.

2. Anisotropic Samples

Continuous sections of wound Nylon "fishing" line are strung through two sides of a mould into which a lattice of holes were drilled. The three layers of holes were staggered by 1.5 mm. The holes were 1.5 mm in diameter and spaced with a 3 mm lattice spacing. Firstly, one end of the Nylon line is placed through a hole until it protrudes outside one of the mould faces. The end is then flared into a mushroom-like head using a fine tipped soldering iron. The other end of the line is placed through the paired lattice hole on the opposite mould face, the end of which is flared out under tension. Each row contains 36 lines and the uniaxial lattice is built up in this fashion both straight and parallel. The other two sides of the mould are then closed off to produce an open topped mould. Into this box degassed epoxy resin is poured until the top layer of line is covered. After cure, excess resin on the top surface is cut away with a fly cutter. A 0.5 mm layer of epoxy resin covers the top and bottom layers of line. Each layer of line is separated by epoxy 1.5 mm thick. The entire sample is approximately 108 $\times$ 108 mm and 5.5 mm thick. This sample of "infinite" line has been measured.

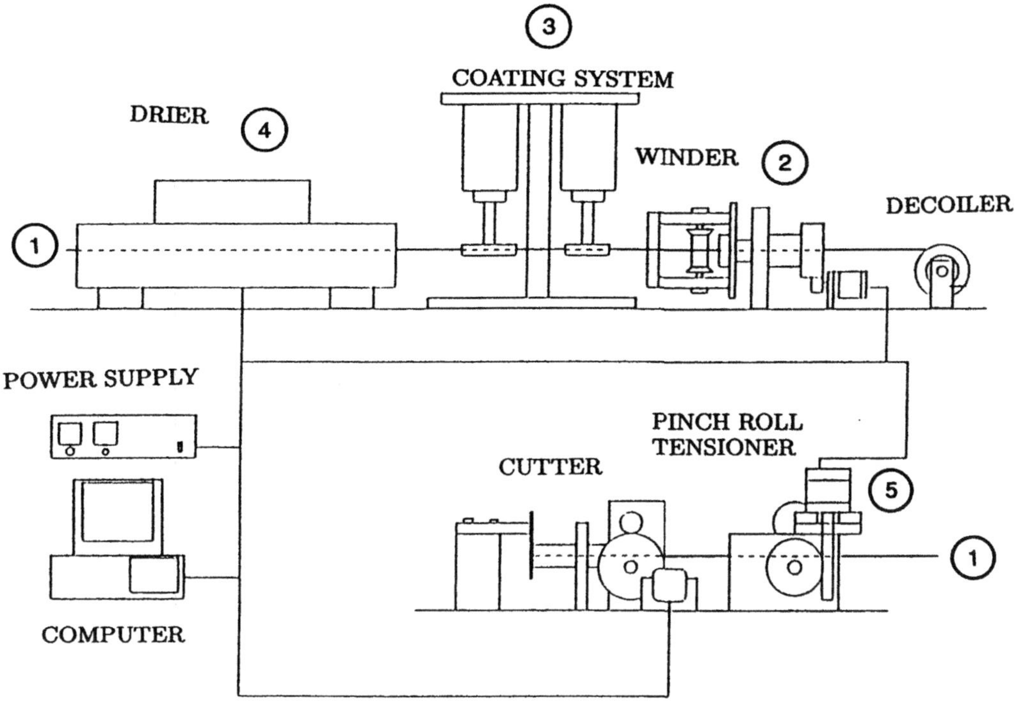

Figure 1. Machine for computer controlled manufacture of SMFH [2] chiral material.

Also, to understand the physics of the media when the helices are cut shorter, whereby the contour length of the conductive line approaches a half wavelength of the microwave radiation within the material and the so called Cotton effect resonance [3], the parent sample was split across the uniaxial direction to produce a multitude of panels containing a lattice work of embedded helices of specific length. If panels 3 mm thick are cut, the helix concentration of about 70/cm3 is achieved ($7 \cdot 10^7\,\mathrm{m}^{-3}$). These panels can be reconfigured in a sample holder using 1 mm dielectric spacers between each panel to preserve the lattice spacing.

The figures below show data accompanied by brief comments for a number of samples prepared in this fashion.

Polarisation dependence of a semi-infinite uniaxial slab (Figure 2)

The data indicates that pronounced rotation is observed in a uniaxial slab when the incident microwave electric field is both parallel and perpendicular to the uniaxial direction. The rotation and ellipticity have similar values at the short wavelength limit for this configuration. An interesting but not well understood frequency dependence is evident at the long wavelength

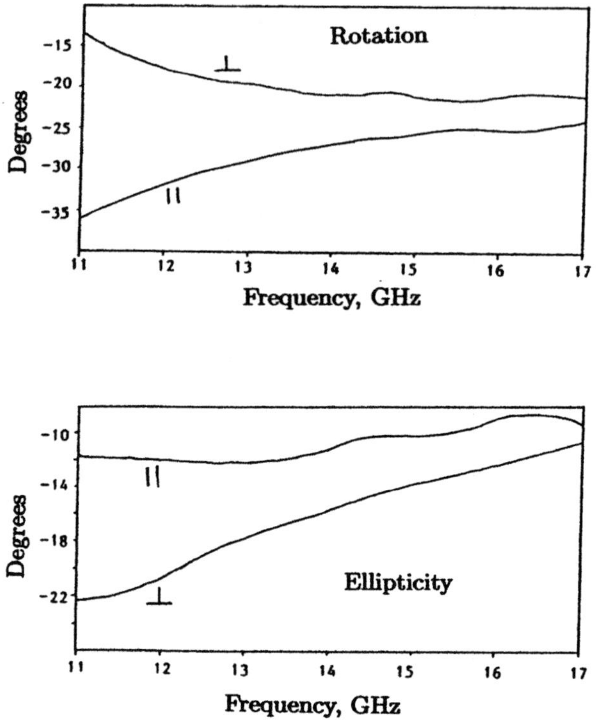

Figure 2. **Rotation and ellipticity measured on a continuous slab of uniaxial lines in a parallel and orthogonal orientations. The sample employed 1 mm diameter Nylon line and was wound with 60 micron Copper wire with a 1.5 mm pitch.**

limit.

Wire Diameter Study: 13 to 140 microns (Figure 3)
This data indicates that coarser solid wire seems to produce more rotation and circular dichroism than fine wire. We do not expect that there is any appreciable effect due to the different intrinsic conductivities of the wires used. However, it should be noted that even though the helices have the same geometrical configuration and are suspended to the same concentration, the outer diameters get larger when thicker wire is used. Thus the helices might exhibit different resonance frequencies. The wire diameters are all larger than the skin depth, and therefore the scattering should not depend on the cross sectional area of the wires. This invites the question as to the dependence of the rotation on helix diameter when the contour length is kept constant.

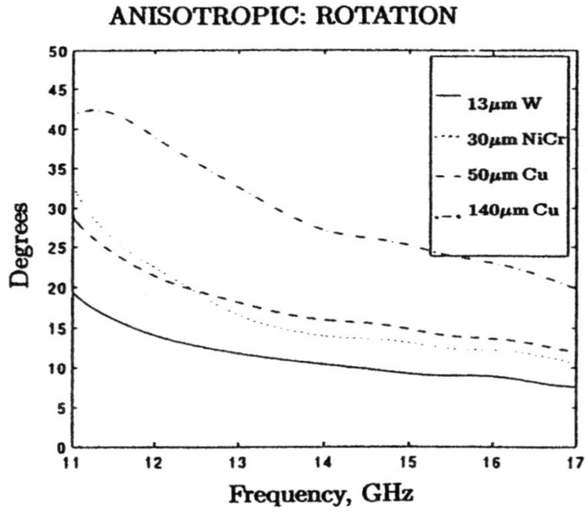

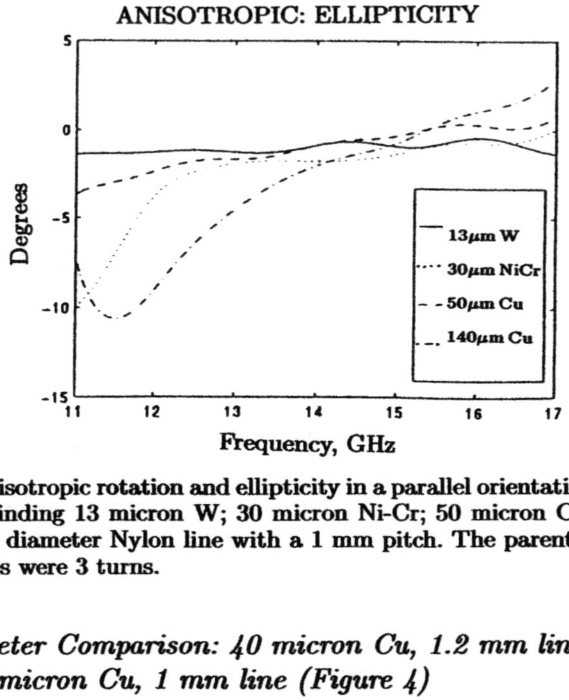

Figure 3. **Anisotropic rotation and ellipticity in a parallel orientation for a chiral material made from winding 13 micron W; 30 micron Ni-Cr; 50 micron Cu and 140 micron Cu around 1 mm diameter Nylon line with a 1 mm pitch. The parent samples were split so that all helices were 3 turns.**

Wire Diameter Comparison: 40 micron Cu, 1.2 mm line versus 140 micron Cu, 1 mm line (Figure 4)

In this case a 140 micron wire sample at 3 turns must be compared to the 40 micron wire at 2.6 turns. The 140 micron wire at 2.6 turns must be compared to the 40 micron wire at 2.2 turns. As the contour length becomes shorter, half-wave resonance is moved upwards onto the 11–17 GHz measurement band.

The data in Figure 4 indicates that the fine wire is just as effective in producing rotation as the coarse wire outside of resonance. If the contour

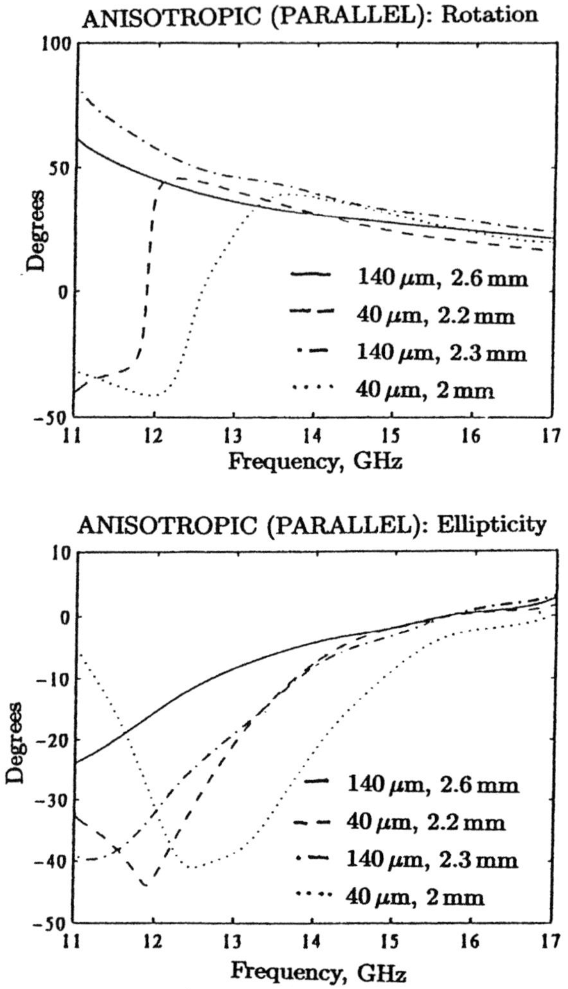

Figure 4. **A comparison between 40 micron Cu wound around a 1.2 mm diameter Nylon line and 140 micron Cu wound around 1 mm Nylon line. The data are for the parallel polarization. The samples all have a 1 mm pitch and are cut to the same contour lengths.**

lengths of these two samples are identical, then they should exhibit resonance at the same frequency barring the end effect. Referring also to Figure 6a, one sees that the 140 micron wire wound around a 1 mm line becomes resonant at the same frequency as the 40 micron wire, 1.2 mm mandrel line a when both are cut to 2 turns. Figure 5a shows that the same is true in that example (50 micron wire) also when the helices are 2 turns. Thus one can conclude that the magnitude of rotation at resonance is essentially independent of wire diameter. Outside of resonance, one can get with finer wire rotations similar to coarser ones if contour length adjustments are

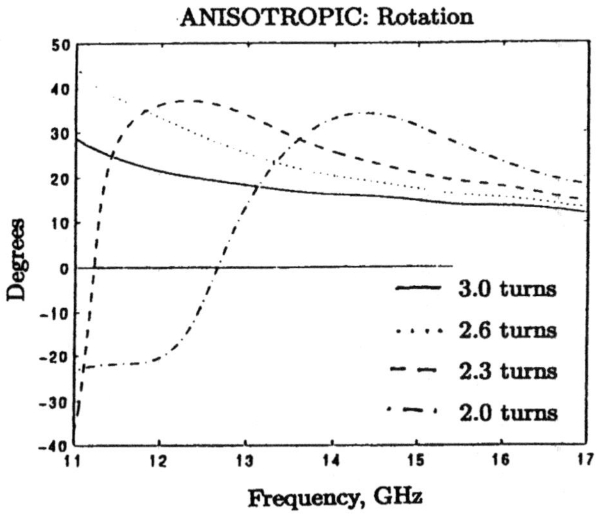

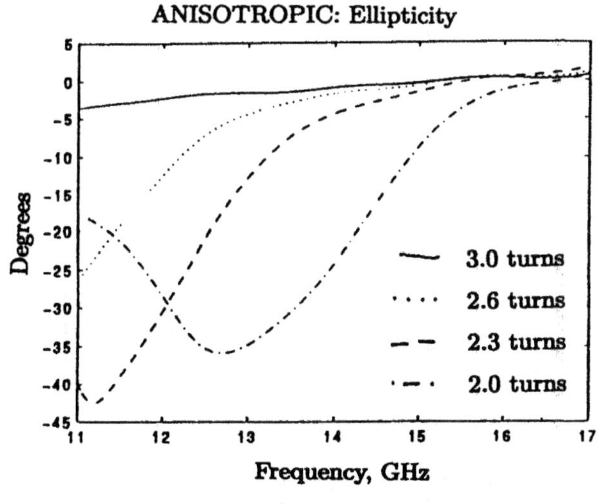

Figure 5. **Anisotropic rotation and ellipticity for 50 micron Cu wire wound around 1 mm Nylon line with a 1 mm pitch. The parent sample was first parted to produce 3 turn helices. After measurement, the individual panels were milled down to 2.6 turns; 2.3 turns and then 2 turns.**

taken into account. The rotation seems to be very sensitive to the helix diameter and current distribution as are the magnetic dipole moments.

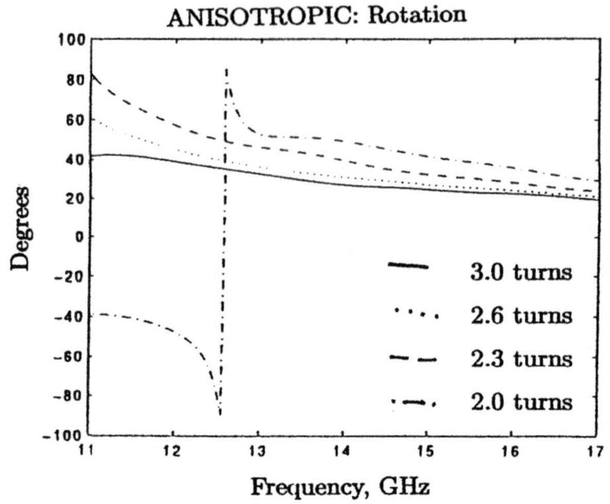

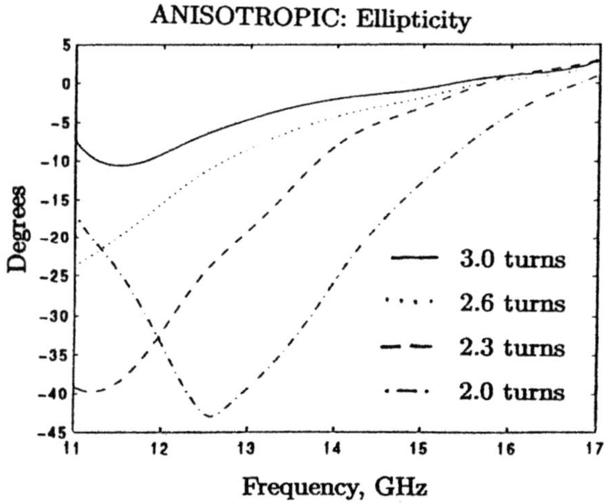

Figure 6. **Anisotropic rotation and ellipticity for 140 Cu wire wound around a 1 mm diameter Nylon line with a 1 mm pitch. As for the 50 micron wire study, samples were milled sequentially from 3 turns to 2 turns.**

Helix Length Study: Contour length of 50 micron wire (Figure 5)

As the contour length becomes closer to the half-wavelength within the material, a Cotton effect [3] resonance is observed along with strong circular dichroism.

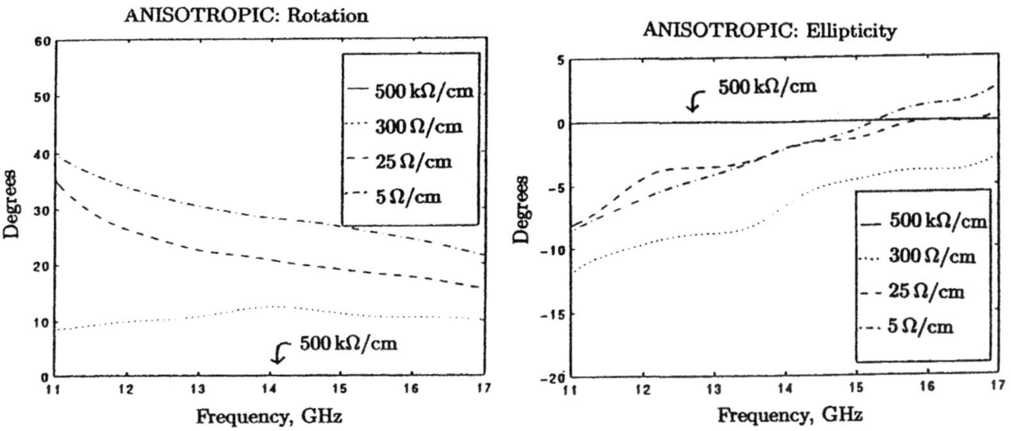

Figure 7. Anisotropic rotation and ellipticity of resistive filament wound around 1 mm diameter Nylon line with a 1 mm pitch. The resistive filament used were BASF Resistat carbon-coated 50 micron Nylon with a resistance of 500 kOhm/cm, Sauquoit Industries single filament Silver-coated 50 micron Nylon with a resistance of 300 Ohm/cm, a 10 filament bundle of these single filaments with a resistance of approximately 25 Ohm/cm and a multifilament thread of these single filaments with a resistance of 5 Ohm/cm. The Silver-coated onto the Nylon filament is reported to be between 60 and 125 Angstroms thick, well below the skin depth at our measurement frequencies.

Helix Length Study: Contour length of 140 micron Cu wire (Figure 6)

Compared to the 50 micron wire study, the 140 micron wire exhibits higher rotations with about the same ellipticity. Again the Cotton effect is observed when the contour length approaches the resonance condition. The resonance frequency is the same as for the 50 micron wire sample. Further comparisons are frustrated due to the samples' different outer diameters. The sharp structure of the Cotton effect resonance curve for the 2 turn sample is a presumed to be an inversion algorithm anomaly.

Resistive Filament Study: 500 kOhm/cm to 5 Ohm/cm (Figure 7)

The data indicates that helically wound insulators exhibit no chirality, presumably because there is no electromagnetic contrast at least up to resistances of 500 kOhm/cm. Higher conductivities produce stronger interactions.

Metal and Metal Coated Filament Comparison (Figure 8)

The data again indicates that higher conductivity filament or simply more conductive material produces stronger interaction. The resonance frequencies are exactly the same for the conductive and semiconductive filaments.

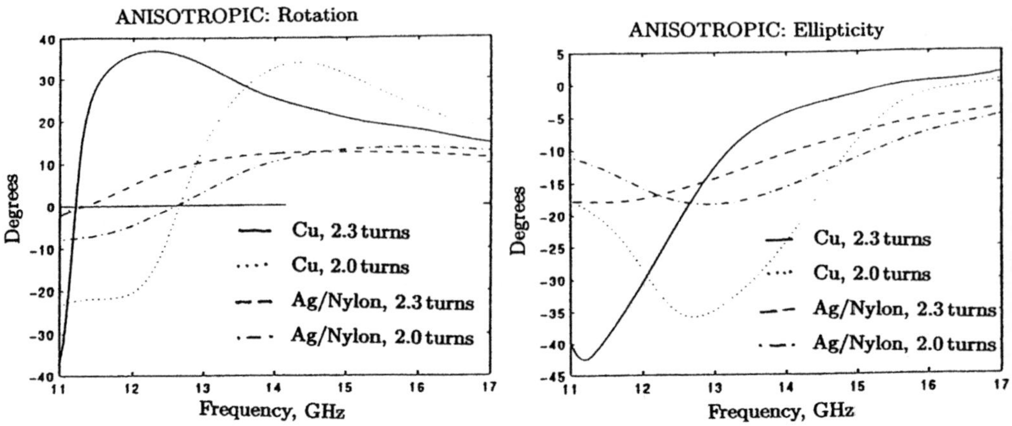

Figure 8. **Anisotropic rotation and ellipticity comparing similar helix geometries with different filament conductivities. 50 micron Cu wire is compared to Sauquoit Industries 50 micron Silver coated Nylon. The samples were measured at 2.3 turns, then 2.0 turns, so that resonance behaviour can be observed.**

Presumably a solid metal wire would lose its rotatory power if its thickness went below the skin depth.

3. Isotropic Samples

Continuous sections of wound line are passed through the chopping section of the SMFH device. The line is fed through brass guide with a two roll pinch drive cylinder which is run by a stepper motor. On the exit side of the brass guide, an overlocker blade is positioned and this is connected to a mild steel support which in turn is mounted to a rigid platform through a spring steel sheet. A solenoid is positioned as in Figure 1 under the mild steel support. The line feed and solenoid are controlled by a computer and engaged sequentially. The line can be chopped at a rate of 120 per minute. For sample manufacture, chopped wound line and chopped blank Nylon line are required. Lossless isotropic samples are simply an intimate random mixture of wound segments and blank Nylon segments. It should be noted that the volume fraction of void in such a sample is about 20%, thus reducing the overall dielectric constant. This mixture is poured into a PVC ring 6 or 8 mm thick with a diameter of 108 mm. The ring is supported between two Perspex sheets which have 108 mm diameter holes cut into them. A 10 micron Mylar sheet is skinned over these holes and tightened with a hot air gun to partially heat shrink the film. The random mixture is well supported by the Mylar while the surfaces remain flat. The

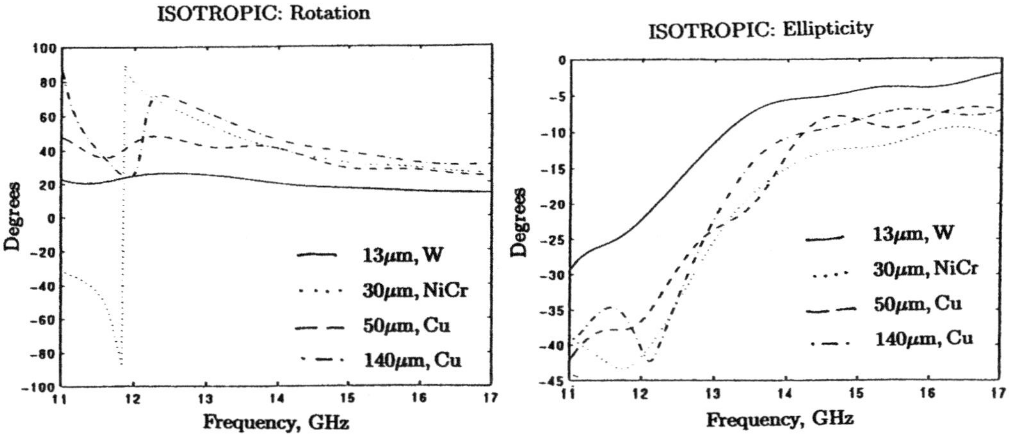

Figure 9. **Rotation and ellipticity for 3 turn helices diluted randomly in 3 mm blank Nylon to a number concentration of 70/cm3. This sample duplicates what was presented in Figure 3 but now as an isotropic dispersion.**

concentration of chiral inclusions was kept to 70/cm3 in all cases. Lossy chiral samples are made by mixing segmented inclusions with our castable lossy material. The mixture is packed into a 10 mm thick PVC ring with a diameter of 108 mm. Pressure is applied with glass plates to insure that the surfaces are flat and parallel. After cure, the sample is demoulded and a free standing lossy sample is produced containing chiral inclusions.

Isotropic Wire Diameter Study: 13 to 140 microns (Figure 9)

The data in figure 9 verifies the same trend as presented in Figure 3. Higher overall rotations are observed in this isotropic sample because the medium dielectric constant is lower due to the void contribution thus making the contour length of each helix closer to the resonance condition in the measurement band. This is evidenced by the ellipticity results. Also, a reduction of rotation below that of the anisotropic samples could have been expected by between 1/3 to 2/3 depending upon the principal values of the chirality tensor contributing to the scattering of these isotropic or anisotropic dispersions. However, the dielectric constants of these samples are not the same. The 30 micron Ni-Cr is unusual. This sample's resonance frequency has shifted upward out of proportion to the others in this series. However, we do not believe that this measurement should be taken seriously.

Resonance Behaviour: Electromagnetic Scaling (Figure 10)

Figure 10 shows rotation and ellipticity results on two samples of different geometries which can be compared through electromagnetic scaling [4] of

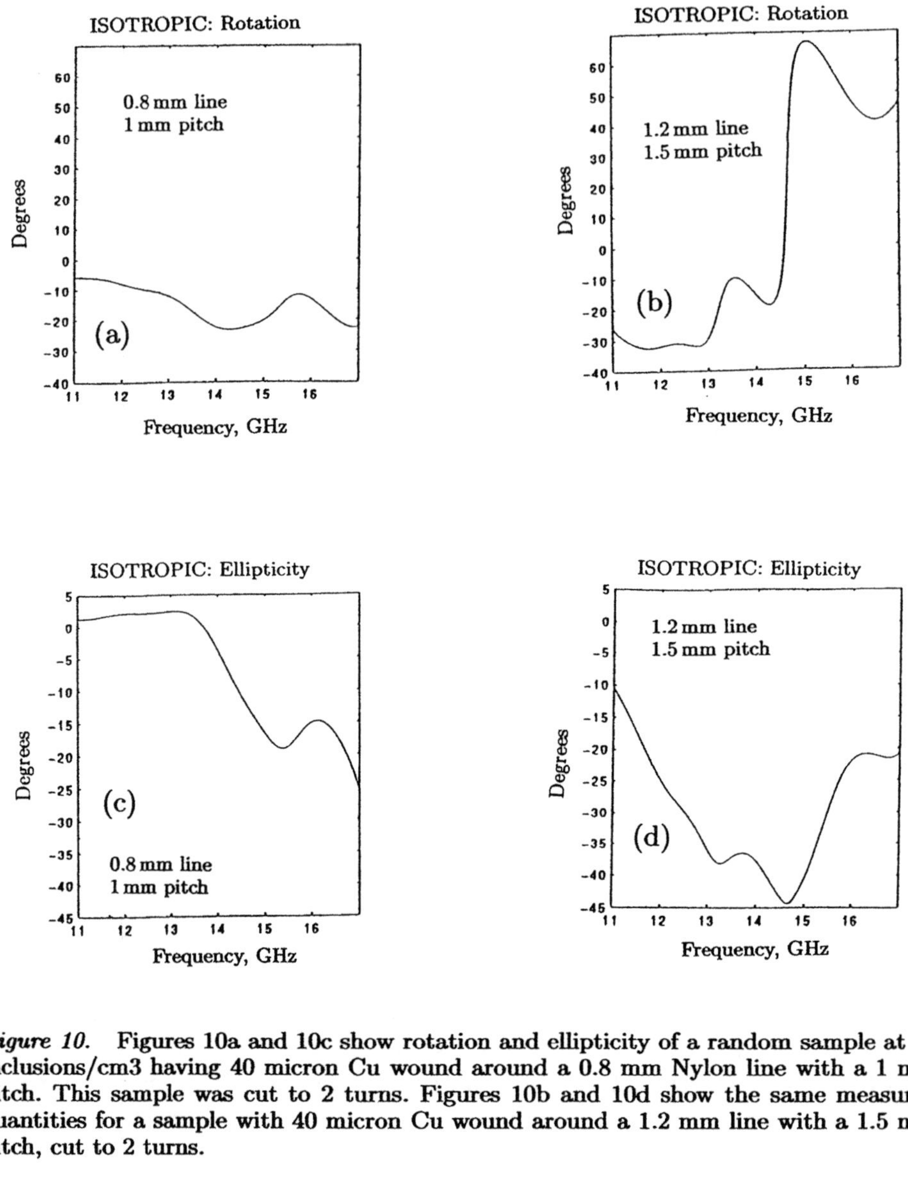

Figure 10. **Figures 10a and 10c show rotation and ellipticity of a random sample at 70 inclusions/cm3 having 40 micron Cu wound around a 0.8 mm Nylon line with a 1 mm pitch. This sample was cut to 2 turns. Figures 10b and 10d show the same measured quantities for a sample with 40 micron Cu wound around a 1.2 mm line with a 1.5 mm pitch, cut to 2 turns.**

these respective contour lengths relative to wavelength. The ellipticity results indicate that resonance is being approached at approximately 18 GHz even though the rotation is not increasing as in the anisotropic samples. This was found to be a result of a 20% variation in the cut length in this sample. A widening of the resonance profile is observed. To see what the profile looks like at resonance and above, a new sample was prepared after scaling the geometry to achieve resonance frequency close to 11 GHz. The geometry chosen was 40 micron Cu wound around a 1.2 mm line with a 1.5 mm pitch cut to 2 turns. The rotation and ellipticity for a random sample is shown in Figure 10b and 10d. Through electromagnetic scaling arguments, one could view this new sample as if it were the same as the first one but measured between 17 and 25 GHz. The complete half-wave resonance curve is now seen. Fine structure is observed in rotation and ellipticity due to the variation in cut length.

4. Lossy Media Containing Chiral Inclusions: Does Chirality Have an Effect on Absorption?

Absorbing or lossy samples containing chiral inclusions were prepared as described above. We chose to make 4 samples. These samples include a normal absorber made from our castable material, the same castable material containing 70/cm3, 2 mm blank Nylon segments, the same absorbing host containing 70/cm3 chiral inclusions made by winding 40 micron Cu wire around a 0.8 mm mandrel with a 1 mm pitch cut to 2 turns, and a racemic sample containing 35/cm3 RH and 35/cm3 LH inclusions. The geometry of the inclusions were chosen to exhibit the Cotton effect in our measurement band. The absorber dielectric constant and thickness were chosen so that a "quarterwave" Dallenbach [5] interference peak is also exhibited in this same band.

Figure 11a shows the rotation of these absorbing samples. The non-chiral absorber and absorber with Nylon have no rotation. The racemic sample has a finite rotation possibly as a result of improper mixing. The chiral sample exhibits mild rotation with a resonance frequency at 12.2 GHz. The permittivities are listed in Table 1 below. There was observed a small frequency dependence in the real part of the permittivity. Reported are average values.

In comparing the absorbing sample with Nylon to the samples containing chiral inclusions, it is apparent that an increase in bandwidth has occurred. Analysis of the permittivities of these samples shows that the imaginary part of the permittivity is larger after the introduction of the metal containing inclusions. However, achiral based theoretical predictions using these new permittivities show that the apparent increase in bandwidth obtained with the chiral inclusions is not entirely explained by the changes in permittivity. We note that the chiral and racemic absorbers have similar profiles. We conclude that the loss mechanisms introduced by chirality are mild compared to the dielectric loss provided by the host absorber. Also, chirality is not necessary to produce this loss. A racemic

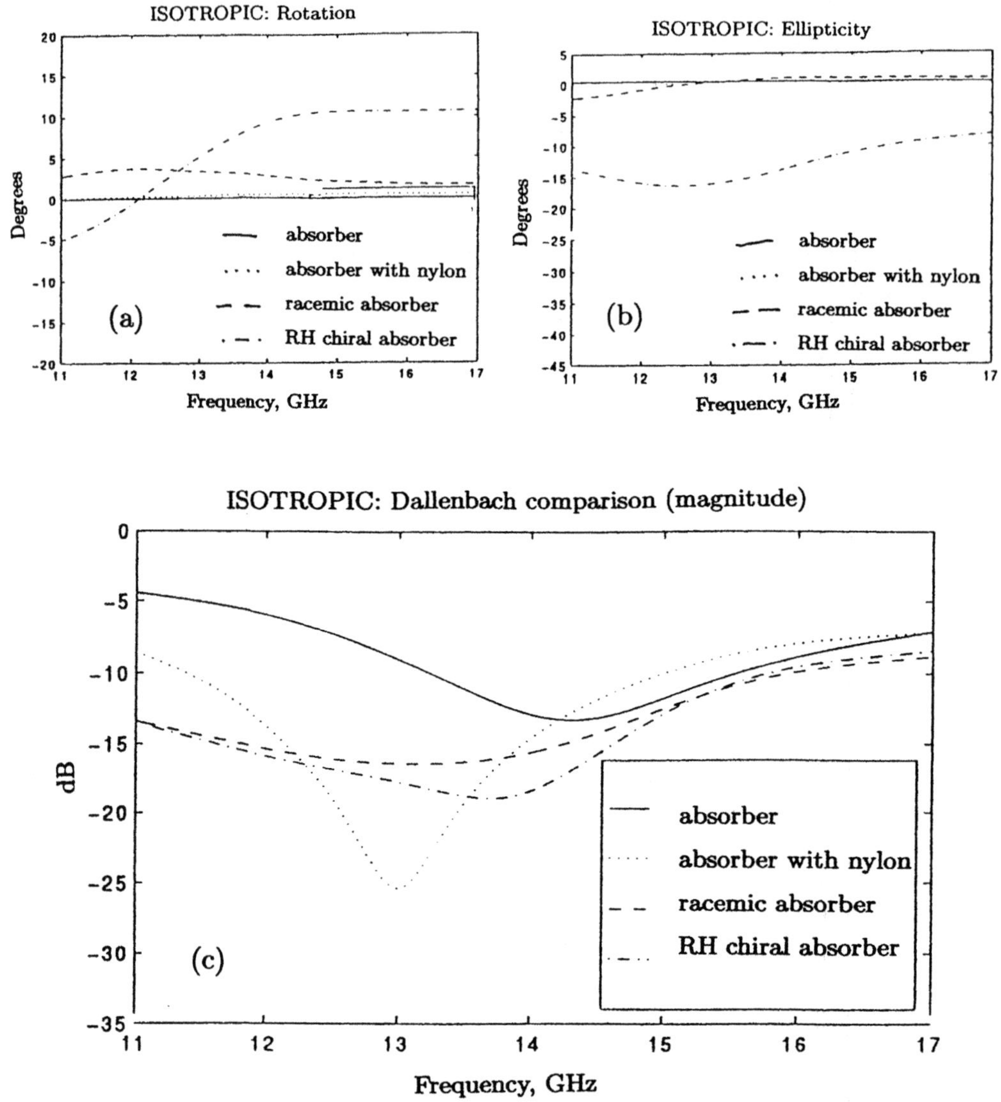

Figure 11. The rotation of the absorbing samples (Figure 11a). Figure 11b shows the ellipticity for these samples while Figure 11c shows the Dallenbach reflection for these samples.

sample produces a summed contribution equivalent to a chiral sample at the same concentration. This result has been recently verified theroretically by Tretyakov [6]. A non-chiral absorber can be made having a similar reflection profile as that obtained with chiral and racemic inclusions.

TABLE 1. Permittivities

	Re (Epsilon)	Im (Epsilon 11 GHz)	Im (Epsilon 17 GHz)
Absorber	2.4	0.4	0.2
Absorber/Nylon	2.7	0.45	0.45
Racemic absorber	2.7	1.0	0.5
Chiral absorber	2.7	0.8	0.4

5. Conclusions and Future Work

The SMFH technique has demonstrated itself as being an efficient method to produce chiral inclusions. The use of fine wire has been proven to be just as efficient in producing rotation in chiral substances as coarse ones as long as the contour lengths are adjusted for outside resonance applications. Rotations seem to be sensitive to the overall helix diameter as evidence points to induced currents flowing on the outside of the wire. The helix diameter will also partially control the helix self-inductance and hence the relative amplitude of the induced current and therefore the magnetic dipole moments. Resistive filament produces less rotation than a solid metal wire of similar diameter. Presumably solid metal wire of a diameter below the skin depth would behave similarly. The chirality tensor appears to be biaxial for the anisotropic samples as pronounced rotations are observed orthogonal to the helix axis in semi-infinite helices and short ones alike. Even a Cotton effect resonance behaviour is observed in the orthogonal direction. Finite rotations are observed for propagation along the helix axis. The 1.2 mm diameter, two turn, 40 micron Cu wire sample shown in Figure 4a produced a rotation down the helix axis of a two layer configuration of 2.25 increasing linearly to 6 degrees at 11 and 17 GHz respectively.

The SMFH device can be used in the production of hypothetical materials such as the artificial magnetic racemic medium [7] and Tellegen [8] medium. The Tellegen material (a non-reciprocal bi-isotropic material) would be made by winding Iron wire around an axially poled ferroelectric PVDF mandrel, then magnetically poling the Iron wire along the helix axis. Coincident and fixed permanent electric and magnetic dipoles could be produced. The Tellegen material could also be rendered chiral depending upon the pitch to diameter ratio the Iron is wound to. The racemic artificial magnetic material could be made by co-winding RH and LH helices around a mandrel and during this process periodically stopping the winding process, engage one of the spindles to wind a circular loop around the line to produce a capacitive coupling. Other methods of producing electrical continuity in fine wire could be found. This continuous line could be parted to segments containing one termination, preferably at the end of a segment. Another material which is under investigation at the moment is a magnetically cored chiral material. This material demonstrates the versatility of the SMFH technique in its ability to manufacture exotic materials. A realm of new science opens up when one can vary the properties of the core. The magnetic core being studied, exhibits a strong permanent magnetic dipole perpendicular to the line axis. This line is wound with 40 micron Cu wire and oriented uniaxially with the north pole of each magnetic core aligned perpendicular to the helix axis and sample faces. Rotation observed in both the parallel and perpendicular orientations was about 13 degrees. We are studying this material as a potential tuneable chiral material when strong DC electric current is driven through the semi-infinite helix windings.

References

1. Kuehl, S.A., Grové, S.S., Smith, A.G., and Theron, I.P. (1995) Manufacture of chiral materials and their electromagnetic properties, in A. Sihvola et al. (eds.) *Proceedings of Chiral'95*, The Pennsylvania State University (11 to 14 October 1995), 13-16.
2. Patent pending.
3. Mason, S.F. (1982) *Molecular Optical Activity and the Chiral Discriminations* Cambridge University Press, Cambridge, p. 20.
4. Sinclair, G. (1948) Theory of models of electromagnetic systems, *Proceedings of the I.R.E.*, November 1948, 1364-1370.
5. The Dallenbach configuration refers to plane wave scattering normal to metal backed flat slab of material.
6. Tretyakov, S.A., Sochava, A.A., Simovski, C.R. (1996) Influence of chiral shapes of individual inclusions on the absorption in chiral composite coatings, *Electromagnetics*, **16**, 113-127.
7. This type of artificial magnetic material consisting of bihelix inclusions was proposed by Alexei Vinogradov at *Chiral'95*.
8. The Tellegen material was referred to by Ari Sihvola at *Chiral'95*.

ON CHIRALITY MEASUREMENTS IN CIRCULAR WAVEGUIDES

G. BUSSE, J. REINERT, M. KLEMT, A. F. JACOB
Institut für Hochfrequenztechnik
Technische Universität Braunschweig
Postfach 3329
D-38023 Braunschweig
FR Germany

1. Introduction

Until now, only free space setups were applied to measure the three scalar constitutive parameters of chiral materials. In principle, the inversion procedure to extract the constitutive parameters from measured scattering data is very simple [1]. But, in practice, difficulties concerning the generation of plane waves, the high accuracy demands, and the limitations regarding the sample thickness lead to expensive setups with limited accuracy [2]. Although waveguide measurement setups are well-known for their stability, reproducability, and accuracy only a few works, for instance [3] and [4], have analysed them in view of chiral material parameter extraction. Recently, a quasi-analytical solution for the inverse problem in circular waveguides was proposed [5]. It requires measurements of chiral slabs of two different lengths and yields, in theory, accurate solutions for the chirality parameter β and the product $\epsilon\mu$, i.e. permittivity times permeability. Here and in the following, the Drude-Born-Fedorov was chosen to define the chiral material.
This study deals with the practical aspects of chirality measurements in circular waveguides. To this end, the experimental setup is described as well as the manufacturing process of a foam-based chiral material. Next, some preliminary experiments are presented that were performed to estimate the behaviour of the chiral material in quality. The results are confirmed in quantity by the chirality parameter, which is calculated using the inversion procedure presented in [5].

2. Setup and Chiral Material

Our experimental setup (Fig. 1) is based on a circular aluminum waveguide of 50mm diameter, which fixes the fundamental mode region from 3.51GHz to 4.59 GHz. To efficiently excite and receive the fields of the fundamental TE_{11}-mode simple linear antennas are used. In their vicinity, several irregularly spaced tuning screws lower the antenna reflection to less than -20 dB over almost the whole considered frequency range. To avoid multiple reflections w.r.t. all polarizations, absorber sections are mounted at the back of each antenna section. They are formed by four 600mm long wedge-shaped staggered absorbers sticked to the waveguide walls. Reflections are then reduced below -30dB at all azimuthal angles. The sample holder, a waveguide section of 600mm, is fixed between both antenna sections using threaded couplings. This ensures reproduceable and

A. Priou et al. (eds.), Advances in Complex Electromagnetic Materials, 333–339.

low loss connections (also to the calibrations standards). In additon, the sample holder can easily be rotated, a useful feature to average out the material inhomogeneities. Last, the antennas are connected to the ports of a Vector Network Analyzer HP8510C via coaxial cables. Before the scattering parameters are measured, a TRL-calibration [6] is performed to cancel out systematic errors and to define suitable reference planes. Then, the measured response is due exclusively to the sample holder and its content.

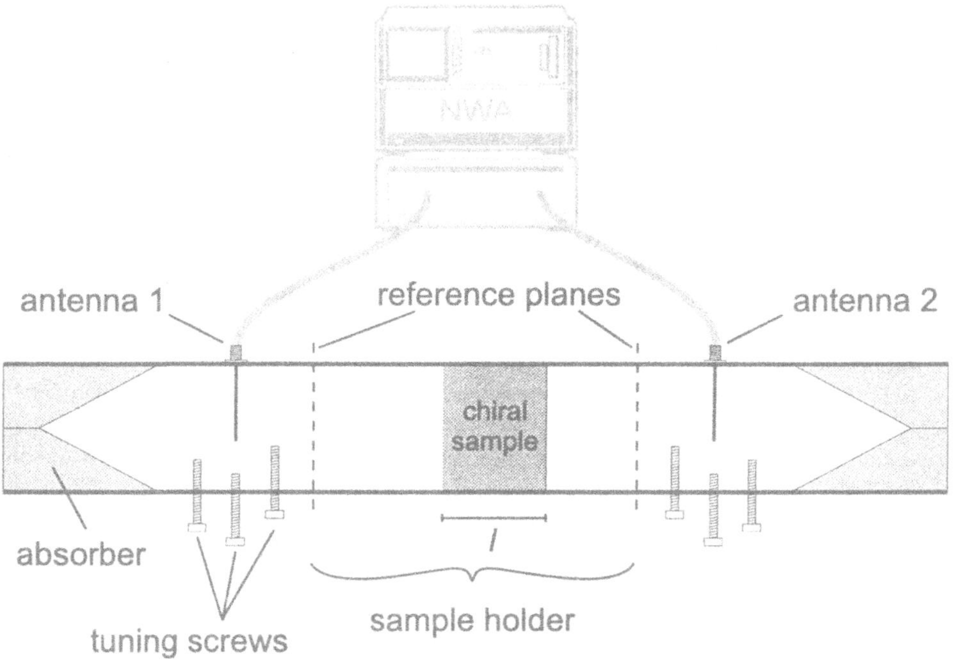

Figure 1. Sketch of the measurement system.

It should be noted, that, in contrast to the theory in [5] where circular rotating waves were considered, it is more convenient, in practice, to handle linear polarized waves. Because of reciprocity forward and backward transmission are identical in this case. Thus, to avoid redundance, forward transmission is measured at different rotation angles of the receiving antenna, for instance at 0° (copolar) and 90° (crosspolar).

The chiral objects of the material are right-handed three turn copper helices of 4mm diameter and height. Although the wire length is 37.5mm, i.e. half wavelength at 4GHz, the geometry of the helix was not optimized in any way. As shown in Fig. 2, the helices were placed in Teflon molds with several injection channels and a matrix of about 2000 hemispherical recesses of 6mm diameter. Six molds were stacked and pressed together, thus forming spherical chambers connected by small channels.

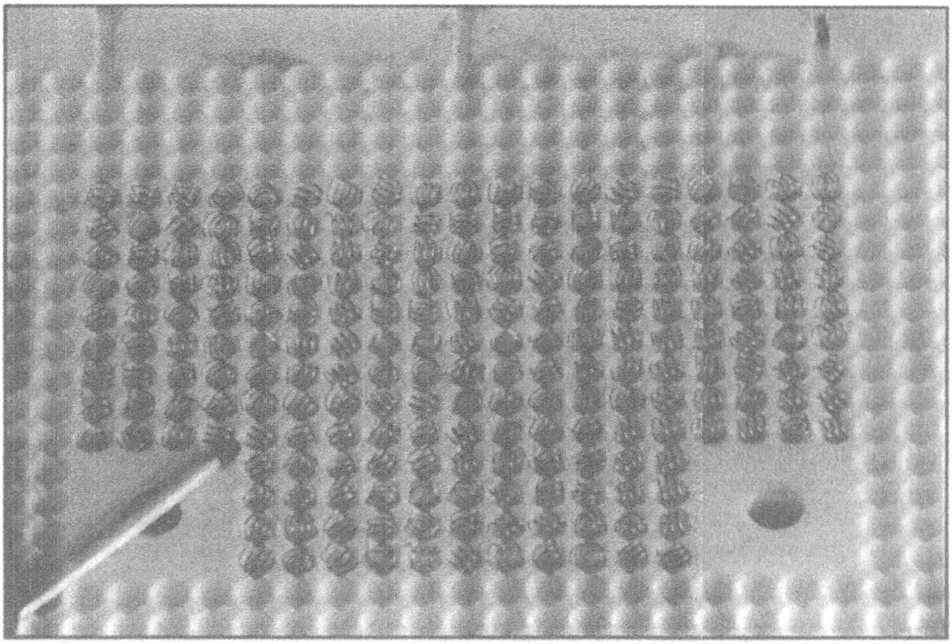

Figure 2. A part of the opened mold with an arrangement of copper helices.

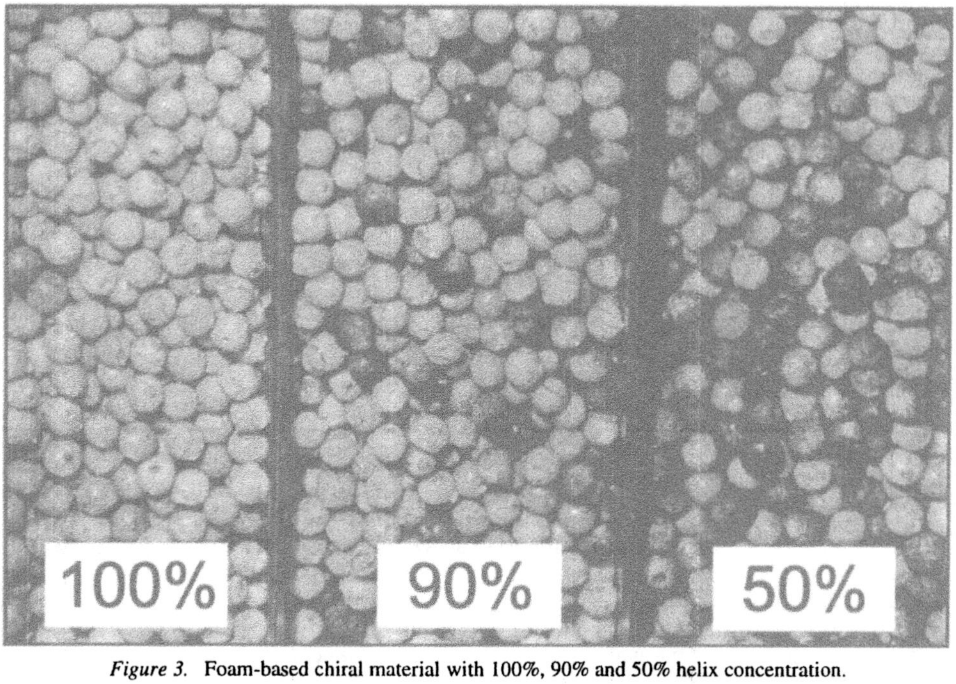

Figure 3. Foam-based chiral material with 100%, 90% and 50% helix concentration.

Then, conventional two components Polyurethan foam, which does not adhere to Teflon, was injected. After a curing time of two hours, the foam wafer could be easily separated from the mold and cut into its elements. In addition, dark colored foam spheres without inclusion were produced as well. By mixing chiral and empty spheres material with reduced chiral concentration is feasible (see Fig. 3). In comparison to the commonly used epoxy-resin-based media, this material with its discrete structure has some important advantages for laboratory purposes: It is lightweight, cheap, can be easily changed w.r.t. shape and chiral concentration, and is easy to handle.

3. Measurement and Results

To qualitatively check the properties of the chiral material, in a first step, the rotation angle produced by a material with 100% chiral spheres was measured as a function of the slab length at 3.85 GHz, 4.0 GHz and 4.3 GHz according to [3]. Refering to Fig. 4 the rotation increases, as expected, almost linearly with slab length, but decreases with frequency. The latter result suggests that the resonance of the material is below 3.85 Ghz. The chirality of the sample was confirmed, at 4 GHz, by the linear dependence (not shown here) of the rotation angle on chiral concentration[1].

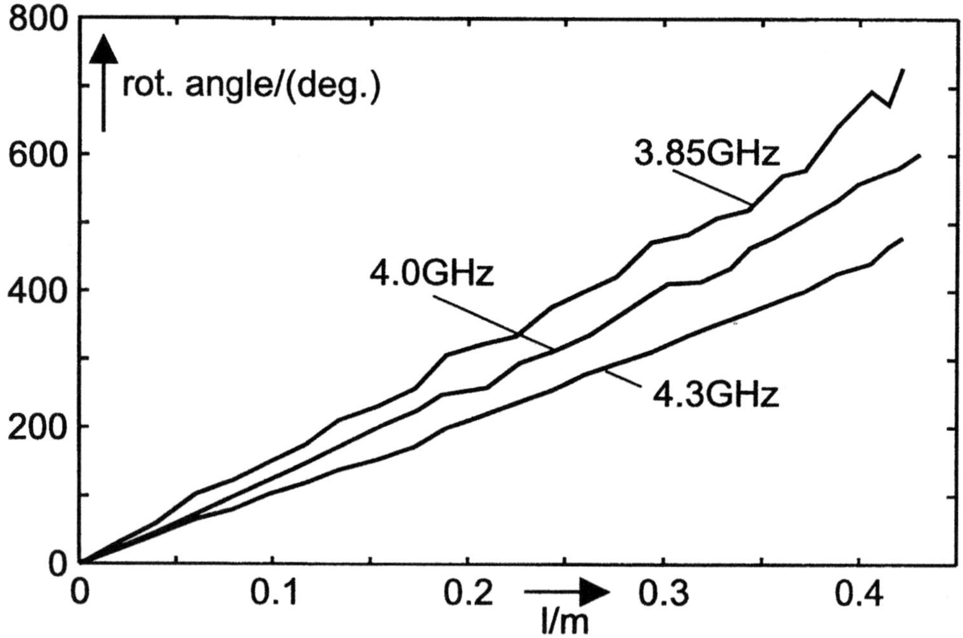

Figure 4. Measured rotation angle versus slab length for 100% chiral concentration at 3.85GHz, 4.0GHz, and 4.3GHz.

Caveat: Because of the waveguide dispersion this dependence is not necessarily linear.

In a second step, as a verification of the measurement procedure, the constitutive parameters of wood shavings, whose low dielectric constant of about ϵ_r=1.1 ensures fundamental mode propagation, were measured. Three slabs of 10, 20, and 30cm length and the same density were placed in the sample holder and, then, the scattering parameters

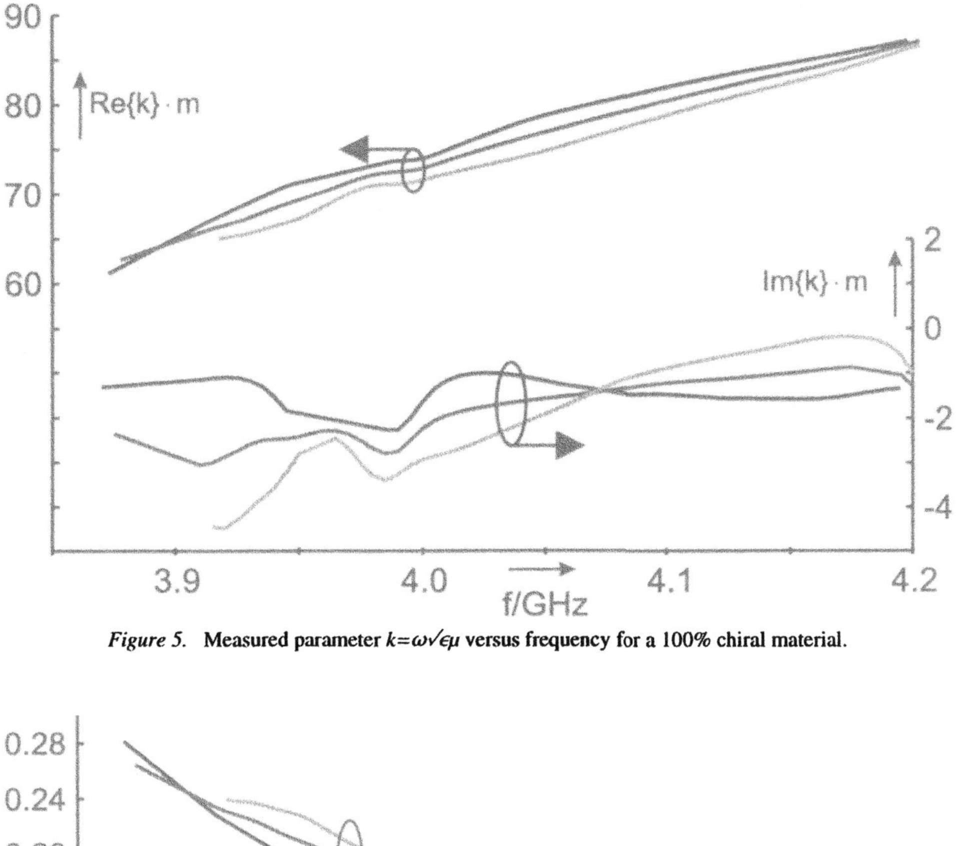

Figure 5. Measured parameter $k=\omega\sqrt{\epsilon\mu}$ versus frequency for a 100% chiral material.

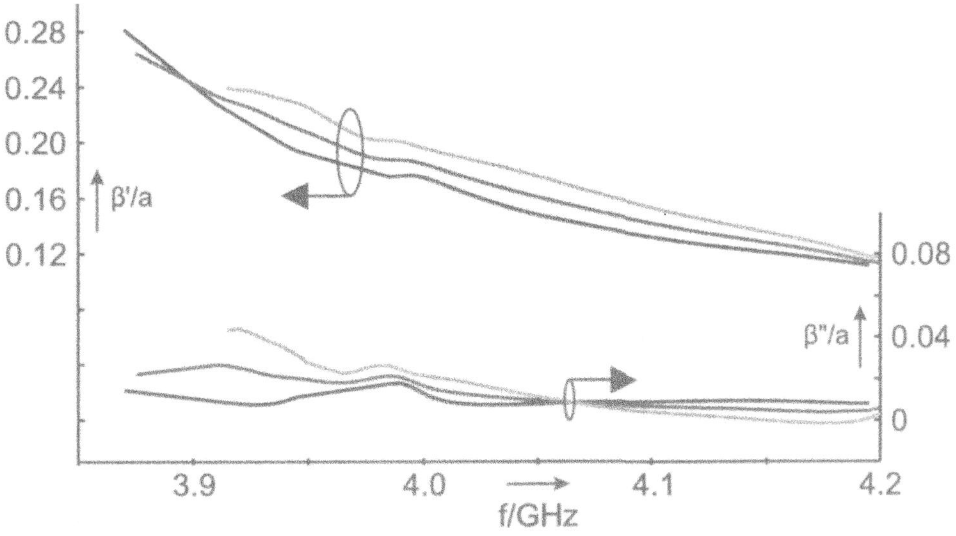

Figure 6. Normalized chirality parameter β/a versus frequency for a 100% chiral material (a=0.025m).

were measured at both co- and crosspolar antenna orientations. From the latter, the chirality parameter and the parameter k, which is related to the product $\epsilon\mu$ by $k=\omega\sqrt{\epsilon\mu}$, were extracted using the inversion of [5]. While, as expected, the chirality parameter is about zero, the parameter k is in excellent agreement with the values obtained by accurate, conventional resonator measurements. Deviations for the imaginary part of k are caused by the lossy wall of the aluminum waveguide, which was not considered for the inversion. The results obtained for all three combinations of slab lengths are remarkably close - a hint to the accuracy of the method.
Finally, the co- and crosspolar scattering parameters of 15, 20, and 25cm long slabs containing 100% chiral spheres were measured. Here, on average, 61% of the waveguide volume were filled with helix loaded spheres corresponding to 5.45 spheres per cubic centimeter. Six measurements were taken at different rotation angles of the sample holder to average out inhomogeneities. Results of the inversion are shown in Figs. 5 and 6 for the different combinations of slab lengths. The normalized chirality parameter confirms in quantity what has been suggested earlier: We are on the back of the chiral resonance and, thus, chirality decreases between 3.85 and 4.2 GHz. In detail, the real part diminishes roughly from 0.3 to 0.1 where the deviations of the curves are within a few percents. The positive imaginary part decreases as well, however, additional measurements should be done to reduce the uncertainties.

Last, the same measurements were done for a chiral material of 60% chiral concentration. As can be seen from Fig. 7, the normalized chirality is reduced approximately from 100% to 60%. In addition, the absolute value of k is increased and, thus, cutoff of this 60% chiral material is shifted towards lower frequencies.

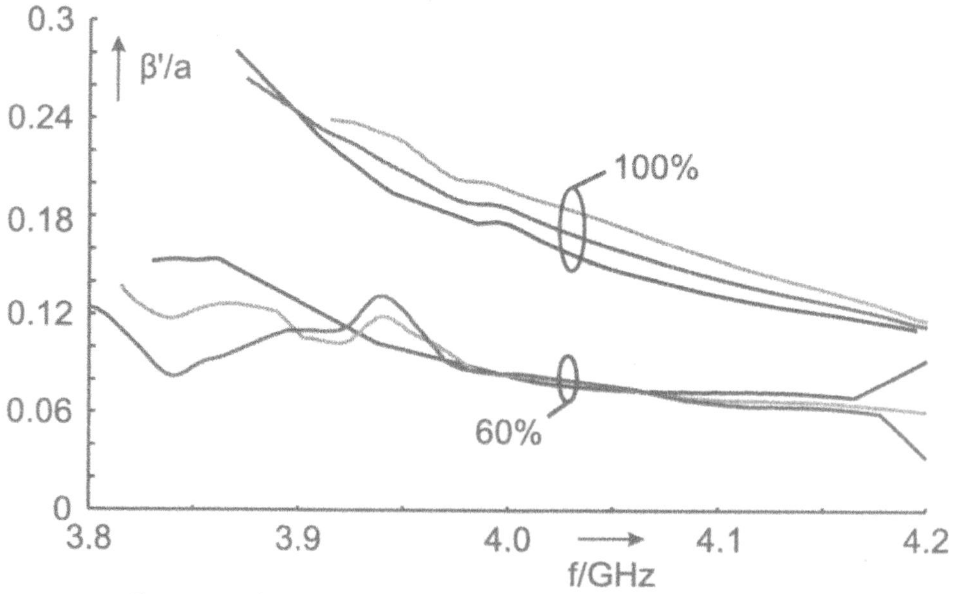

Figure 7. Real part of the normalized chirality parameter β'/a versus frequency for 60% and 100% chiral concentration.

4. Conclusion

The chirality parameter of a foam-based chiral material of discrete structure was extracted from a rugged circular waveguide setup. The simple measurement procedure used leads to plausible and reproducible results, which confirm some qualitative findings of preliminary experiments. The influence of the chiral concentration could be assessed and the material resonance be revealed.

References

1. Cloete, J. H. and Smith, A. G. (1992) The constitutive parameters of a lossy chiral slab by inversion of plane-wave scattering coefficients, *Microwave and Optical Technology Letters*, Vol.5, No. 7, 303-306.

2. Smith, A. G., Cloete, J. H., and Theron, I. P. (1995) The accuracy of a free-space system for measuring the electromagnetic properties of microwave chiral materials, *Proc. 4th Intern. Conf. on Chiral Bi-isotropic and Bi-anisotropic Media*, Pennsylvania State University, 13-16.

3. Hollinger, R. D. Varadan, V. V. and Varadan, V. K. (1992) Experimental characterization of isotropic chiral composites in circular waveguides, *Radio Science*, Vol. 27, No. 2, 161-168.

4. Mariotte, F. and Engheta, N. (1993) Reflection and transmission of guided electromagnetic waves at an air-chiral interface and at a chiral slab in a parallel-plate waveguide, *IEEE Trans. MTT*, Vol. 41, No. 11, 1895-1906.

5. Reinert, J., Busse, G. and Jacob, A. F. (1995) A procedure to extract the chirality parameter from waveguide measurements, *Proc. 4th Intern. Conf. on Chiral Bi-isotropic and Bi-anisotropic Media*, Pennsylvania State University, 9-12 .

6. Hewlett-Packard (1987) Applying the HP 8510B TRL calibration fon non-coaxial measurements, *Product Note 8510-8*, Palo Alto, CA.

EXPERIMENTAL INVESTIGATION OF RESPONSE OF CHIRAL MEDIA AND "CHIRAL MEDIA–FERRITE" STRUCTURES TO MICROWAVE RADIATION AND GOVERNING MAGNETIC FIELD

G.A. KRAFTMAKHER AND YU.N. KAZANTSEV
Institute of Radioengineering and Electronics
Russian Academy of Sciences
11, Mokhovaya str., 103907, Moscow, Russia

Abstract. In this paper, the results of experimental investigations of different chiral structures are reported. Individual inclusions are measured in resonant cavities and in waveguides, reflection and transmission coefficients of planar composite samples are measured in free space. The emphasis is on novel, more complex inclusions (multigone spirals, planar spirals, etc.) and on ferrite-controlled composites.

1. Introduction

Microwave properties of two kinds of chiral objects are under investigation. First, we investigate microwave properties of hollow cylinders whose side walls consist of isolated and oriented helices sandwiched between two dielectric films. The helices form a multistart helical chiral structure.

The second example is a two-dimensional structure. The samples are planar composites on a dielectric substrate with oriented helices formed of thin copper films. Similar samples with single planar helices are also investigated. Frequency dependencies of the reflection and transmission coefficients of the samples under investigation in the standing-wave regime with moving short and in the traveling-wave regime with matched load in a rectangular waveguide as well as in free space are studied.

It has been found that there are resonances which are effectively excited in the standing-wave regime by microwave magnetic fields. They are considered as resonances of magnetic permeability in resonant-cavity mea-

A. Priou et al. (eds.), Advances in Complex Electromagnetic Materials, 341–358.

surements, in analogy with measurements of small samples of conventional magnetic materials.

Besides, there are resonances which can be excited in the standing-wave regime by microwave electric fields. These resonances can be, similarly, thought as resonances of the permittivity of dielectric samples in the resonant-cavity measurements.

It was shown that one can control the resonant frequency and intensity of the resonances in our chiral samples, using combinations with ferrite elements in external constant magnetic fields. Microwave properties of different structures "chiral medium–ferrite" are investigated. The influence of external constant magnetic field on the frequency dependencies of structures "chiral medium–ferrite," such as cylindrical samples with ferrite cores inside and of planar samples with thin ferrite layer at different conditions of excitation are studied. Resonance phenomena in integrated systems with magnetized ferrite elements and chiral inclusions medium are investigated.

The possibilities of identification, separation and excitation of these resonances are discussed. The results can be important for the design of new chiral particles and new artificial composites containing helical inclusions.

2. Investigated Samples and Measurement Methods

We have manufactured several samples of artificial chiral and chiro-ferrite media which are well suited for observations and investigations of electrically controlled microwave resonance phenomena. These samples are also designed so that it is possible to study their microwave by the direct resonant-cavity method.

Figures 1 and 2 show the geometry of the investigated samples. We have prepared the samples in the form of hollow cylinders whose side walls consist of many isolated and oriented helices (one- or one-half turn) sandwiched between two dielectric films (Fig. 1). The cylinders are 2 – 2.5 mm in diameter. The sample length is $L = 5 - 50$ mm, the tungsten wire diameter is 0.015 mm, δ is the distance separating starting positions of neighboring helices. $\delta = 0.200 - 0.070$ mm, which is smaller than the pitch of a single helix. The winding angle α is the angle between the tangent to the helix and the helix axis. The helix axis, which coincides with the sample axis, is marked as O-axis. When α decreases, the helix extends (the helix pitch increases), and it turns into a straight line at $\alpha = 0$. When α approaches 90°, the helix squeezes into a circle. The winding density of the samples equals 40 turns per centimeter, which corresponds to 0.2% metal volume fraction.

We reported about such samples in [1]–[4], where we investigated various samples of the same diameter but with different winding angles α, *i.e.*,

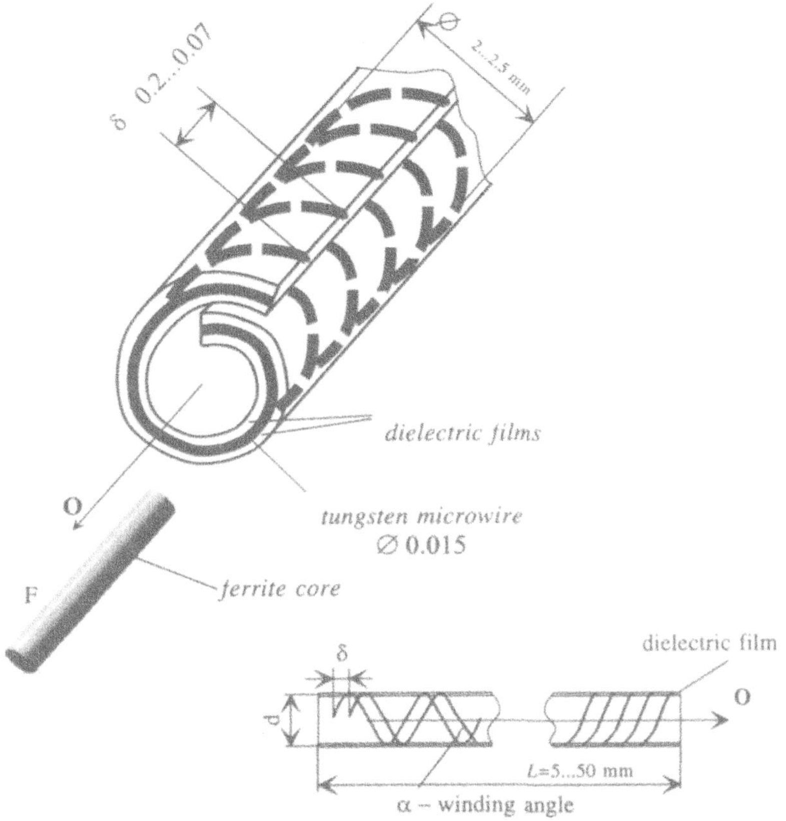

Figure 1. Cylindrical sample with oriented multistart helical structure formed by 1.5-turn helices

with different pitches and lengths of the helical turns. In [1, 2] we reported the results of measurements of the real part of the permeability μ', ($\mu' - 1 \gg 0$). In that measurements we detected giant magnetic losses (high μ'' values), which take place in microwave electromagnetic fields. In that work we used direct measurement methods. We investigated possibilities to control the permeability components μ' and μ'' in structures with ferrite elements (cylindrical samples with a ferrite core inside) by changing the applied constant magnetic field $\mathbf{H_0}$ [3, 4].

In this report we present the results of our latest investigations as development of our previous studies.

We also investigate different samples (Fig. 2), which are arrays of planar elements positioned on a dielectric substrate. The individual elements are oriented helices made of thin copper films (helix diameter is 3 mm, the

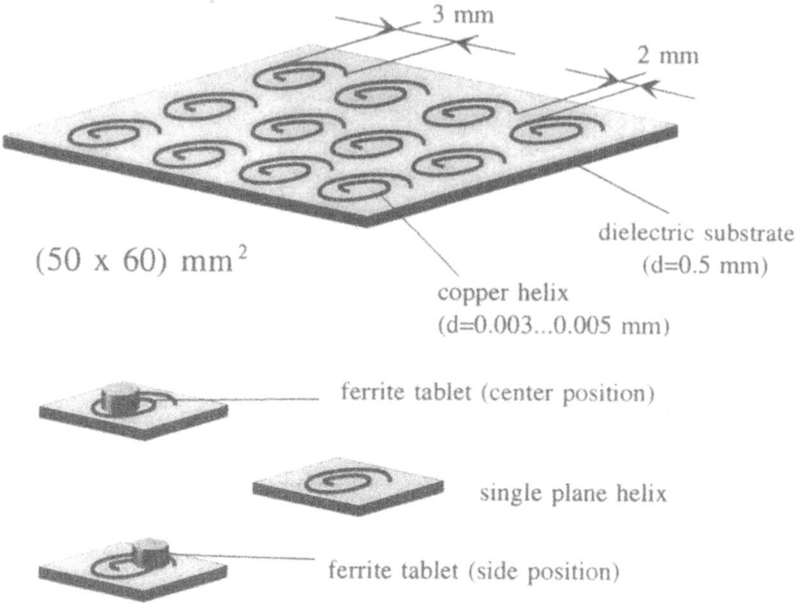

Figure 2. Planar samples

thickness of the copper film is 0.003 – 0.005 mm, the width of the wire layer is 0.1 – 0.05 mm, the distance between the helices is 2 mm). Similar samples with a single plane helix are also investigated.

To study possibilities for electrical control of the particle excitation, we study structures "chiral medium-ferrite" (see Fig. 2). For that purpose, we position a small ferrite tablet on the surface of a planar helix, and study the excitation effects for different positions of the tablet with respect to the helix. In particular, we distinguish the center position and the side position of the ferrite element (Fig. 2).

The measurements are made in rectangular waveguides, cavities and in free space, using swept-frequency network analyzers in the frequency band from 3 to 17 GHz. The measurement technique is similar to that described in Refs. [2, 5]. For discussions of the problems of the methodology of chiral media measurements we refer to Refs. [6]-[12].

In this report we measure the reflection and transmission coefficients of the samples as functions of the frequency, the distance l between the samples and a moving short, and of the constant magnetic field $\mathbf{H_0}$. Standing-wave and traveling-wave regimes with a short end and a matched load are used; various positions and orientations of the samples and of the external constant magnetic field $\mathbf{H_0}$ are tested.

3. Experimental Results

3.1. CHIRAL SAMPLES

Fig. 3 shows the transmission T and reflection R coefficients of the cylindrical sample with the winding angle $\alpha = 70°$ and the cylinder length $L = 5$ mm as functions of the frequency in the resonance area (8 – 12 GHz), the traveling-wave regime in the waveguide being used.

At Fig. 3 we observe resonance effects for two orientations of the sample on the plane which is perpendicular to the wave vector **k**. In the first case, the microwave electric field **E** is perpendicular to the O-axis of the sample, the microwave magnetic field **h** is parallel to the O-axis. In the second experiment, **E** is parallel to the O-axis, and **h** is perpendicular to that axis. For different orientation, we observe resonance phenomena at the same resonant frequency, but the resonance values of the measured coefficients depend on the sample orientation.

Similar dependencies are observed for such samples with different winding angles α at their own resonant frequencies. The dependence of the reflection and transmission coefficients at the resonant frequency on the sample orientation is strong for small α, which corresponds to very extended turns, and for α approaching to 90°, which corresponds to highly compressed turns. The intensity of the resonant effects and the width of the resonance area do not practically depend on rotation of the sample around the helix axis.

In Fig. 4, the frequency dependencies of the transmission and reflection coefficients of a single planar helix are depicted in the resonance area. Waveguide measurements were made in the traveling-wave regime. The geometry of the helix is also shown in Fig. 4. The axes x, y, z refer to the helix position.

For both the cases when the wave vector **k** is perpendicular to the helix axis (and **h** points in the axis direction), and when **k** is parallel to the O-axis (**h** is orthogonal to that), the resonance response reaches its maximum when the field **E** is parallel to the z-axis. The resonance is practically invisible if **E** is parallel to the x-axis (orthogonal to the z-axis). So, the resonance values of the reflection and transmission coefficients (as well as the absorbed energy) depend on helix rotation around the helix axis, and do not depend on the rotation about the z-axis. Hence, when the **E**-field orientation with respect to the z-axis remains the same, the resonance phenomena are the same for all possible orientations of the microwave magnetic field **h**.

It is necessary to note such an interesting peculiarity of the resonance of a plane helix as the existence of a maximum and a minimum of the reflection coefficient curve.

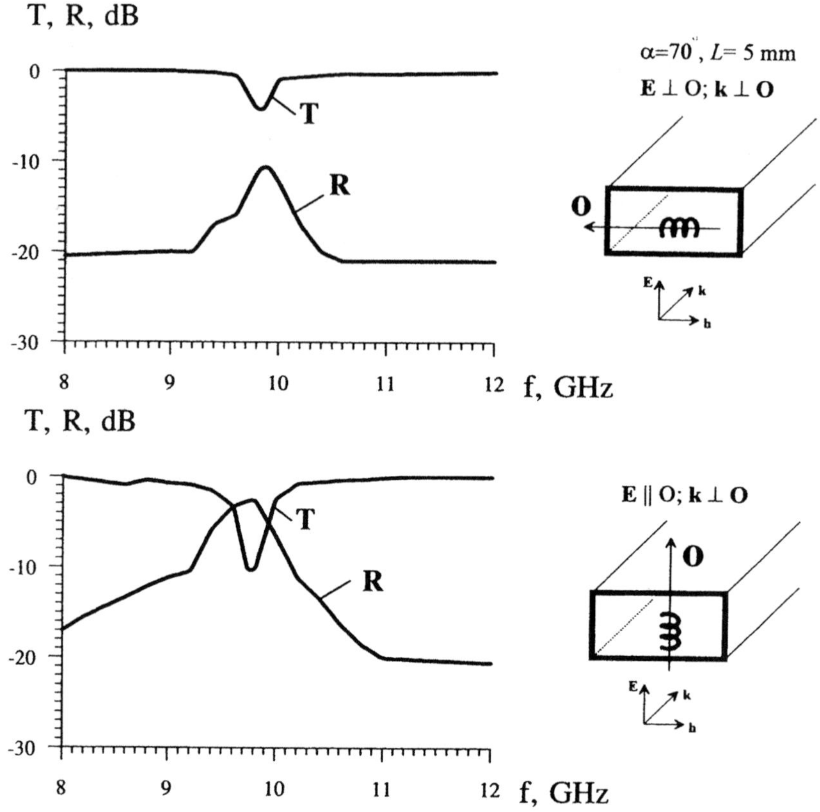

Figure 3. **Transmission T and reflection R coefficients of the cylindrical sample**

In Fig. 5 one can see the transmission T and reflection R coefficients of the planar sample measured in free space as functions of the frequency. We observe the same resonance effects connected with excitation of particles when the microwave electric field **E** is parallel to the z-axis. There is no resonance when **E** is parallel to the x-axis. We see a minimum of the transmission coefficient and the corresponding maximum of the reflection coefficient at the resonant frequency. Besides, there is a minimum of the reflection coefficient, similarly to the case of a single planar helix (Fig. 4). Perhaps, that is due to a combination of the microwave resonance properties of a helix and the microwave properties of thin conductive films.

We have also investigated the reflection coefficient R_m as a function of distance l between the sample and a moving short for various orientations and positions of the samples: in the center of the waveguide cross-section, and at a side wall. In these measurements, the standing-wave regime in the

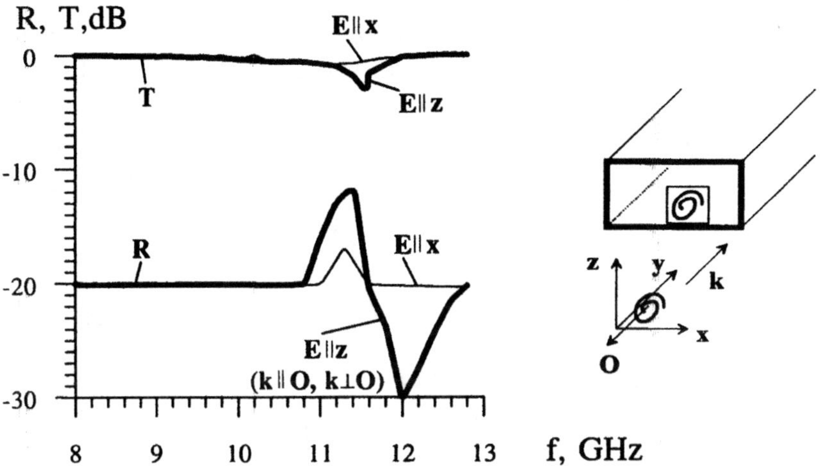

Figure 4. Transmission T and reflection R coefficients of a single planar helix

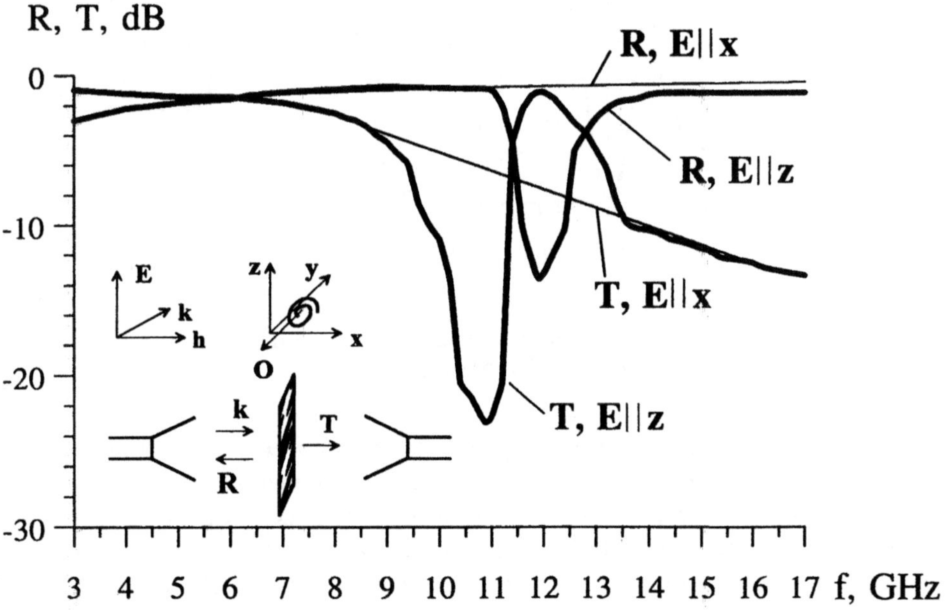

Figure 5. Transmission (T) and reflection (R) coefficients of the planar sample in free space as functions of the frequency

waveguide was realized.

In Fig. 6, dependencies of the reflection coefficient on the distance l

for the cylindrical sample with $\alpha = 70°$, $L = 5$ mm are shown. We see periodical variations of the absolute value of the reflection coefficient, which can be attributed to the distribution of the microwave fields **E** and **h**.

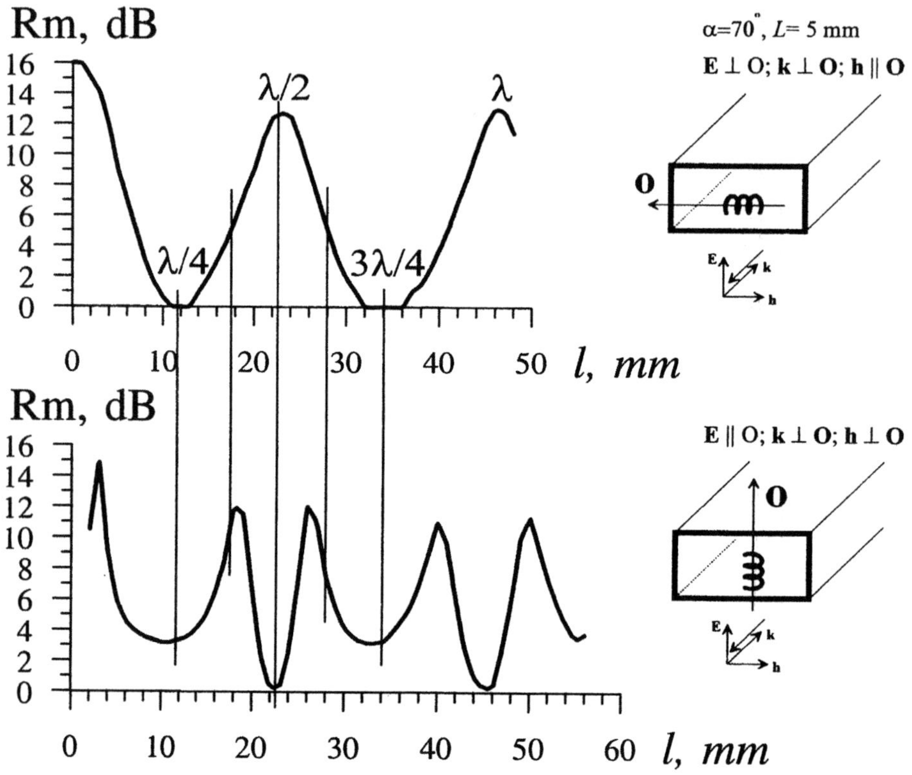

Figure 6. **The reflection coefficient R_m from the cylindrical sample as a function of the distance l between the sample and the moving short**

If the electric field **E** is orthogonal to the helix axis, and **h** is parallel to that (the case depicted in Fig. 6a), the reflection coefficient reaches maxima at $l = 0, \lambda/2 \cdots$, *i.e.*, when the sample is in a maximum of the microwave magnetic field **h**. The reflection is small ($R_m = 0$) at $l = \lambda/4$, when the sample is in the maximum of the microwave electric field **E**. We conclude that the sample is effectively excited resonance is excited in standing-wave regime by the microwave magnetic field **h** parallel to the inclusion axis. This effect was displayed as a resonance of the effective permeability in the cavity measurements [2, 3].

In the case when **E** is parallel to the O-axis, and **h** is orthogonal to that (Fig. 6b), we also see periodical variations of the reflection coefficient, connected with distribution of the microwave fields **E** and **h**. The reflection is small ($R_m = 0$) at $l = 0, \lambda/2 \cdots$ (in the maxima of the **h**-field), and it reaches its maximum at $l = \lambda/4 + \lambda/8$. Besides, there is a resonance at $l = \lambda/4$ (in the maximum of the **E**-field).

To conclude, in this case the particle is excited by the microwave electric field **E**, when this field is parallel to the O-axis. This phenomenon is also displayed as a resonance of the effective permittivity in the cavity measurements.

Thus, the observed resonant effects in cylindrical samples with oriented multistart helical structures (formed by 1.5-turn helices) can be caused either by the microwave magnetic field **h** or by the electric field **E**. The different effects can be observed at the same frequencies, but for different orientations of the samples.

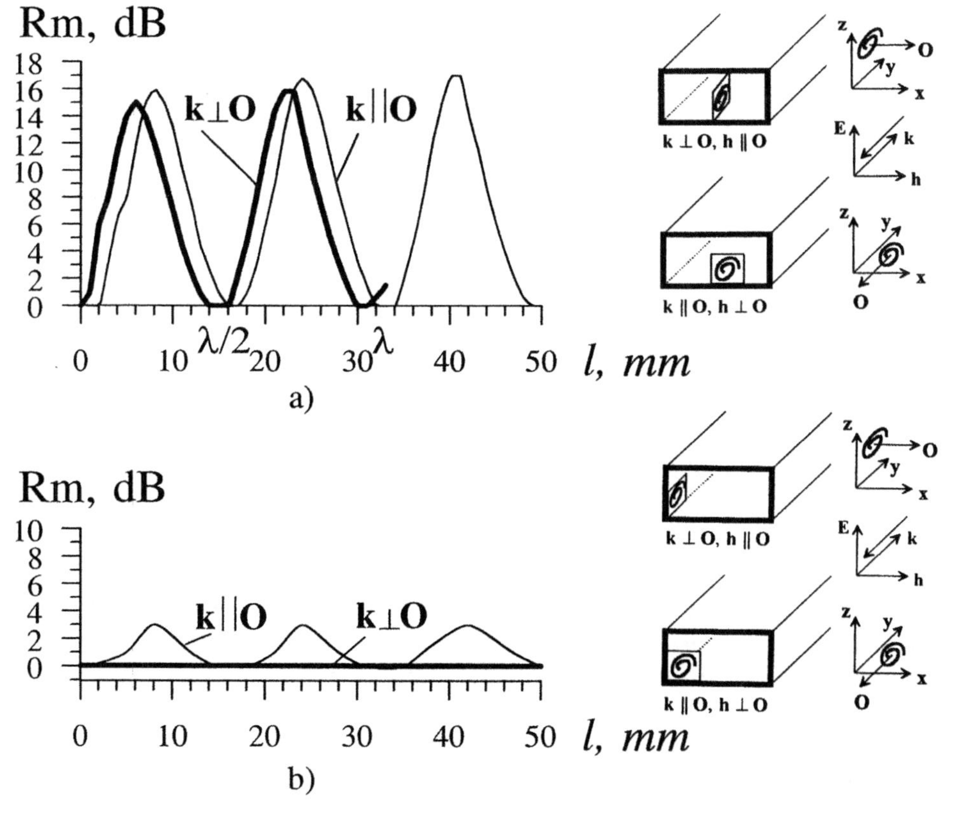

Figure 7. The reflection coefficient R_m from a single planar helix as a function of the distance l between the sample and the moving short (a – center position, b – side position)

When the angle α is close to 90°, the particle can be excited by the microwave magnetic field **h** which is parallel to the O-axis. When the electric field $\mathbf{E}$ is parallel to the sample axis (and $\mathbf{h}$ is perpendicular to the O-axis), the particle can be effectively excited by the $\mathbf{E}$-field at the same frequency, but this effect is rather weak.

The reflection coefficient R_m from a single planar helix as a function of the distance l between the sample and the moving short for the case when $\mathbf{E}$ is parallel to the z-axis is shown in Fig. 7. Various orientations of the helix axis with respect to the magnetic field vector $\mathbf{h}$ ($\mathbf{h}$ parallel and orthogonal to the O-axis), and various positions of the sample are tested.

We see that R_m reaches maxima at $l = \lambda/4$, $3\lambda/4$ (in the maxima of the $\mathbf{E}$-field), independently of the helix axis orientation with respect to the $\mathbf{h}$-field and $\mathbf{k}$-vector. $R_m = 0$ at $l = 0$, $\lambda/2$, when the sample is in a maximum of the $\mathbf{h}$-field (*i.e.*, in a minimum of the $\mathbf{E}$-field). When the sample is positioned at a waveguide side wall, where the $\mathbf{E}$-field is very small (Fig. 7b), we observe that R_m becomes small for the case when the wave vector $\mathbf{k}$ is parallel to the O-axis, and we do not observe resonances when $\mathbf{k}$ is orthogonal to the O-axis, independently of l.

Thus, the observed resonance effect in planar single helices is caused by the microwave electric field $\mathbf{E}$, when it is parallel to the z-axis. This effect can not be excited by microwave magnetic fields $\mathbf{h}$, independently of the particle orientation.

3.2. CHIRO-FERRITE SAMPLES

Next, we investigate microwave properties of different "chiral medium–ferrite" structures. The interest in such elements is mainly due to a possibility of electrical control of the magnetoelectric interactions. Chiro-ferrite media and similar problems are discussed in [13, 14].

In Figs. 8a,b the frequency dependencies of the reflection coefficient R_m from the cylindrical sample with a ferrite core positioned inside, as well as the evolution of these dependencies with variations of the constant magnetic field $\mathbf{H_0}$ are depicted. The measurements were performed in the frequency range from 8 to 12 GHz, and the standing-wave regime when the sample is in a waveguide near the metal short was utilized. These characteristics are investigated for two orientations of the sample: when the microwave magnetic field $\mathbf{h}$ is parallel to the axis of the cylindrical sample (while the microwave electric field $\mathbf{E}$ is perpendicular to the axis), and when $\mathbf{h}$ is orthogonal to the sample axis ($\mathbf{E}$ is parallel to that direction). Also, two different directions of $\mathbf{H_0}$ are realized for both the sample orientations: namely, $\mathbf{H_0}$ orthogonal to the O-axis, and $\mathbf{H_0}$ parallel to that. Fig. 8a corresponds to the case when $\mathbf{h}$ is parallel to the axis and $\mathbf{E}$ is perpendicular to that. We observe resonances of R_m, starting our investigation from $\mathbf{H_0} = 0$. Next, increasing the constant magnetic field, we observe that the resonant frequency shifts, and the resonance reflection coefficient changes. Considering the case when $\mathbf{H_0}$ is orthogonal to $\mathbf{h}$ (and also to the O-axis), we see a strong influence of the constant magnetic field $\mathbf{H_0}$ on the resonance frequency, the widths of the resonance areas, and on the resonance values of the reflection coefficient (Fig. 8a).

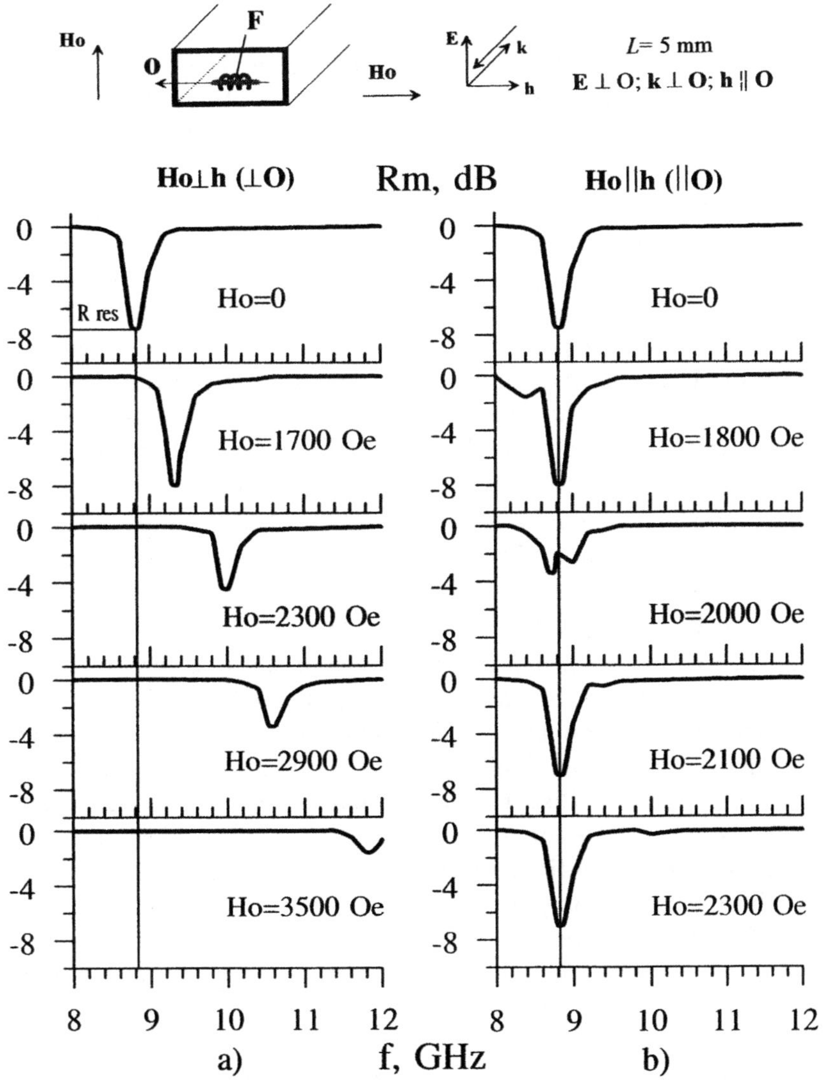

Figure 8. ***(a,b)*** **Influence of the constant magnetic field $\mathbf{H_0}$ on the frequency dependencies of the reflection coefficient in the standing-wave regime for the cylindrical sample with a ferrite core inside**

In contrast, when $\mathbf{H_0}$ is parallel to the inclusion axis and to $\mathbf{h}$ (Fig. 8b) the main (chiral) resonance of R_m practically does not depend on the constant magnetic field.

When we rotate the sample for 90°, so that $\mathbf{E}$ becomes parallel to the O-axis (and $\mathbf{h}$ becomes orthogonal to that, see Figs. 8c,d), a strong influence of $\mathbf{H_0}$ is observed, if $\mathbf{H_0}$ is also rotated by 90°, so that $\mathbf{H_0}$ would now be parallel to $\mathbf{h}$ and orthogonal to the inclusion axis. This can be seen in Fig. 8c. In the case when $\mathbf{H_0}$ is orthogonal to $\mathbf{h}$ and parallel to the axis, the main resonance of R_m remains always on the same frequency while $\mathbf{H_0}$ increases (Fig. 8d), but we see small moving peaks on the slope of the main resonance curve. That are the FMR (ferromagnetic resonance) curves which are usually observed for a single ferrite sample (without chiral inclusions).

In Figs. 8e,f, and 8g,h, the frequency dependencies of R_m are depicted for the case when the ferrite core is outside the cylindrical sample (but near that) for the same orientations as in the case, when the ferrite is inside. We see that the main (chiral) resonance of R_m does not depend on the constant magnetic field $\mathbf{H_0}$ for any orientation of the bias field. However, we see small moving peaks of the FMR, when $\mathbf{H_0}$ is orthogonal to $\mathbf{h}$ (see Fig. 8e for $\mathbf{E}$ perpendicular to the O-axis and $\mathbf{h}$ parallel to that, and Fig. 8h for $\mathbf{E}$ parallel to the O-axis, and $\mathbf{h}$ orthogonal to that). When the bias and microwave magnetic fields are parallel, the FMR is not observed, and we see only the main resonance (Figs. 8f and 8g).

Thus, if the ferrite core is inside the cylindrical cavity of the chiral inclusion, a strong influence of the bias field is observed when $\mathbf{H_0}$ is orthogonal to the inclusion axis for any orientation of the bias field with respect to the microwave field, *i.e.,* independently of the FMR excitation (Fig. 8a).

Thus, we can observe strong changes of the resonant frequency and reflection coefficient at the chiral resonance the FMR is not excited (Fig. 8c), and, vice versa, when the FMR is excited, we do can not achieve effective control of the main (chiral) resonance (Fig. 8d). It may be interpreted taking into account that the ferrite core when positioned inside a chiral element is subject to the action of both the external microwave magnetic field $\mathbf{h}$ and the magnetic field generated by the currents in the chiral inclusion (there exists a magnetic moment $\mathbf{m}$ induced by the microwave fields). This effect takes place when the directions of $\mathbf{m}$ and $\mathbf{h}$ do not coincide (for example, when the field $\mathbf{E}$ is parallel to the helix axis). We note also, that the conditions of effective resonant interaction of the particle currents which create the magnetic moment $\mathbf{m}$ and the ferrite magnetization can differ from the conditions of the ferromagnetic resonance (*i.e.*, excitation of the ferrite core by the external field $\mathbf{h}$). When the directions of $\mathbf{m}$ and $\mathbf{h}$ are the same (the case when $\mathbf{h}$ is parallel to the O-axis), the frequency spectra of the main resonance and that of the FMR overlap.

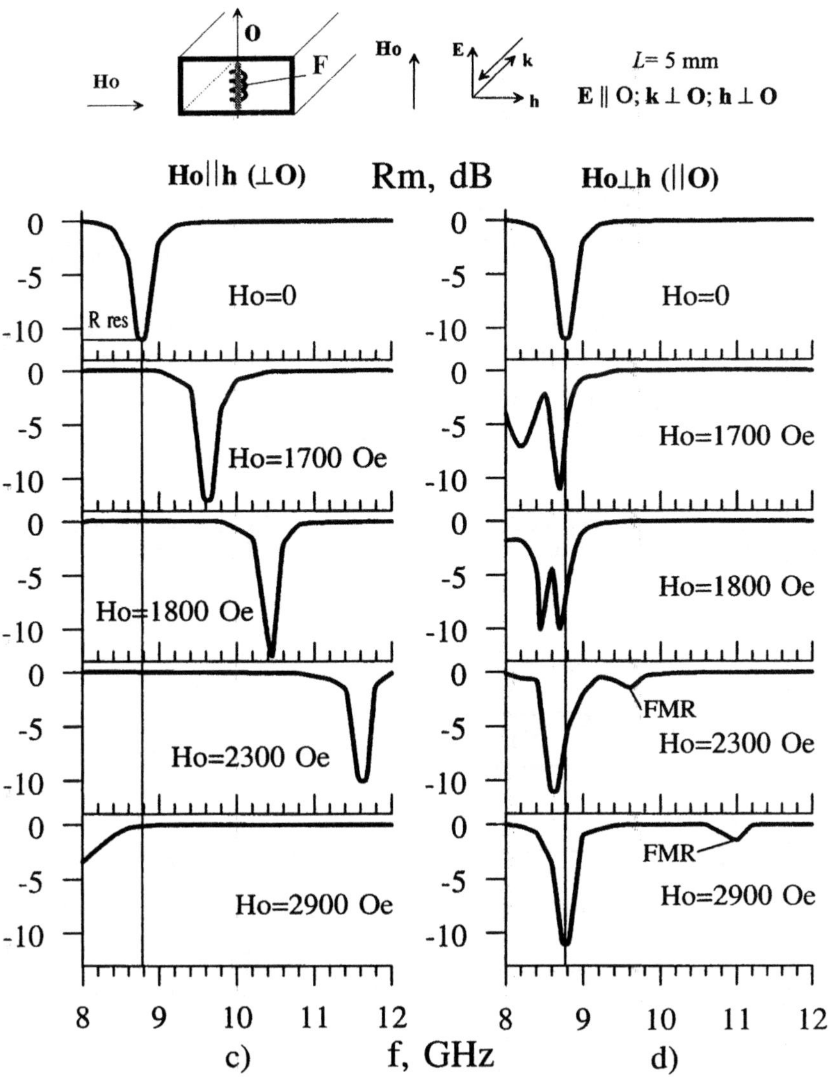

Figure 8. *(c,d)* **Influence of the constant magnetic field $\mathbf{H_0}$ on the frequency dependencies of the reflection coefficient in the standing-wave regime for the cylindrical sample with a ferrite core inside**

The influence of the constant magnetic field $\mathbf{H_0}$ on the reflection coefficient R_m of the planar single helix with a thin ferrite tablet is shown in Figs. 9,10 (in the resonance area). The frequency dependencies as well as the evolution of these dependencies with the constant magnetic field $\mathbf{H_0}$ are depicted for the 8 – 12 GHz range and the standing-wave regime. Ef-

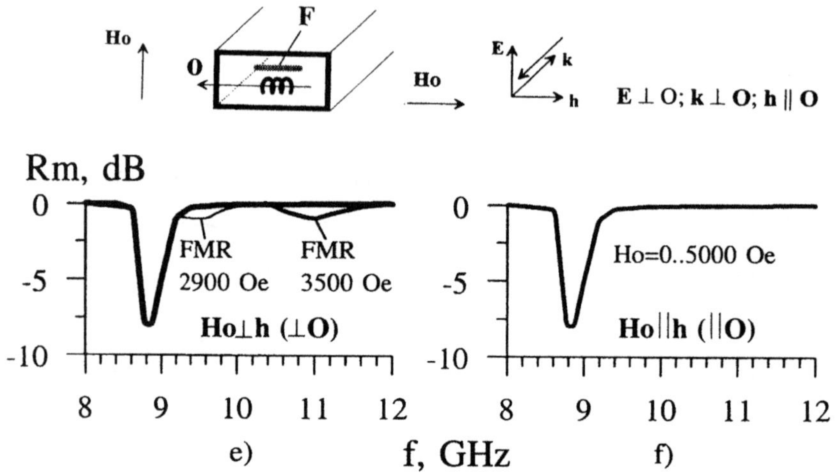

Figure 8. (e,f) **Influence of the constant magnetic field H_0 on the frequency dependencies of the reflection coefficient in the standing-wave regime for the cylindrical sample with a ferrite core outside**

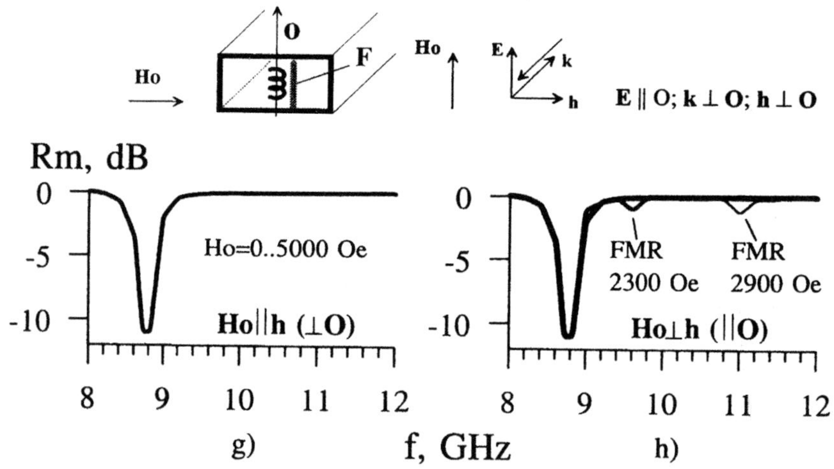

Figure 8. (g,h) **Influence of the constant magnetic field H_0 on the frequency dependencies of the reflection coefficient in the standing-wave regime for the cylindrical sample with a ferrite core outside**

fective interactions are observed when the electric field of the guide mode is parallel to the z-axis of the sample.

Fig. 9 corresponds to the case when the ferrite tablet is moved off from

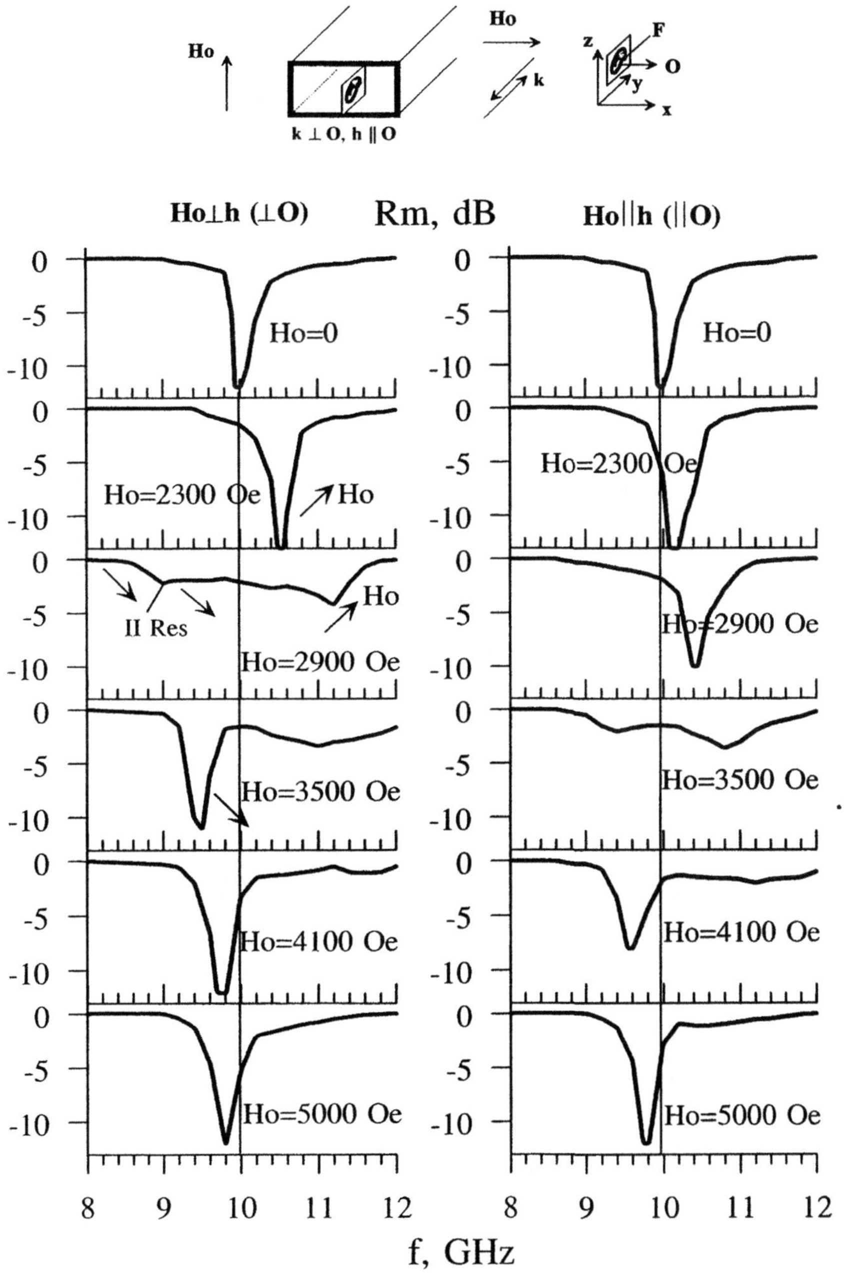

Figure 9. **Influence of the constant magnetic field $\mathbf{H_0}$ on the frequency dependencies of the reflection coefficient in the standing-wave regime for a single plane helix with a ferrite tablet (side position)**

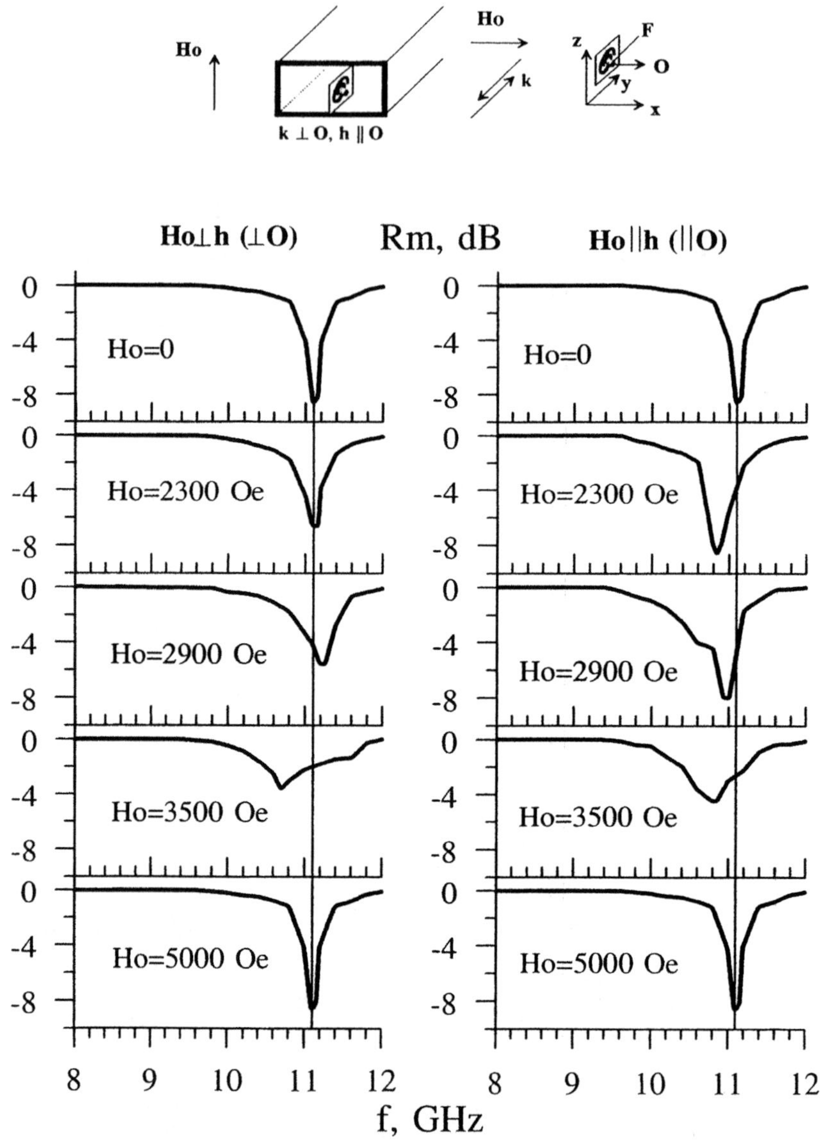

Figure 10. **Influence of the constant magnetic field $\mathbf{H_0}$ on the frequency dependencies of the reflection coefficient in the standing-wave regime for a single plane helix with a ferrite tablet at the center of the helix**

the helix center. The **E**-field is parallel to the z-axis, **h** is in the O-axis direction. Fig. 10 corresponds to the case when the same ferrite tablet is located at the helix center for the same helix orientation. Reflection coefficient curves are shown for the two orientations of $\mathbf{H_0}$: orthogonal to **h** (and to the particle axis), and parallel to both **h** and the axis.

We see strong changes of the intensity of the resonance effects and of the resonant frequency governed by the bias field $\mathbf{H_0}$, when that field is orthogonal to the O-axis, provided the ferrite element is in the side position (Fig. 9). In this case an adjustment of the external constant magnetic field $\mathbf{H_0}$ direction is necessary to ensure effective control of the main resonance, similarly to the case of a cylindrical sample with a ferrite core inside.

If the ferrite plate is positioned at the helix center, the main (chiral) resonance is practically invariable, independently of the vector $\mathbf{H_0}$ direction (Fig. 10). Weak changes of the resonance curve for $\mathbf{H_0}$ orthogonal to $\mathbf{h}$ may be attributed to the influence of FMR on the chiral effects near the main resonant frequency. One can not observe the FMR of such a small and thin ferrite tablet ($1 \times 1 \times 0.3$ mm^3) without a helix in the reflection and transmission regimes.

Thus, we observe strong interactions of the ferrite element with the magnetic moment $\mathbf{m}$ induced in the chiral inclusion when the ferrite tablet is moved off the helix center. The interaction is weak when the ferrite inclusion is at the helix center. This probably reflects the distribution of the local magnetic field of the chiral element.

4. Conclusion

1. Microwave properties of new samples of chiral inclusions and chiral composites containing oriented helices are investigated in waveguides and in free space in the standing-wave and the traveling-wave regimes. Investigated samples are cylinders with oriented multistart helical structures formed of one- and one-half turn helices, planar samples with helices made of thin copper films, and similar structures with magnetized ferrite inclusions. The samples are designed for the observation and investigation of electrically-controlled microwave resonance effects. Besides, the magnetic properties of the cylindrical samples can be studied by the direct resonant-cavity method.

2. It has been found that there are samples which can be excited by the microwave magnetic field $\mathbf{h}$ only, by the microwave electric field $\mathbf{E}$ only, and by both the $\mathbf{E}$- and $\mathbf{h}$-fields, depending on the orientation of the microwave fields. These resonances are also displayed as resonances of the effective permeability or of effective permittivity in the resonant-cavity measurements.

3. It has been shown that one can observe and study resonant interactions of the microwave magnetic field $\mathbf{h}$ with a magnetized ferrite inclusion in the vicinity of a chiral element, that is, the FMR in chiro-ferrite structures. Besides, one can observe and study the resonant interactions of the magnetic moment $\mathbf{m}$ of a chiral inclusion (which is induced by the microwave fields) with a magnetized ferrite element, that is, the controlled

chiral resonance in chiro-ferrite structures. It has been shown that one can possibly utilize the studied effects in chiral structures and in chiro-ferrites structures for comparing properties of different chiral particles and their induced electric and magnetic moments. The results can be important for the design of new chiral particles and new artificial composites containing helical inclusions, as well as novel nonreciprocal composite chiro-ferrite media.

Acknowledgment. The work was supported by the Russian Fund of Basic Research, Grant 95-02-06070a.

References

1. Kazantsev, Yu.N., and Kraftmakher, G.A., (1993) Giant microwave electromagnetism in artificial chiral media not possessing static magnetic properties, *Tech. Phys. Lett.* **19**, November, 665-667.
2. Kazantsev, Yu.N., Kraftmakher, G.A., (1994) Chiral resonance of magnetic permeability of media consisting of aligning multigone helices, Proceedings of *Chiral'94*. Périgueux, France, 215-222.
3. Kazantsev, Yu.N., Kraftmakher, G.A. (1994) Magnetic resonance in chiral medium-ferrite structure, Proceedings of *Chiral'94*, Périgueux, France, 347-357.
4. Kazantsev, Y.N, Kraftmakher, G.A. (1995) "Chiral medium-ferrite" and "electric dipole-ferrite" structures: chiral-ferromagnetic, dipole-ferromagnetic and ferromagnetic resonances, Proceedings of *Chiral'95*, Pennsylvania, USA, 161-170.
5. Tretyakov, S.A. and Viitanen, A.J. (1995) Waveguide and resonator perturbation techniques measuring chirality and non-reciprocity of biisotropic materials, *IEEE Trans. Microw. Theory Techn.* **43**, 222-224.
6. Reinert J., Busse, G., Jacob, A.F. (1995) A procedure to extract the chirality parameter from waveguide measurements, Proceedings of *Chiral'95*, Pennsylvania, USA, 9-12.
7. De Daran F., Vigneras-Lefebvre, V., Parneix, J.-P., Mariotte, F., and Garat, J. (1994) Scattering measurement on chiral objects in the microwave range by using free-space techniques, Proceedings of *Chiral'94*, Périgueux, France, 417-422.
8. Ghodgaonkar, D.K., Varadan, V.V., Varadan, V.K., (1990) Free-space measurement of complex permittivity and complex permeability of magnetic materials in microwave frequencies, *IEEE Trans. Instrum. Meas.* **39**, 387-394.
9. Tretyakov, S.A. and Haliulin D.Ya., (1993) Free-space techniques for biisotropic media parameter measurement, *Microwave and Optical Technology Letters* **6**, 513-515.
10. Tretyakov, S.A. and Viitanen, A.J. (1992) Perturbation theory for a cavity resonator with a biisotropic sample: applications to measurements techniques, *Microwave and Optical Technology Letters* **5**, 174-177.
11. Ougier, S. Chenerie, I., Bolioli, S. (1993) How to tailor and orient metallic helices to get microwave chirality, Proceedings of *Bianisotropics'93*, Gomel, Belarus, 54-59.
12. Lindell, I.V., Sihvola, A.H., Viitanen A.J., (1990) Reflection and transmission of plane waves at a planar interface between isotropic and bi-isotropic media, Helsinki University of Technology, Electromagnetics Laboratory, Report **72**.
13. Varadan, V.K., Varadan, V.V., Ma, Y., and Lakhtakia, A., (1988) Ferrite and chiral polymer composites: Electromagnetic Applications. *Proc. SPIE*, **927**, 213.
14. Krowne, C.M. (1993) Electromagnetic properties of nonreciprocal composite chiral-ferrite media, *IEEE Trans. Antennas Propagat.* **41**, 1289-1295.

MAGNETOSTATICALLY CONTROLLED BIANISOTROPIC MEDIA: A NOVEL CLASS OF ARTIFICIAL MAGNETOELECTRIC MATERIALS

E.O. KAMENETSKII
Tel-Aviv University,
Faculty of Engineering,
Dept. of Electrical Engineering - Physical Electronics
Tel Aviv 69978, Israel

Abstract. Recently an idea of a novel class of artificial bianisotropic materials has been introduced. Theoretical analysis sufficiently demonstrates magnetoelectric effect in proposed media. Now it becomes clear that this idea would make it possible to realize (both on the theoretical and experimental levels) many different bianisotropic material parameters that we have not been able to create before. Further development of these materials may be very productive in terms of new linear and nonlinear effects. For practical realization of such artificial media we can use planar technology well developed for magnetostatic wave (MSW) devices.

Proposed media are based on a composition of MSW resonators with metallic strips on their surfaces. Each MSW resonator with metallization is considered as a separate particle which exhibits a combined effect of electric and magnetic polarization. Because of small sizes of MSW resonators (in comparison with the wavelength), a concentration of particles may be very high. The resonance frequency of MSW resonators can be tuned via bias magnetic field. Nonlinear effects may be displayed due to possible nonlinearity of ferrites. The known experimental results demonstrate possible strong coupling between planar MSW waveguides. Such a phenomenon makes it possible to obtain a combined effect of all the particles in the lattice.

In modern electromagnetics and material science the macroscopic magnetoelectric interaction of the fields is one of the most interesting subjects. Proposed magnetoelectric materials promise to give a series of new fundamental problems and applications.

A. Priou et al. (eds.), Advances in Complex Electromagnetic Materials, 359–376.

1. Introduction

In up-to-date electromagnetics and material science, the macroscopic magnetoelectric interaction of the fields is one of the most interesting subjects. For this reason, artificial bi(an)isotropic media attract considerable attention in view of their potential usefulness in a variety of applications. A lot of journal papers, numerous reviews, many workshops and seminars related to bi(an)isotropic microwave materials have appeared and have been held in the last few years. Concerning this subject, one can use, for example, a detailed list of references in [1].

Strong interest in microwave bi(an)isotropic problems has a tendency to descend now. Despite many theoretical papers on artificial chiral and bianisotropic media published in the last decade, there are only a few experimental works. We cannot see real developments and applications of new microwave devices based on such materials, and for people who work out new devices for the market, now the real question is: does the area of chiral and bianisotropic media have enough practical applications on microwaves to warrant pursuing ?

From our point of view, there are two reasons for such kind of questions. The first reason is the development of reasonable microwave bi(an)isotropic materials. The second reason (which is closely connected with the first one) concerns more clear formulation of general ideas and general principles for microwave applications of bi(an)isotropic materials.

Artificial chiral materials have been developed initially to demonstrate the effect of electromagnetic activity on microwaves, analogous to the effect of optical activity. Microwave chiral media are based on a random distribution of small metallic or dielectric helices in a host material [2,3]. Together with artificial chiral materials an idea of microwave bianisotropic media based on omega or chiro-omega particles has been introduced [4,5]. A special feature of the known artificial chiral and bianisotropic materials is an effect of the first order in ak ($ak < 1$), where a is the scale of inhomogeneity, k is the wavenumber of a wave in the host material. In materials suitable for experimental investigations the effect of chirality is exhibited for dilute random distribution of particles [2,6]. Obviously, applications of such kind of artificial materials has to be very specific and sometimes problematic.

A principally new class of artificial bianisotropic materials has been proposed recently and some aspects of microwave applications have been considered [7,8]. Proposed media are based on a composition of magnetostatic wave (MSW) resonators [9,10] with metallic strips on their surfaces. Each MSW resonator with metallization may be characterized as a separate particle which exhibits combined effect of electric polarization and magnetization. Any scale of inhomogeneity in such media may be very small in terms of the wavelength in the host material. Lakhtakia suggested to name these media as "magnetostatically controlled bianisotropic media" [11]. From some point of view proposed artificial bianisotropic materials may be classified as a microwave analog of natural magnetoelectric crystals [12]. Moreover, due to special geometry of metallization, the lack of symmetry of a MSW resonator may be obtained, and such a medium will exhibit chiral properties [2,13].

The main aim of this paper is to call attention of the electromagnetic community to the novel class of artificial media in which one would make it possible to realize many different bianisotropic effects that we have not been able to create before [14]. We shall demonstrate a real possibility of manufacturing these media and shall show what kind of new physical effects and microwave applications one can expect in proposed materials.

2. On Phenomenological Description of Magnetoelectric Interaction in Media

To describe phenomenologically electromagnetic waves in continuous media one has to average positions of particles in a region which has sizes much larger than the particle dimensions and, at the same time, smaller than the electromagnetic wave wavelength. Such assumptions make it possible to introduce constitutive relations [15].

For the case of nonmagnetic materials and with an assumption of the linear response function, the constitutive relation has the form of an integral relation:

$$D_i(\vec{r},t) = \int_{-\infty}^{t} dt' \int d\vec{r}\,'\epsilon_{ij}(t,\vec{r},t',\vec{r}\,')E_j(\vec{r}\,',t') \tag{1}$$

Here only the causality principle (that the displacement field $\vec{D}$ at the time t is determined by the electric field $\vec{E}$ at times $t' < t$) is taken into account. For a time-invariant and spatially homogeneous medium the constitutive relation (1) has the form of temporal and spatial convolution [15,16]. For magnetic materials, the constitutive relation has a similar form of an integral relation which correlates the vectors $\vec{B}$ and $\vec{H}$ [11].

At optical frequencies it is convenient to combine all induced effects in a renormalized electric dipole moment and to characterize a medium with temporal and spatial dispersion by the tensor of permittivity with parameters depended on the frequency ω and the wave vector $\vec{k}$. In such a consideration the effect of optical activity is introduced as a particular case of the spatial dispersion phenomenon which is characterized by the first order (in a/λ, where a are atomic dimensions, λ is a wavelength) assumption [15,16]. In another approach, magnetization current is taken into account, and one can describe the effect of optical activity by constitutive relations with chirality parameter [17]. Chirality is considered as a purely geometric notion which refers to the lack of symmetry of an object [13]. It was shown that the distinction between these two approaches in description of chiral media may be important for electromagnetic boundary conditions [18].

In crystal optics any scale of material inhomogeneity is very small in terms of a wavelength:

$$\frac{a}{\lambda} << 1 \tag{2}$$

For this reason, the spatial dispersion phenomenon and, as a particular case, the optical activity is not so strongly exhibited in crystal optics [16]. In artificial microwave chiral media, a collection of helices is used. The first resonance in a helix may be characterized as an effect of the first order in a/λ, where a is the characteristic dimension of a helix and λ is the wavelength in the host material [2,3]. Real microwave chiral composites, applied for experimental investigations, are considered as a gas of scattering centers and may be characterized by constitutive relations for a dilute mixture [6]. Man-made chiral materials display chiral effects with orders of magnitude higher than natural media [2].

Whatever approach we use to describe chiral media, we virtually describe nonlocal effects. Nonlocal effects are the main reason of magnetoelectric coupling in chiral materials. This statement becomes clear if we consider the Drude-Born-Fedorov constitutive relations for chiral media [2,6]:

$$\vec{D} = \epsilon\,[\vec{E} + \beta \bigtriangledown \times \vec{E}], \quad \vec{B} = \mu\,[\vec{H} + \beta \bigtriangledown \times \vec{H}] \tag{3}$$

One can see that the chirality parameter β is, a measure of nonlocal effects. The more nonlocal effects, the more chiral parameters we have, and it is clear that there is a limit in a phenomenological description by constitutive relations if we want to increase chirality of a medium.

Magnetoelectric interaction in magnetoelectric crystals has another nature than that in chiral media. If a crystal is to show the magnetoelectric effect, it must have no centre of symmetry and must also be time-asymmetric [12,15]. The effects in magnetoelectric crystals were not analysed in the frequency domain. Moreover, it was suggested that magnetoelectric material cannot be used in time-varying fields. This problem is a subject of strong discussions [19,20].

Magnetoelectric materials proposed in [8] exhibit a principally new type of magnetoelectric interaction in media. Because of very small sizes of MSW resonators in comparison with the wavelength in the host material and due to temporal dispersion of permeability in ferromagnetics, one has virtually local magnetoelectric coupling in time-varying field. Essential magnetoelectric interaction in a medium may be realized when any scale of material inhomogeneity is very small in terms of electromagnetic wavelength (see relation (2)).

Such bianisotropic media may be described by two possible forms of generalized constitutive relations [21]:

$$\begin{aligned}
D_i(\vec{r},t) &= \int_{-\infty}^{t} dt' \int d\vec{r}\,'\alpha_{ij}(t,\vec{r},t',\vec{r}\,')E_j(\vec{r}\,',t') \\
&+ \int_{-\infty}^{t} dt' \int d\vec{r}\,'\beta_{ij}(t,\vec{r},t',\vec{r}\,')B_j(\vec{r}\,',t') \\
H_i(\vec{r},t) &= \int_{-\infty}^{t} dt' \int d\vec{r}\,'\gamma_{ij}(t,\vec{r},t',\vec{r}\,')E_j(\vec{r}\,',t') \\
&+ \int_{-\infty}^{t} dt' \int d\vec{r}\,'\vartheta_{ij}(t,\vec{r},t',\vec{r}\,')B_j(\vec{r}\,',t') \qquad (4)
\end{aligned}$$

$$\begin{aligned}
D_i(\vec{r},t) &= \int_{-\infty}^{t} dt' \int d\vec{r}\,'\epsilon_{ij}(t,\vec{r},t',\vec{r}\,')E_j(\vec{r}\,',t') \\
&+ \int_{-\infty}^{t} dt' \int d\vec{r}\,'\xi_{ij}(t,\vec{r},t',\vec{r}\,')H_j(\vec{r}\,',t') \\
B_i(\vec{r},t) &= \int_{-\infty}^{t} dt' \int d\vec{r}\,'\zeta_{ij}(t,\vec{r},t',\vec{r}\,')E_j(\vec{r}\,',t') \\
&+ \int_{-\infty}^{t} dt' \int d\vec{r}\,'\mu_{ij}(t,\vec{r},t',\vec{r}\,')H_j(\vec{r}\,',t') \qquad (5)
\end{aligned}$$

It should be noted that ϵ_{ij} in (5) is not the same as the permittivity tensor in (1). For time-invariant and spatially homogeneous bianisotropic medium, constitutive relations (4) and (5) have a form of temporal and space convolution [21,22]. In such a case one can consider the effect of spatial dispersion as an additional phenomenon in magnetoelectric media [21]. The temporal convolution form of constitutive relations in bianisotropic media analysed in [23], is a particular case of relations (4) and (5). A special feature of the media proposed in [8] is that due to small sizes of particles, each component of electric and magnetic field in the right-hand sides of Eqns. (4) and (5) can be generated independently at a specific fixed position $\vec{r}$.

3. MSW Resonator as a Bianisotropic Particle

Because of strong temporal dispersion of permeability in ferromagnetics, quasistatic oscillations of magnetization may take place in a certain frequency region [10,15]. One can introduce a magnetostatic approximation in such a case ($\vec{H} = -\nabla\psi$) and use the following equation (Walker's equation) for the magnetostatic potential ψ:

$$\nabla \cdot (\overleftrightarrow{\mu} \nabla\psi) = 0 \tag{6}$$

where $\overleftrightarrow{\mu}$ is the lossless permeability tensor [10,15]. To obtain magnetostatic oscillation at frequency ω a characteristic dimension of a ferromagnetic body l has to be much less than a wave length λ of electromagnetic wave at the same frequency [10,15].

Magnetostatic waves can propagate, for example, in liquid-phase epitaxial yttrium iron garnet (YIG) films. The almost ideal dielectric substrate for YIG film growth is gadolinium gallium garnet (GGG). Depending on the bias field orientation and the propagation direction, one has three sets of pure MSW modes in unbounded YIG film: magnetostatic surface waves (MSSW), magnetostatic forward volume waves (MSFVW), and magnetostatic backward volume waves (MSBVW) [9,10]. A MSW resonator formed by a piece of YIG film has properties of a magnetic dipole. In order to have properties of an electric dipole as well, we can use a region of metallic coating of a certain form on the YIG surface. In such a case an individual MSW resonator with surface metallization can be considered as a bianisotropic particle. Induced electric and magnetic dipole moments $\vec{p}_e$ and $\vec{p}_m$ of the particle are related to external electric and magnetic fields $\vec{E}$ and $\vec{H}$ as:

$$\vec{p}_e = \overleftrightarrow{\alpha}_{ee} \cdot \vec{E} + \overleftrightarrow{\alpha}_{em} \cdot \vec{H}, \qquad \vec{p}_m = \overleftrightarrow{\alpha}_{me} \cdot \vec{E} + \overleftrightarrow{\alpha}_{mm} \cdot \vec{H} \tag{7}$$

Corresponding dyadic polarizabilities are defined using the theory of MSW resonators. Because magnetostatic waves travel in ferromagnetic films with velocities which are two-four orders of magnitude slower than electromagnetic waves [9,10], sizes of every particle are very small in comparison with the wavelength of electromagnetic wave (see inequality (2)).

Different forms of ferromagnetic resonant bodies and forms of surface metallization are possible (Fig. 1). One can see that a ferromagnetic particle is an example of a rather complicated system. We shall use here an analysis of a parallelepiped-form MSW cavity with metallic strip (Fig. 1a), to demonstrate the effect of magnetoelectric coupling in proposed media. This analysis was made in [8].

MSW resonator is considered as a standing-wave structure composed by waveguide sections with and without a metallic strip. It is supposed that the spectrum of every section is known. To find dyadic polarizabilities of a resonator we have to know a distribution of magnetostatic potential $\psi(x, y, z)$ in the presence of the external $\vec{E}$ and $\vec{H}$ fields. When excitation of a resonator takes place, the potential in every waveguide section may be written as:

$$\psi(x, y, z) = \sum_{q=1}^{\infty} a_q(z)\widehat{\psi}_q(x, y), \tag{8}$$

where $\widehat{\psi}_q(x, y)$ is a membrane function of mode q.

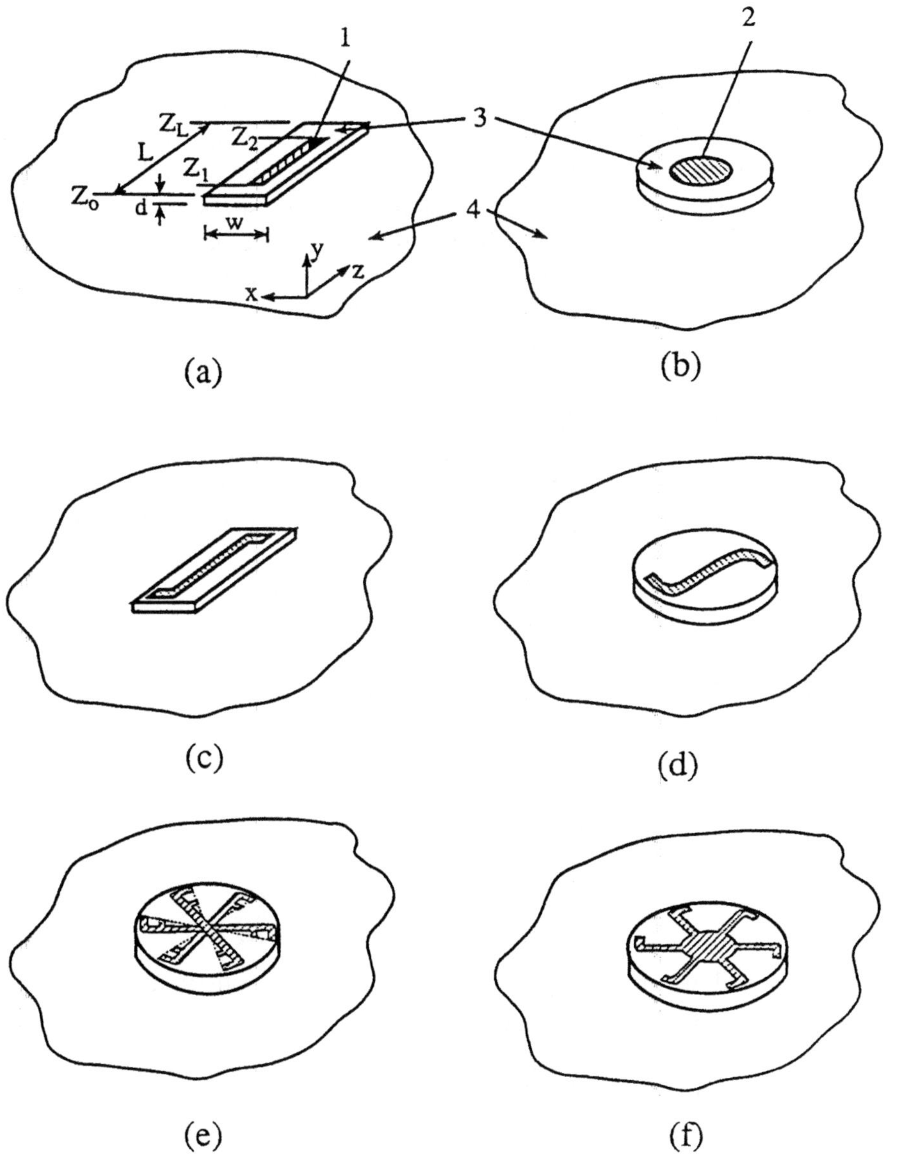

Figure 1. **MSW resonators with surface metallization regions.**
1-Metallic strip, 2-Surface metallization, 3-Ferromagnetic film, 4-Dielectric substrate

The excitation equation for waveguide sections with metallic strips has a form [7,8]:

$$\frac{da_q(z)}{dz} + \gamma_q a_q(z) = \frac{1}{N_q}\left[i\omega \int_S \vec{H}^{exc} \cdot \vec{\hat{B}}^*_{\tilde{q}} ds + \int_C \vec{E} \cdot (\vec{j}^{\,e}_{s\tilde{q}})^* dc\right], \qquad (9)$$

where S is the waveguide cross section which includes the cross sections of the ferrite region and the half-infinite region of a dielectric surrounding a ferrite layer; C is the contour of the metallic strip on the cross section, N_q is the norm of mode q [7], $\hat{B}_{\tilde{q}}$ and $\vec{j}^{\,e}_{s\tilde{q}}$ are, respectively, the membrane function of the magnetic flux density and the surface electric current of mode $\tilde{q}$ of the conjugate problem. The exciting magnetic field $\vec{H}^{exc}$ inside the ferromagnetic film differs from the external magnetic field $\vec{H}$ because of the demagnetization effect. For waveguide sections without metallic strips we have only the first integral on the right-hand side of Eq. (9).

Now one has a problem concerning solving a system of ordinary differential equations (9) for a number of modes in every waveguide section with correspondent boundary conditions at $z = z_0, z_1, z_2, z_l$ (see Fig. 1a). To solve this problem the shooting technique can be used [24].

The magnetization of ferrite is related to the magnetostatic potential as [10]:

$$\vec{m} = -(\overleftrightarrow{\mu} - \overleftrightarrow{I}) \nabla \psi \qquad (10)$$

where $\overleftrightarrow{I}$ is the unit matrix. For structures with transverse tangential magnetization (the bias magnetic field is directed along the x axis), one obtains

$$m_x = 0,$$

$$m_y = -(\mu - 1)\frac{\partial \psi}{\partial y} + i\mu_a \frac{\partial \psi}{\partial z}, \qquad (10a)$$

$$m_z = -i\mu_a \frac{\partial \psi}{\partial y} - (\mu - 1)\frac{\partial \psi}{\partial z},$$

where μ and μ_a are, respectively diagonal and off-diagonal, components of the tensor $\overleftrightarrow{\mu}$. Because of the demagnetization effect, these components depend on the x coordinate. One obtains the dyadic polarizabilities $\overleftrightarrow{\alpha}_{me}$ and $\overleftrightarrow{\alpha}_{mm}$ and, thus, the dipole moment $\vec{p}_m$ (see Eq. (7)) as a result of integration of $\vec{m}(x, y, z)$ over the volume of ferrite.

The dyadic polarizabilities $\overleftrightarrow{\alpha}_{ee}$ and $\overleftrightarrow{\alpha}_{em}$ and, as a result, the dipole moment $\vec{p}_e$ (see Eq. (7)) are found by integrating the surface charge density τ_s over the surface of the metallic strip. The electric charge density is related to the surface electric current density as:

$$\nabla_s \cdot \vec{j}^{\,e}_s = i\omega\tau_s \qquad (11)$$

where $\nabla_s\cdot$ is the two-dimensional divergence (on the plane x, y). We have the following expression to determine the surface current density:

$$\vec{n}_c \times [(\nabla_s \psi)^{(+)} - (\nabla_s \psi)^{(-)}] = \vec{j}_s^e, \tag{12}$$

where $\vec{n}_c$ is the unit normal vector to the contour C and $(\nabla_s \psi)^{(\pm)}$ are surface gradients of the potential ψ above and below the contour C.

Mutual orientation of the vectors $\vec{p}_e$ and $\vec{p}_m$ depends on the geometry of the resonator and on the orientation of the bias magnetic field $\vec{H}_o$. When a parallelepiped-form MSW resonator with $L >> d, w$ and a narrow metallic strip (Fig. 1a) is used, we have a parallel orientation of induced electric and magnetic dipole moments ($\vec{p}_e \parallel \vec{p}_m$) in every particle. For a cylindrical MSW cavity with a circular form of metallization region (Fig. 1b) one obtains a perpendicular mutual orientation of the vectors $\vec{p}_e$ and $\vec{p}_m$ ($\vec{p}_e \perp \vec{p}_m$) if the bias field $\vec{H}_o$ is oriented normally with respect to the plane of a ferromagnetic film.

MSW resonators shown in Fig. 1c,d,e,f exhibit planar structures with the lack of symmetry due to the geometry of metallization. In synthetic chiral composites the basic handed structures are usually three-dimensional. This results in chirality in all directions of observation. Such particles may be synthesized as isotropic media (randomly oriented inclusions) or anisotropic media (aligned inclusions). A step towards planar structures exhibiting magnetoelectric coupling was made with the introduction of omega-shaped particles [4]. A novel class of planar structures based on thin-wire swastika-shaped particles was considered in [25]. The structures based on thin-film MSW resonators [8] one can consider as a further step in development of planar magnetoelectric composites.

Swastika-shaped metal strips on surfaces of MSW cavities (Fig. 1c,d,e,f) have 2^n coplanar legs with angular separation $\gamma = \pi/2^{n-1}$. We have $n = 1$ for the resonators in Fig. 1c,d and $n = 3$ for the resonators in Fig. 1e,d. One can obtain two enantiomorphic metallizations for such resonators [25].

Normally magnetized cylindrical MSW resonator in Fig. 1b is an isotropic particle in the plane of the ferromagnetic film. Normally magnetized cylindrical resonator for swastika-shaped metallization with large number n is also an isotropic particle in the plane of the film. For example, the resonator shown in Fig. 1e may be considered as a quasi-isotropic particle in a plane of a film. The resonator in Fig. 1f, which is a combination of two resonators depicted in Fig. 1b and 1e, is also a quasi-isotropic particle.

4. Bianisotropic Composites Based On MSW Resonators

Composites based on thin-film MSW resonators are planar structures exhibiting magnetoelectric interaction. Induced electric and magnetic dipole moments of every resonator depend on material ferrite properties, geometrical effects (geometry of a resonance body and a surface metallization) and the bias magnetic field $\vec{H}_o$. For randomly distributed MSW resonators with anisotropic geometry (Fig. 1a,c,d) the bias field $\vec{H}_o$ has to be oriented normally to the surface of a ferromagnetic film. In such a case every MSW resonator will have the same electric and magnetic dipole moments. For aligned structures any orientation of $\vec{H}_o$ is possible. Resonators depicted in Fig. 1b,e,f are isotropic (Fig. 1b) or quasi-isotropic (Fig.1e,f) particles in the plane of the film. On the basis of these resonators one can realize randomly distributed (or aligned) structures with normal or tangential orientation of $\vec{H}_o$.

Fig. 2 shows an example of bianisotropic composite with randomly distributed MSW resonators. The structure is performed as a slab constructed by a pile of GGG substrates with YIG bodies. MSW resonators are depicted schematically without a detailed definition of the form of a resonance body and surface metallization. To obtain effective parameters of the composite one can use the modelling based on ellipsoidal approach for bianisotropic particles. The system may be considered as a mixture with very elongated or planar degenerated ellipsoidal inclusions and with random distribution or aligned orientation [2]. Calculation of these effective parameters is beyond the scope of this paper and we restrict ourselves to qualitative description of composites for some particular cases.

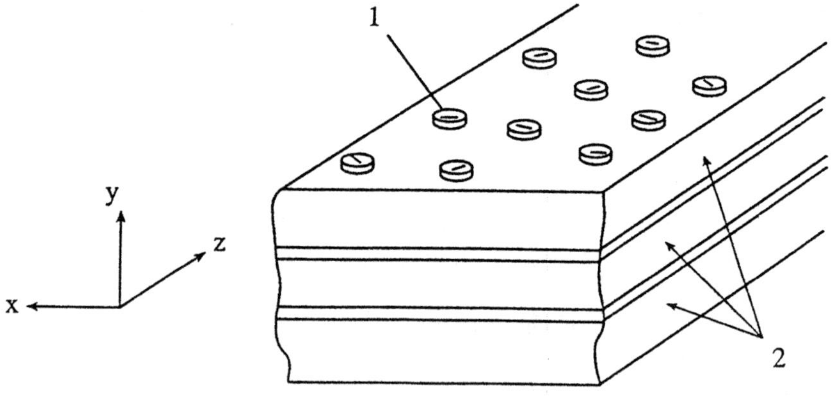

Figure 2. Bianisotropic composite with randomly distributed MSW resonators. 1-MSW resonator, 2-Dielectric substrates

Let a medium be described by constitutive relations:

$$\vec{D} = \overleftrightarrow{\epsilon}\,\vec{E} + \overleftrightarrow{\xi}\,\vec{H} \tag{13}$$

$$\vec{B} = \overleftrightarrow{\mu}\,\vec{H} + \overleftrightarrow{\zeta}\,\vec{E} \tag{14}$$

For some examples of composites we will describe constitutive tensors in (13) and (14) on the basis of geometrical properties of structures and taking into account the conditions for lossless (and in some examples, reciprocal) media [26,27].

a) The structure shown in Fig.2 is composed by randomly distributed very elongated parallelepiped-form ($L >> d, w$; see Fig. 1a) resonators with the normal orientation of the field $\vec{H}_o$ ($\vec{H}_o$ is parallel to the y-axis). For lossless and reciprocal media the constitutive tensors have forms:

$$\overleftrightarrow{\epsilon} = \begin{bmatrix} \epsilon_1 & 0 & 0 \\ 0 & \epsilon_2 & 0 \\ 0 & 0 & \epsilon_1 \end{bmatrix}, \qquad \overleftrightarrow{\mu} = \begin{bmatrix} \mu_1 & 0 & 0 \\ 0 & \mu_2 & 0 \\ 0 & 0 & \mu_1 \end{bmatrix} \tag{15}$$

$$\overleftrightarrow{\xi} = i \begin{bmatrix} c & 0 & 0 \\ 0 & 0 & 0 \\ 0 & 0 & c \end{bmatrix}, \qquad \overleftrightarrow{\zeta} = -i \begin{bmatrix} c & 0 & 0 \\ 0 & 0 & 0 \\ 0 & 0 & c \end{bmatrix} \tag{16}$$

b) The structure in Fig.2 is composed by cylindrical resonators depicted in Fig. 1b. The field $\vec{H}_o$ is parallel to the y-axis. When nonhomogeneous magnetostatic oscillations in resonators take place, the constitutive tensors may be described as:

$$\overleftrightarrow{\epsilon} = \begin{bmatrix} \epsilon_1' & 0 & ia \\ 0 & \epsilon_2' & 0 \\ -ia & 0 & \epsilon_1' \end{bmatrix}, \qquad \overleftrightarrow{\mu} = \begin{bmatrix} \mu_1' & 0 & ib \\ 0 & \mu_2' & 0 \\ -ib & 0 & \mu_1' \end{bmatrix} \tag{17}$$

$$\overleftrightarrow{\xi} = i \begin{bmatrix} 0 & 0 & d \\ 0 & 0 & 0 \\ d & 0 & 0 \end{bmatrix}, \qquad \overleftrightarrow{\zeta} = -i \begin{bmatrix} 0 & 0 & d \\ 0 & 0 & 0 \\ d & 0 & 0 \end{bmatrix} \tag{18}$$

Here we used the condition for the constitutive relations of lossless media [26].

c) The structure is composed by randomly distributed anisotropic resonators depicted in Fig. 1c,d or quasi-isotropic resonators depicted in Fig. 1e,f. For lossless media with $\vec{H}_o$ parallel to y-axis, forms of tensor $\overleftrightarrow{\epsilon}$ and $\overleftrightarrow{\mu}$ are similar to that of tensors (17). For tensors $\overleftrightarrow{\xi}$ and $\overleftrightarrow{\zeta}$ we have the following forms:

$$\overleftrightarrow{\xi} = \begin{bmatrix} e & 0 & f \\ 0 & 0 & 0 \\ f & 0 & e \end{bmatrix}, \qquad \overleftrightarrow{\zeta} = \begin{bmatrix} e^* & 0 & f^* \\ 0 & 0 & 0 \\ f^* & 0 & e^* \end{bmatrix} \tag{19}$$

Structures based on magnetized resonators with swastika-shaped metal strips may be characterized as planar *gyrotropic – chiral structures*. In these media gyrotropic effects are combined with geometric effects caused by the lack of symmetry of the surface metallization. Such structures are principally different from chiroferrite or chiroplasma media which are made from objects immersed in a magnetically biased ferrite or plasma [28].

5. New Physical Problems

A large variety of physical effects in macroscopic interaction of electromagnetic waves with media takes place due to different material properties, a field structure and a character of the field variation [15]. We will use a general description to characterize some new physical effects which one can expect in proposed media.

5.1. TAKING INTO ACCOUNT EXCHANGE INTERACTION

Exchange interaction in ferrites plays an essential role when an attempt to decrease at least one dimension of a resonance body is made. One has to take into account exchange interaction for magnetostatic waves with a large wave number. This phenomenon may be characterized as an effect of spatial dispersion for magnetostatic waves. To describe boundary problems additional (exchange) boundary conditions have to be introduced. Magnetostatic waves with exchange interaction are named spin waves [10].

Spin wave spectrum in ferromagnetic films is a subject of many investigations (see, for example, [29]). A linear excitation of spin waves takes place in nonhomogeneous alternative magnetic field. In ferromagnetic films spin waves can be excited by electric currents in microstrip transducers [29,30].

At present we do not have a rigorous analysis of spin-wave oscillations in a three-dimensional ferromagnetic body. Even more, we do not have such an analysis of spin wave oscillations in a three-dimensional resonator with a certain geometry of a surface metallization. One cannot predict now certain properties of these resonators as a bianisotropic particle. Nevertheless, future application of spin wave resonators as a bianisotropic particle may be very attractive. The sizes of these resonators on microwaves may be very small: in order of units of microns. Temporal dispersion of a bianisotropic medium based on spin-wave resonators may differ essentially from a medium constructed by MSW resonators.

5.2. NONLINEAR EFFECTS

Certain attention is now shifted to the theoretical analysis of nonlinear phenomena in chiral media. A theory of the self-action of waves in nonlinear chiral media based on an introduction of nonlinear constitutive parameters tensor was developed in [31]. In [32] it was shown that unique spatial solitons (chirosolitons) can propagate in media that exhibit the chirality and the Kerr nonlinearity.

In proposed magnetostatically controlled bianisotropic media we can predict the existence of a new type of nonlinear effects. These effects are due to nonlinearity of bianisotropic particles caused by intensive magnetostatic wave propagation in a ferromagnetic film. Nonlinear processes in ferromagnetic films are induced by intensive external magnetic fields and therefore nonlinear terms in the constitutive relations depend on the magnetic field only. In the general case, the constitutive relations may be written as:

$$\vec{D} = \overleftrightarrow{\epsilon}\, \vec{E} + \overleftrightarrow{\xi}\, \vec{H} + \vec{D}^{NL}, \qquad \vec{B} = \overleftrightarrow{\mu}\, \vec{H} + \overleftrightarrow{\zeta}\, \vec{E} + \vec{B}^{NL} \tag{20}$$

Here the constitutive tensors are linear tensors. The nonlinear term $\vec{D}^{NL}$ may be written as [15,33]:

$$D_i^{NL}(t) = \int_0^\infty \int_0^\infty \xi_{ijk}(\tau_1, \tau_2) H_j(t-\tau_1) H_k(t-\tau_2) d\tau_1 d\tau_2 +$$

$$\int_0^\infty \int_0^\infty \int_0^\infty \xi_{ijkl}(\tau_1, \tau_2, \tau_3) H_j(t-\tau_1) H_k(t-\tau_2) H_l(t-\tau_3) d\tau_1 d\tau_2 d\tau_3 + \cdots \tag{21}$$

where ξ_{ijk}, ξ_{ijkl} are nonlinear constitutive tensors. The nonlinear term B^{NL} is written by analogous expression with nonlinear constitutive tensors μ_{ijk}, μ_{ijkl} and correspondent components of magnetic field.

Quadratic nonlinearity as well as cubic nonlinearity may take place in ferromagnetics. This leads to different nonlinear phenomena in ferromagnetic samples, such as three-wave interaction, four-wave interaction, self-action, solitons [34]. For example, a frequency doubling of the alternative magnetization in ferromagnetic resonator is possible. One can realize intensive frequency doubling in thin cylindrical resonators with tangential constant magnetization. Mode interactions in ferromagnetic resonators caused by parametric instabilities may take place [10].

Different nonlinear effects in ferromagnetic resonators will provide a large variety of nonlinear processes in proposed bianisotropic media.

5.3. SPATIAL DISPERSION IN BIANISOTROPIC MEDIA WITH CRYSTAL LATTICE STRUCTURE

Constitutive relations (4) and (5) describe the effect of spatial dispersion in bianisotropic media. This effect implies an assumption that electric and magnetic dipole moments are defined not only by the fields at the given point, but also by the fields at the vicinity of the point.

The known experimental results demonstrate possible coupling between planar MSW waveguides [35]. This makes it possible to obtain a combined effect of all planar MSW resonators in the lattice. When a parameter of this lattice becomes comparable with the wavelength of electromagnetic wave, a medium may be described by constitutive tensors that depend on the frequency ω and the wave number $\vec{k}$. One can carry out an analysis similar to the analysis of spatial dispersion in crystal optics [16] or magnetoelectric crystals [22].

Constitutive relations (13) and (14) in such a case may be written as [16,22]:

$$\vec{D} = \overleftrightarrow{\epsilon}(\omega, \vec{k})\vec{E} + \overleftrightarrow{\xi}(\omega, \vec{k})\vec{H} \tag{22}$$

$$\vec{B} = \overleftrightarrow{\mu}(\omega, \vec{k})\vec{H} + \overleftrightarrow{\zeta}(\omega, \vec{k})\vec{E} \tag{23}$$

Calculation of the constitutive parameters of an artificial bianisotropic medium with crystal lattice structure is a subject of a special microtheory. We have to consider a perfect crystal structure composed by magnetostatically linked ferrite resonators and acted upon by a monochromatic plane wave of the frequency ω and the wave vector $\vec{k}$.

Without any calculations of constitutive parameters, phenomenological analysis of the role of spatial dispersion may be realized. For example, it was shown in [21] that spatial dispersion in bianisotropic media provides additional power flow in Poynting's equation similarly to such effects in anisotropic media. Fig. 3 demonstrates an example of aligned structure. Such a structure may be considered as an artificial crystal lattice.

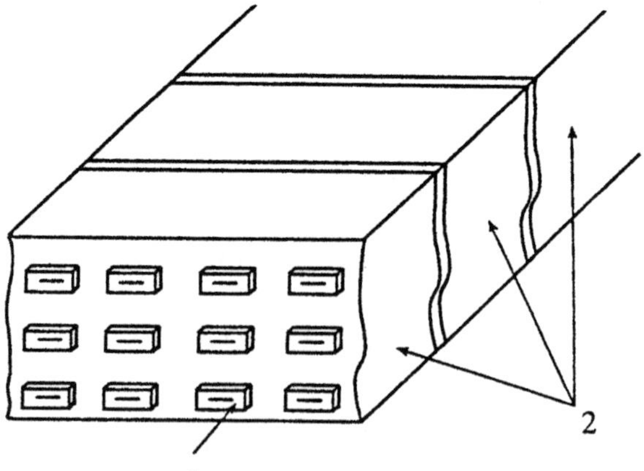

Figure 3. Bianisotropic composite with aligned MSW resonators. 1-MSW resonator, 2-Dielectric substrates

6. Technology and Applications in Microwaves

At present a lack of experimental works in microwave bianisotropics is because of the lack of reasonable microwave bianisotropic materials. Without such materials one cannot verify numerous theoretical results and cannot realize new microwave devices. Proposed magnetostatically controlled bianisotropic media not only promise to give a series of new theoretical problems. There is also a basis for manufacturing such artificial materials using planar technology well developed for modern MSW devices [9]. A medium is composed by GGG slabs with MSW resonators [8]. The resonators are realized by the method of photolithography of YIG films with surface metallization.

Proposed bianisotropic media have potential applications in microwave circuits and microwave antennas. Because of small sizes of MSW resonators in comparison with the wavelength, compact three-dimensional patterns may be realised on microwaves.

One of the applications of patterns based on the proposed material are different insertions in waveguides. Let us consider longitudinal bianisotropic insertion positioned in a hollow (without an insertion) cylindrical waveguide. Obviously, one has to take into account the properties of hollow waveguide to use the magnetoelectric effect of material more efficiently. Fig. 4a shows a case of noneffective usage of bianisotropic insertion. Because of spatial separation of electric and magnetic field in a cross section of a double-ridged waveguide, the pattern exhibits properties of dielectric and magnetic bodies in different parts of a waveguide.

Another situation one has in TEM or quasi-TEM waveguide structures (such as coaxial or strip-line waveguides). In these waveguides we do not have spatial separation of maxima of electric and magnetic fields in a waveguide cross section and, therefore, an application of bianisotropic insertions will be effective. Fig. 4b demonstrates an example of such the application of a bianisotropic pattern.

In the general case of longitudinal bianisotropic insertions in cylindrical waveguides, an analysis of properties of generalized S-matrix for cascade waveguide structure (Fig. 4c) may be very important. Usage of mode matching method [36] demands the knowledge of a spectrum of a waveguide section with bianisotropic insertion. For a medium with complex constitutive parameters such an analysis may be very complicated. The theory developed in [37,38] may give another way to calculate S-matrix, suitable for waveguides with complex media. In this theory material and geometrical properties of bianisotropic insertion are estimated on the basis of an analysis of mode interaction in a regular hollow waveguide.

General properties of multiport systems with magnetized bianisotropic

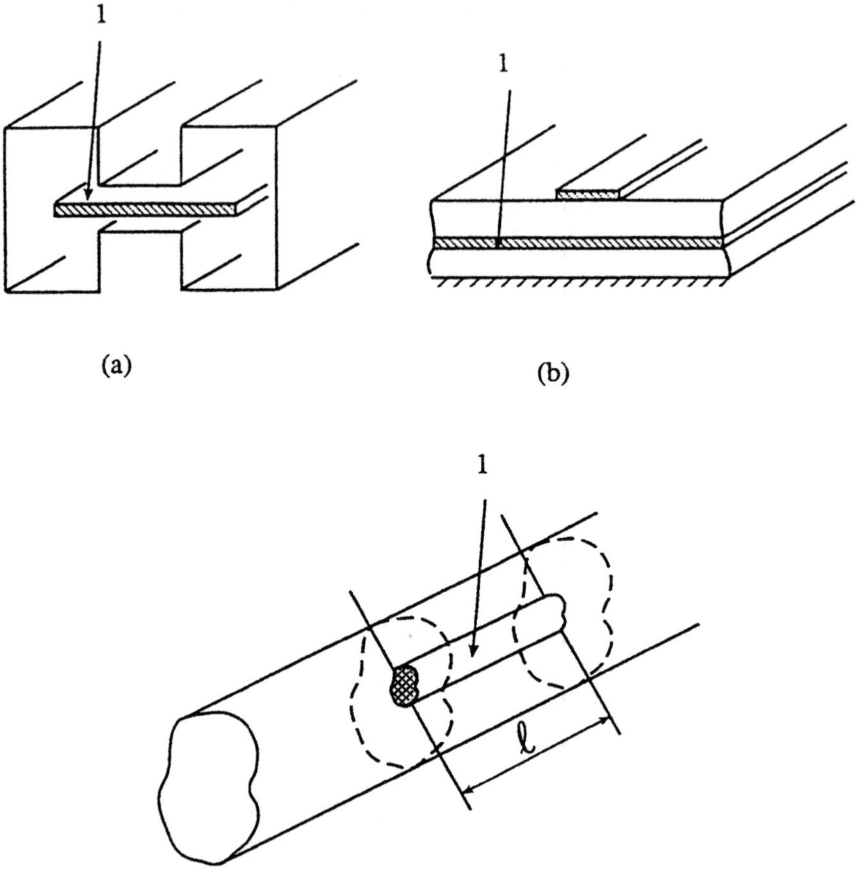

Figure 4. **Waveguides with bianisotropic insertions.**
1-Bianisotropic insertion

patterns have to be a subject of a symmetry analysis that take into account a geometrical symmetry and a symmetry of magnetization. A general procedure has to be developed similarly to an analysis given for symmetrical multiports with gyromagnetic media [39].

7. Conclusion

In this paper we have described a novel class of artificial bianisotropic materials. It was shown that proposed materials exhibit principally a new type of magnetoelectric interaction in media. One has virtually local magnetoelectric coupling in time-varying field due to use of MSW resonators as a bianisotropic particle. Magnetostatically controlled bianisotropic media make it possible to realize principally new material parameters. For example, gyrotropic-chiral structures may be realized. Some new physical effects (taking into account exchange interaction, nonlinear effects, spatial dispersion, etc.) one can expect in magnetostatically controlled media.

Proposed bianisotropic media not only promise to give a series of new theoretical problems. We also have a real basis for manufacturing such artificial materials using the methods of modern technology and, therefore, have a real possibility to extend the number of experimental investigations in this field. New potential applications in microwave circuits and microwave antennas become clear.

This paper may be considered as a general program for analysis, practical realization and applications of a novel class of artificial materials.

Acknowledgment

The author is grateful to A. Lakhtakia, A.H. Sihvola, M. Koshiba, G.Ya. Slepyan and S.A. Tretyakov for useful discussions and comments.

This work was partly supported by the Ministry of Science in Israel.

References

1. Cory, H. (1995) Chiral devices – an overview of canonical problems, *J. Electromagn. Waves & Appl.*, **9**, 805-829.
2. Lindell, I.V., Sihvola, A.H., Tretyakov, S.A., and Viitanen, A.J. (1993) *Electromagnetic Waves in Chiral and Bi-Isotropic Media*, Artech House, Boston-London.
3. Mariotte, F., Guerin, F., Bannelier, P., and Bourgeade, A. (1995) Numerical computations of electromagnetic field scattered by complex chiral bodies, *J. Electromagn. Waves & Appl.*, **9**, 1459-1485.
4. Saadoun, M.M.I. and Engheta, N. (1992) A reciprocal phase shifter using novel pseudochiral or Ω medium, *Microw. & Opt. Technol. Lett.*, **5**, 184-188.
5. Lindell, I.V., Tretyakov, S.A., and Viitanen A.J. (1993) Plane-wave propagation in an uniaxial chiro-omega medium, *Microw. & Opt. Technol. Lett.*, **6**, 517-526.
6. Lakhtakia, A., Varadan, V.K., and Varadan, V.V.(1990) Dilute random distribution of small chiral spheres, *Appl. Opt.*, **29**, 3627-3632.
7. Kamenetskii, E.O. (1996) Excitation of magnetostatic waves in ferromagnetic films by various sources, *Microw. & Opt. Technol. Lett.*, **11**, 99-102.
8. Kamenetskii, E.O. (1996) On the technology of making chiral and bianisotropic waveguides for microwave propagation, *Microw. & Opt. Technol. Lett.*, **11**, 103-107.
9. Ishak, W.S. (1988) Magnetostatic wave technology: a review, *Proc. IEEE*, **76**, 171-187.
10. Gurevich, A. and Melkov, G.(1994) *Magnetic Oscillations and Waves*, [in Russian], Nauka, Moscow.
11. Lakhtakia, A. (1996) Private communication to Kamenetskii, E.O.
12. O'Dell, T.H. (1970) *The Electrodynamics of Magneto-Electric Media*, Amsterdam: North-Holland.
13. Jaggard, D.L.., Mickelson, A.R., and Papas, C.H. (1979) On electromagnetic waves in chiral media, *Appl. Phys.*, **18**, 211-216.

14. Sihvola, A. (1996) Private communication to Kamenetskii, E.O.
15. Landau, L.D. and Lifshitz, E.M. (1984) *Electrodynamics of Continuous Media*, 2nd ed. (revised and enlarged, with Pitaevskii, L.P.) Pergamon Press, Oxford.
16. Agramovich, V.M. and Ginzburg, V.L. (1984) *Crystal Optics and Spatial Dispersion and Excitons*, 2nd ed., Springer-Verlag, New York.
17. Condon, E.U. (1937) Theories of optical rotatory power, *Rev. Mod. Phys.*, **9**, 432-457.
18. Silverman, M.P. (1986) Reflection and refraction at the surface of a chiral medium: comparison of gyrotropic constitutive relations invariant or noninvariant under a duality transformation, *J. Opt. Soc. Am. A*, **3**, 830-837.
19. Sihvola, A.H. (1995) Are nonreciprocal bi-isotropic media forbidden indeed ?, *IEEE Trans. on Microw. Theory Techn.*, **43**, 2160-2162.
20. Lakhtakia, A. and Weiglhofer, W.S. (1995) Comment on "Are nonreciprocal bi-isotropic media forbidden indeed ?" and Author's Reply by Sihvola, A.H. *IEEE Trans. on Microw. Theory Techn.*, **43**, 2722-2724.
21. Kamenetskii, E.O. (1996) Energy balance equation for electromagnetic waves in bianisotropic media, *Phys. Rev. E* **54**, (accepted for publication).
22. Hornreich, R.M. and Strikman, S. (1968) Theory of gyrotropic birefringence, *Phys. Rev.*, **171**, 1065-1074.
23. Karlsson, A. and Kristensson, G. (1992) Constitutive relations, dissipation and reciprocity for Maxwell equations in the time domain, *J. Electromagn. Waves & Appl.*, **6**, 537-551.
24. Ilyinsky, A.S., Slepyan, G.Ya., and Slepyan, A.Ya.(1993) *Propagation, Scattering and Dissipation of Electromagnetic Waves*, Peter Peregrinus Ltd., London.
25. Arnaut, L.R. and Davis, L.E. (1995) On planar chiral structures, *Progress in Electromagnetic Research Symposium (PIERS 1995)*, 24-28 July 1995, Seattle, WA, 165.
26. Kong, J.A. (1990) *Electromagnetic Theory*, New York, Wiley.
27. Lindell, I.V., Sihvola, A.H., and Suchy, K. (1995) Six-vector formalism in electromagnetics of bi-anisotropic media, *J. Electromagn. Waves & Appl.*, **9**, 887-903.
28. Engheta, N., Jaggard, D.L., and Kowatz, M.W. (1992) Electromagnetic waves in faraday chiral media, *IEEE Trans. on Antennas Propag.*, **40**, 367-374.
29. Kalinikos, B.A. (1981) Spin wave spectrum and linear excitation of spin waves in ferromagnetic films, [in Russian] *Izvestiya VUZ'ov, Ser. Physica*, **24**, No. 8, 42-56.
30. Kalinikos, B.A. and Dmitriev, V.F. Self-consistent calculation of radiation resistance of microstrip transducer of spin waves in a perpendicularly magnetized ferromagnetic film, [in Russian], *Zh. Tekh. Fiz.* **58**, 248.
31. Slepyan, G.Ya., Maksimenko S.A., Bass, F.G., and Lakhtakia, A. (1995) Nonlinear electromagnetics in chiral media: self-action of waves, *Phys. Rev. E*, **52**, 1049-1058.
32. Hayata, K. and Koshiba, M. (1995) Chirosolitons: unique spatial solutions in chiral media, *IEEE Trans on Microw. Theory Techn.*, **43**, 1814-1818.
33. Shen, Y.K. (1984) *The Principles of Nonlinear Optics*, Wiley, New York.
34. L'vov, V.S. (1987) *Nonlinear Spin Waves*, [in Russian], Nauka, Moscow.
35. Kamenetskii, E.O. and Solov'ev, O.V. (1990) Magnetostatic modes in a system of coupled spin waveguides in tangential transverse magnetic field, *Pisma Zh. Tekh. Fiz.*, **16**, 28-32 [*Sov. Tech. Phys. Lett.*, **16**, 653-654].
36. Eleftheradies, G.V., Omar, A.S., Katehi, L.P.B., and Rebeiz, G.M. (1994) Some important properties of waveguide junction generalized scattering matrices in the context of the mode matching technique, *IEEE Trans. on Microw. Theory Techn.*, **42**, 1896-1903.
37. Kamenetskii, E.O. (1996) Induced polarization effects in coupling processes of waveguide modes, *IEEE Trans. on Microw. Theory Techn.*, **44**, 572-584.
38. Kamenetskii, E.O. (1996) Mode interactions in chirowaveguides, *J. Appl. Phys.*, **79** 8899-8905.
39. Dmitriev, V.A. (1995) Generators of magnetic groups of symmetry and communication relations, *IEEE Trans. on Microw. Theory Techn.*, **43**, 2668-2675.

LIST OF CONTRIBUTORS

I.N. AKHRAMENKO
Department of Physics, Gomel State University
Sovyetskaya Str. 104, Gomel, 246699, Belarus

L.R. ARNAUT
University of Manchester Institute of Science and Technology
Department of Electrical Engineering and Electronics
P.O. Box 88, Manchester M60 1QD, United Kingdom

L.M. BARKOVSKII
Department of Theoretical Physics, Byelorussian State University
Fr. Skaryny avenue 4, Minsk, 220050, Belarus

V.V. BELYAEV
Central R&D "Cometa"
Velozavodskaya st. 5, Moscow, 109280, Russia

L. BIDEO
CERT/ONERA, 2 Avenue E. Belin, 31055 Toulouse, France

M. BINGLE
Department of Electrical and Electronic Engineering
University of Stellenbosch, Stellenbosch, South Africa

S. BOLIOLI
CERT/ONERA, 2 Avenue E. Belin, 31055 Toulouse, France

A.N. BORZDOV
Department of Theoretical Physics, Byelorussian State University
Fr. Skaryny avenue 4, Minsk, 220050, Belarus

G.N. BORZDOV
Department of Theoretical Physics, Byelorussian State University
Fr. Skaryny avenue 4, Minsk, 220050, Belarus

G. BUSSE
Institut für Hochfrequenztechnik, Technische Universität Braunschweig
Postfach 3329, D-38023 Braunschweig, Germany

J.H. CLOETE
Department of Electrical and Electronic Engineering
University of Stellenbosch, Stellenbosch, South Africa

P.F. COMBES
Université P. Sabatier, 118 route de Narbonne, 31062 Toulouse, France

Y.V. DEVYATKIN
Microcosm Ltd, Moscow

A.G. DMITRENKO
Tomsk State University, Siberian Physical & Technical Institute
1 Revolution Sq., Tomsk, 634050, Russia

T. EGE
University of Gaziantep
Electrical and Electronics Engineering Department
27310 Gaziantep, Turkey

E.A. EVDISCHENKO
Institute of Crystallography, Russian Academy of Sciences
Leninsky pr. 59, Moscow, 117333, Russia

V.V. FISANOV
Tomsk State University, Siberian Physical & Technical Institute
1 Revolution Sq., Tomsk, 634050, Russia

E.B. GRAHAM
Physics Department, University of Natal
P/Bag X01, Scottsville, Pietermaritzburg, 3209, South Africa

S.S. GROVÉ
Institute for Polymer Science, University of Stellenbosch
Stellenbosch, South Africa

A. JACOB
Institut für Hochfrequenztechnik, Technische Universität Braunschweig
Postfach 3329, D-38023 Braunschweig, Germany

E.O. KAMENETSKII
Tel Aviv University
Department of Electrical Engineering–Physical Electronics
Tel Aviv 69978, Israel

Yu.N. KAZANTSEV
Institute of Radioengineering and Electronics, Russian Academy of Sciences
11, Mokhovaya str., 103907, Moscow, Russia

S.A. KHAKHOMOV
Department of Physics, Gomel State University
Sovyetskaya Str. 104, 246699, Gomel, Belarus

M. KLEMT
Institut für Hochfrequenztechnik, Technische Universität Braunschweig
Postfach 3329, D-38023 Braunschweig, Germany

A.O. KOCA
University of Gaziantep
Electrical and Electronics Engineering Department
27310 Gaziantep, Turkey

A.F. KONSTANTINOVA
Institute of Crystallography, Russian Academy of Sciences
Leninsky pr. 59, Moscow, 117333, Russia

M.V. KOSTIN
Institute of Radioengineering and Electronics, Russian Academy of Sciences
11 Mokhovaya str., Moscow, 103907, Russia

G.A. KRAFTMAKHER
Institute of Radioengineering and Electronics, Russian Academy of Sciences
11, Mokhovaya str., Moscow, 103907, Russia

E. KUEHL
Institute for Polymer Science, University of Stellenbosch
Stellenbosch, South Africa

S.A. KUEHL
Institute for Polymer Science, University of Stellenbosch
Stellenbosch, South Africa

A.N. LAGARKOV
Scientific Center for Applied Problems in Electrodynamics
Russian Academy of Sciences
Izhorskaja 13/19, Moscow, 127412, Russia

A. LAKHTAKIA
Department of Engineering Science and Mechanics
The Pennsylvania State University
University Park, PA 16802-1401, USA

S.A. MAKSIMENKO
Institute of Nuclear Problems, Belarus State University
11 Bobruiskaya Str., Minsk, 220050, Belarus

D.A. MARAKASOV
Tomsk State University, Siberian Physical & Technical Institute
1 Revolution Sq., Tomsk, 634050, Russia

G.S. MITYURICH
Department of Physics, Gomel State University
Sovyetskaya Str. 104, Gomel, 246699, Belarus

A.I. MUKOMOLOV
Tomsk State University, Siberian Physical & Technical Institute
1 Revolution Sq., Tomsk, 634050, Russia

B.V. NABATOV
Institute of Crystallography, Russian Academy of Sciences
Leninsky pr. 59, Moscow, 117333, Russia

V.G. NAZARENKO
Institute for Physics of NASU
46 pr. Nauki, Kiev, 252650, Ukraine

M. NORGREN
Department of Electromagnetic Theory, Royal Institute of Technology
S-10044 Stockholm, Sweden

G.B. NOSOV
Organic Intermediates and Dyes Institute
1/4 B. Sadovaya st., Moscow, 103787, Russia

R.E. RAAB
Physics Department, University of Natal
P/Bag X01, Scottsville, Pietermaritzburg, 3209, South Africa

J. REINERT
Institut für Hochfrequenztechnik, Technische Universität Braunschweig
Postfach 3329, D-38023 Braunschweig, Germany

I.V. SEMCHENKO
Department of Physics, Gomel State University
Sovyetskaya Str. 104, 246699, Gomel, Belarus

V.V. SHEVCHENKO
Institute of Radioengineering and Electronics, Russian Academy of Sciences
11 Mokhovaya str., Moscow, 103907, Russia

A.H. SIHVOLA
Helsinki University of Technology, Electromagnetics Laboratory
Otakaari 5 A, 02150 Espoo, Finland

C.R. SIMOVSKI
Radiophysics Department, St. Petersburg State Technical University
Polytekhnicheskaya 29, St. Petersburg, 195251, Russia

G.Ya. SLEPYAN
Institute of Nuclear Problems, Belarus State University
11 Bobruiskaya Str., Minsk, 220050, Belarus

A.A. SOCHAVA
Radiophysics Department, St. Petersburg State Technical University
Polytekhnicheskaya 29, St. Petersburg, 195251, Russia

A.S. SONIN
Organic Intermediates and Dyes Institute
1/4 B. Sadovaya st., Moscow, 103787, Russia

E.G. STARODUBTSEV
Department of Physics, Gomel State University
Sovyetskaya Str. 104, Gomel, 246699, Belarus

H. TORRES-SILVA
Facultad de Ingeniería, Dpto. de Electrónica
Casilla 6-D, Av. 18 de Septiembre 2222, Arica-Chile

S.A. TRETYAKOV
Radiophysics Department, St. Petersburg State Technical University
Polytekhnicheskaya 29, St. Petersburg, 195251, Russia

A.Yu. TRONIN
Institute of Crystallography, Russian Academy of Sciences
Leninsky pr. 59, 117333 Moscow, Russia

S.G. VASHTALOV
Department of Radiophysics, Tomsk State University
36 Lenin Ave., Tomsk, 634050, Russia

A.P. VINOGRADOV
Scientific Center for Applied Problems in Electrodynamics
Russian Academy of Sciences
Izhorskaja 13/19, Moscow, 127412, Russia

Index

Batch number: 10755284

Printed by Printforce, the Netherlands